2018 IEEE 36th VLSI Test Symposium (VTS 2018)

San Francisco, California, USA
22-25 April 2018

IEEE Catalog Number: CFP18029-POD
ISBN: 978-1-5386-3775-3

**Copyright © 2018 by the Institute of Electrical and Electronics Engineers, Inc.
All Rights Reserved**

Copyright and Reprint Permissions: Abstracting is permitted with credit to the source. Libraries are permitted to photocopy beyond the limit of U.S. copyright law for private use of patrons those articles in this volume that carry a code at the bottom of the first page, provided the per-copy fee indicated in the code is paid through Copyright Clearance Center, 222 Rosewood Drive, Danvers, MA 01923.

For other copying, reprint or republication permission, write to IEEE Copyrights Manager, IEEE Service Center, 445 Hoes Lane, Piscataway, NJ 08854. All rights reserved.

*** This is a print representation of what appears in the IEEE Digital Library. Some format issues inherent in the e-media version may also appear in this print version.

IEEE Catalog Number: CFP18029-POD
ISBN (Print-On-Demand):   978-1-5386-3775-3
ISBN (Online):   978-1-5386-3774-6
ISSN: 1093-0167

Additional Copies of This Publication Are Available From:

Curran Associates, Inc
57 Morehouse Lane
Red Hook, NY 12571 USA
Phone: (845) 758-0400
Fax: (845) 758-2633
E-mail: curran@proceedings.com
Web: www.proceedings.com

TABLE OF CONTENTS

GROUP DELAY MEASUREMENT OF FREQUENCY DOWN-CONVERTER DEVICES USING CHIRPED RF MODULATED SIGNAL 1
P. Sarson, T. Yanagida, K. Machida

A COHERENT SUBSAMPLING TEST SYSTEM ARRANGEMENT SUITABLE FOR PHASE DOMAIN MEASUREMENTS 7
Y. Cho, G. Roberts, S. Aouini, M. Parvizi, N. Ben-Hamida

AN OSCILLATION-BASED TEST TECHNIQUE FOR ON-CHIP TESTING OF MM-WAVE PHASE SHIFTERS 13
M. Margalef-Rovira, M. Barragan, E. Sharma, P. Ferrari, E. Pistono, S. Bourdel

SPECIAL SESSION: RECENT DEVELOPMENTS IN HARDWARE SECURITY 19
R. Commarota, N. Karimi, S. Garg, J. Rajendran

INNOVATIVE PRACTICES ON MEMORY TEST PRACTICE 20
M. Casarsa, G. Harutyunyan, K. Chen, R. Sharma, G. Podichetty, M. Keim, S. Chakravarthy

ATPG-BASED COST-EFFECTIVE, SECURE LOGIC LOCKING 21
A. Sengupta, M. Nabeel, M. Yasin, O. Sinanoglu

MODELING AND TEST GENERATION FOR COMBINATIONAL HARDWARE TROJANS 27
Z. Zhou, U. Guin, V. Agrawal

MODELING ATTACKS ON STRONG PHYSICAL UNCLONABLE FUNCTIONS STRENGTHENED BY RANDOM NUMBER AND WEAK PUF 33
J. Ye, Q. Guo, Y. Hu, H. Li, X. Li

SPECIAL SESSION: HOW APPROXIMATE COMPUTING IMPACTS VERIFICATION, TEST AND RELIABILITY 39
L. Sekanina, Z. Vasicek, A. Bosio, M. Traiola, P. Rech, D. Oliveria, F. Fernandes, S. Di Carlo

INNOVATIVE PRACTICES ON QUALITY LEVELS OF A/MS DEVICES 40
W. Dobbelaere, M. Violante, J. Rearick, P. Sarson

HARDWARE TROJAN ATTACKS IN EMBEDDED MEMORY 41
T. Hoque, X. Wang, A. Basak, R. Karam, S. Bhunia

HIGH EFFICIENT LOW COST EEPROM SCREENING METHOD IN COMBINATION WITH AN AREA OPTIMIZED BYTE REPLACEMENT STRATEGY WHICH ENABLES HIGH RELIABILITY EEPROMS 47
G. Schatzberger, F. Leisenberger, P. Sarson, A. Wiesner

TEST CHALLENGES AND SOLUTIONS FOR EMERGING NON-VOLATILE MEMORIES 53
M. Khan, S. Ghosh

SPECIAL SESSION ON RELIABILITY AND VULNERABILITY OF NEUROMORPHIC COMPUTING SYSTEMS 59
S. Yu, C. Liu, W. Wen, Y. Chen

IP SESSION ON ISO26262 EDA 60
A. Schaldenbrand, Y. Zorian, S. Sunter, P. Sarson

AN INTER-LAYER INTERCONNECT BIST SOLUTION FOR MONOLITHIC 3D ICS 61
A. Koneru, K. Chakrabarty

A BUILT-IN SELF-TEST TECHNIQUE FOR TRANSMITTER-ONLY SYSTEMS 67
M. Shafiee, J. Kitchen, S. Ozev

EXPLOITING BUILT-IN DELAY LINES FOR APPLYING LAUNCH-ON-CAPTURE AT-SPEED TESTING ON SELF-TIMED CIRCUITS 73
O. Hasib, D. Crepeau, T. Awad, A. Dulipovici, Y. Savaria, C. Thibeault

SPECIAL SESSION ON BRINGING CORES CLOSER TOGETHER: THE WIRELESS REVOLUTION IN ON-CHIP COMMUNICATION 79
T. Mak, H. Matsutani, P. Pande

INNOVATIVE PRACTICES ON FUNCTIONAL TESTING AND FAULT SIMULATION FOR FUSA 80
A. Krishnan, J. Van Gelder, M. Bhattacharya, S. Chakravarty, P. Goteti

BROADCAST-BASED MINIMIZATION OF THE OVERALL ACCESS TIME FOR THE IEEE 1687 NETWORK 81
Z. Zhong, G. Li, Q. Yang, J. Qian, K. Chakrabarty

EFFICIENT PARALLEL TESTING: A CONFIGURABLE AND SCALABLE BROADCAST NETWORK DESIGN USING IJTAG 87
S. Gupta, J. Wu, J. Dworak

SECURING IJTAG AGAINST DATA-INTEGRITY ATTACKS 93
R. Elnaggar, R. Karri, K. Chakrabarty

SPECIAL SESSION ON OVERCOMING RELIABILITY AND ENERGY-EFFICIENCY CHALLENGES WITH SILICON PHOTONICS FOR FUTURE MANYCORE COMPUTING 99
S. Pasricha, D. Bertozzi, H. Li

INNOVATIVE PRACTICES ON TEST IN JAPAN 100
J. Matsushima, Y. Maeda, K. Hatayama

EFFICIENT GENERATION OF PARAMETRIC TEST CONDITIONS FOR AMS CHIPS WITH AN INTERVAL CONSTRAINT SOLVER 101
F. Neubauer, J. Burchard, P. Raiola, J. Rivoir, B. Becker, M. Sauer

ENHANCED HOTSPOT DETECTION THROUGH SYNTHETIC PATTERN GENERATION AND DESIGN OF EXPERIMENTS 107
G. Reddy, C. Xanthopoulos, Y. Makris

STAGGERED ATPG WITH CAPTURE-PER-CYCLE OBSERVATION TEST POINTS 113
Y. Liu, J. Rajski, S. Reddy, J. Solecki, J. Tyszer

SPECIAL SESSION ON INTELLIGENT SENSOR NODES 119
K. Basu, S. Sen

INNOVATIVE PRACTICES ON SILICON PHOTONICS 120
R. Meade, W. Kim, R. Otte, E. Atwood

SYSTEMATIC *b*-ADJACENT SYMBOL ERROR CORRECTING REED-SOLOMON CODES WITH PARALLEL DECODING 121
A. Das, N. Touba

CIRCUIT-LEVEL RELIABILITY SIMULATOR FOR FRONT-END-OF-LINE AND MIDDLE-OF-LINE TIME-DEPENDENT DIELECTRIC BREAKDOWN IN FINFET TECHNOLOGY 127
K. Yang, T. Liu, R. Zhang, L. Milor

ON-LINE MONITORING AND ERROR CORRECTION IN SENSOR INTERFACE CIRCUITS USING DIGITAL CALIBRATION TECHNIQUES 133
S. Heinssen, T. Hillebrand, M. Taddiken, S. Paul, D. Peters-Drolshagen

SPECIAL SESSION ON BIST/CALIBRATION OF A/MS DEVICES 139
H.-M. Von Staudt, J. Izon, S. Ozev, P. Sarson

INNOVATIVE PRACTICES ON MACHINE LEARNING FOR EMERGING APPLICATIONS 140
K. Madkour, Z. Zhang, A. Crouch, P. Levin, E. Hunter, Y. Huang

IC LAYOUT WEAK POINT EFFECTIVENESS EVALUATION BASED ON STATISTICAL METHODS 141
F. Lin, A. Ahmadi, K. Sekar, Y. Pan, K. Huang

ANALYZING AND MITIGATING THE IMPACT OF PERMANENT FAULTS ON A SYSTOLIC ARRAY BASED NEURAL NETWORK ACCELERATOR 147
J. Zhang, T. Gu, K. Basu, S. Garg

IR DROP PREDICTION OF ECO-REVISED CIRCUITS USING MACHINE LEARNING 153
S.-Y. Lin, Y.-C. Fang, Y.-C. Li, Y.-C. Liu, T.-S. Yang, S.-C. Lin, C.-M. Li, E. Fang

SPECIAL SESSION ON MACHINE LEARNING FOR TEST AND DIAGNOSIS 159
K. Chakraborty, L.-C. Wang, G. Veda, Y. Huang

INNOVATIVE PRACTICES ON CHALLENGES, OPPORTUNITIES, AND SOLUTIONS TO HARDWARE SECURITY 160
S. Aftabjahani, J. Oberg, M. Chen, H. Li

FAST FAULT COVERAGE ESTIMATION OF SEQUENTIAL TESTS USING ENTROPY MEASUREMENTS 161
S. Tanwir, M. Hsiao

REAL-TIME MONITORING OF TEST FALLOUT DATA TO QUICKLY IDENTIFY TESTER AND YIELD ISSUES IN A MULTI-SITE ENVIRONMENT 167
Q. Khasawneh, J. Dworak, P. Gui, B. Williams, A. Elliot, A. Muthaiah

ONLINE INFORMATION UTILITY ASSESSMENT FOR PER-DEVICE ADAPTIVE TEST FLOW 173
Y. Li, E. Yilmaz, P. Sarson, S. Ozev

SPECIAL SESSION ON QUANTUM SYSTEMS: NEXT CHALLENGES IN DESIGN, TEST, INTEGRATION 179
C. Reita, J. Baugh, G. Poulin-Lamarre, B. Kaminska, B. Courtois

NOIDA: NOISE-RESISTANT INTRA-CELL DIAGNOSIS 180
S. Mittal, R. Blanton

MULTI-FACETED MICROARCHITECTURE LEVEL RELIABILITY CHARACTERIZATION FOR NVIDIA AND AMD GPUS 186
A. Vallero, S. Tselonis, D. Gizopoulos, S. Di Carlo

RF CIRCUIT AUTHENTICATION FOR DETECTION OF PROCESS TROJANS... 192
F. Karabacak, R. Welker, M. Casto, J. Kitchen, S. Ozev

SPECIAL SESSION ON MACHINE LEARNING: HOW WILL MACHINE LEARNING TRANSFORM TEST? 198
Y. Makris, A. Nahar, H. Stratigopoulos, M. Hutner

INNOVATIVE PRACTICES ON DESIGN & TEST FOR FLEXIBLE HYBRID ELECTRONICS 199
T.-C. Huang, J. Marsh

Author Index

PROCEEDINGS

2018 IEEE 36th
VLSI Test Symposium
(VTS)

—— VTS 2018 ——

ISBN: 978-1-5386-3774-6

April 22nd – 25th 2018

San Francisco, California (USA)

PROCEEDINGS

2018 IEEE 36th
VLSI TEST SYMPOSIUM
(VTS)

Foreword

Welcome to VLSI Test Symposium (VTS) 2018, the thirty-sixth in a series of annual symposia that focuses on innovation in the field of testing of integrated circuits and systems.

The VTS 2018 program starts with a plenary keynote address from Philip Gadd, who is the Vice President of Data Center Group and General Manager Silion Photonics Product Division at Intel. He will share the challenges in design and test of silicon photonics and its applications in the Data Center Era. This will be followed by an invited keynote address by a luminary of the VTS community from University of Iowa Foundation, Distinguished Professor Reddy M. Sudhakar on "Test Drivers - Past, Present, and Future".

The core of VTS 2018, the three day technical program, responds to the many trends and challenges in the semiconductor design and manufacturing industries, with papers and presentations in the research paper sessions covering the core set of test topics: Analog, Mixed-Signal & RF Test; Hardware Security and Trust; Memory Test and Reliability; Built-in Self Test; Test Standards; Test Generation; Reliability; Applications of Machine Learning in Test and Test Data Analysis. VTS also hosts the E.J. McCluskey Doctoral Thesis Competition and a PhD research forum to showcase the exciting student research spanning all of the above topics. VTS continues a student-focused initiative following last year's success – Student Activities Program – to encourage Masters and early Ph.D. students to participate in the conference.

VTS also continues its tradition of drawing the leading test practitioners and researchers in both industry and academia to contribute to the innovative practices (IP), special sessions, and new topic sessions, enabling it to be the venue where future technology trends and test challenges are debated, test practices are shared, and test research roadmap is charted. In particular, this year's program features the intersection of test and reliability in many emerging and neighboring topics, in order to stimulate discussions and exchange of ideas beyond the traditional test community. We have a rich offering of diverse topics in the IP and special sessions: Recent Developments in Hardware Security, Intelligent Sensor Nodes, how Machine Learning transform test, Reliability and Vulnerability of Neuromorphic Computing Systems, Silicon Photonics for Future Manycore Computing, Wireless Revolution in On-Chip Communication, Impacts of Approximate Computing in Verification, Test and Reliability of Integrated Circuits, BIST/Calibration of A/MS devices, ISO26262 EDA, Test in Asia, Design & Test for Flexible Hybrid Electronics, Memory Test Practice, Functional Testing and Fault Simulation for Functional Safety. A fact worth mentioning is two hot topics, machine learning and silicon phtonics, are offered in both IP and special sessions. One new topic session is offered with focus on Quantum Systems to study its challenges in design, test, and integration.

The highly popular Monday Evening Wine-and-Cheese Panel discusses how new advances in Artificial and Machine Intelligence can impact jobs in Test and EDA. Prior to the start of the conference program, two half-day tutorials will also be offered on Machine Learning and Its Applications in Test and Learning Techniques for Reliability Monitoring, Mitigation and Adaptation.

The social program at VTS provides an opportunity for informal technical discussions among participants. VTS 2008 attendees will experience the beauty, history and infamy of Alcatraz Island on San Francisco Bay and enjoy delectable local cuisine with a stunning San Francisco Bay view at Bristo Boudin.

VTS is the result of the work of many dedicated volunteers: the reviewers, the best paper award judges, the Program Committee, the Organizing Committee, and the Steering Committee. We wholeheartedly thank them all. We also wish to thank all the authors who submitted their work to VTS 2018, and the program participants for their contributions to the Symposium.
We thank the IEEE Computer Society, the IEEE Philadelphia Section and the IEEE Computer Society Test Technology Technical Council (TTTC) for the continued technical sponsorship and support. Furthermore, we are indebted to ams AG, Dialog Semiconductor, and AdvanTest, the Premier Corporate Supporters for VTS 2018, as well as our Corporate Supporter, Mentor Graphics, for their partnership and continued support of this symposium.

We hope that you will find VTS 2018 enlightening, thought-provoking, rewarding, and enjoyable. We wish you all a fun-filled and productive week in the San Francisco area and we hope that you will keep making VTS a success by actively participating in it, assisting in its organization, and letting us always know how we can improve the symposium experience and increase its value for its audience.

Thank you all for coming!

General Chair
Chen-Huan Chiang

Program Co-Chairs
Amit Majumdar
Mehdi Tahoori

VTS 2018 ORGANIZING COMMITTEE

General Chair

Chen-Huan CHIANG
Alcatel-Lucent, US

Program Co-Chair

Mehdi TAHOORI
Karlsruhe Inst. of Technology, DE

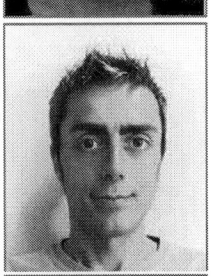

Vice General Co-Chair

Lorena ANGHEL
Grenoble Alpes University, TIMA Laboratory, FR

Vice Program Co-Chair

Stefano DI CARLO
Politecnico di Torino, IT

New Topics Co-Chair

Bernard COURTOIS
BC Consulting, FR

Special Sessions Co-Chair

Haralampos STRATIGOPOULOS
LIP6, FR

Program Co-Chair

Amit MAJUMDAR
Xlinx, US

Past Chair

Yiorgos MAKRIS
University of Texas at Dallas, US

Vice General Co-Chair

Srivaths RAVI
Texas Instruments, US

Vice Program Co-Chair

Sule OZEV
Arizona State University, US

New Topics Co-Chair

Bozena KAMINSKA
Simon Fraser U., US

Special Sessions Co-Chair

Pete SARSON
Dialog Semiconductor, UK

Innovative Practices Track Co-Chair

Suriya NATARAJAN
Intel, US

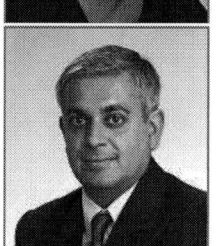

New Initiatives Co-Chair

Rohit KAPUR
Synopsys, US

Finance Chair

Jennifer DWORAK
Southern Methodist University, US

Publications Member

Elena Ioana VATAJELU
Grenoble Alpes University, TIMA Laboratory, FR

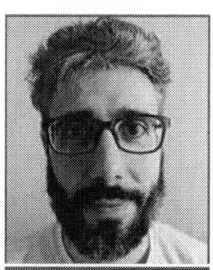

Publicity Co-Chair

Alessandro SAVINO
Politecnico di Torino, IT

Local Arrangements Chair

William EKLOW
Cisco, US

Innovative Practices Track Co-Chair

Peilin SONG
IBM, US

New Initiatives Co-Chair

X. LI
Chinese Academy of Sciences

Publications Chair

Giorgio DI NATALE
LIRMM/CNRS, FR

Publicity Co-Chair

Mango CHAO
NCTU Taiwan, TW

Registration Chair

Chintan PATEL
Univ. of Maryland Baltimore County, US

Audio/Visual Chair

Ke HUANG
San Diego State University, US

Corporate Support Co-Chair

Arani SINHA
Intel, US

Student Activities

Naghmeh KARIMI
*University of Maryland,
Baltimore County, US*

Ex-Officio

Yervant ZORIAN
Synopsys, US

Corporate Support Co-Chair

Vivek CHICKERMANE
Cadence, US

Asian Initiative Co-Chair

Kazumi HATAYAMA
Gunma University, JP

VTS 2018 STEERING AND PROGRAM COMMITTEES

Steering Committee

M. Abadir - Abadir & Associates
J. Figueras - Universita Politecnica de Catalunya
A. Ivanov - University of British Columbia
M. Nicolaidis - Grenoble Alpes University, TIMA Laboratory
P. Prinetto - Politecnico di Torino
A. Singh - Auburn University
P. Varma - Real Intent
Y. Zorian - Synopsys

Program Committee

J. Abraham – University of Texas at Austin
V. Agrawal – Auburn University
C. Argyrides – AMD
M. Barragan – TIMA Laboratory
B. Becker – Universität Freiburg
R. Blanton – Carnegie Melon University
J. Colburn – Nvidia
J. Carulli – Global Foundries
K. Chakrabarty – Duke University
T. Chakrabarty – Qualcomm
A. Chatterjee – Georgia Tech
D. Chen – Iowa State University
H. Chen – Mediatek
K. Chung – Qualcomm
C. Dixit – Broadcomm
D. Gizopoulos – University of Athens
S. Gupta – University of Southern California
I. Hartanto – Xilinx
S. Hellebrand – University of Paderborn
E. Larsson – Lund University
H. Li – Chinese Academy of Sciences

T.M. Mak – Independent
S. Makar –
H. Manhaeve – Ridgetop
P. Maxwell – ON Semiconductor
M. Michael – University of Cyprus
S. Mir – TIMA Laboratory
Z. Navabi – WPI
A. Orailoglu – University of California, San Diego
R. Parekhji – Texas Instruments
J. Rajski – Mentor Graphics
S. Reddy – University of Iowa
M. Renovell – LIRMM
M. Richetti – Synopsys
S. Shoukourian – Synopsys
O. Sinanoglu – NYU Abu Dhabi
S. Sunter – Mentor Graphics
M. Tehranipoor – University of Florida
R. Tekumalla – Broadcom
C. Thibeault – ETS Montreal
L. C. Wang – University of California, Santa Barbara
H.J. Wunderlich – University of Stuttgart

VTS 2017
Best Paper Award

2.A.2: Methodology of Generating Dual-Cell-Aware Tests

Yu-Hao Huang, Ching-Ho Lu, Tse-Wei Wu, Yu-Teng Nien (National Chiao Tung University),

Ying-Yen Chen, Max Wu, Jih-Nung Lee (Realtek Semiconductor Corp.),

Mango Chao (National Chiao Tung University)

The VTS 2017 Best Paper Award selection committee is listed below. VTS extends special thanks to these individuals for reviewing the papers and offering invaluable comments.

TM Mak, *Independant*
Salvador Mir, *TIMA Laboratory*
Arani Sinha, *Intel*

VTS 2017
Best Innovative
Practices Award

Each year, VTS recognizes the organizers and presenters of the Best Innovative Practices Session at the previous year's symposium. The selection is based entirely on audience feedback, as recorded on the attendee feedback forms.

For VTS 2017, the Best Innovative Practices Session Award goes to:

IP Session 4.C: Data Analytics in Test

Organizer:

Suriya Natarajan (Intel Corporation)

Moderator:

Abhijit Sathaye (Intel Corporation)

Presenters:

- **Big Data Analytics Engines for End-to-End Supply Chain and Quality Control**
  Thomas Harper, Paul Simon (Qualtera)
- **Data Mining of Defective Parts Investigation in Test**
  Rahima Mohammed (Intel Corporation)
- **Intelligent Data Driven Test Eco-system**
  Amit Nahar (Texas Instruments)

**VTS 2017
Best Special
Session Award**

Each year, VTS recognizes the organizers and presenters of the Best Special Session at the previous year's symposium. The selection is based entirely on audience feedback, as recorded on the attendee feedback forms.

For VTS 2017, the Best Special Session Award goes to:

Panel Session 8.A: Hot Topic: Future Extensions of IEEE Test Standards

Organizer:

Jennifer Dworak (Southern Methodist University)

Moderator:

Yu Huang (Mentor Graphics)

Presenters:

- **From 1687 to 1687.1**
  Martin Keim (Mentor Graphics)
- **Extending IEEE 1687 for Use on Analog / Mixed-Signal Chips**
  Jeff Rearick (AMD)
- **IEEE Std P1838: DFT Up and Down the Stack**
  Adam Cron (Synopsys)

**VTS 2018
Opening
Keynote**

High volume manufacturing and test of high-speed silicon photonics devices for next generation data center deployments

Philip Gadd

Intel

Vice President of Date Center Group and General Manager of Silicon Photonics Product Division

Speaker Bio:
Mr. Philip Gadd has been the Intel Vice President of Data Center Group and General Manager of Silicon Photonics Product Division since 2017. Previously, he served as Vice president and General Manager of Foxconn Interconnect Technology since October 2015 when Foxconn acquired Avago Technologies optical module division. He served as Senior Vice President and General Manager of Fiber Optics Products Division at Avago Technologies Limited since September 2009. Mr. Gadd served as Senior Vice President at Avago and has been its General Manager of Fiber Optics Products Division since September 2009. Mr. Gadd served as Vice President Avago Technologies Limited since September 2009. Mr. Gadd

oversees the overall development, production, marketing and sales of Avago's fiber optic products used in high-speed wired datacom and telecom applications. Mr. Gadd served as Vice President of Marketing of Quellan LLC. He held senior marketing and marketing management positions at Avago and Hewlett Packard/Agilent. Mr. Gadd served as Senior Director of Marketing for the Wireless Semiconductor Group at Avago until 2008. He also held technical positions at Avantek and Wessex Electronics. He has over 20 years of Radio Frequency and high-speed interconnect Product Marketing and Engineering experience. Mr. Gadd holds a B.Sc. in Communications Engineering from Plymouth Polytechnic and holds an MBA from Cranfield School of Management in the United Kingdom.

**VTS 2018
Invited
Keynote**

Test Drivers - Past, Present, and Future

Sudhakar Reddy

University of Iowa Foundation Distinguished Professor

Speaker bio:
Professor Sudhakar M. Reddy received undergraduate degrees in Physics and Electronic Communication Engineering from Osmania University, India, the M.E. degree from the Indian Institute of Science, India, and the Ph.D. degree from The University of Iowa, USA. Since 1968, he has been a faculty member with ECE Department, The University of Iowa, where he is currently a University of Iowa Foundation Distinguished Professor. He served as the Chair of the ECE Department from 1981 to 2000. He has

authored over 600 papers in archival journals and conferences. He received the von Humboldt Senior Research Fellow Award in 1995 and the first Life Time Achievement Award from the International Conference on VLSI Design. He served as an Associate Editor and twice as a Guest Editor of the IEEE Transactions on Computers and as an Associate Editor of the IEEE Transactions on Computer-Aided Design of Integrated Circuits and Systems. Professor Reddy is an IEEE Life Fellow.

Group Delay Measurement of Frequency Down-Converter Devices Using Chirped RF Modulated Signal

Peter Sarson

ams AG
Premstaetten, 8141, Austria
peter.sarson@ams.com

Tomonori Yanagida, Kosuke Machida

Division of Electronics and Informatics
Gunma University
Kiryu, Gunma 376-8515, Japan

Abstract—**This paper discusses how to measure group delay of radio frequency (RF) frequency-converting devices using typical RF Automated Test Equipment (ATE) when conventional S-parameter measurements cannot be done due to the difference between a device's input and output frequencies. We propose using a chirp waveform to modulate an RF generator to sweep the input frequency to an RF device, and describe how to produce an optimal modulation waveform with digital signal processing. We discuss how to implement the test based on our previous work in the baseband domain and how to understand the measurement errors in this RF context. Group delay measurements results are presented for an RF mixer, measured through pre- and post-down-convert RF/IF filters.**

Keywords—Group Delay, RF, Automotive, Frequency Converter, Automated Test Equipment, Chirp, Modulation.

I. INTRODUCTION

For most applications, group delay specifications of radio frequency (RF) receivers are not tested in production due to the prohibitively high cost of the tests in terms of equipment and test time. As a result, in most cases these specifications are guaranteed by design. However, automobiles are beginning to incorporate more RF receivers and it is anticipated that many will be used in autonomous vehicles, so testing the group delay specifications is likely to become mandatory.

Conventional techniques for measuring group delay use a vector network analyzer (VNA), such as the Aeroflex 6480 Microwave System Analyzer, to provide phase or group delay data relative to a golden device [1][2]. Some equipment suppliers, such as Keysight Technologies, have developed calibration standards for down-converters using three broadband standards: a power meter as a magnitude standard, a comb generator as a phase standard, and an S-parameter calibration kit [2][3]. Other techniques, such as using a transmitter with the same characteristics as the receiver-under-test, allow a VNA to measure the group delay directly [1][4].

There have been efforts to measure the group delay of a system without having the system phase-locked by using a comb generator [3], but this is not feasible because calibrated standards are not available for use in ATE. Using a two-tone stimulus signal has been proposed, where measuring the phase change

between the two tones at both baseband and RF indicates the group delay through the device [4], but it is very difficult to measure the phase of an RF input signal. The user is restricted to down-converting the input by means of the tester's RF measurement hardware, which prevents calibration of part of the signal path within the RF ATE's infrastructure (Fig 1). In addition, using the two-tone option of an ATE would be very slow to sweep across a typical communications band.

The scheme in [4] can be a built upon for measuring and characterizing group delay in production test. The scheme's fundamental idea is that it is not necessary to measure the phase of any individual frequency, just the phase delta between pairs of frequencies. Modulating an RF signal source with a lower frequency two-tone signal would allow control of the phase and would require only one RF source and arbitrary waveform generator (AWG), where the digital-to-analog converter (DAC) resides. This would reduce cost and add flexibility; however, generating a two-tone signal across a large bandwidth could introduce a large error.

The technique in [4] can be further improved by using a high-resolution chirp to modulate the RF source. Using such standards and techniques [2][3][4] depends on the paths within an ATE architecture accessible to a test engineer, for the majority of the ATE testers available today. One solution would be to use of a golden sample correlation method.

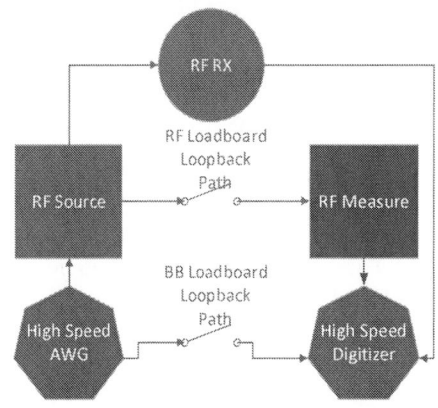

Fig. 1. Inaccessible calibration path (shown in Red) in typical RF tester

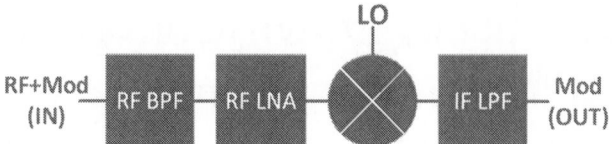

Fig. 2. RF receiver block diagram

A major benefit of RF ATE is that all its instruments are phase locked to a 10 MHz reference clock. A useful feature of its RF source is that it is possible to apply waveforms to its analog IQ modulation inputs. For a limited frequency sweep, such as 25 MHz, it is possible to chirp [5]-[8] the RF source, hence sweeping its output frequency quickly compared to using a two-tone setup [4].

We propose exploiting these benefits of RF ATE, and our experimentally proven technique, to create a production group delay test for RF devices. This would produce RF devices with higher average out-going quality for autonomous vehicles, where a slight error in communication could result in a fatality.

We discuss how to implement a group delay test on a very high frequency (VHF) or ultra-high frequency (UHF), zero intermediate frequency (IF) receiver using typical RF ATE. Fig. 2 shows a simplified block diagram of an RF receiver. The device consists of an RF band-pass filter (BPF), RF low noise amplifier (LNA), mixer, and an IF low-pass filter (LPF). The two principal direct contributors that could cause a group delay failure of the system would be the LPF and BPF. These components' primary function is to alter the phase at specific frequencies; therefore, a defect that changed this characteristic could go undetected if only output magnitude is tested.

This paper is organized as follows: Section II describes the basics of mixer theory, and how a chirp, as described in [7][8], can be used to sweep an RF down-converter's input. Section III discusses how to generate a discrete linear chirp. Section IV details some measurements using the RF tester in loopback mode to verify the technique, and conclusions are drawn in Section V.

II. RF FREQUENCY SWEEP

An RF source of an ATE typically has analog IQ modulation inputs that can be modulated by a high speed AWG of the tester, physically close to AWG's digital input. By applying either the same signal to both IQ inputs (i.e., not in quadrature), or just to the I channel, with DC for the Q channel, a single frequency, double sideband, amplitude modulated (AM) signal (Fig 3) is generated. When applying the chirp waveform described in [7][8], for every step of chirp waveform we would observe the frequency of the RF signal increment, up to the maximum frequency in the chirp waveform. A side effect of the mixing process is creation of an image, f_1-f_2, below the primary RF signal frequency, as well as, f_1+f_2, above the primary frequency, as shown in (1) and Fig. 4, where $cos(\omega_1)$ and $cos(\omega_2)$ are the time-domain representations of f_1 and f_2. The factor ½ shows the input amplitude is divided between the two output terms; in

practice, this represents a 6 dB conversion loss. This side effect of the mixing process can be used as an advantage and to save AWG bandwidth. By placing the RF frequency f_1 in the center of the measurement band, it is possible to sweep either side of the RF carrier frequency.

$$cos\,\varpi_1\,cos\,\varpi_2 = \frac{cos[\varpi_1+\varpi_2]+cos[\varpi_1-\varpi_2]}{2} \quad (1)$$

Therefore, to sweep to the maximum possible frequency using this technique with the typical tester configuration, we need to create a chirp that can be measured by the maximum digitizer rate of 100 MHz. The Nyquist sampling theorem dictates that to oversample a signal correctly, we need at least two points per cycle of the highest frequency of interest, as shown in (2); however, in reality, a slightly higher sampling rate is need to avoid the Nyquist frequency itself. A better rule is four points per cycle. This means we are limited to measuring a maximum frequency of 25 MHz. Since doubling of the signal bandwidth occurs due to the mixing, we can generate just a 0 to 12.5 MHz chirped signal.

$$F_s > 2F_t \text{ or } F_s = 4F_t \quad (2)$$

III. CHIRP CREATION

A typical high speed AWG has a sample rate of 250 MHz, therefore, it is possible to generate a very high fidelity chirp signal from 0 to 12.5 MHz that exhibits very high phase resolution. This is needed to perform a phase unwrap of the measured signal [7][8]. Equation (3) describes a chirp mathematically:

$$F(t) = f_{start} + kt \quad (3)$$

where $k = \frac{f_{start}-f_{stop}}{T}$

f_{start} is the starting frequency (at time t=0), f_{stop} is the final frequency, and T is the time taken to sweep from f_{start} to f_{stop}. k is called the rate of frequency increase or chirp rate.

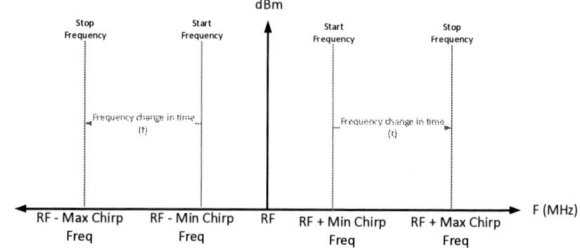

Fig. 3. Frequency spectrum of a chirped RF source without quadrature signal on the IQ inputs.

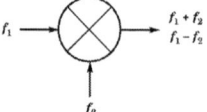

Fig. 4. Frequency mixer schematic diagram

Next we discuss digital data generation for the AWG. Equation (4) is the coherence condition and by using this, the Fourier frequency can be derived as (5). For this example, this is the frequency resolution of the waveform. Hence, each frequency step in the chirp will increase by F_f. Discrete frequency steps are needed to ensure the number of data values needed fits within ATE memory. The unit test period (UTP) of the chirp signal, defined in (6), is the reciprocal of the Fourier frequency. The phase resolution, θ_r, of the waveform is calculated using (7). M is the number of cycles, N is the number of samples, F_s is the sampling frequency of the AWG that will modulate the RF source, and F_t is the discrete modulation frequency of the chirp at any time.

$$F_s/F_t = N/M \qquad (4)$$
$$F_f = F_s/N \qquad (5)$$
$$UTP = 1 / F_f \qquad (6)$$
$$\theta_r = 2\pi/N \qquad (7)$$

By rearranging (4), the bin number, M, can be calculated for the start and stop frequencies defined in (3) as m_{start} and m_{stop}. The frequency resolution, F_f, is dictated by choosing an appropriate array length, N, for the chosen sample frequency, F_s, as defined in (5). A discrete version of (4) can be made using (8), where d_n is the discrete chirp. Because the change in signal frequency is made by incrementing the frequency by F_f, the waveform remains coherent with a measurement instrument that uses the same sampling conditions. Using small frequency increments ensures the phase change for each increment will not be large and allows a phase unwrap of the captured signal.

Reference [10] discusses the issues pertaining to phase reversal using double sideband-suppressed carrier (DSB-SC). Care needs to be taken when generating the chirp, as defined in (8), because when both the mixing signal and the chirp signal pass through zero, phase reversal can occur which could cause the phase unwrap procedure to fail.

$$d_n = \cos\left(\theta_r\big(m_{start} + k(n-1)\big)(n-1)\right), \; n = 1,2,\dots,N$$
$$(8)$$

where $k = \dfrac{1+(m_{start}-m_{stop})}{2N}$

An AWG typically has a programmable LPF, and for this example we would set it to 50 MHz. The effect of this filter on the generated 12.5 MHz chirp waveform would be negligible.

Depending on the difference ΔF, shown in (9), between the start and stop frequencies of the chirp, and the time taken transitioning between these two frequencies, $N\times1/f_f$, will produce a chirp waveform exhibiting sidelobes. $\Delta F\times1/f_f$ is often referred to as the time-bandwidth product (tbp) of the chirp, as defined in (10). As the sidelobes fall away at 3 dB over a tbp interval, the roll off can be described as (11); therefore, it is possible to calculate, at any frequency, what the sidelobe power would be and any undesired influence on a system using this scheme. The

longer the transition from the start to the stop frequency, the larger is tbp, the larger is the side lobe, and the shallower is the roll off.

$$\text{linear chirp bandwidth (lcbw)} = \Delta F \, (3\text{dB}) \qquad (9)$$
$$tbp = \Delta F \frac{1}{f_f} \qquad (10)$$
$$\text{sidelobe roll off} = 3 \text{ dB}/tbp \qquad (11)$$

where $\Delta F = F_{stop} - F_{start}$, chirp -3 dB bandwidth

A key requirement of a chirp for measuring group delay is a small phase change between frequency steps, which usually requires a small frequency change between points, which in turn requires a large number of data points, N. As can be seen in Fig. 5, there is a sharp fall-off of the side lobes at the stop frequency. To setup chirp creation for the 12.5 MHz frequency sweep, a time-bandwidth product of 25,000 can be calculated because there are 200,000 points with a time per sample of 10 ns. By observing the generated chirp in dBc, it is possible to see the 3 dB bandwidth point.

The zoom-in plot of Fig. 6 shows the side lobe occupies approximately 25 kHz of spectrum bandwidth: 25 kHz at the -3 dB points of the start and stop frequencies of the 12.5 MHz chirp signal means the response is flat from 25 kHz to 12.475 MHz. This is confirmed by using (9)-(11): the result for tbp=25,000 is a roll-off starting at 12.475 MHz, finishing 3 dB down at the stop frequency of 12.5 MHz. If a smaller tbp is used, the sidelobe bandwidth occupies more bandwidth.

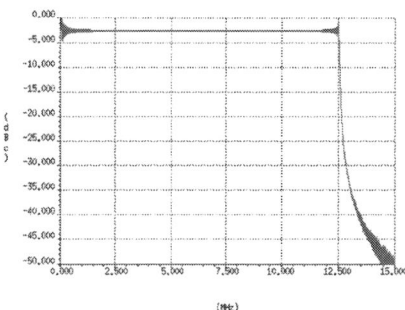

Fig. 5. Frequency domain representation of a discrete linear chirp

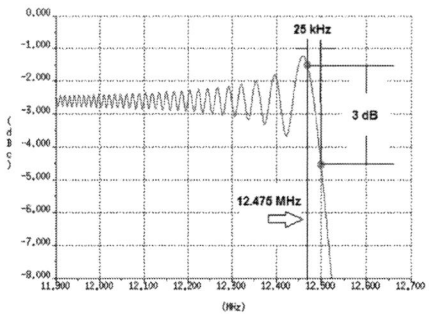

Fig. 6. Zoom in on the chirp upper sidelobe

978-1-5386-3775-3/18 $31.00 © 2018 IEEE

IV. MEASUREMENT SETUP

To verify this technique, the tester can be configured to replicate an RF receiver by using the tester's RF measure module to measure a chirp-modulated RF source.

An RF measure module port is designed for a large range of input power, for example 120 dB, and this is achieved by using appropriate amplifier and attenuator settings within the port module to deliver approximately 1 V amplitude to the digitizer. Amplifiers can introduce distortion, so to maximize accuracy when measuring the reference signal, the setting that uses the fewest active components should be selected. To further improve linearity, only half of the digitizer's range should be used. For this example, the input level to the RF source is set to deliver enough power to produce 0.5 V at the digitizer input.

To cover more than twice the AWG's bandwidth, keeping the IQ waveforms of Fig. 7 in phase produces a DSB AM signal which, when demodulated, produces twice the bandwidth of the generated chirp signal, Fig. 8. To be able to calculate the group delay of the RF subsystem, the tester must be modeled accurately to produce an ideal response to compare against the measured signal. This can be expressed mathematically as (12) and is shown in Fig. 8.

$$s(t) = d(t)\cos(\varpi_1 t) \tag{12}$$

where $\varpi 1$ is the carrier frequency and $d(t)$ is the DAC output for input d_n. $s(t)$ is a discrete modulated chirp, as would be seen at the output of an RF generator when modulated according to (8) by an AWG. Using the RF measure module to down-convert and measure the amplitude response shows the double modulation bandwidth.

In analog/mixed-signal circuit test, it is very common to trigger an AWG and digitizer to start simultaneously [7][8]. Unfortunately, in an RF system this is usually not possible because the digitizer is part of the RF subsystem, thus another method for capturing the data is needed. If the digitizer is setup such that twice as many points are captured compared to the chirp signal, but with sampling rate F_s, then two cycles of the chirp should be present. Locating the position where the chirp stops allows extracting the signal for post-processing and thus solves the triggering issue. To be able to mix down the RF modulated signal shown in Fig. 9, the local oscillator (LO) of the RF measure module must be set at the carrier frequency of the modulated signal, $\varpi_1 \pm \text{Chirp}_{max\_freq}$. In this case, the LO frequency was set to the carrier frequency, Chirp_{max_freq}.

Because the chirp signal is a user-defined signal with its own bandwidth definitions, the RF measure must be setup to allow capturing this signal and to allow easy manipulation of the signal in digital signal processing (DSP) after it is digitized. The RF measure module of a tester will usually have a user-defined range that allows setting the IF that the RF will mix the frequency down to, the sampling frequency, and the number of points the digitizer will capture of the IF. By following (2) and setting the IF to $F_s/4$, verification of (12) can be achieved by multiplying the

digitized signal by a cosine at a multiple of the IF to extract the cosinusoidal component of the chirp from the captured signal. If there is no sinusoidal component, multiplying with a sinusoid at a multiple of the IF should result in zero.

The group delay, T_g, as defined in (13), is a useful measure of time distortion, and is calculated by differentiating, with respect to frequency, the phase response of the device under test (DUT); the group delay is a measure of the slope of the phase response at any given frequency. Variations in the group delay cause signal distortion, just as deviations from linear phase cause distortion.

$$T_g = -d\varphi/d\varpi \tag{13}$$

where $d\varphi$ is the change in phase and $d\varpi$ is the frequency aperture or frequency change. In this case, the frequency aperture is the frequency step size of the chirp, which is, for this example, the Fourier frequency of the signal, F_f (defined in (5)) or the frequency resolution.

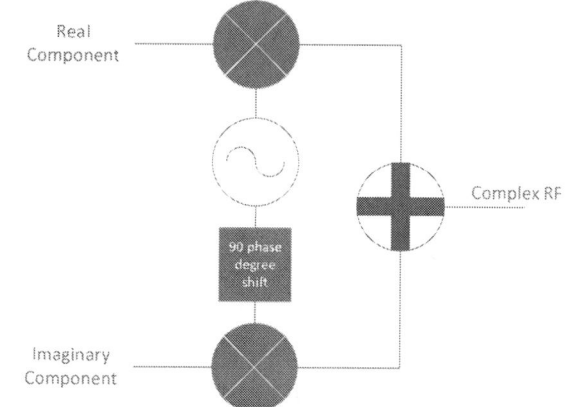

Fig. 7. RF generator with IQ modulation

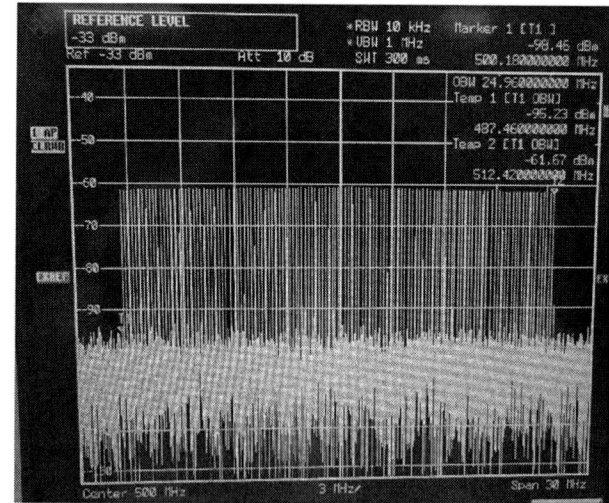

Fig. 8. DSB-SC chirp modulation waveform at 500 MHz carrier frequency

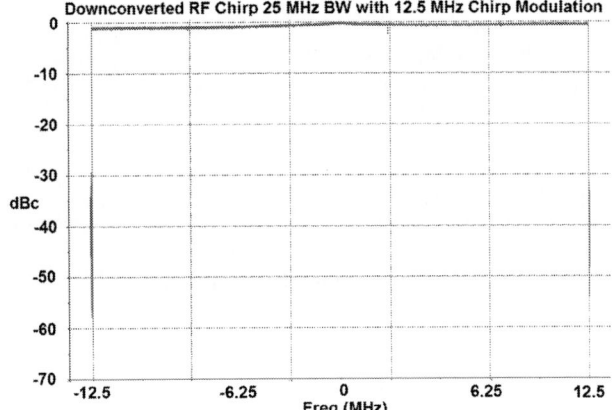

Fig. 9. Demodulated DSB-SC chirp modulation waveform

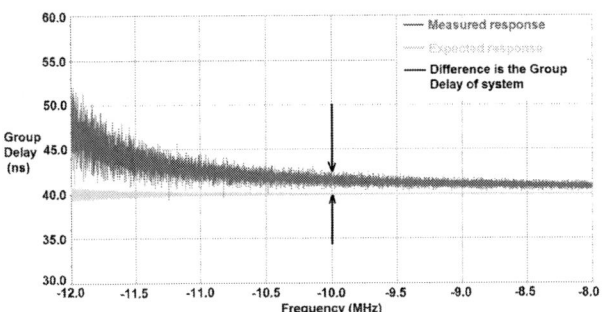

Fig. 10. Zoom in of the measured group delay profile of the measurement vs expected profile using a chirp signal

For an RF system measurement, the group delay measured for the direct signal must be subtracted from the group delay measured via the RF path.

Chirp phase reversing at $Chirp_{max\_freq}/2$ produces phase distortion. As a result, a small part of the chirp bandwidth is unusable and must be corrected. Fig. 10 shows the zoomed in part of the group delay capture, at the end of the frequency sweep, where the effect would be the most apparent compared to the reference. This shows the group delay of the RF measure module becoming increasingly worse above 10 MHz. This makes sense because the useable bandwidth of the Xcerra RF subsystem is approximately 20 MHz, i.e., ±10 MHz, thus proving this technique is viable.

Subtracting the measured group delay of the tester from the group delay of the reference signal (Fig. 11) gives the group delay of the Xcerra Fusion MX subsystem. As no filter is in the path, the only explanation for the roll off observed across the band shown in Fig. 11 is the group delay of the mixer that down-converts the signal.

If an accurate representation of a filter is required, then this characteristic must be calibrated out. This is easily achieved by subtracting the group delay with the chosen RF subsystem filter from the group delay without any filter. Before this can be done,

the measurement accuracy of the system must be known. Measurement error can be assessed by measuring twice the IF path with no filter and subtracting the two responses. Fig. 12 shows the measurement error, and from -10 to +10 MHz, it is less than 25 ps, thus 10x better than the required measurement accuracy for the filter specifications targeted by this technique.

To test any RF receiver, the preceding discussion must be kept in mind. If an RF BPF is to be tested, then some way of switching this filter out of the RF path is essential during design-for-test (DFT) or else the influence of the down-convert mixer will always be present in the measured result. The same consideration is necessary for the LPF post-mixer. If a filter in the IF path of the RF measure port is switched in, the characteristic of this filter should be clearly identifiable.

Fig. 13 shows a typical group delay characteristic of a LPF. In this case, a 10 MHz bandwidth is specified and the filter's 3 dB point is evident at 9.75 MHz, but the group delay starts to degrade at 7.5 MHz where the filter attenuation begins. This is equivalent to the LPF shown in Fig. 2, and thus this technique is capable of accurately testing such a filter if the DFT is adequate. If not, it is still possible to test the filter but with the influence of the down-converter included.

The 10 MHz LPF is a case where the group delay changes slowly across the band due to the width and type of filter. If a more complex filter is switched into the IF path, such as a BPF, where the phase is changing across the whole bandwidth, then this would be more challenging to measure.

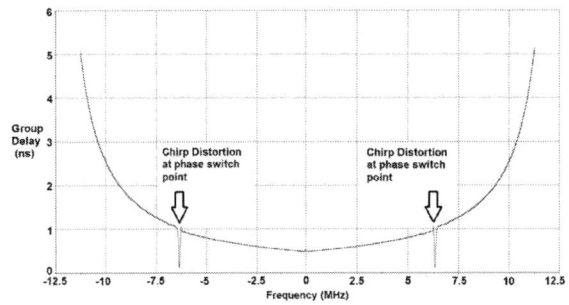

Fig. 11. Group delay of Xcerra RF measure port.

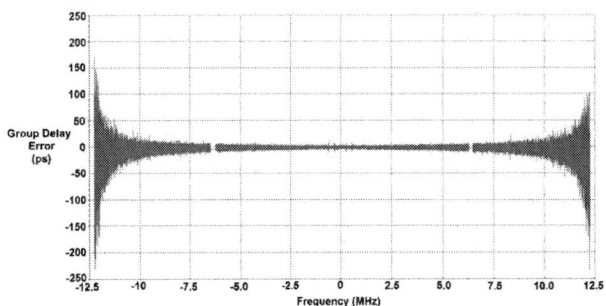

Fig. 12. Measurement error of group delay of IF path of the RF measure port module.

978-1-5386-3775-3/18 $31.00 © 2018 IEEE

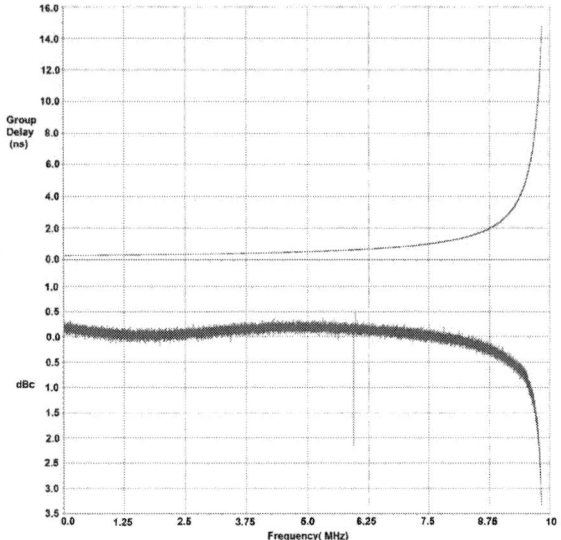

Fig. 13. Group delay and Amplitude Response of 10 MHz LPF in the IF path of the Xcerra RF measure port module.

In Fig. 14, the datasheet amplitude response and group delay of the Mini Circuits BPF-B48+ bandpass filter are shown in the top right corner. Comparing the measured result to the curves from the datasheet [11], it can be observed that the chirp technique achieves a good approximation of the expected result. The expected curve should have a starting group delay of 167 ns at 46 MHz, a minima of 127 ns at 48.5 MHz, and then a return to 155 ns at 50.5 MHz. The measured curves fit this closely except for a 5 ns error at the maximum frequency of 50.5 MHz. Using these two examples covers all the components shown in Fig. 2 for an RF receiver.

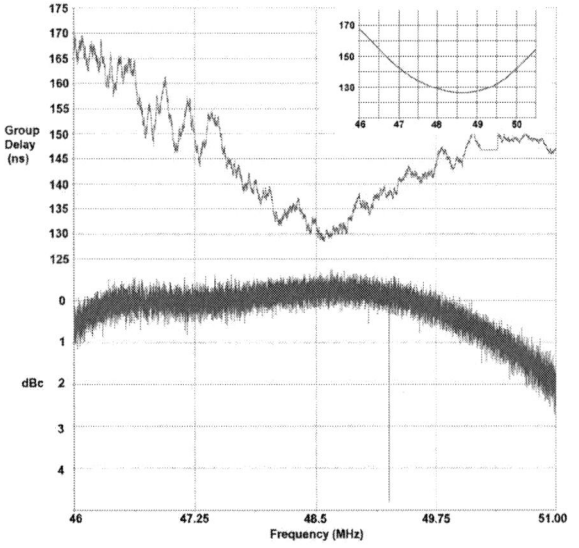

Fig. 14. Group delay and Amplitude Response of 2 MHz BPF in the RF path of the Xcerra RF measure module port (Datasheet top right insert) .

V. CONCLUSION

The range of a typical RF tester that can produce DSB modulation can be extended, even if the system AWG used for modulation has a limited bandwidth, thus allowing older test equipment to test state-of-the-art devices. Using the technique described produces a very accurate and fast method of characterizing RF receiver devices' group delay. By subtracting the measured response from the ideal, an accurate comparison can be made. In most cases today, only a few frequencies of a filter are compared against a threshold value, potentially allowing defects into the field. With the proposed technique, the quality level of tested RF receivers, with any RF or IF filter, can be significantly increased.

REFERENCES

[1] C.J.Clark; A.A.Moulthrop; M.S.Muha; "Transmission Response Measurements of Frequency-Translating Devices Using a Vector Network Analyzer," *IEEE Transactions on Microwave Theory and Techniques*, Volume: 44, Issue: 12 (Dec 1996)

[2] Joel Dunsmore, "A New Calibration Method for Mixer Delay Measurements That Requires no Calibration Mixer", *Microwave Conference (EuMC), 2011 41st European*

[3] J.Scott, M.Hoy, "Group-Delay Measurement of Frequency-Converting Devices Using a Comb Generator," *IEEE Transactions on Instrumentation and Measurement* (Volume: 59, Issue: 11, Nov. 2010)

[4] T.Bednorz, "Group Delay and Phase Measurements at Converters and Multistage Converters Without LO Access," *Microwaves, Communications, Antennas and Electronics Systems*, 2009. COMCAS 2009. IEEE International Conference on

[5] P.Sarson,"RF Filter Characterization Using a Chirp," 2014 *9th International Design and Test Symposium* (IDT), Year: 2014, Pages: 1–5

[6] P. Sarson, "Group Delay Filter Measurement Using a Chirp," 2016 *21st IEEE European Test Symposium (ETS)*, Year: 2016, Pages: 1 – 2

[7] P. Sarson, "Test Time Efficient Group Delay Filter Characterization Technique Using a Discrete Chirped Excitation Signal," *2016 IEEE International Test Conference (ITC)* ,Year: 2016, Pages: 1 – 6

[8] P.Sarson, "An ATE Filter Characterization ToolKit Using a Discrete Chirped Excitation Signal as Stimulus," *Journal of Electronic Testing: Theory and Applications*, Springer (2017), DOI : 10.1007/s10836-016-5633-x

[9] H.Kitayoshi, S.Sumida, K.Shirakawa, S.Takeshita, "DSP Synthesized Signal Source for Analog Testing Stimulus and New Test Method," *Proc. International Test Conference*, pp.825-834, 1985.

[10] B.Kanmani, "The 'Phase-Reversal' in DSB-SC: A Comment," *Digital Signal Processing Workshop and 5th IEEE Signal Processing Education Workshop*, 2009. DSP/SPE 2009. IEEE 13th.

[11] Mini Circuits BPF-B48+ Datasheet , https://ww2.minicircuits.com/pdfs/BPF-B48+.pdf

A Coherent Subsampling Test System Arrangement Suitable for Phase Domain Measurements

Young Gouk Cho‡ †, Gordon W. Roberts‡ Sadok Aouini†, Mahdi Parvizi† and Naim Ben-Hamida†

‡ *McGill University,*
Montreal Quebec, Canada H3A 0E9

† *Ciena Canada,*
Ottawa, Ontario, Canada K2K 2P5

Abstract— This paper presents a coherent subsampling test system arrangement that can be configured to measure the intrinsic noise and jitter transfer characteristic response of a time-based channel. Such as a phase-locked loop using a sampling rate that is established based on the incoming signal bandwidth rather than the highest frequency contained in the measured signal. These test systems are applicable to a wide range of applications from RF to optical communication systems and can be constructed using off-the-shelf components. Experiments conducted on a Teradyne Flex tester are used to validate the proposed subsampling principles.

Key words— Coherent subsampling, phase domain, phase measurement, jitter transfer characteristic

I. INTRODUCTION

THIS paper will present a test system that can be used to coherently subsample the intrinsic noise as well as the jitter transfer characteristic components of a signal. The traditional method of sampling data, is to use the well-known Nyquist-Shannon sampling theorem - which states that the sampling rate must be greater than twice the highest frequency contained in the signal. On the other hand, today's technology is operating at extremely high data rates, in excess of 100 Gbps in fiber optic channels, and these rates are expected to increase further in time. With such signal rates, a traditional coherent sampling system would require digitizers with extremely high sampling rates, roughly ten-times the incoming clock rates. A test system operating at such rate would be extremely expensive to build, if at all possible with today's technologies. In this paper, a solution to this problem is proposed based on a coherent subsampling method, whereby the sampling rate is determined based on the desired signal bandwidth instead of the highest frequency contained in the signal, allowing for a much lower sampling rate. Proposed test system is relatively inexpensive to construct and can be built today using off-the-shelf components.

Previous work on coherent subsampling was performed by Okawara [1]. Although our method and Okawara's method leads to identical outcomes, differences exist in the way in which the data is processed. In Okawara's method, additional post-processing is need to re-shuffle the captured data set, as well as detect which image was captured. Furthermore, depending on the image type (odd or even), re-ordering of its time sequence are necessary to recover the original signal information. Conversely, our method does not need any post-processing DSP procedure to recover the original signal information; thus, less steps are involved during the subsampling process.

To illustrate the principle of subsampling, consider the diagram shown in Fig. 1 involving a sine wave with period T_T. Here, the sinewave has been sampled with a period greater than a single period, specifically one sample every $p8+1$ seconds apart where p is an arbitrary integer. For sake of illustration, this sinewave has been sampled at a rate that results in 8 points being collect over a single period T_T of this waveform. Here these points are labeled 1, 2, 3, … 8. On account of the timing between the sampling rate and the signal frequency, the sampling of the next cycle of the sinewave results in the exact same sample set, and the next and the next and so on. Because of the periodic extension property of this sample set, the sample points can be collected at a much slower rate, and in any order, without loss of signal information.

One of the objectives of this paper is to identify the sampling conditions for a subsampling process that is synchronized with

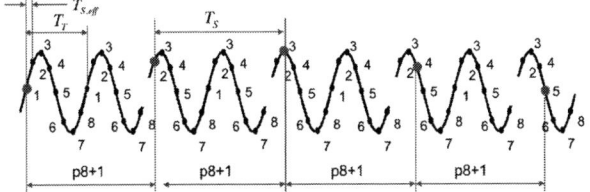

Figure 1: **Highlighting the principle of subsampling on a sine wave signal consisting of 8 sample points whose resulting sample set is periodic with period $T_T=8T_{s,eff}$ (black dots) or $8(p8+1)$ (blue dots).**

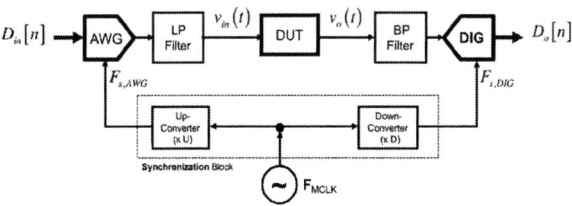

Figure 2: **Proposed coherent subsampling test system using a bandpass digitizer and a phase-locked synchronization network.**

the incoming test signal, herein referred to as a coherent subsampling system. Another objective of this work is to demonstrate how such a test system can be used to measure a wide range of analog and sampled channels that use phase-modulation (PM) type signals such as phase-locked loop and delay-locked loops.

In addition, this paper will also demonstrate how intrinsic noise measurements can be performed with this test system. A phase noise measurement attempts to gain information on the random phase shifts that occurs in a periodic signal. The most direct way in which to obtain phase noise information is to use a spectrum analyzer and observe its frequency-domain characteristic [2]. Digital Signal Processing (DSP) techniques, along with time-to-digital converters (TDCs) can also be used to measure a phase noise [3]. In this work, a DSP-based technique called coherent phase extraction [4, 5] will be used to obtain the phase domain characteristic within a test signal using a subsampling approach.

The paper is organized as follows: section II describes the general sampling considerations for a coherent subsampling system of phase-modulated noise signals. Section III describes a similar set of conditions for measuring intrinsic noise signals. Section IV describes the clock synchronization requirements for coherency. Section V will provide simulation results using MATLAB/Simulink and Section VI will include experimental results using an ATE. Finally, the paper concludes in Section VII.

II. General Sampling Considerations

The proposed coherent subsampling test system is shown in Fig. 2 where an arbitrary waveform generator (AWG) takes a digital signal D_{in} from the source memory of the ATE and produces a base-band test signal over some bandwidth BW_{AWG}, denoted by $v_{in}(t)$, that is used to excite the device-under-test (DUT). This response, described by $v_o(t)$, is subsequently filtered with a digitizer (DIG) having a bandpass response with bandwidth BW_{DIG} and whose sample values D_o are delivered to the capture memory of the ATE. The AWG and DIG is assumed to be clocked with a frequency $F_{S,AWG}$ and $F_{S,DIG}$, respectively. The sampling clocks are derived from a synchronization network involving a master reference clock with frequency F_{MCLK}, and an up- and down-frequency converter. By design, the sampling rate of the AWG is much greater than the sampling rate of the digitizer,

$$F_{S,AWG} \gg F_{S,DIG} \qquad (1)$$

so that the DIG subsamples or under-samples the incoming signal $v_o(t)$.

The AWG is assumed to generate a base-band signal according to the principles of coherency, i.e.,

$$F_T = \frac{M_{AWG}}{N_{AWG}} F_{S,AWG} \qquad (2)$$

where N_{AWG} represents the number of samples values used in the source memory of the AWG and M_{AWG} represents the number of cycles a sinusoidal will complete in the time period $N_{AWG}/F_{S,AWG}$. In contrast, the DIG is assumed to sample an aliased version of this tone, i.e.,

$$F_{S,AWG} > F_T > F_{S,DIG} > F_{T,alias} \qquad (3)$$

then, according to the principles of coherency, i.e.,

$$F_{T,alias} = \frac{M_{DIG}}{N_{DIG}} F_{S,DIG} \qquad (4)$$

where N_{DIG} represents the number of samples values used in the capture memory of the DIG and M_{DIG} represents the number of cycles a sinusoidal will complete in the time period $N_{DIG}/F_{S,DIG}$. As both the AWG and the DIG are sampled systems, the spectral behavior in both cases are periodic in frequency, but the DIG exhibits a much smaller period. Of further consideration is the fact that any one period of the frequency spectrum consists of an even and odd part that are fold-symmetric [6].

It is also important to note the response of the AWG is low-pass in nature, whereas the DIG has a bandpass response. Here it is shown that the bandwidth of the DIG is set equal to $F_{S,DIG}/2$, it need not be. It can, in principle, be made much larger than the AWG bandwidth without violating the conditions of Eqn. (3).

Using a DIG sampling rate that is much lower than its incoming signal frequency can be advantageous when trying to capture signal information associated with high-frequency (HF) analog/mixed-signal systems. Coherent subsampling test arrangements for measuring RMS value of the individual phase frequency components of a phase-modulated (PM) signal will be described as well as its sampling conditions. Although, only the PM signal is described in this paper, the subsampling method can also be applied to amplitude based signals.

A. Subsampling a Phase-Modulated (PM) Signal

Often the frequency response behavior of a time-sensitive channel such as phase-locked loop (PLL) is to be measured. To do so, requires the AWG to generate a multi-tone signal embedded in the phase of a sinewave signal acting as the carrier. The proposed coherent subsampling system can be used to perform this measurement.

Consider a sinusoidal signal with amplitude A_{in} that is coherently generated by the AWG without any phase signal component as follows

$$D_{in}[n] = A_{in} \cdot \sin\left(2\pi \frac{M_{AWG}}{N_{AWG}} n\right) \qquad (5)$$

As Eqn. (5) must satisfy the AWG coherency requirements, we can rewrite the above equation in terms of the test frequency and AWG sampling rate as

$$D_{in}[n] = A_{in} \cdot \sin\left(2\pi \frac{F_T}{F_{S,AWG}} n\right) \qquad (6)$$

Now, assume that this signal is applied to the DUT input and the output signal is sampled by the DIG. Assuming the frequency of the output signal remains the same and the bandwidth of the BP filter is sufficient to pass the test signal with little attenuation, the DIG output signal can be written as

$$D_o[n] = A_o \cdot \sin\left(2\pi \frac{F_T}{F_{S,DIG}} n\right) \qquad (7)$$

where A_o represents the amplitude of the DUT output signal. As the frequency of the output signal is related to the coherency parameters of the AWG described by Eqn. (2), one can re-write Eqn. (7) as

$$D_o[n] = A_o \cdot \sin\left(2\pi \frac{M_{AWG}}{N_{AWG}} \frac{F_{S,AWG}}{F_{S,DIG}} n\right) \quad (8)$$

The relationship between the original tone frequency F_T and the aliased tone frequency $F_{T,alias}$ can be summarized as

$$F_T = \begin{cases} p \cdot F_{S,DIG} + F_{T,alias} & \text{even image} \\ p \cdot F_{S,DIG} - F_{T,alias} & \text{odd image} \end{cases} \quad (9)$$

where p is an arbitrary integer representing the number of cycles of the digitizer spectrum encompassed or nearly encompassed by the incoming tone frequency. Substituting the AWG and DIG coherency relationships, specifically, Eqns. (2) and (4), into Eqn. (9) leads to

$$\frac{M_{AWG}}{N_{AWG}} F_{S,AWG} = p \cdot F_{S,DIG} \pm \frac{M_{DIG}}{N_{DIG}} F_{S,DIG} \quad (10)$$

or, when re-arranged,

$$\frac{F_{S,AWG}}{F_{S,DIG}} = \begin{cases} \dfrac{p + M_{DIG}/N_{DIG}}{M_{AWG}/N_{AWG}} & \text{even image} \\ \dfrac{p - M_{DIG}/N_{DIG}}{M_{AWG}/N_{AWG}} & \text{odd image} \end{cases} \quad (11)$$

The above relationship identifies the synchronization condition between the sampling rates of the AWG and the DIG.

With above relationship in mind, let us consider the following PM signal equation used to produce the samples for the AWG source memory,

$$D_{in}[n] = A_{c,in} \sin\left(\begin{array}{l} 2\pi \dfrac{M_{AWG,c}}{N_{AWG}} n \\ + \sum_{i=1}^{P} A_{\phi,in,i} \cdot \sin\left(2\pi \dfrac{M_{AWG,\phi,i}}{N_{AWG}} n\right) \end{array} \right) \quad (12)$$

Here $A_{c,in}$ represents the amplitude of the carrier signal, and $A_{\phi in,i}$ are the amplitudes of the individuals phase tones. Here it is assumed that P-tones are used. The carrier and phase tone frequencies are assumed to be coherent with AWG sampling rate $F_{S,DIG}$ satisfying

$$F_c = \frac{M_{AWG,c}}{N_{AWG}} F_{S,AWG} \quad (13)$$

Likewise, the frequencies of the individual phase tones are given by

$$F_{\phi,i} = \frac{M_{AWG,\phi,i}}{N_{AWG}} F_{S,AWG} \quad (14)$$

Correspondingly, the samples collected by the digitizer have the familiar form,

$$D_o[n] = A_{c,o} \sin\left(\begin{array}{l} 2\pi \dfrac{M_{AWG,c}}{N_{AWG}} \dfrac{F_{S,AWG}}{F_{S,DIG}} n \\ + \sum_{i=1}^{P} A_{\phi,o,i} \cdot \sin\left(2\pi \dfrac{M_{AWG,\phi,i}}{N_{AWG}} \dfrac{F_{S,AWG}}{F_{S,DIG}} n\right) \end{array} \right) \quad (15)$$

where $A_{c,o}$ represents the amplitude of the carrier signal and

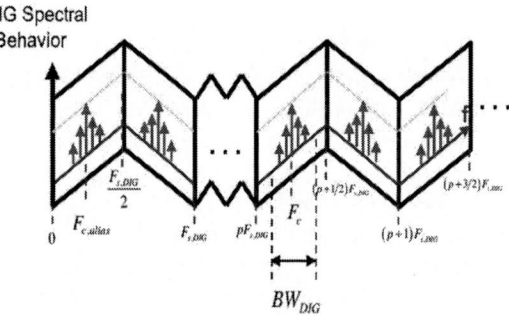

Figure 3: Illustrating the aliasing effect of a PM signal with carrier frequency F_c.

$A_{\phi,o,i}$ are the amplitudes of the individuals phase tones seen at the DIG. Equation (11) provides the subsampling rule for a single sinusoidal, whereas for a multi-tone signal, there are many choices in which to select the subsampling frequency $F_{S,DIG}$. The easiest choice, and one with the easiest book keeping, is to select the subsampling frequency based on the primitive frequency (i.e., smallest frequency) of the AWG and DIG. This occurs when $M_{AWG} = M_{DIG} = 1$. Subsequently, the subsampling constraint becomes

$$\frac{F_{S,AWG}}{F_{S,DIG}} = \begin{cases} \dfrac{m + 1/N_{DIG}}{1/N_{AWG}} & \text{even image} \\ \dfrac{m - 1/N_{DIG}}{1/N_{AWG}} & \text{odd image} \end{cases} \quad (16)$$

where m is a positive integer that relates the number of cycles of the DIG spectrum that is encompassed by the original tone, i.e., $p=mM_{AWG}$. Substituting the sampling constraint provided in Eqn. (16) to Eqn. (15) leads to the final result,

$$D_o[n] = A_{c,o} \sin\left(\begin{array}{l} 2\pi \dfrac{M_{AWG,c}}{N_{DIG}} n \\ + \sum_{i=1}^{P} A_{\phi,o,i} \cdot \sin\left(2\pi \dfrac{M_{AWG,\phi,i}}{N_{DIG}} n\right) \end{array} \right) \quad (17)$$

Both the carrier and its sidebands are mapped to the DIG spectrum provided $M_{AWG,c}$ and $M_{AWG,\phi I}$ for all i less than $N_{DIG}/2$. Moreover, the bandwidth of the DIG must allow for enough sidebands to pass to the DIG unchanged, but remove any other higher frequency components to avoid these aliasing into the Nyquist band. Figure 3 depicts the sample spectrum of the PM signal just described.

III. SAMPLING CONSIDERATIONS FOR INTRINSIC NOISE MEASUREMENTS

A. Voltage-Based Noise Measurement

Voltage-based intrinsic noise measurements (no input signal present) of a DUT can be performed with a subsampling digitizer with arbitrary subsampling frequency $F_{S,DIG}$, provided the DIG bandwidth satisfies

$$BW_{DIG} < \frac{F_{S,DIG}}{2} \quad (18)$$

and the center frequency f_o of the BP filter meets the condition

$$f_o = p \cdot F_{S,DIG} \pm \frac{F_{S,DIG}}{4} \qquad (19)$$

By doing so, the base-band spectral coefficients of the DIG sample set will correspond one-to-one with those that fall within the frequency region bounded by the BP filter.

Alternatively, if the DIG bandwidth is much greater than its sampling frequency $F_{S,DIG}$ or the center frequency is not aligned with the sampling frequency then aliasing will occur. In the case of the center frequency being aligned according to Eqn. (19), but the DIG bandwidth is much greater $F_{S,DIG}$, then the noise power will increase on account of the aliases present by the following factor,

$$\gamma = \frac{BW_{DIG}}{F_{S,DIG}/2} \qquad (20)$$

Any other situation will depend on the specifics and no general answer can be provided *a priori*.

B. Phase-Based Noise Measurement

Phase-based intrinsic noise measurements, such as those associated with a PLL, or similar type of device, can be measured using the coherent subsampling approach described in Section II.A along with coherent phase extraction from [6]. A reference signal would be generated by the AWG operating at a clock rate of $F_{S,AWG}$ without PM modulation and applied to the DUT. The DUT output would then be digitized at a subsampling frequency $F_{S,DIG}$ given by Eqn. (16) and the corresponding sample set collected. These samples would take on the same form as those described by Eqn. (17).

An alternative approach is one that removes the AWG as the source of the reference signal and, instead, takes the reference signal directly from the up-converter block as shown in Fig. 4. In this case, the input signal to the DUT can be described as

$$v_{in}(t) = \text{sqwave}\left[\sin\left(2\pi \cdot U \cdot F_{MCLK} \cdot t\right)\right] \qquad (21)$$

where the *sqwave* operation simply maps the sine wave signal into a square wave signal with the same frequency, $U \times F_{MCLK}$. Here F_{MCLK} is the frequency of oscillation of the master clock and U is the up-conversion factor.

On passing the DUT output through the BP filter and subsampling the output at a sampling rate of $F_{S,DIG}$, the DIG samples appearing in the capture memory can then be described as

$$D_o[n] = \sin\left(2\pi \cdot U \cdot \frac{F_{MCLK}}{F_{S,DIG}} n\right) \qquad (22)$$

With the sampling constraint,

$$\frac{F_{MCLK}}{F_{S,DIG}} = \begin{cases} m + 1/N_{DIG} & \text{even image} \\ m - 1/N_{DIG} & \text{odd image} \end{cases} \qquad (23)$$

Eqn. (22) can be reduced to

$$D_o[n] = \sin\left(2\pi \frac{U}{N_{DIG}} n\right) \qquad (24)$$

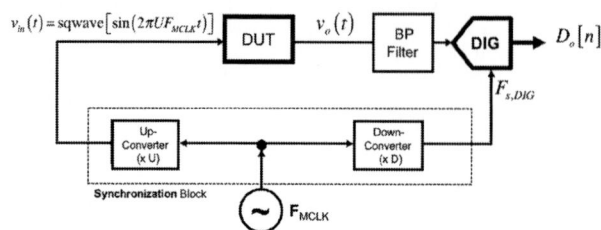

Figure 4: Proposed intrinsic noise measurement test setup with input reference signal taken directly from the up-converter block.

Here the reference clock signal will appear in FFT bin U provide $U < N_{DIG}/2$. All other bins of the DIG FFT will therefore contain components of the noise.

IV. Clock Synchronization Details

A coherent subsampling system relies directly on the ability of the clock network to generate the appropriate sampling clocks for the AWG and DIG. In Fig. 2, the synchronization subsystem consists of two components, an up-converter and a down-converter. The up-converter takes the incoming master clock signal and scales its output frequency by an integer factor U to drive the AWG, i.e.,

$$F_{S,AWG} = U \cdot F_{MCLK} \qquad (25)$$

Conversely, the down-converter takes the incoming master clock signal and scales its output frequency by some factor D that can be expressed as the ratio of two factors, β and α, to drive the DIG, i.e.,

$$F_{S,DIG} = D \cdot F_{MCLK} = \frac{\beta}{\alpha} F_{MCLK} \qquad (26)$$

The rationale for this choice lies with the circuits used to realize the up and down conversion blocks. The up-conversion clock component would be realized using a PLL with a frequency divider with gain of U. The down-converter clock component would be realized using a fractional-N PLL, which can set the output frequency as the ratio of two integers, e.g., Eqn. (26).

As the ratio of the AWG to DIG sampling frequency for the multi-tone PM test case must satisfy Eqn. (16), one can write

$$\frac{F_{S,AWG}}{F_{S,DIG}} = \frac{U \cdot F_{MCLK}}{\frac{\beta}{\alpha} F_{MCLK}} = \frac{p + \frac{M_{DIG}}{N_{DIG}}}{\frac{M_{AWG}}{N_{AWG}}} \qquad (27)$$

Solving for β and α, one finds

$$\beta = U \cdot N_{DIG} \qquad (28)$$

and

$$\alpha = N_{AWG}(mN_{DIG} + 1) \qquad (29)$$

Following a similar line of reasoning, the PLL parameters corresponding to test conditions for the two proposed subsampling systems are summarized in Table I.

978-1-5386-3775-3/18 $31.00 © 2018 IEEE

Table I: Synchronization Block Parameters

Function	PLL	FPLL	
		β	α
multi-tone PM with AWG, TS 1	U	$U N_{DIG}$	$N_{AWG}(mN_{DIG}+1)$
intrinsic noise, TS 2	U	N_{DIG}	$mN_{DIG}+1$

V. SIMULATIONS

The subsampling coherent test systems shown in Figs. 2 and 4 were investigated using MATLAB/Simulink simulations. The focus is to verify that the proposed sampling considerations leads to the correct output signal conditions.

A. Ideal DIG Behavior

Assuming a DIG is implemented with an ADC with infinite resolution, i.e., no quantization error, below is an investigation of the two test conditions involving a multi-tone with PM and an intrinsic noise measurement.

i. Subsampling a Multi-Tone PM Signal

Samples from a multi-tone PM signal consisting of six equal-amplitude sine waves were loaded into the AWG having parameters $M_{AWG,\phi}$ parameters: 1, 7, 11, 17, 23 and 29, N_{AWG} =256 with $F_{S,AWG}$ = 40 MHz. The carrier frequency is set to 2.305 MHz. The amplitude of the carrier was set to 1 V and the rms value of each phase tone was set to 0.01 rads. The DIG set-up has N_{DIG} = 1024 and m=1, resulting in $F_{S,DIG}$ = 156.097 kHz. A bandpass filter precedes the DIG with a 78 kHz bandwidth centered at 9.248 MHz. Using the analytic phase extraction method described in [6], the spectral coefficients of the instantaneous phase of the captured signal is shown in Fig. 5. As is evident, six tones in the appropriate FFT bins are present, with almost equal phase levels of about -43 dB. There is a slight roll-off present on account of the front-end filter. Through a calibration procedure, this error can be removed (not shown). Additionally, jitter transfer characteristic can also be measured by generating a PM signal that has equal powers in the band of interest. Figure 6 was obtained by generating a PM signal consists of 50μ rads tones from bin 1 to 29 and all other parameters are kept identical to simulation in Fig. 5.

ii. Subsampling a PM Noise Signal

A reference signal produced by a PLL acting as the DUT was simulated using MATLAB/Simulink. The PLL produced an output frequency of 40 MHz according to the test setup shown in Fig. 4, using a master clock reference of 10 MHz. A wide range of white noise was injected into the voltage control port of the PLL to simulate the noise that would be carried by the PLL output signal. The result of this simulation is shown in Fig. 7 together with a linear regression line to account for the repeatability of the noise measurement. As is evident, the system is behaving quite linear.

VI. EXPERIMENTAL VALIDATION

In this section the proposed coherent subsampling principles

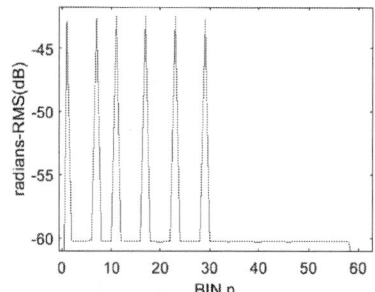

Figure 5: Magnitude of the spectral coefficients of the PM signal.

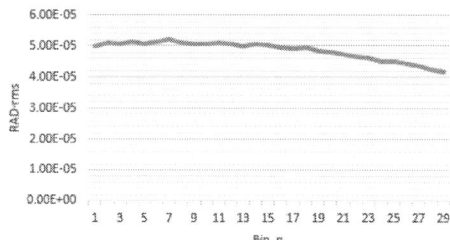

Figure 6: Jitter transfer characteristic using PM signal of 50u rads from bin 1 to 29.

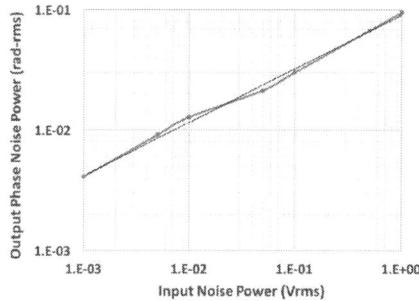

Figure 7: Input-output phase noise transfer characteristic of the coherent subsampling test system.

described in this paper will be experimentally investigated. Measurements will be performed using a Teradyne Flex Tester. The Flex tester uses an 18-bit DAC in the AWG and a 16-bit ADC in the DIG. The sampling frequencies of each block is derived from the internal clocking section of the ATE. The test setup is similar to that shown in Fig. 2 for measuring multi-tone PM signals. For measuring PM noise signals, setup described in Fig. 8 was used. The filter after the AWG is a low-pass anti-imaging filter whose characteristics are set by the signaling conditions. This is internally set by the ATE. A BP filter is inserted in the signal path in front of the DIG; this is separate from the anti-aliasing incorporated in the DIG of the ATE. To account for the frequency response behavior of these filters, a calibration procedure is run prior to the start of any measurement. In our tests, a short circuit between the AWG and the DIG acts as the DUT. In this way, the proposed theory can be verified without any DUT concerns.

A. Subsampling a Multi-Tone PM Signal

This test involves a six-tone multi-tone PM signal. Using the same clocking parameters as for the Section V, a PM signal was

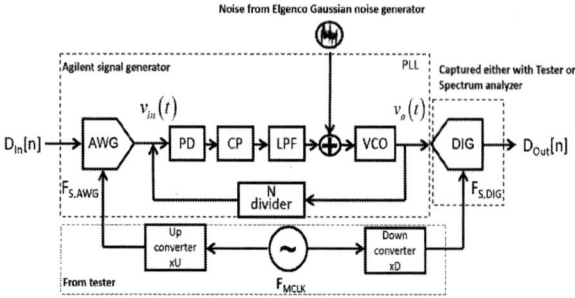

Figure 8: Test setup for testing intrinsic noise where signals are captured using Teradyne Tester.

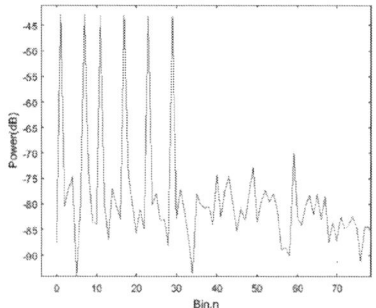

Figure 9: The measured spectral coefficients of a multi-tone PM signal with phase components in bins 1, 7, 11, 17, 23 and 29.

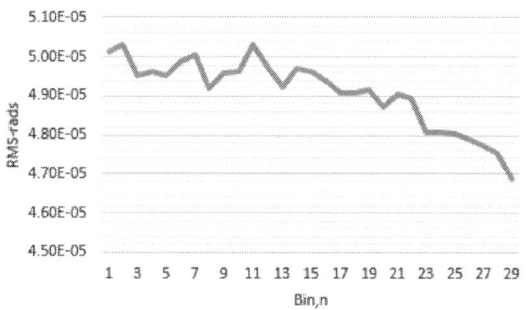

Figure 10: The measured jitter transfer characteristic of a a PM signal consisting of 50µ rads tones from bin 1 to 29.

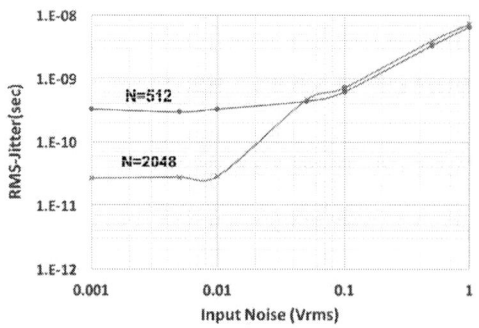

Figure 11: Output rms phase noise power (jitter) versus input noise power injected into VCO port of an Agilent 33220A voltage-controlled oscillator.

loaded into the source memory of the AWG with carrier bin 59 and phase tones located in bins: 1, 7, 11, 17, 23 and 29. An

analysis of the spectral properties of the instantaneous phase of the DIG captured signal is shown in Fig. 9. As is evident, all six phase components are clearly visible in their appropriate bins. A small carrier-feedthrough is present in bin 59. This leakage is quite small as the ADC resolution used by the DIG is equal to 16 bits. Similarly, PM signal with 50µ rads tones has been generated for bin 1 to 29 and loaded onto the memory of a tester and captured with DIG. Figure 10 shows the result of the jitter transfer characteristic of Teradyne FLEX tester.

B. Subsampling an PM Noise Signal

An Agilent 33220A signal generator was set up to produce a PM signal with a carrier of 20 MHz. A random noise signal from a Type 603-A noise generator made by Elgenco Inc. was used to modulate the carrier with rms levels from 1 mV to 1 V. The corresponding output PM signal was then applied to the coherent subsampling system for measurement. Two different data lengths were used. One with a sample set of 512 points and another consisting of 2048 points. The results for these two cases are displayed in Fig. 11. For the longer data set, linear operation can be observed from about two decades of change in the input/output noise level. For inputs less than 10 mV, the output measurement saturates at an output noise level of about 30 ps. For shorter data lengths, the output saturation noise level is much higher at 300ps and the linear range of operation reduces to a single decade change. The change in the saturation limit/linear range of operation is attributed to jitter-induced quantization noise effects associated with the digitizer.

VII. CONCLUSION

In this paper, coherent subsampling test system was proposed to measure PM type signals. This system was shown to be able to measure multi-tone PM signal, jitter transfer characteristic and intrinsic noise. Simulation and experimental measurement results were used to validate the proposed method of coherent subsampling method. Limitations of the proposed system includes BP filter and lower jitter clock for obtaining accurate results. Although the simulation and experiment was conducted at low frequency due to limitations of the available instruments, these test method can be extended to mmWave range measurements.

REFERENCES

[1] H. Okawara, "DSP-Based Testing – Fundamentals 27, Multi-tone under-sampling conditioning," Verigy Japan, July 2010.

[2] Hewlett & Packard, "Phase Noise Measurement Seminar". http://hparchive.com/seminar_notes/HP_PN_seminar.pdf, 1985

[3] S.D.Grignot, F.Azais, L.Latorre, F.Lefevre, (2014). Low-Cost Phase noise Testing of Complex RF ICs using Standard Digital ATE. Presented at International Test Conference.

[4] C. M. Miller and D. J. McQuate, "Jitter Analysis of High-Speed Digital Systems," Hewlett-Packard Journal, pp. 49-56, Feb. 1995.

[5] T. J. Yamaguchi, M. Soma, M. Ishida, T. Watanabe and T. Ohmi, "Extraction of instantaneous and RMS sinusoidal jitter using an analytic signal method," IEEE Transactions on Circuits and Systems II: Analog and Digital Signal Processing, Volume 50, Issue 6, pp. 288 – 298, June 2003.

[6] G. W. Roberts, F. Taenzler and M. Burns, An Introduction to Mixed-Signal IC Test and Measurement, 2nd Edition, Oxford University Press, New York, USA, 2011.

2018 IEEE 36th VLSI Test Symposium (VTS)

An Oscillation-Based Test technique for on-chip testing of mm-wave phase shifters

M. Margalef-Rovira[†], M. J. Barragan[†], E. Sharma[‡], P. Ferrari[†], E. Pistono[‡], and S. Bourdel[‡]

[†]*Univ. Grenoble Alpes, CNRS, Grenoble INP*, TIMA F-38000 Grenoble, France*
[‡] *Univ. Grenoble Alpes, CNRS, IMEP-LAHC, F-38000 Grenoble, France*
marc.margalef-rovira@univ-grenoble-alpes.fr

Abstract—**Beam-forming techniques using phased arrays are one of the most promising solutions for the practical implementation of future high-data-rate point-to-point communication protocols. The functionality of phased arrays is based on the use of phase shifters that should provide an accurate and controllable phase difference between the different paths of the array. However, the integration of phase shifters in current nanometric technologies is prone to imperfections that may affect the intended phase shift and degrade the performance of the antenna array. This requires extensive testing and calibration and represents a bottleneck in the production line of these system. In this work, we propose a simple Oscillation-Based Test technique that may be suitable for Built-In Self-Test applications of phase shifters. The technique is demonstrated on a Reflection-Type Phase Shifter implemented in a 55 nm BiCMOS technology. Electromagnetic and electrical simulation results show the feasibility of the proposed technique.**

Index Terms—**Phase shifters, phased arrays, Oscillation-Based Test, BIST, mm-wave integrated circuits.**

I. INTRODUCTION

The ever-increasing need of transmitting large volumes of data has led the semiconductor industry to the millimeter-wave (mm-wave) frequency band. High operating frequencies provide larger bandwidths for high-data-rate systems. In this evolution toward higher data rates, it is becoming much more difficult to improve spectral efficiency using traditional time and frequency domain techniques. A promising solution consists in taking advantage of the spatial dimension by spatially directing the antenna beam pattern.

Steering the antenna beam pattern, usually called beamforming, requires the use of a phased antenna array, or phased array. Such a system allows shaping and directing electronically the antenna beam pattern; it brings the advantages of improved Signal-to-Noise Ratio (SNR), spatial multiplexing and spatial interference cancellation. Phased arrays are used in high-end communication equipment (e.g., military communication systems) and are expanding into the consumer electronic market (e.g., car radar, high-speed communications) [1]. Indeed, beamforming has been proposed as a key enabling technology for future communication standards such as 5G cellular communications [2], [3].

A phased array is a set of multiple antennas working together as a single highly-directive antenna. In this system, the response of each individual antenna is shifted and combined

*Institute of Engineering Univ. Grenoble Alpes

in order to shape and steer the resulting radiation pattern. A phased array is composed of two main elements: the antennae and the phase shifters that interface with the front-end of the transceiver. The phase of each individual channel in the array is electrically controlled in such a way that the transmitted (or received) radio waves interfere constructively in one particular direction. This controlled phase-shift is provided by phase shifters that adjust the phase difference between the different channels in the array to combine the different signals coherently.

Even if advanced technology nodes allow the integration of phase shifters, they are prone to imperfections, such as process and mismatch variations, unintended coupling between adjacent elements, supply voltage gradients, etc., which may produce variations on the provided phase shift. Temperature drifts must be also considered. This is a key issue since small variations on a phase shift may produce large degradations in the performance of the phased array [4]–[6]. For this reason, phase shifters have to be tested and calibrated in the production line to ensure that they provide the appropriate phase shift for a correct operation. Let us notice that this does not solve the temperature drift issue. Moreover, testing and calibrating phase shifters relies on direct functional measurements using dedicated and costly mm-wave testers. In this scenario, characterizing and calibrating a complete phased array becomes a costly and time-consuming procedure that may represent a bottleneck in the production line. Developing Built-In Self-Test (BIST) strategies for phase shifters may be a promising solution to overcome these issues. BIST applications move some of the functionality of the tester to the Device Under Test (DUT) itself, in such a way that the DUT becomes self-testable, signal manipulations remain internal, and the cost of the test equipment is greatly reduced. Moreover, BIST applications may enable self-calibration during in-field operation.

In this line, a variety of test techniques have been proposed for the characterization of phased arrays and phase shifters. Thus, the works in [6]–[8] propose direct functional test techniques based on the excitation of the phased array with an external radiofrequency wave and the processing of the array response. The contact-less characterization procedures described in [7], [8] are not suitable for production line testing since they require manual adjustment and a long measurement time using dedicated test equipment. On the other hand, in [6]

978-1-5386-3775-3/18 $31.00 © 2018 IEEE

13

a simplification of contact-less tests allows measuring under near-field conditions, reducing this way the physical size of the test setup.

BIST strategies for phased arrays have been also proposed in [4], [9]–[12]. In [9]–[11] a practical system-level test for phased arrays in an RF system is demonstrated: it allows characterizing and calibrating phase and amplitude inaccuracies in each individual channel of the array. The proposed test circuitry requires the integration (or re-use) of a frequency synthesizer for exciting the different channels in the array and a dedicated receiver block to extract the test information from the response signals. In the same line, the work in [4] proposes a system-level BIST and calibration circuitry for the phased array of a receiver section. BIST circuitry makes use of an LC oscillator for test stimulus generation and a simplified receiver consisting in a mixer in a homodyne configuration for test response acquisition.

The approach in [12] takes advantage of code-modulation techniques for multiplexing different test signals into orthogonal codes that are then applied to the phased array under test. An on-chip demodulator is then used to recover the test information of each individual channel.

In this work, we propose to adapt the classic Oscillation-Based Test (OBT) technique to enable self-test applications for integrated mm-wave phase shifters. To our knowledge, this is the first time that OBT is employed for the test of these mm-wave circuits. OBT, firstly introduced in [13], consists in reconfiguring the DUT into an oscillator, in such a way that the parameters of the resulting oscillation (i.e., frequency, amplitude) are directly correlated to the DUT functionality and/or performance.

Compared to previous test strategies, enabling OBT for phase shifters has a two key advantages:

1) The need of a dedicated test stimulus generator is eliminated since the DUT itself generates the test signal.
2) The interpretation of the test results is simplified since the phase shift introduced by the phase shifter is naturally encoded in the amplitude and frequency of the resulting oscillation.

In this work, it is shown that a phase shifter can be easily reconfigured as an oscillator whose frequency and amplitude are correlated to its phase shift. The proposed OBT strategy allows then to replace a complex phase measurement by simpler magnitude and frequency signatures. In addition, the proposed test strategy opens the door to a full BIST implementation due to its simplicity and low area overhead.

The rest of the paper is organized as follows. In section II, we present the proposed test approach based on OBT for the characterization of phase shifters, focusing on the Reflection-Type Phase Shifters (RTPS) as a case study. Section III describes the practical implementation of the proposed OBT architecture for a selected case study in a 55 nm BiCMOS technology and presents some relevant results obtained by electrical and electromagnetic simulations. Finally, section IV summarizes the main contributions of this work.

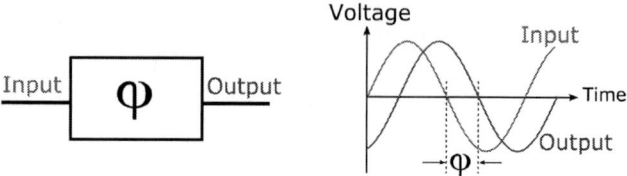

Fig. 1. Ideal phase shifter.

II. PROPOSED APPROACH

As it was mentioned above, OBT was first introduced in [13] as a structural Design-for-Test approach based on reconfiguring the DUT as an oscillator. The key idea behind OBT is that any defect in the DUT would have an impact in the frequency of the resulting oscillator. This test technique has been extended to a wide variety of analog and mixed-signal applications, including the static test of Analog-to-Digital Converters [14], testing analog integrated filters [15], testing Dual-Tone Multi-Frequency receivers [16], etc.

Adapting the OBT principles to phase shifters requires to reconfigure the phase shifter itself into an oscillator in test mode, in such a way that any imperfection that may cause a deviation of the intended phase shift, would cause as well a deviation of the oscillation characteristics (i.e., frequency and/or amplitude).

An ideal phase shifter, as the one conceptually represented in Fig. 1, is a two-port device producing a controlled phase shift φ between its input and output signals without power loss at a given frequency. Several topologies of phase shifters have been presented in the literature with different phase shift control techniques. This paper focuses on electronic phase shifters based on passive components and, as a proof-of-concept case study, we will particularize the methodology to a Reflection-Type Phase Shifter (RTPS) with varactor loads. This classic phase shifter architecture, first presented in [17], offers a high precision continuous tuning of the phase shift with a good performance in terms of return and insertion loss compared to other architectures, which makes it an interesting design choice in practical mm-wave systems. It should be noted, however, that the proposed OBT technique could be applied to any other kind of electronically controlled phase shifter based on passive devices.

Without loss of generality, a RTPS with varactor loads can be electrically modeled at a particular frequency as a lossy LC-tank, as it is schematically depicted in Fig. 2, where L_{eq} and C_{eq} represent the inductive and capacitive behavior of the circuit, respectively, and R_p represents the ohmic losses of the circuit. Note that C_{eq} is a tunable capacitor, since it includes the contribution of the varactor loads.

As it is well-known, LC-tanks are lossy resonators that can be set to sustain oscillations by adding an active element that compensates the ohmic losses of the parasitic resistor [18]. This active element acts as an equivalent negative resistor and can be practically implemented using a pair of nMOS transistors in a crossed-coupled pair configuration. Figure 3

978-1-5386-3775-3/18 $31.00 © 2018 IEEE

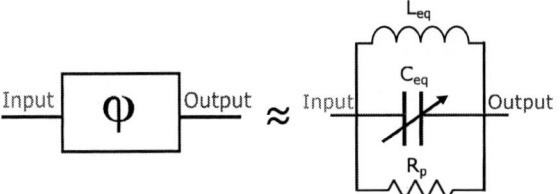

Fig. 2. Reflection-type phase shifter equivalent circuit as a lossy LC-tank. Notice that C_{eq} includes the contribution of the varactor loads of the phase shifter.

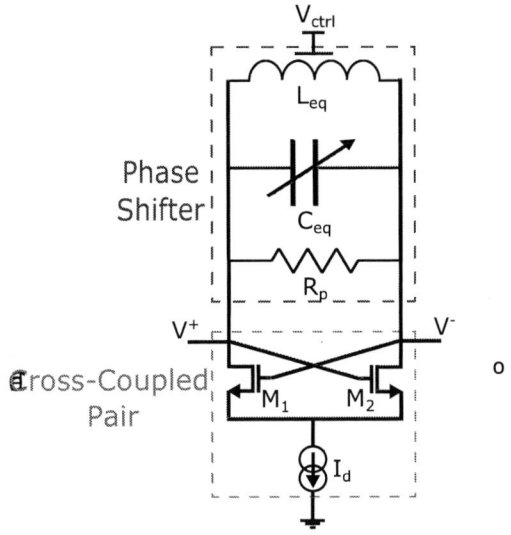

Fig. 3. Reconfiguration of a reflection-type phase shifter with varactor loads as an oscillator.

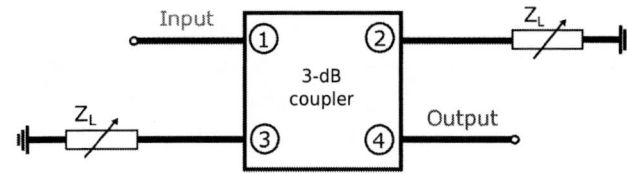

Fig. 4. General architecture of a RTPS with tunable loads. Port 1 is the input port, port 2 is the forward port, port 3 is the coupled port, and port 4 is the isolated port of the 3-dB coupler.

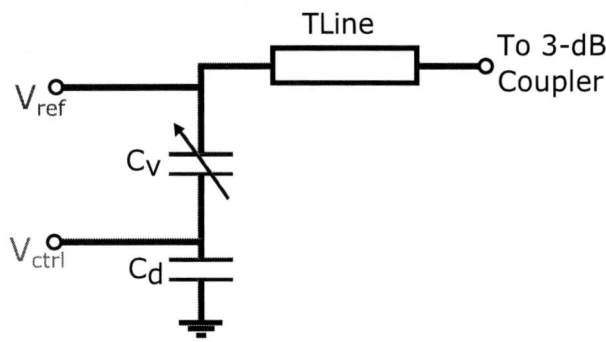

Fig. 5. Tunable loads for the considered RTPS case study.

shows a conceptual block diagram of the proposed reconfiguration during the test mode, in which the phase shifter under test –represented in Fig. 3 as its equivalent lossy LC-tank– is connected to a cross-coupled pair. Note that the V_{ctrl} power supply of the resulting oscillator corresponds to the biasing voltage of the varactor loads in the RTPS. In order to produce sustained oscillations, the Barkhausen criteria has to hold. A closed-loop system as the one in Fig. 3, consisting in a lossy complex impedance $Z(j\omega_0)$ (i.e., the lossy LC tank) and a block that exhibits a gain g_m (i.e., the transconductance of the nMOS transistors in the cross-coupled pair), will sustain steady-state oscillations only at frequencies ω_0 for which:

- The loop gain is equal to unity in absolute magnitude, i.e. $\left| g_m \cdot Z(j\omega_0)^2 \right| = 1$.
- The phase shift around the loop is zero or integer multiple of 2π, i.e. $\angle g_m \cdot Z(j\omega_0) = 2\pi n$, $n = 0, 1, 2, \dots$

To fulfill the Barkhausen criteria in practice, the transconductance g_m of the transistor has to meet the condition $g_m > 2/R_p$. Under this constraint, it is easy to demonstrate that the system may oscillate at a frequency F_{osc} given by,

$$F_{osc} = \frac{\omega_0}{2\pi} = \frac{1}{2\pi\sqrt{L_{eq}C_{eq}}} \qquad (1)$$

using the same notation previously defined.

In the following, we will demonstrate that this oscillation frequency is a function of the phase shift introduced by the phase shifter during normal operation, in such a way that the measurement of the phase shift can be replaced by the measurement of F_{osc}. Let us evaluate the phase shift introduced by a RTPS. Figure 4 presents the general topology of an ideal RTPS. It is composed of an ideal 3-dB coupler and two variable loads. If we denote Γ as the reflection coefficient at ports 2 and 3, it can be demonstrated that the phase shift φ between ports 1 and 4 can be written as

$$\varphi = \pi/2 + \arg(\Gamma). \qquad (2)$$

Multiple structures have been proposed to implement the loads, including switched capacitors, switched stubs, varactors, etc. In our case study we will consider varactor loads. Figure 5 shows a practical implementation of these varactor loads, which consists in a transmission line, a MOS varactor (usually implemented as an array of varactors in parallel) and one decoupling capacitor C_d. The capacitance of the varactor, C_v, is controlled by the DC voltage difference applied to its ports ($V_{ref} - V_{ctrl}$). In this configuration, equation (2) can be developed as

$$\varphi = \frac{\pi}{2} - 2\arg\left(\frac{Z_0 C_v \omega}{1 - L_c C_v \omega^2}\right) \qquad (3)$$

where, in the sake of simplicity, it has been assumed that the varactors and the transmission line are ideal devices without ohmic losses, Z_0 represents the input impedance of the phase shifter, and L_c is the equivalent inductance of the transmission line.

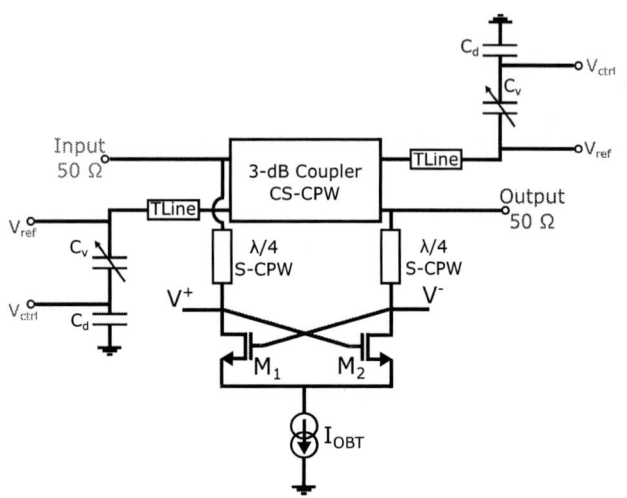

Fig. 6. Proposed RTPS with built-in OBT.

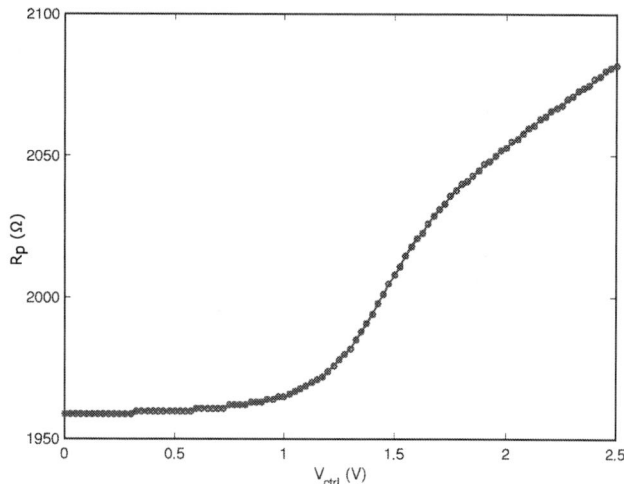

Fig. 7. Ohmic losses, R_p, in the considered reflection-type phase shifter at 60 GHz, obtained by electromagnetic simulation, as a function of the varactors bias voltage V_{ctrl}.

As it can be observed, the phase shift introduced by the phase shifter at a given frequency is a function of the varactor capacitance C_v. If we go back to the expression of the oscillation frequency in test mode given by equation (1), it is easy to see that the oscillation frequency is also a function of the varactor capacitance (being C_{eq} a function of the intrinsic impedance Z_0 and the varactor capacitance). Hence, it can be concluded that the phase shift introduced by the phase shifter in normal operation is correlated to the oscillation frequency in the proposed OBT configuration, as we intended to demonstrate. Moreover, it is interesting to notice that a similar analysis can be carried out for the amplitude of the oscillation: given that the impedance of the RTPS is a function of C_v and ω, variations in the oscillation frequency also result in variations in the oscillation amplitude.

III. PRACTICAL IMPLEMENTATION AND RESULTS

To validate the proposed OBT strategy, a proof-of-concept phase shifter with built-in OBT was designed in STMicro-electronics 55-nm BiCMOS technology. The selected device under test is a RTPS with varactor loads employing a 3-dB coupler based on a Coupled Slow-wave CoPlanar Waveguide architecture [19]. The selected DUT has a maximum phase shift tuning range of 55° at a central frequency of 60 GHz.

Figure 6 shows a block diagram of the DUT with added OBT circuitry. A nMOS cross-coupled pair has been co-designed together with the phase shifter in order to enable oscillations during the test mode. The cross-coupled pair is connected to the input and output nodes of the phase shifter using quarter-wavelength slow-wave CoPlanar Waveguides (CPWs) [20]. Transistors were sized in order to a) comply with the Barkhausen criteria to ensure sustained oscillation in test mode, and b) minimize the impact on the phase shifter performance during normal operation. It is important to note that in the proposed architecture no RF switches are needed to alternate between normal and OBT operation. Test mode is

activated simply by turning on the current source I_{OBT} and normal operation is recovered by turning off I_{OBT}.

To meet the conditions in the Barkhausen criteria, the cross-coupled pair has to compensate the ohmic losses in the equivalent LC-tank. Thus, the equivalent negative resistor seen from the ports of the cross-coupled pair ($V+$ and $V-$ in Fig. 6), that is, $R_{neg} = -2/g_m$, where g_m is the transconductance of the transistors in the pair, has to comply with the design equation,

$$|R_{neg}| \leqslant R_p \implies g_m \geqslant \frac{R_p}{2}, \qquad (4)$$

where R_p represents the ohmic losses in the phase shifter equivalent LC-tank seen from the $V+$ and $V-$ nodes. Figure 7 shows this resistance R_p as a function of the biasing voltage V_{ctrl} applied to the phase shifter, obtained by electromagnetic simulation of the system with ANSYS HFSS. Note that for this analysis, in the OBT configuration proposed in Fig. 6, the quarter-wavelength transmission lines are included in the tank, that is, the equivalent resistance is computed from the output nodes of the resulting oscillator. To comply with the Barkhausen criteria, nMOS transistors were sized to $W = 300\ \mu m$, $L = 60$ nm, which yields a value of $|R_{neg}| = 89\ \Omega$ when I_{OBT} is set to 6 mA.

Under the described conditions, and again according to the Barkhausen criteria, the system produces sustained oscillations at frequencies for which the total phase shift around the loop is 2π, that is, when the phase shift at the nodes of the phase shifter is π (neglecting layout parasitics). Figure 9 represents the phase shift between the nodes of the phase shifter when it is configured as an oscillator as a function of the biasing voltage V_{ctrl}, obtained by electromagnetic simulations. As it can be observed, the π phase shift can be achieved for the complete tuning range of V_{ctrl}, and the output frequencies compatible with the oscillation criteria are in the range from 14 GHz to 17 GHz.

978-1-5386-3775-3/18 $31.00 © 2018 IEEE

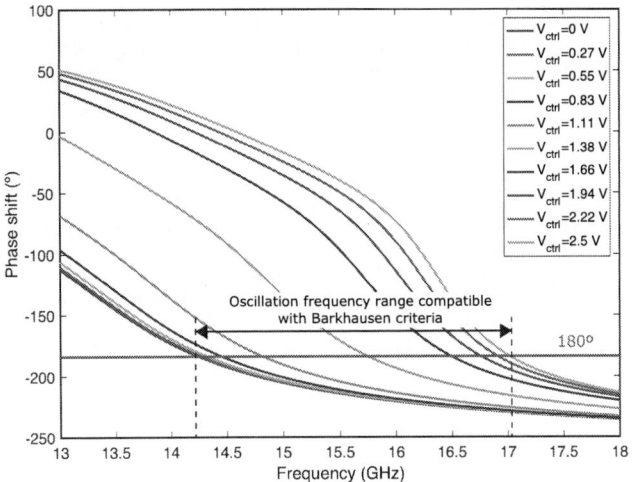

Fig. 8. Oscillation frequencies compatible with the Barkhausen criteria as a function of the bias voltage V_{ctrl}, obtained by electromagnetic simulation.

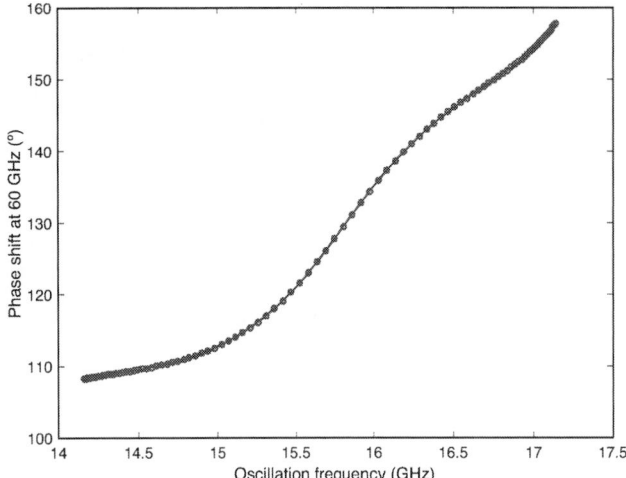

Fig. 9. Phase shift introduced by the phase shifter at 60 GHz as a function of the oscillation frequency in OBT mode.

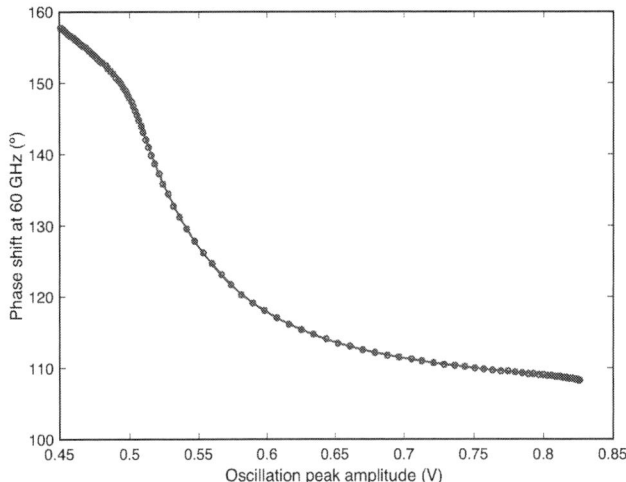

Fig. 10. Phase shift introduced by the phase shifter at 60 GHz as a function of the oscillation amplitude in OBT mode.

Regarding the impact of the added test circuitry on the performance of the original phase shifter without test circuitry, electromagnetic simulations show that the worst-case insertion loss for the original phase shifter without the OBT circuitry is of 1 dB, while the insertion loss of the phase shifter with built-in OBT during normal operation is 1.4 dB. The degradation of the insertion loss is only 0.4 dB, which is negligible for most applications. In the same line, the worst-case return loss is 28 dB for the original phase shifter, while the introduction of the OBT circuitry gives a worst-case return loss of 20 dB in normal operation mode, which still represents a good adaptation of the phase shifter.

During OBT mode, the oscillation frequency of the resulting oscillator is a signature of the phase shift introduced by the phase shifter. Figure 9 shows the phase shift introduced by the phase shifter at 60 GHz as a function of the oscillation frequency in OBT mode, obtained by electromagnetic simulation. As it can be observed, the measurement of the phase shift can be replaced by the measurement of the oscillation frequency. In addition, the amplitude of the resulting oscillation in OBT mode is another signature typically employed in classical OBT strategies [15] to complement the frequency signature. Fig. 10 shows the phase shift introduced by the phase shifter at 60 GHz as a function of the oscillation amplitude in OBT mode. As it can be observed, the measurement of the oscillation amplitude is also a simple signature that can be used to replace the complex measurement of the phase shift. The 3D plot in Fig. 11 illustrates the equivalence of the proposed measurements and represents the phase shift at 60 GHz as a function of both the oscillation frequency and amplitude in OBT mode. In other words, it is shown that we can replace the measurement of phase shift at 60 GHz by measurements of frequency and/or amplitude in the frequency range of 14 GHz to 17 GHz.

These results show that applying OBT to mm-wave phase shifters is not only a feasible test approach, but it opens the door to simple on-chip characterization of phase shifters with reduced hardware resources. Future work will include the extension of the proposed OBT architecture to integrate on-chip amplitude and/or frequency detectors that may enable on-chip self-test and calibration applications. Moreover, it is particularly interesting to consider the extension of the proposed technique to complete phased arrays. Phase shift differences between the different branches of a phased array will appear as differences in the oscillation frequency and amplitude in OBT mode. Measuring frequency and amplitude differences is a much simpler and precise task than measuring phase differences, which opens the door to future BIST applications for phased arrays based on the proposed OBT strategy.

IV. CONCLUSIONS

In this paper, it has been shown for the first time that an Oscillation-Based Test technique could be suitable for the

978-1-5386-3775-3/18 $31.00 © 2018 IEEE

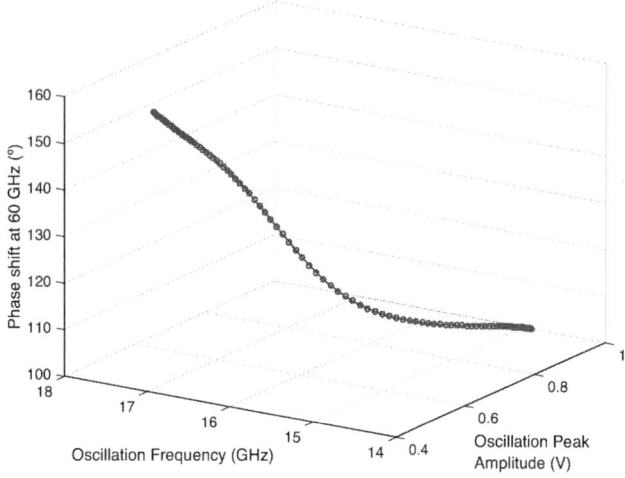

Fig. 11. Phase shift introduced by the phase shifter at 60 GHz as a function of both the oscillation frequency and amplitude in OBT mode.

characterization of integrated mm-wave phase shifters. The proposed OBT technique is based in the reconfiguration of the phase shifter under test as an oscillator during test mode and it enables the replacement of complex phase shift measurements at high frequencies by simpler signatures such as the frequency and amplitude of the resulting oscillator.

The fundamentals of the proposed test strategy have been analytically explored, and the proposed technique has been validated in a proof-of-concept prototype designed in STMicroelectronics 55-nm BiCMOS technology. The designed prototype consists in a reflection-type phase shifter with varactor loads with a built-in nMOS cross-coupled pair to enable oscillations during OBT mode. Electromagnetic and electrical simulations show the feasibility of the proposed technique with a negligible impact on the performance of the phase shifter during normal operation. The proposed OBT technique requires very few hardware resources and may be a promising solution for practical BIST applications of phased arrays.

ACKNOWLEDGEMENTS

This work has been partially funded by the TARANTO project.

REFERENCES

[1] A. Hajimiri, H. Hashemi, A. Natarajan, X. Guan, and A. Komijani, "Integrated Phased Array Systems in Silicon," *Proceedings of the IEEE*, vol. 93, pp. 1637–1655, Sept 2005.

[2] W. Roh, J. Y. Seol, J. Park, B. Lee, J. Lee, Y. Kim, J. Cho, K. Cheun, and F. Aryanfar, "Millimeter-wave beamforming as an enabling technology for 5G cellular communications: theoretical feasibility and prototype results," *IEEE Communications Magazine*, vol. 52, pp. 106–113, February 2014.

[3] J. Bae, Y. S. Choi, J. S. Kim, and M. Y. Chung, "Architecture and performance evaluation of mmwave based 5g mobile communication system," in *2014 International Conference on Information and Communication Technology Convergence (ICTC)*, pp. 847–851, Oct 2014.

[4] J. W. Jeong, J. Kitchen, and S. Ozev, "A self-compensating built-in self-test solution for RF phased array mismatch," in *2015 IEEE International Test Conference (ITC)*, pp. 1–9, Oct 2015.

[5] C. Y. Kim, D. W. Kang, and G. M. Rebeiz, "A 44-46-GHz 16-Element SiGe BiCMOS High-Linearity Transmit/Receive Phased Array," *IEEE Transactions on Microwave Theory and Techniques*, vol. 60, pp. 730–742, March 2012.

[6] M. Shafiee and S. Ozev, "Contact-less near-field measurement of RF phased array antenna mismatches," in *2017 22nd IEEE European Test Symposium (ETS)*, pp. 1–6, May 2017.

[7] X. Guan, H. Hashemi, and A. Hajimiri, "A fully integrated 24-GHz eight-element phased-array receiver in silicon," *IEEE Journal of Solid-State Circuits*, vol. 39, pp. 2311–2320, Dec 2004.

[8] S. Jeon, Y. J. Wang, H. Wang, F. Bohn, A. Natarajan, A. Babakhani, and A. Hajimiri, "A Scalable 6-to-18 GHz Concurrent Dual-Band Quad-Beam Phased-Array Receiver in CMOS," *IEEE Journal of Solid-State Circuits*, vol. 43, pp. 2660–2673, Dec 2008.

[9] S. Y. Kim, O. Inac, C. Y. Kim, D. Shin, and G. M. Rebeiz, "A 76-84-GHz 16-Element Phased-Array Receiver With a Chip-Level Built-In Self-Test System," *IEEE Transactions on Microwave Theory and Techniques*, vol. 61, pp. 3083–3098, Aug 2013.

[10] O. Inac, F. Golcuk, T. Kanar, and G. M. Rebeiz, "A 90-100-GHz Phased-Array Transmit/Receive Silicon RFIC Module With Built-In Self-Test," *IEEE Transactions on Microwave Theory and Techniques*, vol. 61, pp. 3774–3782, Oct 2013.

[11] T. Kanar and G. M. Rebeiz, "A 2-15 GHz built-in-self-test system for wide-band phased arrays using self-correcting 8-state I/Q mixers," in *2016 IEEE MTT-S International Microwave Symposium (IMS)*, pp. 1–4, May 2016.

[12] K. Greene, V. Chauhan, and B. Floyd, "Code-modulated embedded test for phased arrays," in *2016 IEEE 34th VLSI Test Symposium (VTS)*, pp. 1–4, April 2016.

[13] K. Arabi and B. Kaminska, "Oscillation-test strategy for analog and mixed-signal integrated circuits," in *Proceedings of 14th VLSI Test Symposium*, pp. 476–482, Apr 1996.

[14] K. Arabi and B. Kaminska, "Efficient and accurate testing of analog-to-digital converters using oscillation-test method," *Proceedings European Design and Test Conference. ED & TC 97*, pp. 348–352, 1997.

[15] G. Huertas, D. Vazquez, E. J. Peralias, A. Rueda, and J. L. Huertas, "Practical oscillation-based test of integrated filters," *IEEE Design Test of Computers*, vol. 19, pp. 64–72, Nov 2002.

[16] G. Huertas, D. Vazquez, A. Rueda, and J. L. Huertas, "Built-in self-test in mixed-signal ICs: a DTMF macrocell," *VLSI Design 2000. Wireless and Digital Imaging in the Millennium. Proceedings of 13th International Conference on VLSI Design*, pp. 568–571, 2000.

[17] B. T. Henoch and P. Tamm, "A 360° Reflection-Type Diode Phase Modulator (Correspondence)," *IEEE Transactions on Microwave Theory and Techniques*, vol. 19, no. 1, pp. 103–105, 1971.

[18] B. Razavi, *RF Microelectronics (2Nd Edition)*. Upper Saddle River, NJ, USA: Prentice Hall Press, 2nd ed., 2011.

[19] Z. Iskandar, J. Lugo-Alvarez, A. Bautista, E. Pistono, F. Podevin, V. Puyal, A. Siligaris, and P. Ferrari, "A 30-50 GHz reflection-type phase shifter based on slow-wave coupled lines in BiCMOS 55 nm technology," *2016 46th European Microwave Conference (EuMC)*, pp. 1413–1416, 2016.

[20] A. Bautista, A. L. Franc, and P. Ferrari, "Accurate Parametric Electrical Model for Slow-Wave CPW and Application to Circuits Design," *IEEE Transactions on Microwave Theory and Techniques*, vol. 63, no. 12, pp. 4225–4235, 2015.

978-1-5386-3775-3/18 $31.00 © 2018 IEEE

Special Session: Recent Developments in Hardware Security

Ro Commarota, Qualcomm Inc., USA
Naghmeh Karimi, University of Maryland Baltimore County, USA
Siddharth Garg, New York University, USA
Jeyavijayan Rajendran, Texas A&M University, USA (Organizer)

I. INTRODUCTION

In this session, we explore some of the recent challenges facing hardware security: (i) Challenges in the implementation of post-quantum crypto algorithms, (ii) Impact of aging on security, and (ii) Security challenges in machine learning.

II. A PATH TOWARD EARLY ADOPTION OF LATTICE-BASED CRYPTOGRAPHY SCHEMES IN HARDWARE (RO COMMAROTA)

The emergence of quantum computing as a plausible computational paradigm threatens to break many classical cryptographic schemes, leading to innovations in public key cryptography that focus on the foundational work and standardization of the so called post-quantum cryptography (PQC) primitives and their protocols that are resistant to quantum computing threats. In particular, lattice-based cryptography is a promising PQC family, both in terms of foundational properties, as well as its application to both traditional and emerging security problems such as encryption (asymmetric, but also symmetric), digital signature, key exchange, homomorphic encryption etc. While lattice-based PQC schemes provide guarantees in theory, they require significant computational resources, making their realization in varying scenarios (e.g., from high-performance servers to resource constrained IoT) challenging. Furthermore, since these schemes are yet to be standardized, there is a critical need for agile deployment in the face of emerging and changing standards.

Both issues (computational pressure and agile deployment) for PQC lattice-based schemes can be alleviated by deploying programmable accelerators with specialized datapath and controllers that can achieve improved performance and energy efficiency. Generally, flexible and customized hardware accelerators are implemented on FPGAs. However, these customized hardware accelerators do not provide programmability to satisfy the variable goals in the security community. By scarifying agility, Application Specific Integrated Circuits (ASICs) are the preferred platform to implement highly customized implementations of lattice-based schemes with the lowest energy and delay. Specialized accelerators have been highly customized to support a single cryptographic scheme or improved variants of the scheme; however a successful attack against any scheme necessitates a complete redesign of the entire custom hardware accelerator. On one hand, the lack of programmability in custom accelerators incurs a large overhead for the tedious redesign of modified accelerators to fix vulnerabilities. On the other hand, programmable accelerators may not achieve the energy efficiency, performance or small footprint of custom accelerators. Therefore there is a need to efficiently explore the design space of architectural solutions spanning the spectrum between general purpose processors and a custom hardware solutions, trading off flexibility for performance, energy, and resilience.

In this work we present our approach to the design of programmable hardware accelerators for lattice-based PQC schemes, and early DSE results for popular schemes for lattice-based key exchange and digital signature using LBC. Our analysis includes both compiler and micro-architectural optimization aspects.

III. DEVICE AGING AND POWER ANALYSIS ATTACKS (NAGHMEH KARIMI)

With the aggressive scale down of process technology in recent years, time-dependent deviation of the electrical properties in CMOS circuits has become more severe. Due to this time-dependent deviation so-called "aging," the electrical specifications of transistors included in a device, and in turn, the timing and power consumption of the underlying device alters over time. For cryptographic devices, aging is not only crucial from the reliability point of view but also needs to be considered from security perspective as aging-related degradations may affect leaking sensitive information through side-channel analysis attacks, and in particular, power analysis attacks. In this talk, we discuss how aging affects the success rate of power analysis attacks, and in particular profiling attacks launched on cryptographic devices. We will focus on profiling attacks and show aging misalignment between the device used for characterization and the target device can make the attack more difficult. Our experimental results show the attack-efficiency decrease due to aging misalignments is especially significant for mismatches of few weeks.

IV. BACKDOORED NEURAL NETWORKS (BADNETS): A NEW CHALLENGE FOR THE TEST COMMUNITY (SIDDHARTH GARG)

Deep neural networks are typically computationally expensive to train, requiring weeks of computation on many GPUs; as a result, many users outsource the training procedure to the cloud or rely on pre-trained models that are then fine-tuned for a specific task. In this talk we will highlight how outsourced training introduces new security risks: an adversary can create a maliciously trained network (a backdoored neural network, or a BadNet) that has state-of-the-art performance on the user's training and validation samples, but behaves badly on specific attacker-chosen inputs. We demonstrate backdoors on a U.S. street sign classifier that identifies stop signs as speed limits when a special sticker is added to the stop sign; we then show in addition that the backdoor in our US street sign detector can persist even if the network is later retrained for another task and cause a drop in accuracy of 25% on average when the backdoor trigger is present. The talk will conclude with a discussion on the role the VLSI test community can play in adapting semiconductor test and verification methodologies to the neural network context.

978-1-5386-3775-3/18 $31.00 © 2018 IEEE

Innovative Practices on Memory Test Practice

M. Casarsa ST Microelectronics and G. Harutyunyan, Synopsys
Kaitlyn Chen and Ramesh Sharma, Intel Corporation, Giri Podichetty and Martin Keim, Mentor Graphics, A Siemens Business
Sreejit Chakravarthy, Intel Corporation

Ramesh Tekumalla, Broadcom (Organizer)

I. INTRODUCTION

In this IP session, there will be 3 presentations focusing on the state-of-the-art memory test practices from tools and industrial implementation perspectives. The 1st presentation will discuss the memory test strategies in the automotive space. The 2nd presentation discusses memory tests strategies for enabling test cost reduction. The 3rd presentation will discuss on array BIST and its presence and usage in various applications of the current generation designs.

II. LEVERAGING EMBEDDED MEMORY TEST AND REPAIR FOR FUNCTIONAL SAFETY (M CASARSA)

Currently the automotive SOCs are one of the fastest growing sectors in semiconductor industry. In addition to their functionality, these SOCs need to satisfy several functional safety requirements mainly mandated by ISO 26262 standard. These requirements bring new challenges and considerations into the SOC design process. Typically, a large ratio of such SOCs is populated with embedded memories. These memories need on-chip infrastructure to ensure functional safety during in-field operation. The main challenges for such infrastructure is to ensure power-on self-test and repair capabilities; execute periodic test capability during the operation; meet the required test coverage with short test time window; optimize test time in-system power dissipation; maintain transparency of memory content; as well as ensure communication with on-chip safety management to provide short notification time. To meet above functional safety challenges, the embedded memory test & repair infrastructure is augmented and leveraged. In this Innovative Practices presentation, the authors will present their recent experience in using an efficient embedded test and repair solution to address the above mentioned functional safety requirements. This solution was validated on a comprehensive automotive SoC, which includes in it considerable amount of embedded memories, along with numerous logic and AMS IP blocks. The obtained results show the advantages and effectiveness of this solution.

III. MEMORY TEST CAPABILITIES FOR ADDRESSING TEST COST REDUCTION AND FUNCTIONAL SAFETY NEEDS (MARTIN KEIM)

Recent developments in MBIST capabilities enable Intel to address some of the challenges related to test cost reduction and enabling memory test needs driven by functional safety requirements. The capabilities for reducing test time for both, pass/fail tests as well as volume diagnosis data collection are key requirements for Intel to adopt usage of industry standard solutions. In addition, the core tool capabilities with richer support for system Verilog and flexible RTL editing has provided a good framework for Intel to develop memory test solution for FuSa that is not available out-of-box in industry. In this presentation we'll discuss key features of new memory test solution and how that help Intel to easily automate the memory test implementation flow for their next generation needs.

IV. IMPROVING ARRAY BIST FOR INFIELD TEST AND REPAIR AND YIELD ANALYSIS (SREEJIT CHAKRAVARTY)

Currently Work on Array BIST over the decades have focused on issues that crop up in High Volume Manufacturing. There is an increased use of array BIST for infield diagnostic applications in the general area of functional safety (automotive, industrial), server and high performance computing. In this presentation we will highlight some of the new demands on array BIST resulting from these new applications.

ATPG-Based Cost-Effective, Secure Logic Locking

Abhrajit Sengupta[†], Mohammed Nabeel[‡], Muhammad Yasin[†], and Ozgur Sinanoglu[‡]

[†] Tandon School of Engineering, New York University, New York, USA
[‡] New York University Abu Dhabi, Abu Dhabi, United Arab Emirates
{as9397, mtn2, yasin, ozgursin}@nyu.edu

Abstract—The globalization of IC supply chain lead to the emergence of hardware security threats such as IP piracy, reverse engineering, overbuilding, and hardware Trojans. Among the techniques developed to mitigate these threats, logic locking offers the most versatile protection and is being actively researched. The most recent locking technique SFLL thwarts with provable and quantifiable security all the state-of-the-art attacks including SAT, AppSAT, and the removal attack. However, the implementation cost of SFLL can sometimes be prohibitive, as it lacks an automated framework that explores cost-effective implementation options. In this paper, we show how VLSI testing principles and tools can be adopted to automate critical steps in SFLL and minimize its cost. We propose "SFLL-fault" that utilizes fault injection driven synthesis to efficiently explore design options and ATPG to assess security levels. Our experimental results confirm the efficacy of our strategy; SFLL-fault can reduce the implementation cost by 35% compared to SFLL without compromising security.

Index Terms—IP piracy, reverse engineering, logic locking, ATPG, VLSI testing

I. INTRODUCTION

In today's ever-growing market of digital electronics, threats such as IP piracy, hardware Trojans, overbuilding, and reverse-engineering are becoming an increasing concern for military and commercial organizations [1]. With globalization of IC fabrication, many companies send their intellectual property (IP) to off-shore foundries for cost-effective access to advanced technology nodes. This way the companies are forced to share their valuable IP with potentially *untrusted* parties, exposing it to the aforementioned threats. In fact, it was estimated that due to IP piracy the semiconductor industry loses $4 billion annually [2].

Design-for-Trust (DfTr) techniques. To protect IP from unauthorized access, several techniques such as IC metering [3], watermarking [4], camouflaging [5], split manufacturing [6], physically unclonable function (PUF) [7], and logic locking [8]–[14] have been proposed. Due to its simplicity and effectiveness, logic locking has gained significant popularity in the research community. It can protect against an attacker located anywhere in the supply chain whereas other DfTr techniques protect only against a subset of malicious entities.

Logic locking. Logic locking was first introduced in EPIC [8]. It is effected at the hardware level by inserting extra *key-gates* in the original netlist to lock its functionality. The *secret key* is stored in a *tamper-proof* memory [15] on the chip as shown in Fig. 1. Without the secret key loaded onto its memory, the IC is rendered practically useless as it will produce corrupted outputs. The threat model for

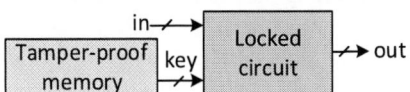

Fig. 1. The output of a logic locked circuit is a function of the key inputs and the primary inputs. The key inputs are driven from a tamper-proof memory.

logic locking is illustrated in Fig. 2. Several other works followed [9], [10] improving the security of logic locking. Nonetheless, a lethal attack based on Boolean satisfiability (SAT) broke all the existing logic locking techniques [11] then. To thwart SAT attack, SARLock [12] and Anti-SAT [16] were proposed. However, both SARLock and Anti-SAT fall short against removal [14] and bypass attack [17]. Later, TTLock was proposed to protect against such attacks [18]. The idea is to make the on-chip hardware implementation different from that of the original design (and thus, hidden/protected from an attacker). The modified design implemented in hardware produces different outputs compared to the original design for a selected set of input patterns/cubes known, accordingly, as *protected input patterns/cubes*[1]. This set of protected input patterns constitutes the secret key. Consequently, a restoration logic, driven by this key, is also implemented on-chip to restore the original functionality for the protected cubes. However, TTLock protects only one input cube, which results in low output corruptibility and is vulnerable to approximate attacks such as AppSAT [13].

The baseline: SFLL. Recently, stripped functionality logic locking (SFLL) was proposed to generalize TTLock, protecting a large number of input patterns using a security-aware CAD framework [19]. One variant of SFLL, referred to as SFLL-flex, modifies the design based on the designer-specified protected patterns using a simulated annealing approach [19]. Though this might work well for a certain class of applications, simulated annealing is extremely slow in practice and may run into scalability issues for general applications.

Note that none of the previous work on logic locking accounts for the cost of a tamper-proof memory during the synthesis process, which might have a significant impact on the area footprint of the design leading to high overhead[2]. Moreover, most existing SAT resilient logic locking techniques rely solely on a designer to specify the security-critical parts

---

[1]Input patterns may be compactly represented using input cubes, which contain don't care entries (x's). We use the two terms interchangeably.

[2]As part of standard power up sequence, the content of non-volatile tamper-proof memory is shifted to chain of flops, thereby, not impacting the timing of a design afterwards.

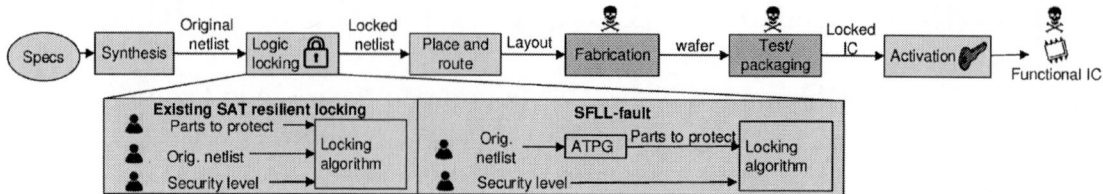

Fig. 2. Threat model for logic locking. The skulls denote the untrusted stages. The inset highlights that the proposed SFLL-fault technique uses ATPG to determine which parts of the netlist are cost-effective candidates for protection; otherwise, the designer is burdened with providing these security specifications.

of a design or the protected input patterns(s) [19]. This leads to additional work for the designers, who are already overburdened with hard deadlines and frequent engineering change orders in design specifications.

Contribution. The discussion above asserts that in order to minimize reliance on the designers, it is imperative to develop design tools that automate the additional security related design decisions. The objective of this work is to provide an automated framework that enables the designers to both 1) identify and modify the parts of a design that constitute cost-effective candidates to be protected using functionality-strip operation, and 2) compute the associated protected patterns. It is crucial that such a framework is built around well-established EDA tools and can be easily integrated with existing IC design flow with minimal changes. Accordingly, our strategy for selecting protected patterns is to leverage the principles of VLSI testing and make use of automatic test pattern generation (ATPG) tools. The contributions of our work are:

1) We develop a fault-injection based heuristic to select the protected input patterns minimizing the overhead.
2) We present a scalable, comprehensive SFLL-fault optimization framework that also accounts for the cost of tamper-proof memory. We achieve on average 35% improvement in area overhead when compared to the baseline SFLL-flex.
3) We conduct a detailed security analysis of the proposed SFLL-fault against the state-of-the-art attacks.

II. PRELIMINARIES

Before delving into further details, we briefly discuss the state-of-the-art attacks against logic locking and elucidate in particular stripped-functionality logic locking (SFLL). A summary of all the existing locking techniques is shown in Table I.

A. Attacks on logic locking

1) SAT attack: SAT attack is the most effective attack against logic locking to date. In each iteration, the attack finds an input pattern, called a *distinguishing input pattern* (DIP), which divides the key space into two classes; correct and incorrect. A DIP is an input pattern for which the circuit produces different outputs for two different sets of keys. The attack iteratively divides the key space by finding a DIP in each iteration. It terminates successfully when the incorrect search space is pruned, that is, all incorrect key values have been eliminated. The computational effort of the attack is

TABLE I
COMPARISON OF LOGIC LOCKING TECHNIQUES WRT ATTACK RESILIENCE AND OPTIMAL OVERHEAD. ✓ DENOTES PRESENCE OF A FEATURE.

Feature	RLL [8]	FLL [10]	SLL [9]	AntiSAT [16]	SARLock [18]	SFLL [19]	Proposed SFLL-fault
SAT resilient [11]	✗	✗	✗	✓	✓	✓	✓
Removal resilient [14]	✓	✓	✓	✗	✗	✓	✓
AppSAT resilient [13]	✗	✗	✗	✗	✗	✓	✓
Cost-effective	✗	✗	✗	✗	✗	✗	✓

determined by the number of DIPs required by the attack to successfully find the correct key.

2) Removal attack: Removal attack exploits structural traces to identify, and ultimately remove the protection logic [14]. Both SARLock and Anti-SAT leave the original circuit unchanged; the protection circuitry is added merely as a wrapper around the original circuit. SARLock hardcodes the secret key in the mask logic, which can be easily extracted by tracing the fan-outs of the key inputs. The symmetric construction of Anti-SAT block allows the identification of the protection logic through signal probability analysis. Even when the symmetry is dissolved using additional obfuscation, Anti-SAT remains vulnerable to slightly more sophisticated AppSAT guided removal attack [14].

3) Bypass attack: The principle of Bypass attack is to select a random key as the correct key, find the patterns that lead to an incorrect output for that particular key, and finally construct a bypass circuit to rectify the incorrect circuit output [17]. The complexity of the Bypass attack is dictated by the size of the bypass circuit. Both SARLock and Anti-SAT are vulnerable to the Bypass attack as well.

B. Stripped-functionality logic locking

SFLL comes in two flavors: 1) SFLL-HD that is suited to protect an arbitrarily large number of cubes, and 2) SFLL-flex that is suitable for applications where the cubes-to-be-protected can be compactly represented [19]. SFLL (and also TTLock, which is a special case of SFLL-HD) introduced the notion of *protected input cubes* to achieve resilience against removal and bypass attacks. It "strips" part of the functionality from the design and implements this modified functionality in hardware; protected input cubes are those that differentiate the outputs of the original and the modified designs. A restoration unit is also implemented on chip, canceling the errors produced by the protected input patterns only when the correct key is in place. Removal attack thus fails as it only delivers a netlist with stripped functionality. While TTLock thwarts removal attacks by introducing only one protected cube [18], SFLL allows for a larger number of protected cubes,

978-1-5386-3775-3/18 $31.00 © 2018 IEEE

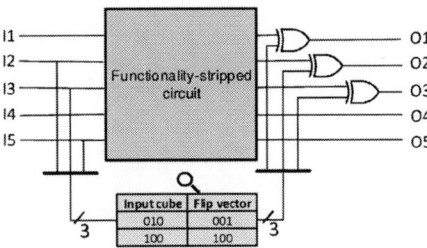

Fig. 3. Architecture for SFLL-flex. The protected input cubes are stored in an LUT along with flip vectors that restore the stripped functionality. Source: [19].

and thus a controllably stronger protection against removal attacks.

As opposed to SARLock and Anti-SAT, SFLL only focuses on the average number of keys eliminated by a DIP and not the worst case number. As such, SFLL allows trade-offs between SAT attack resilience and removal attack resilience. Note that if the DIP corresponds to the protected input cube, then SAT can immediately recover the correct key by eliminating all the incorrect keys in a single iteration. Thus, the resilience against a SAT attack is defined by the probability of finding such a protected input cube by the attacker [18], [19].

SFLL-flex. SFLL-flex strips functionality of a design based on either randomly or designer selected protected input cubes. As demonstrated in Fig. 3, it comprises a functionality-stripped circuit (FSC) and a restore unit, integrated using XOR gates. The restore unit is driven from a tamper-proof look-up table (LUT) that stores the compressed input cubes along with the flip vectors. The stripped functionality is thus restored for the protected input patterns, which act as "keys" of SFLL-flex.

III. PROPOSED SFLL-FAULT

In this section, we illustrate our technique with necessary details. SFLL-flex compactly represents the randomly-generated or designer-selected protected input cubes by first compressing them. It then relies on simulated annealing driven logic synthesis to lower the cost of stripping functionality based on the resulting set of compressed cubes. However, there are two pitfalls of this technique. First, simulated annealing is extremely slow in practice and rarely used for modern, large designs. Second, while SFLL-flex scales well for a small class of applications where protected input cubes are few and easy to extract, it may not scale well for generic applications where cube compression on a poorly selected initial set of protected cubes leads to high overhead in terms of area, power and timing. A question that naturally follows is that can we develop a heuristic to intelligently select the protected patterns such that the overall cost of implementation is minimized without compromising its security? In this paper, we present a new heuristic which is not only suitable for generic applications, but also allows for a cost-effective implementation of the proposed SFLL technique.

To identify the most cost-effective parts to strip and the list of protected input cubes associated with it, we leverage well-understood concepts and effective tools from VLSI testing.

We inject easy-to-enumerate faults[3], one at a time, to strip functionality from a design. Fault injections are traditionally used to model the behavior of physical defects in higher levels of abstraction. In this context, we use faults to enumerate our options to strip functionality from a design. We use ATPG tools to identify the list of patterns that detect a fault; these patterns are indeed the protected (or equivalently "failing") input cubes for the design modified by injecting the fault. This way, we can assess the implementation cost and security levels associated with different design modification options and choose the least-cost implementation with sufficient security.

Example. Let us consider the example of c17 ISCAS benchmark circuit shown in Fig. 4a. A stuck-at-zero fault was injected at node N5 at the output of the NAND gate. This reduces the circuit to that in Fig. 4b. The two circuits differ at the output O23 for the cubes listed in Fig. 4c. It is evident from the figure that the number of gates gets reduced by two for the FSC. Next, a restore circuit is constructed from these failing cubes to recover the original functionality of the circuit as shown in Fig. 4d.

A. Architecture.

The architecture consists of a tamper-proof [15] look-up table (LUT) and a *restore* unit as shown in Fig. 4d.

Cost. The cost[4] of implementation is dictated by the size of the look-up table that is used to store the protected input cubes, in addition to the comparators and XOR gates inserted at the outputs. Suppose an injected fault causes f outputs to fail where output o_1 fails for k_1 patterns each with n_1 bits, o_2 fails for k_2 patterns each with n_2 bits and so on. This implies that the number of bits to be stored in the LUT is $n_1 \cdot k_1 + n_2 \cdot k_2 + \cdots + n_f \cdot k_f$ and its associated cost is:

$$cost_{lut} = (n_1 \cdot k_1 + n_2 \cdot k_2 + \cdots n_f \cdot k_f) \cdot \alpha_{lut} \qquad (1)$$

where α_{lut} denotes the per-bit cost of a tamper-proof LUT.

The restore unit consists of f comparators of size $n_1, n_2,$ and n_f, respectively, along with f XOR gates at the failing outputs. Thus, the cost of the restore circuit is given by:

$$cost_{rest} = (n_1 + n_2 + \cdots n_f) \cdot \alpha_{comp} + f \cdot \alpha_{xor} \qquad (2)$$

where α_{comp} and α_{xor} denotes the cost of one bit comparator and a 2-input XOR gate, respectively. Thus, the total cost of the LUT and the restore unit is $cost_{lut} + cost_{rest}$.

B. Optimization framework

Suppose the cost of the FSC is denoted by $cost_{sf}$, then the problem reduces to the following optimization problem:

$$\text{minimize:} \quad cost_{sf} + cost_{lut} + cost_{rest}$$
$$\text{such that} \quad \max\{n_i - \log_2 k_i\} \geq s; 1 \leq i \leq f \qquad (3)$$

where s is the desired security level and $\max\{n_i - \log_2 k_i\}$ is the security level attained against SAT attack, explained later

---

[3]In this work, we use stuck-at faults, while we note that the fault model can be extended.

[4]Here, we focus on area as our cost metric. Similarly, we can optimize for other metrics such as power or timing of a design.

978-1-5386-3775-3/18 $31.00 © 2018 IEEE

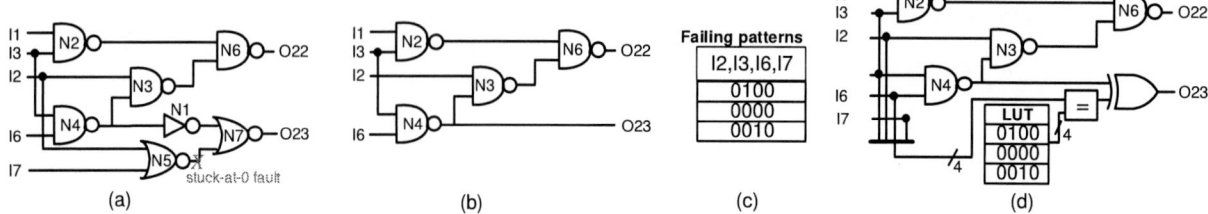

Fig. 4. SFLL-fault applied on c17 circuit. (a) Original circuit. (b) FSC after fault injection at N5. (c) List of failing patterns at O23. (d) Locked circuit including restore unit.

Algorithm 1: Fault injection based synthesis algorithm

Input : Original netlist N, Security level s
Output: Optimized netlist N_{opt}, Set of failing patterns V_{opt}

1   $N_{opt} \leftarrow \phi$
2   $V_{opt} \leftarrow \phi$
3   $X \leftarrow$ parse_list_of_nodes(N)
4   $cost_{min} \leftarrow \infty$
5   **for** *each $x \in X$* **do**
6      $N_{sf} \leftarrow$ inject_fault(N, x)
7      $V_{sf} \leftarrow$ get_failing_patterns(N_{sf}, N)
8   **if** *security_level(V_{sf}) $< s$* **then**
9   select_random_pattern(N_{sf}, s)
10   **end**
11     $cost_{new} \leftarrow cost_{sf} + cost_{lut} + cost_{rest}$
12   **if** *$cost_{new} < cost_{min}$* **then**
13        $cost_{min} \leftarrow cost_{new}$
14        $N_{opt} \leftarrow N_{sf}$
15        $V_{opt} \leftarrow V_{sf}$
16   **end**
17 **end**

---

in this section. Thus, essentially we want to select the FSC which has the least overall implementation cost, accounting also for the restore logic. The synthesis algorithm to find such an optimized FSC is depicted in Algorithm 1.

The input to the algorithm is the original netlist N and a desired security level s. We inject a stuck-at fault at every node in the netlist N and list its corresponding failing patterns. These patterns constitute the protected input patterns for which the circuit produces an incorrect output without the restore operation. Next we compute the security level delivered by the protected input cubes of the fault from Eq. 3. If the desired security level is not achieved, then a random pattern, satisfying the security requirement, is added to the set of protected patterns. The algorithm terminates when all the faults in the design have been explored and produces the overall optimal circuit in terms of area.

C. Security analysis of SFLL-fault

In this section, we analyze the security of our technique against state-of-the-art attacks against logic locking.

SAT attack resilience. We prove the following claim.

Claim 1. *SFLL-fault achieves a security of* $\max\{n_i - \log_2 k_i\}, 1 \leq i \leq f$ *against the SAT attack.*

Proof. Recall from Section II-B that the resilience against a SAT attack is defined by the probability of an attacker to identify a protected input cube. However, as the attacker has no information about the protected cubes, he/she can not do any better than a random guess. Note that in our case, finding one

of the protected cubes does not lead to the recovery of the other protected cubes. Thus, in order to recover the full functionality of the circuit, the attacker is forced to find all the input cubes, implying the SAT attack would be successful if and only it can find all the DIPs corresponding to the protected cubes. Thus, the security is dictated by the DIP having the minimum probability of being detected. Suppose, V_i denotes the set of protected cubes for output o_i and v denotes a randomly selected DIP.

Thus, the probability of success for SAT attack is given by:

$$\Pr[\text{SAT success}] = \min\{\Pr[p \in V_i]\}, 1 \leq i \leq f$$
$$= \min\{\frac{k_i}{2^{n_i}}\}, 1 \leq i \leq f \quad (4)$$

From [19], a logic locking technique is called λ-secure if the probability of finding a protected pattern is not greater than $\frac{1}{2^\lambda}$. In our case, the probability that p belongs to a set V_i (and leads to the recovery of key) is $\frac{k_i}{2^{n_i}}$; the security level attained for the output o_i is $n_i - \log_2 k_i$. Using Eq. 4, the security level for the locked circuit is:

$$s = \max\{n_i - \log_2 k_i\}, \quad 1 \leq i \leq f$$

$\square$

Removal attack resilience. Similar to SFLL-flex, SFLL-fault thwarts the removal attack by making the on-chip FSC implementation different from the original circuit for k input patterns, where $k = max\{k1, k2, \cdots, k_f\}$. While the current SFLL-fault framework optimizes for area, it can also be extended to optimize for removal attack resilience by simply maximizing k.

Bypass attack resilience. Bypass attack terminates by running only one iteration of the SAT attack and outputs an incorrect key. It then constructs a wrapper circuitry around the locked netlist to correct the output for that particular incorrect key [17]. However, for our design, the attacker fails to extract precise information about the protected input cubes, even if he/she creates a bypass circuit around the locked netlist; the circuit would still fail for the protected cubes, delivering resiliency against Bypass attack.

IV. RESULTS

A. Experimental setup

In this section, we present detailed experimental results to establish the effectiveness of the proposed SFLL-fault. All the experiments have been carried out on a 128-core Intel Xeon

978-1-5386-3775-3/18 $31.00 © 2018 IEEE

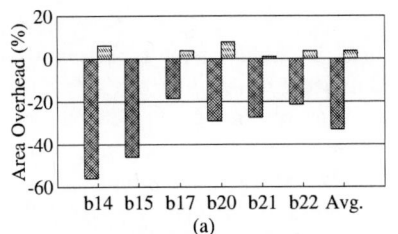 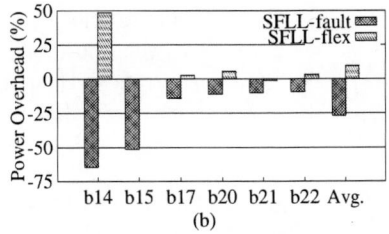 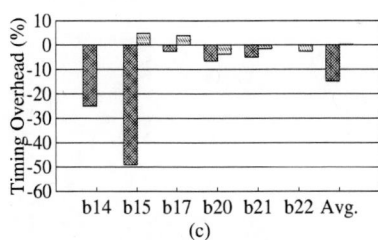

Fig. 5. Overhead comparison of FSC for SFLL-fault and SFLL-flex. (a) Area, (b) Power, and (c) Timing. Note that negative numbers imply gain in terms of cost, underscoring efficacy of SFLL-fault.

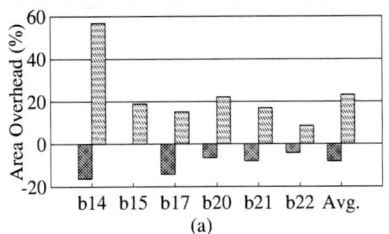 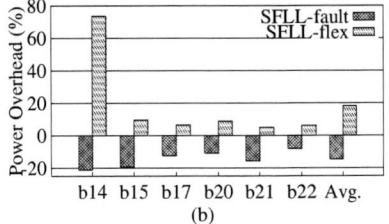 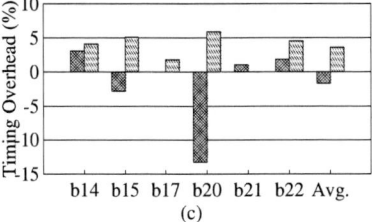

Fig. 6. Overhead comparison of locked circuits for SFLL-fault and SFLL-flex for s=128. (a) Area, (b) Power, and (c) Timing.

TABLE II
DETAILS OF ITC'99 BENCHMARK CIRCUITS

Benchmark	# Inputs	# Outputs	# Gates
b14	277	299	9,767
b15	485	519	8,367
b17	1452	1,512	30,777
b20	522	512	19,682
b21	522	512	20,027
b22	767	757	29,162

processor running at 2.2 GHz with 256 GB of RAM. All the codes have been compiled using GCC compiler 4.4.7 on Cent OS 6.9 (Final). We implemented our technique on ITC'99 benchmark circuits whose details are provided in Table II. We used Synopsys Design Compiler along with Global Foundries (GF) 65nm LPe library to compute the area, power, and timing overhead numbers. We assume that the tamper-proof LUT occupies $1.46\mu m^2$ per bit, the same as the area of a single bit of one-time programmable memory in GF 65nm LPe technology. Atalanta-M ATPG tool is used to compute the failing patterns for the fault-injected circuit[5] and the results were cross verified using the Cadence Conformal for logical equivalence.

B. Overhead analysis

Functionality-stripped circuit. The basic difference between SFLL-flex and SFLL-fault is the underlying method used to effect the functionality-strip operation; SFLL-flex uses simulated annealing, whereas SFLL-fault relies on fault injection. To demonstrate the cost-effectiveness of the functionality-strip operation in SFLL-fault as compared to that in SFLL-flex, we report the area, power, and timing overhead of the FSC circuit for both techniques. Note that negative numbers indicate savings compared to the original design.

As shown in Fig. 5a, SFLL-fault always achieves a significant reduction in area as compared to that of SFLL-flex. On average, the area overhead incurred by SFLL-fault is

[5]Note that Tetramax ATPG tool could not be used as it does not return all the failing patterns for a fault.

only -33.0% compared to 3.7% of SFLL-flex. Thus, the FSC obtained using fault injection occupies 33.0% lesser chip area than the original one. These savings in area can be attributed to the autonomy given to the ATPG tool in deciding both 1) the part(s) of the design to be protected and 2) the set of protected patterns.

As shown in Fig. 5b, the power overhead is -26.8% for SFLL-fault and 9.7% for SFLL-flex on average. As expected the power overhead correlates well with the area overhead. As illustrated in Fig. 5c, SFLL-fault is also able to achieve an average reduction of -14.8% for timing. Overall, SFLL-fault significantly outperforms SFLL-flex in terms of area, power, and timing savings for the FSC circuit, underscoring the benefits of delegating critical design decisions to carefully crafted EDA algorithms.

Logic-locked circuit. Next, we report the overhead of the overall logic-locked circuits with s=128 bits of security in Fig. 6. SFLL-fault is geared towards minimizing the overall cost which comprises three components listed in Eq. 3. In SFLL-flex, the FSC circuit incurs a small area overhead of about 3.7%. Upon addition of the overhead of LUT and restore unit (comparators and XORs), the overall overhead is 23.3%, on average. In contrast, the overall overhead of SFLL-fault is only -8.2% on average. SFLL-fault achieves significant savings in the FSC area which can offset the overhead of the restore unit and LUT, leading to a notable area savings for the locked circuit. Similarly, the average power overhead of the locked circuits is 18.3% and −14.6% for SFLL-flex and SFLL-fault, respectively. Also, SFLL-fault improves timing by -1.7% compared to 3.4% for SFLL-flex. However, for some of the circuits SFLL-fault does incur small timing overhead. This can be attributed to the fact that SFLL-fault protects a larger part of the design compared to SFLL-flex. This inadvertently requires a comparatively larger restore unit.

Runtime. In SFLL-fault, the fault injections are independent of one another, allowing for a parallelized implementation. For faster, exhaustive processing of faults in large circuits such as

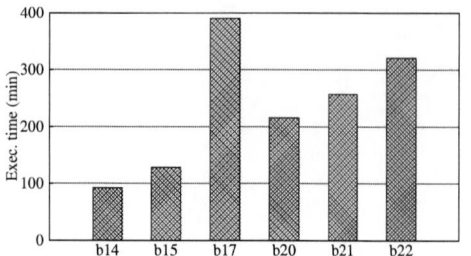

Fig. 7. Execution time for SFLL-fault for different benchmark circuits.

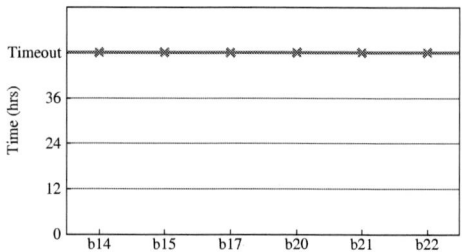

Fig. 8. Timeout for SAT attack for all the locked benchmark circuits.

b17, we simulate up to 100 faults in parallel. As reported in Fig. 7, the execution time of SFLL-fault is in the order of a few hours even for large circuits with >30K gates. The majority of the time is spent on optimizing the FSC using Synopsys DC compiler.

C. Security analysis

Note that the overall architecture of SFLL-fault and the security properties of SFLL-fault are the same as SFLL-flex; SFLL-flex differs mainly in the implementation methodology. As such, the security achieved against the other attacks such as AppSAT would be the same as that achieved by SFLL-flex. Nevertheless, we validate the security of SFLL-fault by launching SAT attack on circuits locked with SFLL-fault based on 128-bit security. Note that each experiment is repeated 10 times to improve the statistical significance of the results. Fig. 8 shows that the attack fails to terminate within the specified time limit of 48 hours for every trial.

D. Discussion

Comparison with SFLL-HD. Although we cannot directly compare SFLL-fault to SFLL-HD that uses a purely combinational logic for restore unit, we present some qualitative differences here. SFLL-HD allows to protect a large number of patterns; however, the functionality-strip operation in SFLL-HD relies on existing logic synthesis tools, which may not be security-aware [19]. Moreover, SFLL-HD has a significantly higher overhead compared to SFLL-fault; the functionality-strip operation itself incurs significant overhead. As an example, for the b14 circuit, SFLL-HD incurs an area overhead of 43% compared to only -16.2% for SFLL-fault.

V. CONCLUSION

We presented an automated synthesis framework by leveraging the principles of VLSI testing towards developing a low cost and secure logic locking technique, seamlessly incorporating security into the IC design flow. SFLL-fault is

able to achieve savings of 33.0%, 26.8%, and 14.8% for area, power, and timing, respectively during the functionality-strip operation. This ultimately results in savings of -8.1%, -14.6%, and -1.7% for area, power, and timing, respectively, for the overall locked circuit. While the platform currently optimizes for area, it can be easily extended to optimize for other metrics such as power, timing or resilience against different attacks. The proposed platform demonstrates the effectiveness of leveraging design automation techniques and advocates their application in emerging fields such as hardware security.

REFERENCES

[1] M. Rostami, F. Koushanfar, and R. Karri, "A Primer on Hardware Security: Models, Methods, and Metrics," *IEEE*, vol. 102, no. 8, pp. 1283–1295, 2014.

[2] SEMI, "Innovation is at Risk Losses of up to $4 Billion Annually due to IP Infringement," 2008. [June 10, 2015].

[3] Y. Alkabani and F. Koushanfar, "Active Hardware Metering for Intellectual Property Protection and Security," in *USENIX Security Symposium*, pp. 291–306, 2007.

[4] A. Kahng, J. Lach, W. H. Mangione-Smith, S. Mantik, I. Markov, M. Potkonjak, P. Tucker, H. Wang, and G. Wolfe, "Watermarking Techniques for Intellectual Property Protection," in *IEEE/ACM Design Automation Conference*, pp. 776–781, 1998.

[5] J. Baukus, L. Chow, R. Cocchi, and B. Wang, "Method and apparatus for camouflaging a standard cell based integrated circuit with micro circuits and post processing," 2012. US Patent no. 20120139582.

[6] F. Imeson, A. Emtenan, S. Garg, and M. V. Tripunitara, "Securing Computer Hardware Using 3D Integrated Circuit (IC) Technology and Split Manufacturing for Obfuscation," in *USENIX Security Symposium*, pp. 495–510, 2013.

[7] R. Pappu, B. Recht, J. Taylor, and N. Gershenfeld, "Physical One-Way Functions," *Science*, vol. 297, no. 5589, pp. 2026–2030, 2002.

[8] J. Roy, F. Koushanfar, and I. L. Markov, "Ending Piracy of Integrated Circuits," *IEEE Computer*, vol. 43, no. 10, pp. 30–38, 2010.

[9] M. Yasin, J. Rajendran, O. Sinanoglu, and R. Karri, "On Improving the Security of Logic Locking," *IEEE Transactions on CAD of Integrated Circuits and Systems*, vol. 35, no. 9, pp. 1411–1424, 2016.

[10] J. Rajendran, H. Zhang, C. Zhang, G. Rose, Y. Pino, O. Sinanoglu, and R. Karri, "Fault Analysis-Based Logic Encryption," *IEEE Transactions on Computer*, vol. 64, no. 2, pp. 410–424, 2015.

[11] P. Subramanyan, S. Ray, and S. Malik, "Evaluating the Security of Logic Encryption Algorithms," in *IEEE International Symposium on Hardware Oriented Security and Trust*, pp. 137–143, 2015.

[12] M. Yasin, B. Mazumdar, J. Rajendran, and O. Sinanoglu, "SARLock: SAT Attack Resistant Logic Locking," in *IEEE International Symposium on Hardware Oriented Security and Trust*, pp. 236–241, 2016.

[13] K. Shamsi, M. Li, T. Meade, Z. Zhao, D. Z., and Y. Jin, "AppSAT: Approximately Deobfuscating Integrated Circuits," in *IEEE International Symposium on Hardware Oriented Security and Trust*, pp. 95–100, 2017.

[14] M. Yasin, B. Mazumdar, O. Sinanoglu, and J. Rajendran, "Removal Attacks on Logic Locking and Camouflaging Techniques," *IEEE Transactions on Emerging Topics in Computing*, vol. 99, no. 0, p. PP, 2017.

[15] P. Tuyls, G. Schrijen, B. Škorić, J. van Geloven, N. Verhaegh, and R. Wolters, "Read-Proof Hardware from Protective Coatings," in *International Conference on Cryptographic Hardware and Embedded Systems*, pp. 369–383, 2006.

[16] Y. Xie and A. Srivastava, "Mitigating SAT Attack on Logic Locking," in *International Conference on Cryptographic Hardware and Embedded Systems*, pp. 127–146, 2016.

[17] X. Xu, B. Shakya, M. M. Tehranipoor, and D. Forte, "Novel Bypass Attack and BDD-based Tradeoff Analysis Against All Known Logic Locking Attacks," in *International Conference on Cryptographic Hardware and Embedded Systems*, pp. 189–210, 2017.

[18] M. Yasin, A. Sengupta, B. Schafer, Y. Makris, O. Sinanoglu, and J. Rajendran, "What to Lock?: Functional and Parametric Locking," in *Great Lakes Symposium on VLSI*, pp. 351–356, 2017.

[19] M. Yasin, A. Sengupta, M. Nabeel, M. Ashraf, J. Rajendran, and O. Sinanoglu, "Provably Secure Logic Locking: From Theory to Practice," in *ACM SIGSAC Conference on Computer and Communications Security*, 2017. To appear.

Modeling and Test Generation for Combinational Hardware Trojans

Ziqi Zhou, Ujjwal Guin, and Vishwani D. Agrawal

Department of Electrical and Computer Engineering

Auburn University, AL 36849, USA

{zhouziq, ujjwal.guin, agrawvd}@auburn.edu

Abstract—Due to globalization of semiconductor manufacturing, appearance of malicious circuitry known as hardware Trojan is now a recognized security threat. A Trojan may be added to the verified netlist without the knowledge of the designer or user causing unexpected malfunction or data theft when the device is in use. In this research we devise tests that would detect a Trojan in a manufactured chip. We recognize that a Trojan must escape manufacturing tests provided with the netlist by the designer. Based on the two parts of a Trojan, namely, a trigger derived as a Boolean function of any set of signals and a payload (typically, an XOR gate) inserted on a signal line, we develop a test generation model. A single-line trigger combined with a single payload line gives a set of $2K \times (K-1)$ Trojans in this model for a circuit with K signal lines. Tests for these are shown to be vectors that detect "conditional stuck-at" faults, for which we give a test generation algorithm using standard ATPG tools. The model allows us to define and measure a Trojan coverage metric for tests. Results show scalability of these tests, besides being more effective in detecting real Trojans than N-detect stuck-at test vectors or random vectors.

Index Terms—Hardware Trojans, modeling, logic testing, N-detect ATPG, verification

I. INTRODUCTION

Ensuring the security of integrated circuits (ICs) becomes a major challenge due to the globalization of the semiconductor industry. Majority of system-on-chip (SoC) design companies outsource their production across the world to fabrication units (fabs or foundries) due to a massive cost (several billion dollars [40]) for building and maintaining such foundries. This creates the threat of *hardware Trojans* (HT), which is a leading security concern for government and industry [3], [9], [19], [28], [30]–[33], [38]. A hardware Trojan is a malicious alteration to the original design to modify its functionality such that an adversary can gain control of the system. An adversary may insert a hardware Trojan into a design to interrupt its normal operation in the field. The Trojan would act like a "silicon time bomb" [19]. It can also create a backdoor in a secure system to give access to critical system functionality or leak secret information to an adversary.

Researchers have proposed numerous techniques to detect and prevent HTs. These techniques are broadly classified into two groups, namely, solutions targeted for the detection of HTs, and solutions designed for preventing an adversary to insert a HT in a design. The detection methods for HTs can further be classified into logic testing [8], [13], [15], [20], [35], and side-channel analysis [4], [6], [7], [21], [22], [25]. Prevention methods can be grouped into design-for-trust measures [12], [23], [26], [29], [39] and split manufacturing [27], [34], [36].

The overall aim is to detect HTs in chips manufactured in an untrusted environment and, thus, prevent Trojan infected devices from getting into the electronics supply chain. Logic testing can be used to detect these Trojans, where we apply stimuli to primary inputs (PIs) and observe responses at primary outputs (POs) [8], [9], [13], [15], [20], [30]. Detection of a HT occurs when there is a mismatch between the observed and expected responses. Such detection of a HT through logic testing does not have any impact from the process and environmental variations. On the other hand, the side-channel analysis uses physical characteristics such as power [37], temperature [24], delay [18], and radiation [16] to detect the HT. Side-channel detection methods primarily rely on the availability of Trojan-free golden circuits, which may not be available in reality. Moreover, process and environmental variations may mask the side channel leakage, if the Trojan circuitry is small. Despite significant research performed on HT, we still lack methods for modeling and test generation to detect them.

A. Contribution

We propose a generalized model of a combinational hardware Trojan. We believe this is the first time such a model for a Trojan is being presented. We then propose a generalized method based on conditional stuck-at faults to detect the modeled Trojans. The contributions of this paper are:

- *Design of a combinational hardware Trojan:* We have proposed a generalized model of hardware Trojan based on the circuit netlist. We call this Type-n Trojan, where n is the number of the trigger inputs. The payload of this Trojan can be delivered to a location where a stuck-at fault (SAF) is detectable by a Trojan activation pattern (TAP). Because TAPs may detect several stuck-at faults, the location of the Trojans is not unique. Any such fault site inserted with payload will result in Trojan behavior for the TAP. We believe this is the first approach to model a generalized Type-n Trojan.

- *Detection based on conditional SAFs:* We have proposed a hardware Trojan detection technique based on conditional detection of SAFs. With reasonable test length, we can detect all Type-1 Trojans. These conditional SAF patterns (CSP) also detect higher order Trojans with reasonable confidence. It is reasonable to assume that an adversary will not have access to the CSP since they would not be included in manufacturing data.

The rest of this paper is organized as follows: Section II describes the generalized model for a combinational hardware Trojan, termed as Type-n Trojan. Section III details our approach for detecting Type-n Trojans. Section IV describes simulation results to demonstrate the effectiveness of our pro-

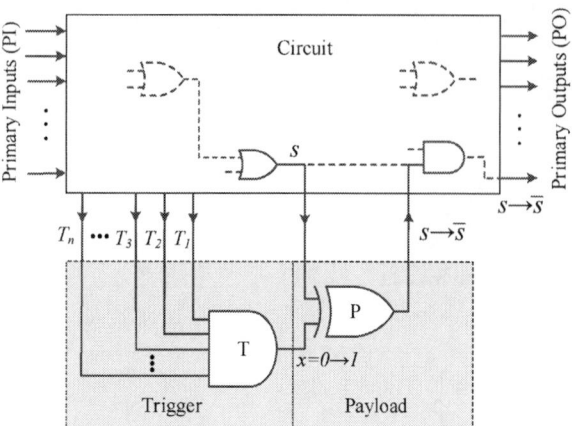

Figure 1. A model for a combinational hardware Trojan.

posed conditional SAF patterns for hardware Trojan detection. Section V concludes the paper.

II. MODELING A HARDWARE TROJAN

A. Hardware Trojans

A hardware Trojan has two parts, namely, a trigger and a payload as shown in Figure 1. The trigger activates the hardware Trojan when a certain condition is satisfied. Inputs to the trigger can directly come from primary inputs (PI) or from internal nets of the circuit. Although shown here as an AND gate, the trigger can be any logic function. When the Trojan is activated, e.g., when the AND gate output becomes 1, it delivers the payload to the circuit by modifying its functionality. A two-input XOR gate with inputs from the trigger and a net in the circuit, can be used for such purpose. The output of the XOR gate is taken back to the circuit.

Trojans, added for malicious purposes, consist of circuitry (trigger and payload) that has been added to a VLSI chip without the knowledge of the designer or the user. A hardware Trojan must have the following properties:

- **Property 1:** A Trojan modifies the logical function of a chip, although the modification may be subtle. For certain inputs, termed as *activation vectors* or *activation patterns*, the output of the chip then deviates from its correct value. This incorrect result may help an adversary to fulfill his/her malicious purpose.
- **Property 2:** A Trojan must not be activated by production (scan-based structural or functional) tests. This leads the Trojan circuitry to remain undetected during production testing of the chip.
- **Property 3:** Although the effect of a Trojan may appear similar to a design error, the Trojan distinctly differs from a design error. In case of a remaining error in a completed design, production tests are generated for the chip with error and these tests aim at preserving the error in the manufactured chip. Since the Trojan is inserted in the chip design after the production tests were generated, the Trojan circuitry is, by design, made transparent to tests. Thus, the design of a Trojan must consider its function (activation inputs and modified outputs), the chip function (typically, a netlist), and the production tests.

B. Hardware Trojan Model

The Trojan circuitry may be designed for malicious purposes, such as, expose some secret key to an adversary, transmit unencrypted data to an unsecured channel, disable a circuitry, or incorrectly execute an intended function. A hardware Trojan modifies the input-output characteristics of the chip and thus provides an adversary to gain undue advantage. In this paper, we assume that the chip is sequential, is implemented with flip-flops and combinational logic, and is tested through the scan technique [11]. Typically, manufacturing tests are generated for single stuck-at faults of the combinational logic. These tests are digital vectors applied to primary inputs (PI) of the combinational logic and the results at primary outputs (PO) are verified against expected responses. Functional tests and delay tests are also performed at the manufacturing site. Without loss of generality, we focus our discussion to stuck-at fault (SAF) manufacturing tests for designing a Trojan and its detection.

Two single stuck-at (SSA) faults, namely, stuck-at-0 ($sa0$) and stuck-at-1 ($sa1$), are modeled on every *signal* or *line*, where a signal can be a primary input (PI), a gate output, or a fanout branch. Thus, the number K of fault sites is given by:

$$K = \#PI + \#Gates + \#Fanout\ branches \qquad (1)$$

A test for a fault on a signal assumes all other signals to be fault-free. The test activates the fault by setting the signal to an appropriate value, for example, 0 for a $sa1$ fault, and propagates the state of the signal to a primary output (PO). In addition, there are specific test sequences to verify the function of the scan shift register [11].

To facilitate the testing of a hardware Trojan, we propose a model shown in Figure 1 with following attributes:

1) Trigger: The objective of a Trojan designer is to evade manufacturing tests, otherwise every chip will fail at the testing site. The trigger circuit must remain quiet (e.g., output of the trigger x remains "0") during the tests. The selection of the trigger inputs (T_i) can be from the primary inputs or internal nets of the circuit.

2) Payload: A net (s) is selected in the circuit to deliver the payload of the Trojan. The original signal s, shown with broken line in Figure 1, is rerouted through a two-input XOR gate whose other input is either the trigger x as shown in Figure 1 or $\overline{x}$. We define this net as Trojan location and assume that it is distinctly different from the set $\{T_i\}$ used to generate the trigger. Two conditions must be satisfied by a vector at PI to activate the Trojan. First, the vector should activate a path from s to a PO, hence it should be a test for either a $sa0$ or $sa1$ fault on s. Second, this vector should place a logic 1 on x (or logic 0 if $\overline{x}$ is connected to the XOR). As a result the PO will experience a signal inversion, changing the true function of the circuit.

Definition 1. A *Type-n Trojan* is defined as a combinational hardware Trojan of order n and has n trigger inputs.

Definition 2. The *location* of a Type-n Trojan is defined as a site (signal or line) in the circuit where the payload is delivered.

978-1-5386-3775-3/18 $31.00 © 2018 IEEE

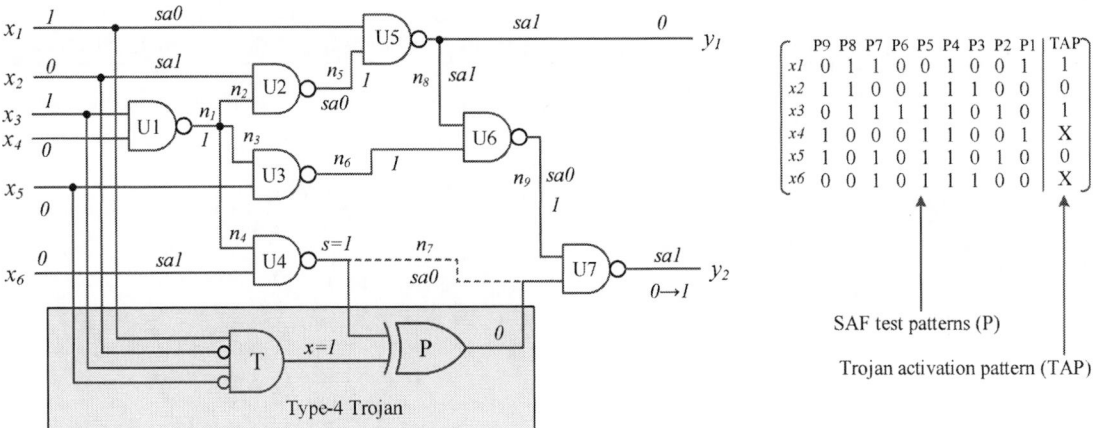

Figure 2. An 18-line ($K = 18$) combinational circuit with Type-4 Trojan ($n_7|x_1, \overline{x_2}, x_3, \overline{x_5}$). For Trojan activation pattern (TAP) 101000, logic states of lines and detectable SAFs are marked on the circuit.

A Type-1 Trojan has only one trigger input that can come from any part of the netlist or a primary input. Similarly, a Type-2 Trojan has two trigger inputs. We note that for lower order Trojans, in general, the trigger inputs may come from low switching nets to keep the Trojans mostly quiet.

Figure 2 shows an example of a Type-4 Trojan inserted in a 6-input, 2-output circuit, where we select trigger inputs directly from PIs. We specify this Trojan as ($n_7|x_1, \overline{x_2}, x_3, \overline{x_5}$), where n_7 is the payload site and $x_1, \overline{x_2}, x_3$ and $\overline{x_5}$ are the trigger input signals. Note that, in general, triggers can be tapped from any line in the circuit. Nine test vectors $\{P1, P2, \cdots, P9\}$ are generated using the ATPG tool Tetra-Max [1] as manufacturing tests to provide 100% stuck-at fault (SAF) coverage. As long as the Trojan is not activated by these test vectors, the circuit will pass the production test. For this example, we have Trojan activation pattern, $TAP = (101X0X)^T \notin \{P1, P2, \ldots, P9\}$, where "X" denotes *don't care* state. As the order of this Trojan ($n = 4$) is less than the number of PI (6), the Trojan may be activated by multiple input patterns. It is thus necessary to verify that the trigger output (x) remains 0 for all test vectors ($P1, P2, \ldots, P9$).

One can deliver the payload at any site, where a SAF is detected by the TAP. Because the two X's can be enumerated in four ways, the TAP 101X0X corresponds to four vectors. Simulating [1] two SAFs at the payload site n_7 we find that 101001 detects $sa1$ and the other three patters, 101000, 101100 and 101101 detect $sa0$ at n_7. Thus, the Trojan in Figure 2 will produce four errors at y_2, $0 \rightarrow 1$ for 101001 input and $1 \rightarrow 0$ for 101000, 101100 and 101101.

In this example, the Trojan delivers the payload at signal n_7, where a SAF fault is detected by one or more TAPs. By simulating any TAP, alternative locations for delivering the payload can be found. Figure 2 shows SAFs detectable at x_1, $x_2, x_6, n_1, n_2, n_3,$ or n_5 the TAP 101000. Thus, any of these lines can be used as an alternative payload site.

C. Finding All Type-n Trojans

An upper bound on the number of Type-n Trojans (T_n) is given by:

$$T_n \leq \binom{K}{n} \times 2^n \times (K - n) \qquad (2)$$

Table I
MODELED HARDWARE TROJANS IN CIRCUIT OF FIGURE 2.

Trojan Category	Type-1	Type-2	Type-3	Type-4
All possible Trojans (Eq. 2)	612	9,792	97,920	685,440
Feasible Trojans	605	8,097	60,905	294,538
Trojans removed by SAF tests	586	6,985	43,852	17,4114
Valid Trojans	19	1,112	17,053	120,424

where K is the number of lines in the Trojan-free circuit. From Equation 1, $K = 18$ for the circuit of Figure 2. Numbers of all possible Trojans of types 1 through 4, computed from Equation 2, are shown in Table I. Note that the number of Trojans goes up by one order for each higher type. However, all Trojan structures (payload and trigger) do not modify the truth table of the circuit; when trigger is active, a path from payload site to PO may or may not be sensitized. Those modifying the truth table are shown as feasible Trojans in Table I. These were determined using the exhaustive set of $2^6 = 64$ patterns, a fault simulator [1] to identify sensitized paths, and a logic simulator [2] to examine the trigger states. We find that a significant number of the feasible Trojans is detectable by the set of nine SAF manufacturing test patters $P1$ through $P9$ shown in Figure 2. We do not consider those as valid Trojans because chips containing them will be eliminated during production testing. Removing them from feasible Trojans gives us the number of valid Trojans.

Although the Type-n Trojan model seems general, even for a small circuit ($K = 18$), the number of valid Trojans grows rapidly with n. For generating tests for Trojan detection and for coverage analysis, we will use Type-1 Trojans assuming their number equals the upper bound of Equation 2:

$$T_1 = 2K(K - 1) \qquad (3)$$

This number of target Trojans is $O(K^2)$, or quadratic in circuit size, K. The number would be $O(K^{n+1})$ for Type-n Trojans, where $n \leq K - 1$. Thus, our methodology parallels SAF whose tests are known to detect multiple stuck-at and many other types of faults [11].

Trigger circuitry of a Type-n Trojan model is an n-inputs AND gate. The payload is delivered through an XOR gate to any site activated by the Trojan activation pattern (TAP).

Algorithm 1: Design of a Type-n Trojan.

Input : Circuit Netlist (C), Manufacturing test patterns (P), Order of a Trojan (n)
Output: Trojan activation pattern (TAP), Trigger Inputs (T)

1 Read the netlist ;
2 Read manufacturing test patterns (P);
3 Select a random pattern as Trojan activation pattern, $TAP \notin P$;
4 Perform logic simulation using P to obtain all internal node values (M_K);
5 Perform logic simulation with TAP to obtain all internal node values (S_T);
6 Select a n random locations of the netlist to form the trigger inputs ;
7 Form a new matrix M_n, $M_n \leftarrow mod(M_K)$;
8 **if** $S_n \in M_n$ **then**
9   Drop the selection as it will activate the Trojan;
10   Go to Step 6;
11 **else**
12   Choose T as trigger input;
13 **end**
14 Perform fault simulation and logic simulation with TAP ;
15 Select a fault site (from Step 14) for delivering the payload, i.e., Trojan location.

Algorithm 2: Conditional stuck-at fault (SAF) pattern generation for hardware Trojan detection.

Input : Circuit Netlist, C
Output: Conditional SAF detection pattern set, CSP

1 Read the netlist C ;
2 Determine number of nets in C, $K \leftarrow \#PIs + \#Gates + \#fanout\_branches$;
3 Initialize empty set of conditional SAF patterns, CSP $\leftarrow \phi$;
4 Initialize count $c \leftarrow 0$;
5 **for** $i \leftarrow 0$ **to** K **do**
6   **for** $j \leftarrow 0$ **to** 1 **do**
7   **for** $k \leftarrow 0$ **to** 1 **do**
8   Initialize $TempNets \leftarrow Nets$;
9   Initialize count $l \leftarrow 0$;
10   **while** $TempNets \neq \phi$ **do**
11          $CSP[c] \leftarrow$ Test pattern for stuck-at j fault with $net_l = k$;
12   Invoke logic simulation with $CSP[c]$ to get internal node values ;
13   Remove nets with signal value from $CSP[c]$, $TempNets \leftarrow update(TempNets)$;
14          $c \leftarrow c + 1$, and $l \leftarrow l + 1$;
15   **end**
16   **end**
17   **end**
18 **end**
19 Report CSP for Trojan detection;

Algorithm 1 designs a Type-n Trojan that will not be activated the manufacturing tests. The inputs are original netlist (C), manufacturing test patterns (P), and the order of the Trojan (n). The algorithm reports the TAP and trigger inputs (T). It reads the Trojan free circuit netlist and manufacturing test patterns (Lines 1-2). A random TAP is selected (Line 3), which is not present the manufacturing test pattern set ($TAP \notin P$). Logic simulation gives the internal node values (M_K) of the circuit for all manufacturing test patterns (Line 4). M_K is a $K \times p$ matrix where K and p denote number of circuit nodes and number of manufacturing test patterns, respectively. We simulate the circuit with TAP for all internal node values (S_T). Here, S_T is a $K \times 1$ vector. Select n locations from K nets, either randomly or by some given criterion, to form an ($n \times 1$) vector S_n from S_T (Line 6). The $mod()$ function returns a new matrix M_n corresponding to the selected n nodes (Line 7). Selection of these n nets is not valid if $S_n \in M_n$, as one of the test pattern will trigger the Trojan (Line 9), so select a new set of n nodes (go to Step 6). Finally, fault simulation with TAP gives possible sites for payload (Lines 14-15).

III. TEST GENERATION FOR TYPE-n TROJANS

Definition 3. CSP-n, CSP-1 or CSP, and CSP-0: For a signal s in a digital circuit, two *type-n conditional stuck-at fault (SAF) patterns* (CSP-n) detect $sa0$ and $sa1$ faults, respectively, while setting specified [0,1] values on n other signals $t_1, \cdots t_n$. CSP-1, or simply CSP, are conditional tests with a condition on a single signal. CSP-0 are tests without any condition and are identical to the classical SAF tests.

Type-n HT's with signal s as payload are detectable by CSP-n we denote as $(s\ sa0\ |\ C_1, \cdots C_n)$ or $(s\ sa1\ |\ C_1, \cdots C_n)$, where $C_i = t_i$ or $\overline{t_i}$. CSP-n is a generalization of the CSP-1 defined in the literature [14].

Clearly, a Type-n Trojan is detectable by any of the two CSP-n's for which s is the payload site and $t'_i s$ are trigger in-

puts. For example, the Type-4 Trojan in Figure 2 is detected by CSP-4 ($n_7\ sa0\ |\ x_1, \overline{x_2}, x_3, \overline{x_5}$) and ($n_7\ sa0\ |\ x_1, \overline{x_2}, x_3, \overline{x_5}$). Considering complexity, we restrict to test generation for CSP-1 and evaluate their coverage for higher type of Trojans.

A. Conditional SAF Pattern (CSP-1 or CSP) Generation

Algorithm 2 generates conditional SAF patterns (CSP) for detecting hardware Trojans. It is necessary to determine the total number of nets (K) according to Equation 1 (Line 2). The algorithm initializes CSP-n as an empty set (Line 3) and then iterations with their signal values ($Nets$) are stored in $TempNets$. A CSP-1 is generated for a specific SAF with signal values for specific nets (Line 11). Note that, a CSP-n generation capability in the ATPG tool TetraMAX [1] is invoked by specifying an SAF target with n other signals and their values. Logic simulation is performed with this pattern to find the internal node values (Line 12). All {net, signal value} pairs corresponding to this pattern are dropped from $TempNets$ (Line 13). Repeat this process until $TempNets$ is empty (Line 10). Once all iterations are complete, the algorithm reports CSP-n's for Type-n Trojan detection.

B. An Example: Circuit of Figure 2.

Once again, considering the high complexity due to large number of higher type of Trojans, explained in Section II C, we generated tests for all 612 Type-1 Trojans using Algorithm 2. As a result, 48 CSP-1 detected 605 feasible Trojans confirming the data in Table I. Without confusion, we simply call them CSP. Assuming that 9 SAF vectors would have already tested for 586 Trojans during production, those were removed leaving as set of 48 vectors that detect all 19 valid Type-1 Trojans. Manufacturing tests are applied during production to all chips

Table II
HT TEST COVERAGE (%) OF VALID TROJANS (V_n).

Trojan type	Circuit	Lines, K Eq (1)	All Trojans T_n, Eq (2)	SAF tests	Valid Trojans V_n, Eq (5)	HT tests	V_n Coverage (%)		
							CSP	N-det.	Random
Type 1	Fig. 2	18	612	9	19	48	100	100	100
	c432	307	187,884	69	16,684	99,991	100	96.86	70.29
	c880	577	664,704	76	152,583	531,846	96.25	95.88	94.81
Type 2	Fig. 2	18	9,792	9	1,112	48	100	99.60	99.73
	c432	307	5.73×10^7	69	1.2710^7	99,991	93.48	89.71	61.93
	c880	577	3.82×10^8	76	1.22×10^8	531,846	93.72	92.98	91.62
Type 3	Fig. 2	18	97,920	9	17,053	48	99.80	98.46	97.64
	c432	307	1.16×10^{10}	69	3.96×10^9	99,991	89.62	83.89	59.16
	c880	577	1.46×10^{11}	76	6.25×10^{10}	531,846	90.08	89.99	88.06
Type 4	Fig. 2	18	685,440	9	120,424	48	99.30	97.40	95.95
	c432	307	1.76×10^{12}	69	7.27×10^{11}	99,991	85.28	78.67	53.11
	c880	577	4.19×10^{13}	76	2.176×10^{13}	531,846	87.87	87.51	85.62

to eliminate defective ones. Assuming that a Trojan remains undetected (we define this as a *valid Trojan*, all passing chips must have the same Trojan. Hence it is sufficient to test just one chip for Trojans and the Trojan tests can be much longer than the manufacturing tests. For any type (n), the quality of Trojan tests is their coverage of valid Trojans, which, in turn depends on the manufacturing tests. Thus,

$$\text{Trojan Coverage} = \frac{\# \ of \ detected \ valid \ Trojans}{\# \ of \ all \ valid \ Trojans} \times 100 \ \% \quad (4)$$

We generated two other sets, each with 48 vectors, an N-detect set and a random set. Valid Trojans of types 1 through 4 (Table I) were simulated. Results are given in Table II (Rows 1, 4, 7 and 10). Coverage of CSP was always higher and dropped slower with increasing n. Four Type-2 Trojans, $(x_5 \mid \overline{x_2}, x_4)$, $(x_5 \mid x_4, n_5)$, $(n_4 \mid \overline{x_1}, x_4)$ and $(n_3 \mid x_1, x_4)$, were only detected by CSP. Their payloads are closer to PI. Besides, they indicate superior capability of CSP in covering trigger combinations.

IV. BENCHMARK CIRCUITS

To study the effectiveness of the proposed conditional SAF patterns (CSP), we used a simulation setup for ISCAS 85 benchmark circuits [10]. TetraMax [1] provided manufacturing tests for each circuit covering 100% of all detectable SAFs. Next, we generate the CSP for each circuit using Algorithm 2. As the number of valid Trojans is large, we perform the coverage analysis of CSP based on four random sample sets of 20,000 Trojans of Type-1 through Type-4, respectively. Some Trojans cannot be triggered from inputs, nor do they affect outputs. Excluding these, we get feasible Trojans. Feasibility within each sampled set was assessed using the TetraMax conditional ATPG capability [1]. Some feasible Trojans are detectable by the manufacturing tests. Excluding those, we get valid Trojans (v_n), any of which an adversary may insert in the netlist. It is economical to estimate the total number of valid Trojans (V_n) in a circuit based on the 20,000-Trojan sample. This sample size is large enough for reasonable accuracy [11]. Total number of valid Trojans is,

$$V_n = \frac{v_n}{20,000} \times T_n \quad (5)$$

Table II shows the results. Trojan sampling was not used for the circuit of Figure 2 (Rows 1, 4, 7 and 10). Next, Rows 2, 5, 8 and 11 show data for c432 benchmark. For $K = 307$, Equation 2 gives $T_1 = 187,884$ Type-1 Trojans (Column 4). This circuit has 712 SAFs detected by 69 manufacturing test patterns (Column 5). Algorithm 2 generated 99,991 CSP, beyond 69 SAF patterns, shown as HT tests in Column 7. From 187,884 Type-1 Trojans, we take a random sample of 20,000 Type-1 Trojans to estimate the Trojan coverage. Among these, 513 Trojans could not be triggered from inputs, leaving 19,487 feasible Trojans. In addition, 17,711 Trojans were detected by 69 SAF patterns. Hence, number of valid Trojans, $v_n = 19,487 - 17,711 = 1,776$. From Equation 5, number of valid Type-1 Trojans, $V_n = 16,684$ (Column 7). Next three columns of Row 2 give Type-1 Trojan coverage by CSP, N-detect patterns, and random patterns, respectively, each containing 99,991 patterns. Similarly, results for Trojans of Type 2 (Row 5), Type 3 (Row 8) and Type 4 (Row 11) were obtained. Notably, the CSP coverages are consistently higher.

Results for c880 benchmark in Rows 3, 6, 9 and 12 were obtained in a similar manner with one exception. The number of HT tests is 531,846 and will grow significantly larger for bigger circuits. We randomly sampled 5,000 patterns from 531,846 HT tests to estimate the coverage of Trojans of Types 1 through 4 [17]. The results are given in Columns 8-10 (rows for c880). Once again, CSP coverages are higher.

V. CONCLUSION

The Type-n Trojan is a generalized model that facilitates test generation and coverage analysis. A Consideration of the complexity issue leads to the Type-1 Trojan model and its test by conditional stuck-at fault patterns (CSP). Thus, the number of Trojans to be modeled is $O(K^2)$ for a circuit with K signal lines. Although the detection coverage is measured over valid Type-1 Trojans, not detectable by manufacturing tests, tests are generated for all Type-1 Trojans. This is because our "real" targets include higher types as well. We find that both Trojan sampling and vector sampling are beneficial for coverage estimates. For larger circuits, CSP generation for a randon sample of Type-1 Trojans may also be used [5].

In the future, the scope of modeling and test generation should be expanded to solve diagnostic problems. Another aspect to explore is the minimization of Trojan tests. Despite the fact that the Trojan tests need not be applied to all chips, the numbers of Type-1 Trojans and their CSP for large circuits can be enormous. A third aspect to explore is the behavior of CSP in detecting Trojans with $n > 4$.

Acknowledgment: This research was supported by an internal grant from the ECE Department at Auburn University.

REFERENCES

[1] "TetraMAX ATPG: Automatic Test Pattern Generation." Synopsys, Inc., 2017.

[2] "VCS: Industrys Highest Performance Simulation Solution." Synopsys, Inc., 2017.

[3] S. Adee, "The Hunt for the Kill Switch," *IEEE Spectrum*, vol. 45, no. 5, pp. 34–39, 2008.

[4] D. Agrawal, S. Baktir, D. Karakoyunlu, P. Rohatgi, and B. Sunar, "Trojan Detection Using IC Fingerprinting," in *Proc. IEEE Symp. Security and Privacy (SP)*, 2007, pp. 296–310.

[5] V. D. Agrawal, H. Farhat, and S. C. Seth, "Test Generation by Fault Sampling," in *Proc. Int. Conf. on Computer Design (ICCD)*, 1988, pp. 58–61.

[6] M. Banga and M. S. Hsiao, "A Region Based Approach for the Identification of Hardware Trojans," in *Proc. IEEE Int. Workshop on Hardware-Oriented Security and Trust*, 2008, pp. 40–47.

[7] M. Banga and M. S. Hsiao, "A Novel Sustained Vector Technique for the Detection of Hardware Trojans," in *Proc. 22nd Int. Conf. VLSI Design*, 2009, pp. 327–332.

[8] M. Banga and M. S. Hsiao, "Odette: A Non-Scan Design-for-Test Methodology for Trojan Detection in ICs," in *Proc. IEEE Int. Symp. Hardware-Oriented Security and Trust*, 2011, pp. 18–23.

[9] S. Bhunia, M. S. Hsiao, M. Banga, and S. Narasimhan, "Hardware Trojan Attacks: Threat Analysis and Countermeasures," *Proc. IEEE*, vol. 102, no. 8, pp. 1229–1247, 2014.

[10] D. Bryan, "The iscas'85 benchmark circuits and netlist format," *North Carolina State University*, vol. 25, 1985.

[11] M. L. Bushnell and V. D. Agrawal, *Essentials of Electronic Testing for Digital, Memory, and Mixed-Signal VLSI Circuits.* Springer, 2000.

[12] R. S. Chakraborty and S. Bhunia, "Security Against Hardware Trojan Through a Novel Application of Design Obfuscation," in *Proc. Int. Conf. Computer-Aided Design*, 2009, pp. 113–116.

[13] R. S. Chakraborty, F. G. Wolff, S. Paul, C. A. Papachristou, and S. Bhunia, "MERO: A Statistical Approach for Hardware Trojan Detection," in *Proc. International Workshop on Cryptographic Hardware and Embedded Systems (CHES)*, LNCS 5747, Springer, 2009, pp. 396–410.

[14] O. E. Cornelia, "Conditional Stuck-At Fault Model for PLA Test Generation," Master's thesis, McGill University, Montreal, Canada, Dec. 1987.

[15] S. K. Haider, C. Jin, M. Ahmad, D. Shila, O. Khan, and M. van Dijk, "Advancing the State-of-the-Art in Hardware Trojans Detection," *IEEE Transactions on Dependable and Secure Computing*, 2017.

[16] J. He, Y. Zhao, X. Guo, and Y. Jin, "Hardware Trojan Detection Through Chip-Free Electromagnetic Side-Channel Statistical Analysis," *IEEE Trans. Very Large Scale Integration Sys.*, vol. 25, no. 10, pp. 2939–2948, Oct. 2017.

[17] K. Heragu, V. D. Agrawal, and M. L. Bushnell, "FACTS: Fault Coverage Estimation by Test Vector Sampling," in *Proc. 12th IEEE VLSI Test Symp.*, 1994, pp. 266–271.

[18] Y. Jin and Y. Makris, "Hardware Trojan Detection Using Path Delay Fingerprint," in *Proc. HOST*, 2008, pp. 51–57.

[19] R. Karri, J. Rajendran, K. Rosenfeld, and M. Tehranipoor, "Trustworthy Hardware: Identifying and Classifying Hardware Trojans," *Computer*, vol. 43, no. 10, pp. 39–46, 2010.

[20] N. Lesperance, S. Kulkarni, and K.-T. Cheng, "Hardware Trojan Detection Using Exhaustive Testing of k-bit Subspaces," in *Proc.*

20th Asia and South Pacific Design Automation Conf. (ASP-DAC), 2015, pp. 755–760.

[21] J. Li and J. Lach, "At-Speed Delay Characterization for IC Authentication and Trojan Horse Detection," in *Proc. IEEE Int. Workshop on Hardware-Oriented Security and Trust*, 2008, pp. 8–14.

[22] Y. Liu, K. Huang, and Y. Makris, "Hardware Trojan Detection Through Golden Chip-Free Statistical Side-Channel Fingerprinting," in *Proc. 51st Design Automation Conf.*, 2014.

[23] X. T. Ngo, S. Bhasin, J.-L. Danger, S. Guilley, and Z. Najm, "Linear Complementary Dual Code Improvement to Strengthen Encoded Circuit Against Hardware Trojan Horses," in *Proc. IEEE Int. Symp. Hardware Oriented Security and Trust*, 2015, pp. 82–87.

[24] A. N. Nowroz, K. Hu, F. Koushanfar, and S. Reda, "Novel Techniques for High-Sensitivity Hardware Trojan Detection Using Thermal and Power Maps," *IEEE Trans. Computer-Aided Design of Integrated Circuits and Systems*, vol. 33, no. 12, pp. 1792–1805, Dec. 2014.

[25] R. Rad, J. Plusquellic, and M. Tehranipoor, "Sensitivity Analysis to Hardware Trojans Using Power Supply Transient Signals," in *Proc. IEEE Int. Workshop on Hardware-Oriented Security and Trust*, 2008, pp. 3–7.

[26] J. Rajendran, J. Jyothi, O. Sinanoglu, and R. Karri, "Design and Analysis of Ring Oscillator Based Design-for-Trust Technique," in *Proc. IEEE 29th VLSI Test Symp.*, 2011, pp. 105–110.

[27] J. J. V. Rajendran, O. Sinanoglu, and R. Karri, "Is Split Manufacturing Secure?," in *Proc. Conf. Design, Automation and Test in Europe (DATE)*, 2013, pp. 1259–1264.

[28] M. Rostami, F. Koushanfar, and R. Karri, "A Primer on Hardware Security: Models, Methods, and Metrics," *Proc. IEEE*, vol. 102, no. 8, pp. 1283–1295, 2014.

[29] H. Salmani, M. Tehranipoor, and J. Plusquellic, "A Novel Technique for Improving Hardware Trojan Detection and Reducing Trojan Activation Time," *IEEE Trans. Very Large Scale Integration Sys.*, vol. 20, no. 1, pp. 112–125, 2012.

[30] O. Sinanoglu, N. Karimi, J. Rajendran, R. Karri, Y. Jin, K. Huang, and Y. Makris, "Reconciling the IC Test and Security Dichotomy," in *Proc. 18th IEEE European Test Symp.*, 2013.

[31] M. Tehranipoor and F. Koushanfar, "A Survey of Hardware Trojan Taxonomy and Detection," *IEEE Design & Test of Computers*, vol. 27, no. 1, 2010.

[32] M. Tehranipoor, H. Salmani, X. Zhang, M. Wang, R. Karri, J. Rajendran, and K. Rosenfeld, "Trustworthy Hardware: Trojan Detection and Design-for-Trust Challenges," *Computer*, vol. 44, no. 7, pp. 66–74, 2011.

[33] M. M. Tehranipoor, U. Guin, and D. Forte, *Counterfeit Integrated Circuits: Detection and Avoidance.* Springer, 2015.

[34] K. Vaidyanathan, B. P. Das, and L. Pileggi, "Detecting Reliability Attacks During Split Fabrication Using Test-Only BEOL Stack," in *Proc. 51st Design Automation Conf.*, 2014, pp. 1–6.

[35] A. Waksman, M. Suozzo, and S. Sethumadhavan, "FANCI: Identification of Stealthy Malicious Logic Using Boolean Functional Analysis," in *Proc. ACM SIGSAC Conf. on Computer & Communications Security*, 2013, pp. 697–708.

[36] Y. Wang, P. Chen, J. Hu, and J. J. Rajendran, "The Cat and Mouse in Split Manufacturing," in *Proc. 53rd Design Automation Conf.*, 2016, pp. 1–6.

[37] S. Wei, S. Meguerdichian, and M.Potkonjak, "Malicious Circuitry Detection Using Thermal Conditioning," *IEEE Trans. Information, Forensics and Security*, vol. 6, no. 3, pp. 1136–1145, Sept. 2011.

[38] K. Xiao, D. Forte, Y. Jin, R. Karri, S. Bhunia, and M. Tehranipoor, "Hardware Trojans: Lessons Learned After One Decade of Research," *ACM Trans. Design Automation of Electronic Systems (TODAES)*, vol. 22, no. 1, p. 6, 2016.

[39] K. Xiao and M. Tehranipoor, "BISA: Built-In Self-Authentication for Preventing Hardware Trojan Insertion," in *Proc. IEEE Int. Symp. Hardware-Oriented Security and Trust*, 2013, pp. 45–50.

[40] A. Yeh, "Trends in the Global IC Design Service Market." DIGITIMES Research, March 2012. http://www.digitimes.com/news/a20120313RS400.html?chid=2.

978-1-5386-3775-3/18 $31.00 © 2018 IEEE

Modeling Attacks on Strong Physical Unclonable Functions Strengthened by Random Number and Weak PUF

Jing Ye, Qingli Guo, Yu Hu, Huawei Li, Xiaowei Li

State Key Laboratory of Computer Architecture, Institute of Computing Technology, Chinese Academy of Sciences
University of Chinese Academy of Sciences
{yejing, guoqingli, huyu, lihuawei, lxw}@ict.ac.cn

ABSTRACT

Physical Unclonable Function (PUF) is a promising hardware security primitive. One important category of PUFs is the strong PUF with numerous Challenge-Response Pairs (CRPs). Since the typical strong PUFs, the arbiter PUF and several its variants, were broken by modeling attacks, many new designs for resisting modeling attacks have been proposed. Do they really achieve their promise, or are they only another pipe dream? This paper targets two PUF designs: the randomized PUF and the obfuscation PUF, which strengthen the arbiter PUF by leveraging the random number and the weak PUF, respectively. A heuristic algorithm is proposed for attacking these PUFs. The algorithm is implemented in CUDA. Some PUFs that cannot be broken in several months by CPU show their vulnerabilities in days by leveraging the GPU acceleration. The experimental results show that, for certain scales of objective PUFs, the prediction accuracy is beyond the reliability of CRPs, indicating successful attacks.

KEYWORDS

Modeling attack, Heuristic algorithm, Randomized PUF, Obfuscation PUF, Reliability

1 INTRODUCTION

The Physical Unclonable Function (PUF) is a promising hardware security primitive [1]. It exploits the process variations to output particular responses for input challenges, which are called the Challenge-Response Pairs (CRPs). Even with the same design, different manufactured PUFs will have different CRPs, which are hard to be predicted before manufacturing, hard to be controlled during manufacturing, and hard to be cloned after manufacturing. The CRPs are the key of PUF security. If attackers obtain or predict the CRPs of a PUF, then the PUF can be counterfeited and is no more secure.

The PUF can be generally divided into two categories: the weak PUF and the strong PUF. In weak PUFs, normally an independent circuit produces only one response bit [2-6]. If attackers can obtain all the response bits of a weak PUF, the weak PUF is no more secure. Currently, photon side-channel [7, 8] and wearout [9] are utilized to break the SRAM PUF [2, 3].

Different from weak PUFs, the strong PUFs normally have many or even numerous CRPs. A typical strong PUF is the arbiter PUF [10]. It compares the delays of two paths to produce a response bit. Each path consists of the path segments which are selected by a challenge. If the challenge contains N_{CB} bits, the number of CRPs is $2^{N_{CB}}$. If attackers can access the CRP interface, it is still impossible for them to directly read out all the CRPs in a reasonable time. However, the large amount of CRPs is achieved by sharing path segments among different CRPs. This leads to a serious security vulnerability: different CRPs of a strong PUF have high correlation, so attackers can model this correlation, collect certain number of CRPs as a training set, and then use machine learning methods to learn this correlation. In this way, other unknown CRPs of the attacked PUF can be predicted by using the learned model. The CRP prediction accuracy of the arbiter PUF is as high as 99% [11, 12].

To resist modeling attacks, there are generally two kinds of design methods: (1) adopting new electronic characteristics and (2) controlling the CRP access interface.

In [13] and [14], the voltage and the current, instead of delay, are adopted to produce CRPs, respectively. The usage of voltage transfer and current mirror increases the complexity of CRP modeling. However, new attack algorithms [15–17] have been proposed to break them with certain scales.

For the other kind of design methods [18, 31], the CRP access interface is strengthened so that attackers cannot directly control the challenge or read out the response of PUF. For example, the XOR arbiter PUF [18] adopts several arbiter PUFs. The challenge is input to them, and each of them produces an internal response bit, which is inaccessible to attackers. Then, these bits are XORed to produce the external response bit, which is accessible. In this way, attackers cannot directly obtain the CRPs of each arbiter PUF. Currently, photon side-channel [19], power side-channel [20, 21], and CRP unreliability [20, 22] are used to break it.

In recent years, new modeling attack resistant strong PUFs are proposed by leveraging the random number [23–26] and the weak PUF [27]. All these designs make attackers not know what internal challenge really produces each observed response bit. The objective of this paper is to analyze their security. Our contributions are followed.

Firstly, this paper provides meaningful new data of successful attacks on the randomized PUF and the obfuscation PUF with certain scale.

Secondly, a heuristic algorithm is proposed for attacking objective PUFs. The algorithm is implemented in CUDA with up to 250X speedup by GPU than CPU. Some PUFs that cannot be broken in several months by CPU show their vulnerabilities in days by leveraging the GPU acceleration. All the PUFs are attacked using the same parameters to make a fair comparison.

It is necessary to notice that, the successful attacks on the PUFs with certain scales do not mean such designs of PUFs are totally insecure. With larger scales, these designs can still be secure enough.

978-1-5386-3775-3/18 $31.00 © 2018 IEEE

The rest of the paper is organized as follows. Section 2 introduces the objective PUFs. Section 3 proposes the heuristic algorithm for modeling attack. Section 4 and 5 show the experimental results of attacks. Section 6 summarizes this paper.

2 OBJECTIVE PUF

This paper takes the randomized PUF and the obfuscation PUF as the attack objectives, which are based on the XOR arbiter PUF. All the attacked PUFs are implemented in two Xilinx KC705 boards. The arbiter PUFs are implemented using different hardware resources in the FPGA chips. The PUF design in FPGA is similar as [30].

2.1 XOR Arbiter PUF

The XOR arbiter PUF consists of several arbiter PUFs. The scale of a XOR arbiter PUF depends on two parameters:

N_{CB}: the number of challenge bits.

N_{XA}: the number of arbiter PUFs in the XOR arbiter PUF.

When N_{XA}=1, the XOR arbiter PUF is the arbiter PUF.

In the two Xilinx KC705 boards, we implement totally 60 XOR arbiter PUFs. N_{CB} is set to 64. N_{XA} is set to 1, 2, 3, 4, 5, and 6. For each configuration of N_{CB} and N_{XA}, 10 XOR arbiter PUFs are implemented. There are two commonly used metrics for evaluating a PUF: the uniformity and the reliability. The uniformity [29] of response bits is the percentage of response bits of 1 of different challenges, the ideal value of which is 50%. If a challenge is input to a PUF for t times, and at most s response bits are the same, then the reliability [29] of this CRP is s/t. For each XOR arbiter PUF, we firstly random generate 10^4 challenges. Each challenge is input to the XOR arbiter PUF for 20 times in the room temperature (20°C) and the normal supply voltage (1v) to obtain its reliable response bit. These challenges and their reliable response bits are used as the testing set for evaluating the CRP prediction accuracy of attack. Next, each challenge is input to the XOR arbiter PUF again for 10 times under each temperature and supply voltage corner shown in Table 1 for evaluating the reliability. Experimental results of uniformity and reliability are shown in Fig.1.

Table 1: Temperature and Supply Voltage Corners for Evaluating Reliability

Temperature(°C)	Supply Voltage	Temperature(°C)	Supply Voltage
-10	1v	50	1v
-10	0.9v	50	0.9v
-10	1.1v	50	1.1v
20	0.9v	20	1.1v

Figure 1: Uniformity and Reliability of XOR Arbiter PUF with N_{CB}=64.

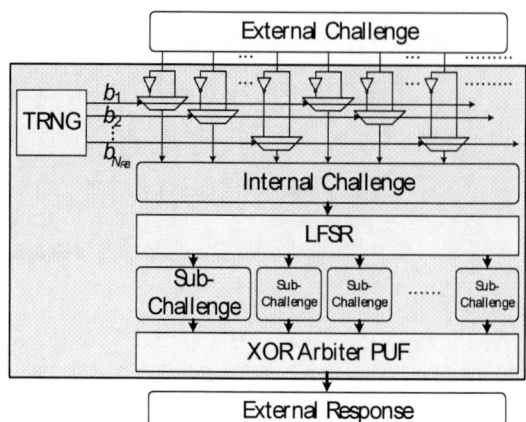

Figure 2: Randomized PUF.

2.2 Randomized PUF

The randomized PUF [26] leverages a TRNG to strengthen the arbiter PUF. The structure of randomized PUF is shown in Fig.2. Besides N_{CB} and N_{XA}, the scale of a randomized PUF also depends on the following parameter:

N_{RB}: the number of random bits for generating internal challenge from external challenge.

The TRNG generates N_{RB} random bits. Every random bit controls N_{CB}/N_{RB} external challenge bits. If the random bit is 0, the corresponding external challenge bits directly go through the multiplexer, or else the external challenge bits are inverted. The internal challenge is extended by LFSR to multiple sub-challenges and produces the multi-bit external response.

For the randomized PUF, an external challenge has $2^{N_{RB}}$ possible internal challenges, so have $2^{N_{RB}}$ possible external responses. Meanwhile, $2^{N_{RB}}$ external challenges will also have the same possible responses, so the valid number of challenges is reduced to $2^{N_{CB}-N_{RB}}$. The trust verification party needs to collect all the possible responses of a challenge, or they can use the same way as noise bifurcation PUF to model each arbiter PUF through one-time access interface. Then during authentication, if a PUF returns any one of the possible responses of the challenge, the PUF passes the authentication.

For attackers, they do not know which of the $2^{N_{RB}}$ possible internal challenges produces the observed response.

The randomized PUF is implemented in FPGA by adding random number generators to the 60 XOR arbiter PUFs explained in Section 2.1. The number of response bits is set the same as N_{CB}=64. The 10^4 response bits of XOR arbiter PUF are used as the testing set. The random bits have negligible effects on the uniformity and the reliability.

2.3 Obfuscation PUF

The obfuscation PUF [27] leverages weak PUF to strengthen the arbiter PUF. The key idea is to generate the internal challenge by calculating the external challenge with the weak PUF response bits. In [27], only the design for strengthening single arbiter PUF is proposed. We extend the obfuscation PUF for XOR arbiter PUF as shown in Fig.3.

978-1-5386-3775-3/18 $31.00 © 2018 IEEE

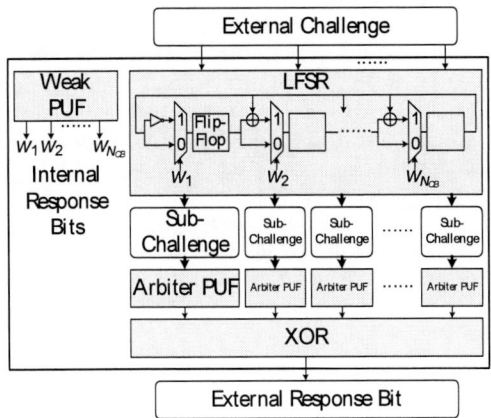

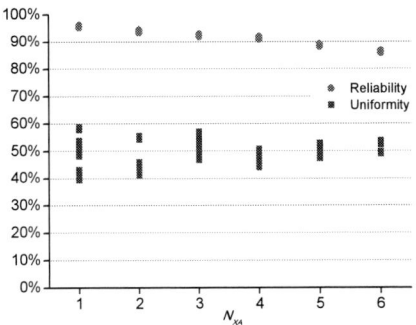

Figure 3: Obfuscation PUF.

Figure 4: Uniformity and reliability of obfuscation PUF with N_{CB}=64.

The external challenge is input to a LFSR. The weak PUF produces N_{CB} internal response bits to determine which bits in LFSR are XORed during shift. The LFSR generates N_{XA} internal sub-challenges, each of which is input to an arbiter PUF. The internal response bits of these arbiter PUFs are XORed to produce the external response bit.

In FPGA implementation, the arbiter PUFs used in obfuscation PUF is still the ones explained in Section 2.1. There are many ways to implement weak PUF in FPGA, while we use the ROs. Two ROs are placed and routed with the same nominal oscillating frequencies in the FPGA, and their actual oscillating frequencies are compared to produce one weak PUF response bit.

Due to the usage of weak PUF, the CRPs of obfuscation PUF are different from those of XOR arbiter PUF. For each obfuscation PUF, we also collect 10^4 reliable CRPs as the testing set for evaluating the CRP prediction accuracy. Their uniformity and reliability is shown in Fig.4.

3 MODELING ATTACK

For a specific PUF, its design may be leaked during supply chain or be reversely engineered. Please notice that the reverse engineering can only derive the circuit design, but cannot precisely measure the electronic characteristics such as delays. Hence, like previous works of attacking PUF, we assume attackers know the objective PUF design, and can obtain the training set that contains pairs of external challenge and external response, abbr. external CRPs.

In our experiments, we randomly generate external challenges and collect their external responses to construct the training set. Every external challenge is input to the objective PUF for only once under randomly selected temperature and supply voltage corner of Table 1. Thus, whether the collected CRPs are reliable or not is unknown. What we can assume is that reliability of training set should be similar as that of testing set.

An attack is considered successful if the CPR prediction accuracy is no less than the CRP reliability [22]. Such CRP prediction accuracy is high enough to counterfeit a PUF.

The arbiter PUF is modeled as previous works. Each path is modeled as the sum delay of path segments. The delays of paths segments are the unknowns to be solved. For the randomized PUF, because attackers do not know the random number, the random number is also considered an unknown. But the random numbers are not shared among different CRPs, so they are not counted in the number of unknowns in the following content.

Many machine learning algorithms have been used to attack PUFs, such as support vector machine, logistic regression, evolution strategy, lattice basis reduction, and so on. This paper proposes a heuristic algorithm. It is based on evolution strategy and simulated annealing. The simulated annealing is used for individual mutant of evolution strategy to improve the efficiency, while the parameters used by simulated annealing and evolution strategy are intelligently adjusted during the heuristic process. Though heuristic algorithm is not as efficient as logistic regression for attacking XOR arbiter PUF [11, 12], it is a more common one for attacking different kinds of PUFs.

Some terminologies used in the proposed heuristic algorithm are explained in the following:

CRP{C->R}: the challenge of CRP is *C* and its response is *R*.

Incorrect CRP: a CRP{*C->R*} of a training set is considered incorrect if it is an unreliable one, or the PUF of the training set does not actually produce *R* under *C*.

Individual: an individual denotes a set of guessed values of unknowns, e.g. delays of path segments of arbiter PUF.

Fit: an individual fits a CRP{*C->R*}, if the individual can produce *R* under *C*.

Fitness: the fitness of an individual to a training set is the percentage of CRPs the individual can fit.

Please notice that, because the arbiter PUF compares delays of two paths to produce a response bit, it is unnecessary to obtain the exact delay value of each path segment. When we guess the path segment delays, the values range from 0 to 1.

The abstract procedure of proposed heuristic algorithm is illustrated in Fig.5. At the beginning, multiple individuals are initialized by randomly guessing the values of unknowns. The individual mutant is operated by iterative simulated annealing. For each individual in every iteration, several unknowns are randomly selected and their values are modified to new randomly guessed values. Then the fitness of the modified individual is calculated. If the fitness becomes larger after modification than before modification, this modification is accepted. Or else, the modification is only accepted with a certain probability. This acceptance probability decreases as iterations increase.

978-1-5386-3775-3/18 $31.00 © 2018 IEEE

If the iteration number of continuous rejection achieves a user-defined threshold, we suppose the individual is stuck in a local optimization solution. In such case, the parameter adjustment is activated. The parameter adjustment is also activated for every hundred or thousand iterations defined by users. Its function is to analyze the history of optimization procedure so to adjust the parameters for achieving higher efficiency of searching solution space. The two key parameters are followed:

(1) The acceptance probability. The history status of individuals is analyzed. If an individual is stuck much more frequently than others, the individual is probably abandoned, or else the acceptance probability can be increased so the individual can jump out the local optimization solution. If the fitness of an individual increases much slower than those of other individuals, its acceptance probability is probably decreased.

(2) The number of individuals. The optimization procedure normally contains two states: one is a fast enhancement of fitness, and the other is a slow one. In the state of fast enhancement, a few individuals are sufficient, while for the other state, more individuals are expected to enlarge the solution searching space and find a better solution. To identify the state, both how many individuals are stuck in recent iterations and how much fitness is improved in recent iterations are analyzed. If only a few individuals are needed, the individuals with lower fitness are probably abandoned. If more individuals are required, new individuals can be generated by crossing two existing individuals or directly copying from an existing individual. In heuristic algorithm, two same individuals can become much different after many iterations, since the mutant and the cross operations are controlled by pseudo random numbers.

The ideal fitness is the percentage of correct CRPs in the training set. From the aspect of attackers, they do not know the percentage of correct CRPs, so the higher fitness, the better.

The proposed heuristic algorithm is implemented in C++ running in Ubuntu 16.04 OS. To accelerate the proposed heuristic algorithm, firstly, a CPU thread is used specifically for parameter adjustment. When this thread is analyzing the history of optimization procedure, all the individuals do not wait but are still mutant in parallel. CUDA programming is used for calculating the fitness of each individual. Whether an individual fits one CRP is calculated by GPU threads in parallel. The program is running on our computer with Intel i7-4790 3.6GHz CPU, NVidia GTX-1080 GPU, and 16G memories. The speed is accelerated 70~250X by GPU than CPU.

All the attacks toward the randomized PUF and the obfuscation PUF use the same initial parameters for a fair comparison, e.g. the initial number of individuals, the initial acceptance probability, and so on.

4 ATTACK ON RANDOMIZED PUF

In the randomized PUF, the random bits determine the internal challenge. For example, if N_{RB}=1, then an external challenge c_e may generate two possible internal challenges c_{i1} and c_{i2}. If the

external response of c_e is r_e, then attackers do not know whether r_e is produced by c_{i1} or c_{i2}. In the heuristic attack, an individual is considered fitting an external CRP $\{c_e\text{->}r_e\}$ in the training set if the individual can produce r_e under either c_{i1} or c_{i2}. In addition, for a fair comparison with other objective PUFs, when counting the number of CRPs in the training set, the external CRP with 64-bit response is counted as 64 CRPs.

As we explained before, the training set contains unreliable CRPs, so the ideal fitness should be the percentage of correct CPRs in training set, which may not be 100%. From this point of view, we define the F_C:

$$F_C = \frac{Fitness\ to\ Training\ Set}{Percentage\ of\ Correct\ CRPs\ in\ Training\ Set}$$

The ideal value of F_C is 100%. When calculating F_C, the percentage of correct CRPs in training set is assumed the same as the reliability of testing set.

Firstly, we set N_{RB}=1, and analyze the relation among F_C, CRP prediction accuracy, and size of training set. The experimental results are shown in Fig.6. When the number of correct CRPs is 200 times of unknowns, the difference between F_C and CRP prediction accuracy is below 5%. Then, we set the number of correct CRPs = 400 × number of unknowns. The experimental results are shown in Table 2. The CRP prediction accuracy of all the randomized PUFs with N_{RB}=1 are no less than their reliability. The runtime is shown in Fig.7. The GPU achieves about 250X speedup than CPU. Finally, we set N_{RB}=2. The experimental results are shown in Table 3. The randomized PUFs with N_{XA}=1, 2, 3, 4 are all successfully attacked with longer runtime.

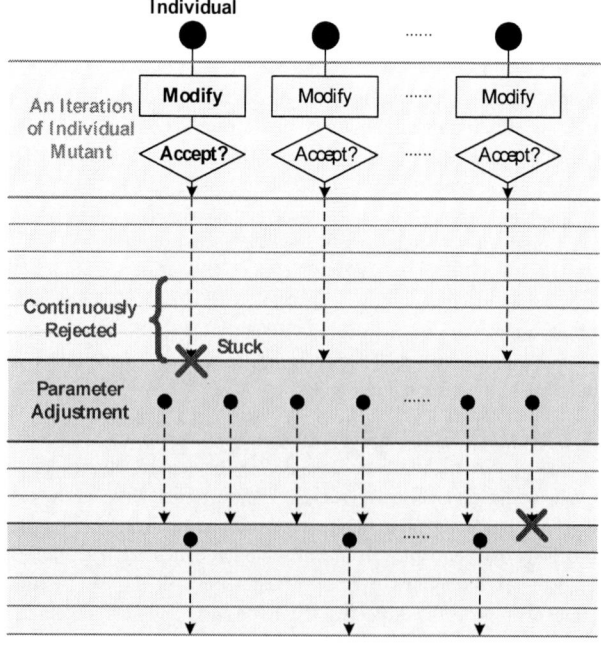

Figure 5: Heuristic algorithm.

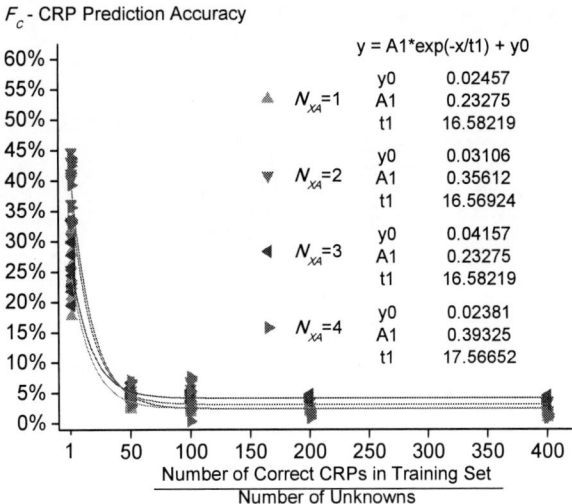

Figure 6: Relation among F_C, CRP prediction accuracy, and size of training set of randomized PUF with N_{RB}=1.

Table 2: CRP Prediction Accuracy of Randomized PUF with N_{RB}=1

N_{XA}	Number of CRPs in Training Set	Average Reliability	CRP Prediction Accuracy		
			Avg.	Min.	Max.
1	$0.27×10^5$	98.1%	98.2%	97.9%	98.6%
2	$0.54×10^5$	96.1%	96.2%	95.9%	97.0%
3	$0.82×10^5$	95.1%	95.2%	94.7%	95.7%
4	$1.11×10^5$	93.5%	93.7%	93.1%	94.3%
5	$1.41×10^5$	92.2%	92.4%	91.8%	92.9%
6	$1.72×10^5$	90.9%	91.1%	90.6%	91.2%

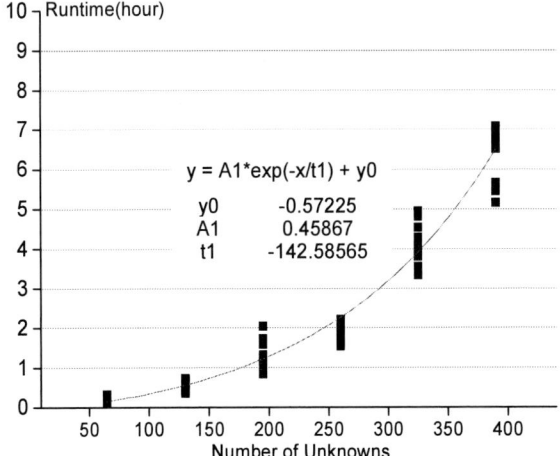

Figure 7: Runtime of randomized PUF with N_{RB}=1 (250X speed up by GPU).

Table 3: CRP Prediction Accuracy of Randomized PUF with N_{RB}=2

N_{XA}	Number of CRPs in Training Set	Average Reliability	CRP Prediction Accuracy			Average Run Time (hour)
			Avg.	Min.	Max.	
1	$0.27×10^5$	98.1%	98.2%	97.8%	98.5%	1.01
2	$0.54×10^5$	96.1%	96.3%	95.8%	96.8%	1.54
3	$0.82×10^5$	95.1%	95.2%	94.7%	95.8%	3.15
4	$1.11×10^5$	93.5%	93.7%	93.3%	94.2%	4.86

5 ATTACK ON OBFUSCATION PUF

The obfuscation PUF leverages weak PUF to strengthen the modeling attack resistance. The response bits of weak PUF are also unknowns, so the number of unknowns is $(N_{CB}+1)×N_{XA}+N_{CB}$. Hence, the number of unknowns of obfuscation PUF with N_{XA} arbiter PUFs is similar as that of XOR arbiter PUF with $N_{XA}+1$ arbiter PUFs.

The relation among F_C, CRP prediction accuracy, and size of training set is shown in Fig.8. When the number of correct CRPs in training set is 400 times of unknowns, the difference between F_C and CRP prediction accuracy is below 5%. Using this size of training set, the experimental results of CRP prediction accuracy are shown in Table 4. All the obfuscation PUFs are successfully attacked. The runtime is shown in Fig.9. When N_{XA} increases from 5 to 6, the runtime increases much more. The GPU achieves about 70X speedup than CPU. Here the speedup is not as good as randomized PUFs, because for different individuals of obfuscation PUF, different sub-challenges are generated. Hence the threads of GPU do not only need to calculate the response under a challenge, but also need to extend the challenge into sub-challenges. The extension of challenge slows down the calculation speed of every GPU thread.

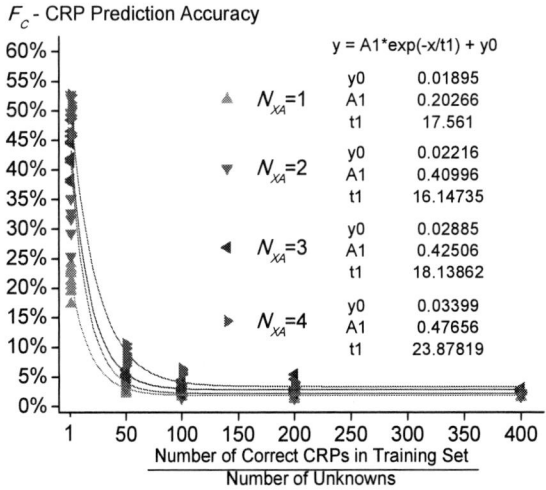

Figure 8: Relation among F_C, CRP prediction accuracy, and size of training set of obfuscation PUF.

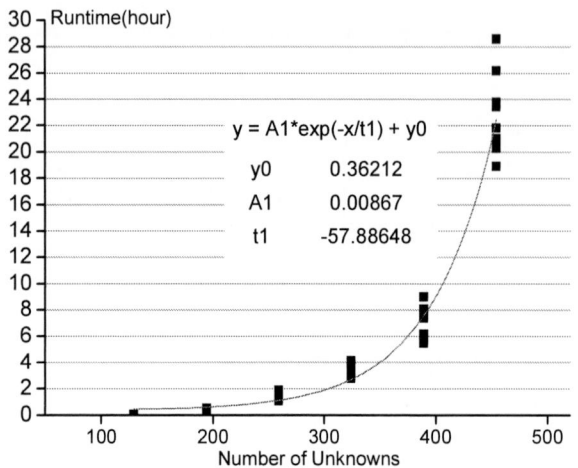

Figure 9: Runtime of obfuscation PUF (70X speed up by GPU).

Table 4: CRP Prediction Accuracy of Obfuscation PUF

N_{XA}	Number of CRPs in Training Set	Average Reliability	CRP Prediction Accuracy Average	CRP Prediction Accuracy Minimum	CRP Prediction Accuracy Maximum
1	0.54×10^5	95.6%	95.7%	95.4%	96.1%
2	0.83×10^5	93.9%	94.1%	93.5%	94.4%
3	1.12×10^5	92.3%	92.3%	92.0%	92.8%
4	1.42×10^5	91.5%	91.5%	91.2%	92.0%
5	1.75×10^5	88.8%	88.8%	88.4%	89.2%
6	2.10×10^5	86.3%	86.5%	86.0%	86.8%

6   SUMMARY

This paper proposes a heuristic algorithm for attacking the randomized PUF and the obfuscation PUF. The algorithm is based on simulated annealing and evolution strategy, with an intelligent parameter adjustment. By using GPU, the PUFs that cannot be broken in several months using CPU show their vulnerabilities in days.

However, this does not mean all the obfuscation PUFs and randomized PUFs are unsecure. It is necessary to understand that, with the increasing of unknowns, the runtime will not be acceptable under the current computational capabilities.

In addition, the proposed attack algorithm is obviously a common one for attacking other PUFs. Because in every iteration, we randomly guess the unknowns, the efficiency may not as good as the attack algorithm based on gradient. However, as a common attack method, it is suitable to fairly compare the security of different PUFs. From this point of view, we will release a version of the algorithm, and hope it can be used by designers to evaluate their PUFs.

7   ACKNOWLEDGE

This paper is supported in part by National Natural Science Foundation of China (NSFC) under grant No. 61532017, 61432017, 61704174, 61376043, and 61521092. The corresponding authors are Xiaowei Li and Yu Hu.

REFERENCES

[1] U. Ruhrmair, D. E. Holcomb, "PUFs at a Glance," DATE, 2014.
[2] J. Guajardo, S. S. Kumar, G.-J. Schrijen, P. Tuyls, "FPGA Intrinsic PUFs and Their Use for IP Protection," CHES, pp. 63-80, 2007.
[3] L. Zhang, C.-H. Chang, Z. H. Kong, C. Q. Liu, "Statistical Analysis and Design of 6T SRAM Cell for Physical Unclonable Function with Dual Application Modes," ISCAS, pp. 1410-1413, 2015.
[4] P. Prabhu, A. Akel, L. Grupp, W. K. Yu, G. Suh, E. Kan, S. Swanson, "Extracting Device Fingerprints from Flash Memory by Exploiting Physical Variations," ICTTC, pp. 188-201, 2011.
[5] S. Rosenblatt, S. Chellappa, A. Cestero, N. Robson, T. Kirihata, S. S. Iyer, "A Self-Authenticating Chip Architecture Using an Intrinsic Fingerprint of Embedded DRAM," IEEE JSSC, vol. 48, no. 11, pp. 2934, 2013.
[6] P. Koeberl, U. Kocabas, et. al., "Memristor PUFs: A New Generation of Memory based Physical Unclonable Functions," DATE, 2013.
[7] C. Helfmeier, C. Boit, D. Nedospasov, J.-P. Seifert, "Cloning Physical Unclonable Function," HOST, 2013.
[8] D. Nedospasov, J.-P. Seifert, C. Helfmeier, C. Boit, "Invasive PUF Analysis," FDTC, pp. 30-38, 2013.
[9] A. Roelke, M. R. Stan, "Attacking an SRAM-Based PUF through Wearout," ISVLSI, pp. 206-211, 2016.
[10] D. Lim, J. W. Lee, B. Gassend, G. E. Suh, M. van Dijk, S. Devadas, "Extracting Secret Keys from Integrated Circuits," IEEE TVLSI, vol. 13, no. 10, pp. 1200-1205, 2005.
[11] U. Ruhrmair, F. Sehnke, J. Solter, G. Dror, S. Devadas, J. Schmidhuber, "Modeling Attacks on Phyiscal Unclonable Functions," CCS, 2010.
[12] U. Ruhrmair, J. Solter, F. Sehnke, X. Xu, A. Mahmoud, V. Stoyanova, G. Dror, J. Schmidhuber, W. Burleson, S. Devadas, "PUF Modeling Attacks on Simulated and Silicon Data," IEEE TIFS, vol. 8, no. 11, 2013.
[13] A. Vijayakumar, S. Kundu, "A Novel Modeling Attack Resistant PUF Design based on Non-Linear Voltage Transfer Characteristics," DATE, pp 653-658, 2015.
[14] R. Kumar, W. Burleson, "On Design of a Highly Secure PUF Based on Non-Linear Current Mirrors," HOST, pp. 38-43, 2014.
[15] A. Vijayakumar, V. C. Patil, C. B. Prado, S. Kundu, "Machine Learning Resistant Strong PUF: Possible or a Pipe Dream," HOST, pp. 19-24, 2016.
[16] Q. Guo, J. Ye, Y. Gong, Y. Hu, X. Li, "Efficient Attack on Non-Linear Current Mirror PUF with Genetic Algorithm," ATS, pp. 49-54, 2016.
[17] J. Ye, Y. Hu, X. Li, "POSTER: Attack on Non-Linear Physical Unclonable Function," CCS, pp. 1751-1753, 2016.
[18] M. Majzoobi, F. Koushanfar, M. Potkonjak, "Lightweight Secure PUFs," ICCAD, pp. 670-673, 2008.
[19] F. Ganji, J. Kramer, J. Seifert, S. Tajik, "Lattice Basis Reduction Attack against Physically Unclonable Functions," CCS, pp. 1070-1080, 2015.
[20] G. T. Becker, R. Kumar, "Active and Passive Side-Channel Attacks on Delay Based PUF Designs," IACR Cryptology ePrint Archive, 2014.
[21] A. Mahmoud, T. Munchen, M. Majzoobi, U. Ruhrmair, F. Koushanfar, "Combined Modeling and Side Channel Attacks on Strong PUFs," IACR Cryptology ePrint Archive, 2014.
[22] G. T. Becker, "The Gap Between Promise and Reality: On the Insecurity of XOR Arbiter PUFs," Springer, vol. 9293, pp 535-555, 2015.
[23] M. Majzoobi, M. Rostami, F. Koushanfar, D. Wallach, S. Devadas, "Slender PUF Protocol: A Lightweight, Robust, and Secure Authentication by Substring Matching," SPW, pp. 33-44, 2012.
[24] M. Rostami, M. Majzoobi, F. Koushanfar, D. Wallach, S. Devadas, "Robust and Reverse-Engineering Resilient PUF Authentication and Keyexchange by Substring Matching," IEEE TETC, 2014.
[25] M. D. Yu, D. M'Raihi, I. Verbauwhede, S. Devadas, "A Noise Bifurcation Architecture for Linear Additive Physical Functions," HOST, 2014.
[26] J. Ye, Y. Hu, X. Li, "RPUF: Physical Unclonable Function with Randomized Challenge to Resist Modeling Attack," AsianHOST, 2016.
[27] J. Ye, Y. Hu, X. Li, "OPUF: Obfuscation Logic Based Physical Unclonable Function," IOLTS, pp. 156-161, 2015.
[28] G. T. Becker, "On the Pitfalls of Using Arbiter-PUFs as Building Blocks," IEEE TCAD, vol. 34, no. 8, pp. 1295-1307, 2015.
[29] M. Gao, K. Lai, G. Qu, "A Highly Flexible Ring Oscillator PUF," DAC, 2014.
[30] M. Majzoobi, F. Koushanfar, S. Devadas, "FPGA PUF using Programmable Delay Lines," WIFS, 2010.
[31] J. Ye, Y. Hu, X. Li, "VPUF: Voter based Physical Unclonable Function with High Reliability and Modeling Attack Resistance," IOLTS, 2017.

Special Session: How Approximate Computing impacts Verification, Test and Reliability

L. Sekanina[1], Z. Vasicek[1], A. Bosio[2], M. Traiola[2], P. Rech[3], D. Oliveria[3], F. Fernandes[3], S. Di Carlo[4]

[1]IT4I, Brno University of Technology, Czechia; [2]LIRMM - France; [3]UFRGS, Brazil; [4]Politecnico di Torino - Italy

I. INTRODUCTION

Two AxC techniques have been successfully applied to hardware components. The first one is the *functional approximation* [1]that modifies the circuit structure replacing the original function F with the function G. G implementation leads to area/energy reduction at the cost of reduced accuracy, meaning that some errors can be observed at the outputs of G. The observed errors are a variation between the output values of F (precise) and G (approximate). The variation is the accuracy loss measured by means of quality metric(s) [1]. The second AxC technique is the *over-scaling based approximation*. Basically, the HW component is forced to work outside its specified operating conditions [1]. The classical example is the reduction of the supply voltage under the minimum value.

In the functional approximation, the circuit netlist is modified to introduce the approximation. Thus we have to guarantee that the approximate circuit does not introduce an error greater than the acceptable one. It is therefore required to adapt the "verification" phase to the AxC design flow. We present an approach in which the approximation problem is formulated as a complex multi-objective design and optimization problem and solved using advanced search methods. In order to **exactly** determine the quality (the average error, the worst-case error etc.) of candidate solutions, symbolic methods based on BDD analysis and SAT problem solving are employed. Several case studies in the area of arithmetic circuit approximation will be presented to demonstrate the efficacy and scalability of the search-based approximation combined with formal error analysis [2]. Once the verification phase is done, the AxC design enters in the manufacturing process, where physical defects may impact the produced AxC. This means that the manufactured device may have a different error w.r.t. the verified one. In this context, the role of "testing" is to ensure that the amount of error (due to manufacturing defects) is not greater than the acceptable error threshold. We propose our **AxC aware ATPG** for generating test vectors targeting only the "critical faults" [3]. The benefits are a lower test set and yield improvement. Experimental results are carried out on a public benchmark suite to prove the efficiency of the proposed approach.

AxC exploits the index property of some applications to tolerate the presence of inaccuracy. This inaccuracy is quantified as an error w.r.t to the precise application and quantified by means of quality metrics. Actually the applications are inherently able to tolerate the presence of a certain amount of error because they intrinsically embed redundancy: a higher number of loop iterations or data precision larger than the required. AxC cut this embedded redundancy and on one hand it gains in terms of overhead reduction, but on the other hand it introduces a reduction of the capability to tolerate errors (because of the accuracy reduction). This means that if a resilient application is able to tolerate a certain amount of errors due to harsh environment or aging problems, the approximated application can tolerate only a reduced fraction of errors. For safety critical applications, it is thus very important to identify a trade off between approximation and reliability. Moreover, the use of the over-scaling technique may exacerbate this problem.

We experimentally evaluate the benefits and challenges in terms of reliability introduced by approximate computing. Faults that only slightly impact the code output could be tolerated if an approximate solution is accepted as correct. We have seen through accelerated neutron beam experiments (i.e., the harsh environment) that a 0.1% approximation of the output value allows the application to tolerate up to 90% of radiation-induced transient errors [4]. It is then fundamental to analyze the application and distinguish between tolerable and critical errors. Unfortunately, approximate hardware is likely to increase the device error rate. Errors in the least significant digits of a 32 bit number are much less severe than those in a 16 bit number. Moreover, a fault in a 32 bit hardware used to execute two 16 bit operations is likely to corrupt both half-precision outputs, thus reducing the application reliability. As a result, while the benefits of approximate hardware are unquestionable it is necessary to carefully evaluate the impact of such an hardware in the system reliability.

This work has been partially supported by the IT4Innovations excellence in science project LQ1602.

REFERENCES

[1] S. Mittal, "A survey of techniques for approximate computing," *ACM Comput. Surv.*, vol. 48, no. 4, pp. 62:1–62:33, Mar. 2016. [Online]. Available: http://doi.acm.org/10.1145/2893356

[2] M. Ceska, J. Matyas, V. Mrazek, L. Sekanina, Z. Vasicek, and T. Vojnar, "Approximating complex arithmetic circuits with formal error guarantees: 32-bit multipliers accomplished," in *Proc. of 36th IEEE/ACM Int. Conf. On Computer Aided Design.* IEEE, 2017, pp. 416–423.

[3] I. Wali, M. Traiola, A. Virazel, P. Girard, M. Barbareschi, and A. Bosio, "Towards approximation during test of integrated circuits," in *2017 IEEE 20th International Symposium on Design and Diagnostics of Electronic Circuits Systems (DDECS)*, April 2017, pp. 28–33.

[4] D. Oliveira, L. Pilla, N. DeBardeleben, S. Blanchard, H. Quinn, I. Koren, P. Navaux, and P. Rech, "Experimental and analytical study of xeon phi reliability," in *Proc. of the Int. Conf. for High Performance Computing, Networking, Storage and Analysis*, ser. SC '17, 2017, pp. 28:1–28:12.

Innovative Practices on Quality Levels of A/MS Devices

Wim Dobbelaere, OnSemi,
Massimo Violante, Turin Polytechnic,
Jeff Rearick, AMD,
Peter Sarson, Dialog Semiconductor (Organizer)

I. INTRODUCTION

In this IP session, there will be 3 presentations focusing on how to increase the in-field quality level of A/MS devices. The 1st presentation will discuss using optical inspection data with electrical test data such that the combined data can be used to further screen out potentially bad devices that would have normally passed electrical testing. The 2nd presentation discuses using a top level fault injection method to determine the ASIL levels in the ISO26262 standard. The 3rd presentation will discuss the current progress of the P2427 Standardized Analogue Test Coverage Working Group and how this standard can be used to increase quality levels of A/MS devices.

II. PRODUCING DEFECT-FREE IC'S BY COMBINING ELECTRICAL TEST DATA AND SILICON INSPECTION DATA (WIM DOBBELAERE)

Historically, electrical testing of integrated circuits has always been executed without any knowledge about the in-fab visual inspection data. The urge to avoid electronic failures in automotive applications is driving the IC industry to research defect-oriented methods that complement the traditional functional tests. Recent research has focused on new electrical methods that are enabled by smart DfT in combination with automatic test generation and statistical post-processing. In this paper, we assess the benefits and complications of combining silicon visual inspection data and electrical test data. An experiment on an automotive product demonstrates that only 51 % of the visual killing defects are caught electrically. This proves that a full defect inspection would lead to a dramatic fault coverage improvement that however needs to be traded of versus unjustified yield loss.

III. USNG FAULT INJECTION TO DETERMINE ASIL LEVEL (MASSIMO VIOLANTE)

As automotive embedded hardware and software are growing in complexity, and they are more and more often responsible for safety critical tasks, meeting the quality levels applications demand is becoming a challenging task for designers. To fulfill such a goal, the automotive industry defined the ISO26262, which is today the best practice for designing electronic and electrical systems, known as items, for automotive. The adoption of such a standard in a design flow could have a significant impact, as many of the established processes shall be reconsidered. Among them, the verification of hardware design is a crucial activity, which aims at analyzing the hardware design of an item to evaluate if it meets the reliability goals the standard prescribes for the different Automotive Safety Integrity Levels (ASILs).

This activity is mostly manual, although supported by available calculation models, as it requires the analysis of the item schematics, and the identification of how faults affecting hardware components propagate through the item (including software), eventually reaching its outputs.

For each of the hardware components saboteurs are used to inoculate the relevant fault models, and simulations are used to evaluate how fault effects propagate through the item. With this approach we analyzed a representative use case where an item responsible for controlling an electric motor has to be designed according to ASIL D requirements. The item elements are an input stage based on discrete components, an automotive microcontroller running a software to convert the input signal through the analog to digital converter the microcontroller embeds and to generate a pulse width modulated signal using the embedded programmable timer, and an output stage consisting of a half bridge used to drive the electric motor.

Given the item functionality and the adopted hardware architecture, thanks to the proposed approach we were able to identify a safety goal violation, and to propose an alternative implementation that meets the ASIL D requirement.

IV. IEEE P2427: PROPOSING THE ESSENTIAL FRAMEWORK FOR MEASURING DEFECT COVERAGE IN ANALOG CIRCUITS (JEFF REARICK)

The informal team that has been working on analog defect coverage measurement and reporting for the last three years has matured into a Working Group (IEEE P2427). This talk will provide an update on the decisions made which crisply define the key elements of the current proposal: defining defect models, deriving the defect universe, specifying detection criteria, and calculating coverage numbers. A running example will be used to illustrate the proposal and highlight the issues still under discussion. Define the key elements of the current proposal: defining defect models, deriving the defect universe, specifying detection criteria, and calculating coverage numbers. A running example will be used to illustrate the proposal and highlight the issues still under discussion.

Hardware Trojan Attacks in Embedded Memory

Tamzidul Hoque[1], Xinmu Wang[2], Abhishek Basak[3], Robert Karam[4], and Swarup Bhunia[1]

[1]University of Florida, Gainesville, FL 32608, [2]Northwestern Polytechnical University, Xi'an, China
[3]Case Western Reserve University, Cleveland, OH 44106, [4]University of South Florida, Tampa, FL 33620

Abstract—**Embedded memory, typically implemented with Static Random Access Memory (SRAM) technology, is an integral part of modern processors and System-on-Chips (SoCs). The reliability and integrity of embedded SRAM arrays are essential to ensure dependable and trustworthy computing. In the past, significant research has been conducted to develop automated test algorithms aimed at comprehensively detecting SRAM faults. While such tests have advanced our ability to detect manufacturing imperfection induced faults, they cannot ensure detection of *deliberately implemented* design modifications, also known as hardware Trojans, in an SRAM array by untrusted entities in the design and fabrication flow. Indeed, these attacks constitute an emerging concern, since they can affect the integrity of fabricated ICs and cause severe consequences in the field. While a growing body of research addresses Trojan attacks in logic circuits, little to no research has explored these attacks in embedded memory arrays. In this paper, for the first time to our knowledge, we propose a new class of hardware Trojans targeting embedded SRAM arrays. The Trojans are designed to evade industry standard post-manufacturing memory tests (e.g. March test) while enabling targeted data tampering after deployment. We demonstrate various forms of Trojan circuits in SRAM that cause diverse malicious effects and have diverse activation conditions while incurring minimal overhead in power, performance, and stability. Further, the proposed layouts preserve the SRAM cell footprint and incur negligible silicon area overhead.**

I. INTRODUCTION

Static Random Access Memory (SRAM) is an essential part of any processor or system-on-chip (SoC) design that bridges the rising gap between execution speed and data transfer latencies. Cache memories can occupy more than 50% of a processor or SoC die area, a number which is expected to increase further as technology advances [1]. SRAM failures can lead to corruption of stored data, which can easily propagate in the system and effect data integrity. Therefore, reliable SRAM arrays are crucial for dependable computing. While existing fault models and associated algorithms improve detection of manufacturing process induced SRAM faults, they cannot adequately assure detection of a malicious design change, also known as hardware Trojan attack, deliberately implemented by an attacker in an untrusted foundry. Moreover, the high cost of testing large SRAM arrays prohibits IC vendors from applying exhaustive combinations of various test algorithms and test stress conditions, which further limits the ability of industry-standard testing to detect such Trojans.

Hardware Trojan attacks have emerged as a serious security concern for integrated circuits (ICs). Numerous untrusted components in the IC life cycle, such as third-party intellectual property (IP) cores, Computer-Aided Design (CAD) tools, and overseas foundries, have made it possible to maliciously modify an IC during design or fabrication. Such attacks can pose severe threats to IC operational reliability and integrity, and can cause system performance degradation, malfunction, or secret information leakage after deployment [2] [3]. While significant research efforts have, with limited success, addressed the detection of Trojans hidden inside the logic portion of ICs [4], the problem of possible Trojan attacks in embedded memory and their effective detection remains unexplored to a large extent. Among the 93 different Trojan inserted benchmarks available on *Trust-Hub* [2] not one involves a memory array. Previous work in SRAM Trojans has sought to leverage data [5] or access patterns [6] for triggering a Trojan, or else implemented a DoS Trojan in SRAM which can be detected by X-propagation in the front end memory design process [7]. However, the possibility of a Trojan residing within the memory array is not explored. Further, as all the above Trojan designs follow the same design methodology as Trojans mounted in general logic circuits, the design overhead and side-channel footprint cannot be minimized sufficiently to avoid affecting the design layout or evade side-channel detection. Das et al. describe a system which can detect and prevent malicious writes to the memory; however, a Trojan circuit within the memory array writing directly to internal cells is not considered, yet this will bypass the proposed safeguards [8]. While optical imaging based techniques [9] could identify structural modifications in SRAM arrays, they are applied mainly in post-deployment failure analysis instead of post-manufacturing tests, hence not effective in proactive Trojan detection, especially in cases where hardware Trojans are inserted in only a small sample of an IC batch. Hence, designing and implementing realistic Trojans in memory arrays is a valuable step forward in hardware Trojan research.

In this paper, we explore the topic of deliberate malicious modification of embedded memories to cause wide variety of functional failures impacting data integrity in memories (e.g. in processor cache), which could be leveraged for various software-oriented, system-level attacks. We consider attacks mounted in untrusted foundries by modifying design layout before IC fabrication. We validate the Trojan design in a compact SRAM layout to demonstrate the feasibility. In particular, the paper makes the following novel contributions:

1) It presents the design of a new class of hardware Trojans in embedded memory arrays. Unlike previous works, the Trojans are inserted in the memory array instead of peripheral logic and designed to cause SRAM failure during deployment. We show that they evade detection from industry standard SRAM testing as well as existing hardware Trojan detection approaches.

2) The proposed Trojans have a unique trigger (activation) and payload (effect) mechanism. They are essentially well-designed low foot-print structures that induce controlled faults under rare internal circuit conditions. They are beyond the coverage of existing test methods and test-time detection using complex trigger mechanisms.

3) The proposed Trojan model can also evade side-channel (e.g. transient/leakage current or path delay based) analysis based Trojan detection since it does not noticeably change the power, performance, or cell stability when inactive. It also does not change the cell footprint or incur any silicon area overhead. This is demonstrated by transistor-level simulation and custom SRAM layout.

978-1-5386-3775-3/18 $31.00 © 2018 IEEE

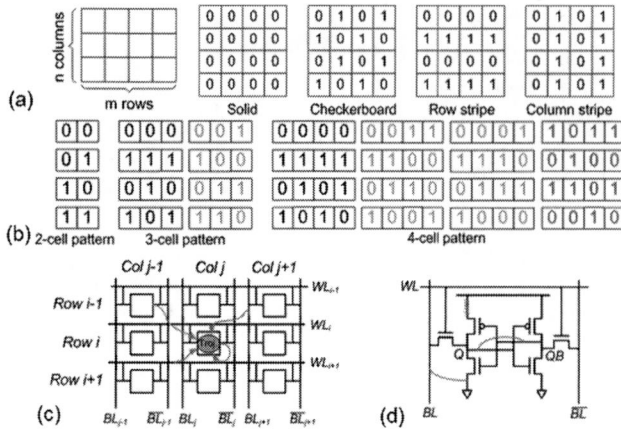

Fig. 1. Possible data patterns in SRAM array: (a) commonly used patterns in testing. (b) patterns that can be used for trigger mechanisms. A Trojan attack in SRAM: (c) a general model; (d) effective defect types.

The rest of the paper is organized as follows. Section II briefly describes common SRAM faults and associated testing algorithms. The proposed Trojan designs are described in Section III. Section IV provides simulation results for Trojan functional verification and side-channel impact characterization. Finally, we conclude the paper in Section V.

II. BACKGROUND

A. SRAM Fault Models

Here, we describe commonly used SRAM functional fault models. While a *stuck-at fault* (SAF) causes the value of a cell to remain stuck at fixed logic (0 or 1), a *Stuck-open fault* (SOF) disables access to a particular cell. With a *transition fault* (TF), the cell fails to undergo a particular type of transition $(0 \rightarrow 1$ or $1 \rightarrow 0)$. There are also faults that compromise cell hold or read stability. A data retention fault (DRF) means a cell fails to hold the value for more than time T, whereas read destructive faults (RDF) causes cell values to flip with a read operation. Faults involving two cells are called *coupling faults* (CFs). The cell causing the faulty behavior is called the *aggressor cell* (a-cell), and the cell exhibiting faulty behavior is called the *victim cell* (v-cell). Coupling faults can be broadly categorized into *state coupling fault* (CFst), *inversion coupling fault* (CFin), and *idempotent coupling fault* (CFid). In CFst, the faulty behavior occurs when a particular state in the a-cell forces a value in the v-cell. CFin and CFid are sensitized by a transition write operation in the a-cell. In CFin, the v-cell value is inverted, while in CFid, the v-cell is forced to a fixed value. Weak faults that cause slight disturbance in SRAM operations may also occur and may lead to more complex fault models like *dynamic faults*, which require multiple read or write operations to appear. In *linked faults*, multiple faults affect the same cell, and the impact of one fault can mask that of another which makes detection of each fault more difficult. Efficient testing for these faults is essential and *march tests* have emerged as the industry standard.

B. SRAM Test Algorithms

March tests consist of multiple march elements. Each march element is a group of operations executed as a unit on every cell and repeated throughout the SRAM array. The March C-[13] test is described by the following equation:

$$\{\updownarrow (w0); \Uparrow (r0, w1); \Uparrow (r1, w0); \Downarrow (r0, w1); \Downarrow (r1, w0); \updownarrow (r0)\}$$

It contains $10n$ operations, where n is the number of cells in the SRAM array. It can detect all simple SAFs and TFs that are unlinked with CFs, because reading both 0 and 1, and exciting both $0 \rightarrow 1$ and $1 \rightarrow 0$ transitions are covered by the test. It can also detect all *CFst*s and unlinked *CFin*s. However, march C- cannot detect linked faults, DRF, or SOF. March SL was developed to cover all simple linked faults [12]. March RAW is designed to detect dynamic faults sensitized by read-after-write operations [11]. Researchers have also considered a more general form of coupling fault, i.e., pattern sensitive fault (PSF), in which multiple a-cells collaboratively cause a faulty behavior in the v-cell. While general pattern sensitive fault testing is considered impossible, the target of existing algorithms is to detect PSFs within a restricted region of the a-cells with respect to the v-cell. Different data backgrounds can improve the coverage of coupling faults, and to some extent facilitate detection of PSFs. The four data backgrounds used in industrial testing are displayed in Fig. 1(a). In this work, the Trojans are evaluated in the context of fast-y march tests with commonly used data backgrounds.

III. TROJAN ATTACKS IN SRAM ARRAY

To evade detection, a cache memory Trojan should: 1) bypass conventional memory testing mechanisms, and 2) minimally impact cell area and side-channel signature (IDDQ, IDDT, access time). Here, we describe designs and trigger mechanisms of Trojans which satisfy these requirements.

A. Trojan Trigger Mechanism

To effectively evade detection by various SRAM testing algorithms, we exploit data patterns in the array that are unlikely to appear during march tests, along with operating conditions denoted by the state of word lines and bit lines (Fig. 1(c)). Considering the four data backgrounds listed in Fig. 1(a), any arbitrary cell group of n rows and m columns only provides a limited or no data patterns as a possible triggering mechanism (Fig. 1(b)). If $m = 2$, all four data combinations are valid patterns in march tests, denoted in black in Fig. 1(b). Therefore, 2-cell patterns are not adequate to establish a safe Trojan trigger condition. Half of the 3-cell patterns are invalid (marked in orange) in march tests when the block is not being accessed and can be used to form the trigger mechanism. However, these patterns are possible when multi-column stripe data backgrounds are applied, e.g., each stripe contains more than one column, which is not always used in industrial tests due to time limitations. As the cell number increases to 4, 12 out of the 16 possible patterns cannot occur except when the block is being accessed. Moreover, out of the 12 invalid patterns, 8 are theoretically possible in tests with multi-column stripe backgrounds (marked in orange), and the remaining 4 patterns are unlikely to happen in all regular pattern data backgrounds (marked in red). Therefore, these patterns can be used to design robust Trojans that can evade any march test. The observations hold true for column-wise patterns as well. Note that these patterns are likely to occur, thus can cause Trojan trigger in field deployment, in which case the data patterns are not restricted and the read/write operations do not occur in a deterministic manner as in March tests. In addition, states of word and bit lines in an SRAM array can be used to trigger a Trojan. For instance, the invalid data patterns previously discussed are not being operated on directly, so the value of their word line can be used instead. If all the

978-1-5386-3775-3/18 $31.00 © 2018 IEEE

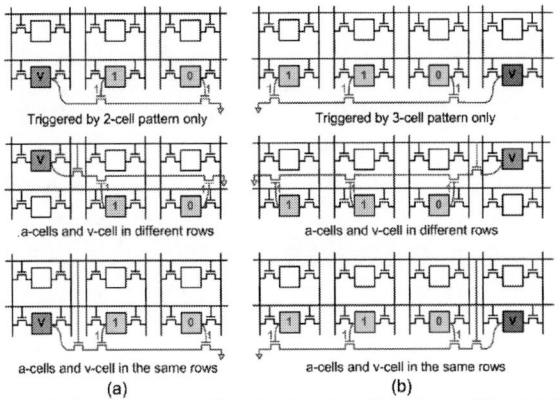

Fig. 2. Trojans causing v-cell node shorted to V_{ss} triggered by: (a) 2-cell data pattern and a word line; (b) 3-cell data pattern and a word line.

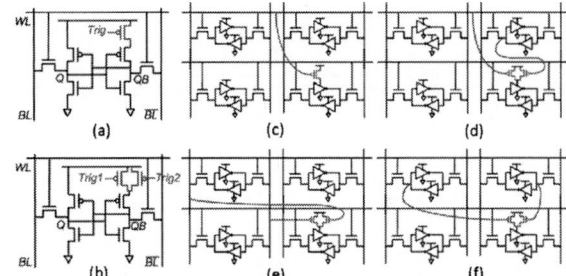

Fig. 3. Trojans causing v-cell pull-up path broken: (a) controlled by one node; (b) controlled by two nodes. (c), (d), (e), (f) shows four examples of Type-2 Trojans leveraging (a) and (b).

trigger cells are in one row, the corresponding word line value can be incorporated to guarantee an impossible condition, i.e., $\overline{WL} \cdot P$, where P is the invalid data pattern.

Although the proposed Trojan design is in the context of bit-oriented SRAMs (BOMs), the Trojan model remains valid in word-oriented SRAMs (WOMs) where memory is accessed by words containing B bits ($B > 1$). Here, testing a word is equivalent to testing B bits in parallel, if the memory block containing each bit has the same data backgrounds as in BOMs. Therefore, the capability of a given Trojan trigger mechanism in evading a march test is the same for a BOM and a bit-interleaved WOM. We leverage this trigger mechanism to design two types of Trojans with different payloads.

B. Trojan Type-1: Resistive Short/Bridge

Trojans can be designed to cause resistive shorts between one circuit node and V_{dd}/V_{ss}, or bridges between two nodes, where a node can be a cell node, a word line, or a bit line. Fig. 2 demonstrates representative designs of Trojans that short a cell node with V_{ss} upon trigger. The trigger condition of a Type-1 Trojan is realized by multiple nMOS pass transistors (PT) concatenated in series. The Gate terminals of the PTs are connected to the trigger nodes. Only when all Gate terminals hold logic 1 the shorting path is activated. Otherwise, the v-cell node sees an extremely high resistance (on the order of GΩ), which does not affect normal v-cell functionality. NMOS PTs are chosen because of their higher mobility hence lower effective resistance. This is further necessitated by the concatenation of multiple PTs, which weaken the path conductivity significantly. Second, nMOS PTs are good at conducting logic 0 (Vss) while a pMOS PT is subject to voltage degradation of V_{thp} when conducting logic 0, which is intolerable in deep sub-micron, especially low-power technologies. Finally, the traditional SRAM layout used in this work prohibits insertion of an arbitrary pMOS transistor, whereas nMOS PTs are implemented in free space introduced by word lines and access transistors. Fig. 2(a) and (b) are implementations of the 2- and 3-cell trigger pattern, respectively, for evading march tests with various data backgrounds (following the methods in Section III-A). A 4-cell trigger pattern can be designed in the same manner to bypass all march tests. The cells requiring a logic 1 in the trigger pattern are connected with the true nodes, while those requiring a logic 0 are connected to the complementary nodes. Trigger cells are chosen from the same row to avoid the need of an extra metal layer for interconnection. Each Trojan can also contain a trigger

node from a nearby non-trigger-cell word line to lower the probability of triggering. If the trigger word line is not that of the v-cell, the Trojan could never be triggered while the v-cell is accessed, making the payload a weak fault that merely tampers cell hold stability.

While a larger number of PTs can provide a lower trigger probability, more PTs in series will incur larger effective on-resistance, degrading the shorting/bridging impact on the v-cell. When the PT number is above a threshold, the Trojan will only cause a weak fault that decreases the v-cell SNM without causing other faulty behaviors. In addition, longer Trojan paths will cause larger parasitic effects on the array. Therefore, we limit the number of PTs to between 2 and 5.

C. Trojan Type-2: Resistive Open

Causing a conditional open requires adding circuitry in the target path, which is difficult in a compact SRAM array. We were able to implement variations of one Trojan type that breaks one pull-up path in a cell. The simplest form uses one extra pMOS PT to control the defect (Fig. 3(a)).

The space between the pull-up transistor and V_{dd} is the only location where an extra pMOS PT can fit (see Section IV). The PT is controlled by *Trig* that can turn off the PT with logic 1. To incorporate more trigger nodes, additional PTs could be connected in parallel. However, the number is limited to two due to space limitations. Fig. 3 illustrates Trojans implemented to cause resistive open defects. The Trojan in Fig. 3(c) is controlled merely by the word line of the previous row. This implementation assures that the Trojan will not be triggered while the v-cell is being operated. During this period, DRF can occur if the v-cell holds a 0 and the trigger word line remains high for sufficient time. Note that the v-cell does not have a problem in holding a 1 because the Trojan can only cause an open in the inverter generating the complementary output. The time required to cause a DRF depends on the pull-down path leakage, which is on the order of order of $100\mu s$ in our framework. Hence DRFs are unlikely to surface in this case during either testing or deployment, as the word line is active only briefly, and remains low during the bit line pre-charge time, refreshing the v-cell every cycle when it holds a 1. Thus, this Trojan will not cause a strong fault and will evade all tests.

The other three Trojans contain two trigger nodes. The Trojan in Fig. 3(d) is modified from Fig. 3(c) by introducing an additional trigger control from another cell node. This lowers the occurrence probability of the soft error by 50%. Fig. 3(e) demonstrates a Trojan using the true and complementary bit lines ($BL_t, \overline{BL_t}$) of another column (t) to establish the trigger condition. The Trojan is turned on when both BL_t and

978-1-5386-3775-3/18 $31.00 © 2018 IEEE

TABLE I
IMPLEMENTED TROJANS OF TYPE-1 AND TYPE-2.

Trojan Type	Trojan Name	Defect Type	Defect Terminals/Spot	# of Triggers	Trigger Lines	Payload Effect	Subject to Tests
Type-1	Ts_Q(B)_Vss_2	Short	Cell node, Vss	2	2 other cells	Reduced hold-SNM, RDF, write failure	non-Solid data background
	Ts_Q(B)_Vss_3	Short	Cell node, Vss	3	3 other cells	Reduced hold-SNM, RDF, write failure	advanced data background
	Ts_Q(B)_Vss_4	Short	Cell node, Vss	4	4 other cells	Reduced hold-SNM, RDF, write failure	None
	Ts_Q(B)_Vss_4_WL	Short	Cell node, Vss	5	WL, 4 other cells	Temporarily reduced hold-SNM	None
	Ts_BL(B)_Vss_2	Short	Bit line, Vss	2	2 cells	Incorrect read	non-Solid data background
	Tb_Q(B)_Q(B)_2	Bridge	Cell node, Cell node	2	2 other cells	RDF, write failure, coupling fault	non-Solid data background
	Tb_Q(B)_Q(B)_3	Bridge	Cell node, Cell node	3	3 other cells	write failure	advanced data background
	Tb_Q(B)_Q(B)_4	Bridge	Cell node, Cell node	4	4 other cells	reduced hold/read-SNM	None
Type-2	To_QB_Vdd_WL	Open	Cell pull-up	1	WL_{i-1}	Temporary neg. hold-SNM	None
	To_QB_Vdd_WL_Q(B)	Open	Cell pull-up	2	$WL_{i-1}, Q(B)1$	Temporary neg. hold-SNM	None
	To_QB_Vdd_BL_BLB	Open	Cell pull-up	2	$BL_t, \overline{BL_t}$	$(w0)(r0)^3$	March RAW
	To_QB_Vdd_Q(B)_Q(B)	Open	Cell pull-up	2	$Q(B)1, Q(B)2$	$(w0)(r0)^2$, DRF	March RAW/SL/ DRF tests w/ checkerboard/ col. stripe DB

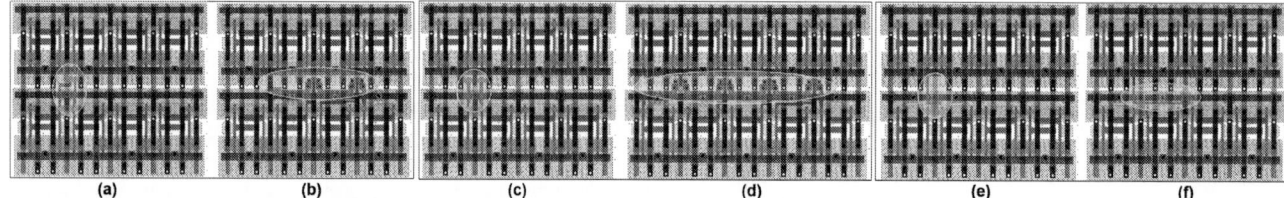

Fig. 4. Layout of Trojans in a compact SRAM, causing short defects: (a) Ts_BL_Vss_2_WL, (b) Ts_QB_Vss_2_WL; bridge defects: (c) Tb_BLB_BL_2_WL, (d) Tb_QB_QB_3_WL; open defects: (e) To_Q(B)_Vdd_WL, (f) To_Q(B)_Vdd_WL_QB.

$\overline{BL_t}$ are high. The payload effect is two-fold. If the trigger condition is satisfied, e.g., after a read operation on a cell in column t, BL_t and $\overline{BL_t}$ can both have floating values close to logic 1, and the v-cell has a read(0)-after-write(0) fault that flips the value with multiple read operations. While not accessed, the v-cell can have a DRF if BL_t and $\overline{BL_t}$ remain high for a long time; however, one of the bit line voltages will always have a slight degradation during the evaluation phase, creating a weakly conducting path that prevents the DRF from occurring. In our framework, the v-cell exhibits $(w0)(r0)^3$ when the Trojan is on. Thus the Trojan will be detected by test algorithms covering read-after-write dynamic faults while maintaining BL_t and $\overline{BL_t}$ high when operating on the v-cell, e.g., March RAW [11]. However algorithms like March SL [12], which keep one of BL_t and $\overline{BL_t}$ low with a write as the last operation in the march element while accessing the v-cell, cannot detect this Trojan. Although March-RAW is only intended to cover $(w0)(r0)^2$ faults, the compromised value after two read 0 operations cannot be recovered by the weakly-on bit line, and eventually will have a flip (see Section IV). Finally, the Trojan in Fig. 3(f) triggers when both trigger cells situated on one side of the v-cell hold different values. Such a Trojan can evade detection in march tests with solid backgrounds. With checkerboard and column stripe, march tests covering read-after-write dynamic faults or DRFs can detect the Trojan. As the advanced march tests are generally not applied with complex data backgrounds, there is a high chance that the Trojan can pass industrial-standard tests. The four implemented Trojans are listed in Table I.

IV. FEASIBILITY VERIFICATION

We have implemented the proposed Trojans in a compact SRAM layout in 45nm CMOS technology. Fig. 4 illustrate each type of Trojan with two representative examples. Trojan insertion does not change the cell footprint or incur extra silicon area. Trojans that are not mounted inside a cell (Type-1) make no modification on the original cell layout, but only add extra connections to certain existing nodes. Although the

locations and structure of SRAM cells are to be preserved, certain objects have some flexibility in their locations. For example, V_{DD}-to-N-Well contacts to assure a uniform potential of the N-well can be slightly shifted while maintaining the necessary density (Figs. 4(b)(d)(f)).

An optimized Trojan design should try to share the source/drain of Trojan transistors with those of SRAM existing transistors as much as possible (Figs. 4(a)(c)(e)). For Trojans containing several transistors or creating bridging defects, contacts are required between active regions and interconnects. Since metal 1 and 2 are extensively used in the SRAM array horizontally and vertically, it is difficult to use these layers for Trojan interconnection. Hence, we leverage polysilicon to build Trojan interconnects, resulting in higher interconnect resistance compared to metal. The actual resistance impact will be elaborated in Section IV. Note that, although the original SRAM elements are guaranteed to have no design rule violations, we allow design rule errors to some extent in Trojan circuitry. These errors can lead to broken Trojan circuitry in some ICs due to process variations and eventually result in benign Trojans. For instance, insufficient active overlap around active contacts in Trojan Type-1 can cause broken contacts and disable the Trojan path. Since the attack scenario considered in this paper is untrusted foundries, who have control over the IC fabrication process, attackers are able to ensure the design rule correctness of the original circuit while preserving the violations in the Trojan circuitry. Inserting Trojans directly in third-party SRAM IP hard cores would be prohibited as the violations will be detected in full-chip physical verification. As no design rule violations occur in the original circuit, the yield of the fabrication process will not be affected, hence no suspicion regarding the IC integrity would arise from this aspect.

HSPICE simulations were performed with 45nm low-power CMOS predictive technology model. The payload effects were analyzed on a 32x64 SRAM array, considering the v-cell in standby mode and during operations. The impact of each idle Trojan on other cells in the array was also characterized.

978-1-5386-3775-3/18 $31.00 © 2018 IEEE

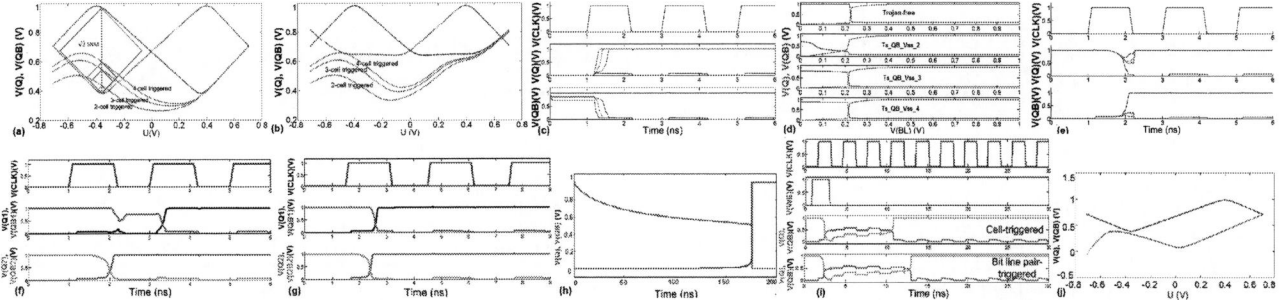

Fig. 5. Figures showing Trojan's impact when Type-1(short) Trojan is activated: (a) Hold-SNM of the v-cell; (b) Read-SNM of the v-cell; (c) RDF in the v-cell during read 0 operation; (d) Shifted write-0 trip point and degraded logic 1 voltage at QB; (e) Write-0 failure (RDF) in the v-cell. Type-1(bridge) Trojan Tb_QB_QB_2 causing: (f) RDF with 2ns clock period; (g) CFid with 3ns clock period. Type-2(open) Trojan To_Q_VDD_2 causes (h) data retention fault; (i) read-after-write dynamic faults. (j) Type-2 Trojans with WL in trigger cause temporary negative SNM in the v-cell.

A. Trojan Type-1: Short

Trojan Ts_QB_Vss_x ($x \in \{2,3,4\}$) introduces a shorting path between the complementary node of a cell and V_{ss}. The trigger is controlled by the data pattern of 2, 3 or 4 other cells. The Trojan payload impacts the v-cell's capability in holding/operating with a logic 0.

Hold-0: While not operated, the v-cell can still hold value 0; however, the static noise margin (SNM) is severely compromised due to the Trojan-induced pull-down path. Fig. 5(a) illustrates the distorted DC transfer characteristics of the shorted inverter. The butterfly curve of the v-cell when the Trojan is untriggered is the same as that of a Trojan-free cell. The red curves correspond to the transfer characteristics of the v-cell QB transistor with an on-state Trojan. A straightforward observation is that a triggered Trojan leads to dramatically reduced SNM for holding 0, while the SNM for holding 1 is improved. This means the cell is extremely sensitive to noise when storing 0, which can be caused by supply voltage or temperature fluctuation, coupling of neighborhood cells, and radiation. Hence, the soft error rate for the v-cell is significantly increased. Also, fewer trigger cells leads to a shorting path with lower series resistance. This aggravates the discharging of QB and makes it more difficult for the v-cell to hold 0. The actual SNM values are provided in Table II.

Read-0: Similarly, the Trojans also reduce the v-cell read-SNM when triggered. Since read-SNM is generally much lower than hold-SNM for a cell, the v-cell read-SNM with an on-state Trojan is negative for read 0. Fig. 5(b) shows that even the Trojan with a 4-cell trigger which has a relatively weak shorting path leads to a non-positive read 0 SNM. This implies that the v-cell cannot perform a read 0 correctly even with no disturbance due to noise. This is proved in Fig. 5(c) which shows the occurrence of RDFs during read 0. For comparison, the blue curves display the correct read 0 operation of Trojan-free cells. Degraded logic 1 voltages due to Trojans can also be seen in the figure, confirming the reduced SNMs from another perspective.

Write-0: Fig. 5(d) shows an on-state Trojan slightly shifts the write-0 trip-point of the v-cell because the shorting path results in a stronger pull-down path in the QB transistor. The Trojans also degrade logic 1 voltages at QB due to the shorting path. More importantly, transient analysis demonstrates remarkably increased write access time due to triggered Trojans, as shown in Table II. The read/write access times are measured from the beginning of the evaluation phase, excluding the bit line conditioning or address decoder propagation delay, where

TABLE II
IMPACT OF Ts_QB_Vss_x ($x \in \{2,3,4\}$) ON A 32X64 SRAM ARRAY.

Parameters	Golden	Trojan Untriggered			Trojan Triggered		
		x=2	x=3	x=4	x=2	x=3	x=4
SNM-hold (V)	0.42	0.42	0.42	0.42	0.04	0.12	0.16
SNM-read (V)	0.24	0.24	0.24	0.24	<0	<0	<0
Read access time (ns)	0.26	0.26	0.26	0.26	-	-	-
Write access time (ns)	0.84	0.85	0.86	0.87	1.45	1.06	0.99
Standby power (nW)	1.43	1.43	1.43	1.43	-	-	-
Read Energy (fJ)	118.29	118.29	118.29	118.29	-	-	-
Write Energy (fJ)	110.95	111.03	111.13	111.22	-	-	-

Trojans do not make a difference. The increased write access time causes write-0 failures in Trojan-compromised v-cells at high operating frequencies. Fig. 5(e) demonstrates write-0 failures in v-cells at the Trojan-free array's nominal frequency. Similar to SNM, the impact on write access time also exacerbates with a reduced number of trigger cells because of stronger shorting paths. Considering a pre-charge phase of 1ns, a Ts_QB_Vss_2 causes a performance degradation from 543 MHz (period of 1.84ns) to 408 MHz (period of 2.45ns).

Untriggered Trojan: Table II shows that untriggered Trojan Ts_QB_Vss_x has virtually no impact on v-cell SNM/read access time, and incurs minimal (<0.03ns) degradation on write access time due to capacitive loading. Considering 1ns pre-charge phase, the performance degradation is below 1.6%, hence negligible for sub 50-nm technologies with large process variations. Moreover, for larger arrays the access time will be much higher due to large word/bit line capacitance, thus the relative performance overhead would be negligible.

The parametric impact of two other Type-1 (short) Trojans is given in Table III. Here, an on-state Trojan Ts_BL_Vss_2 causes incorrect read 1 in the v-cell because the shorting path prevents BL from pre-charging to as high as BLB, which poses a significant effect on the latch-based sense amplifier that only senses the bit line pair voltage difference for a short time. Trojan Ts_QB_Vss_4_WL incurs temporal hold-SNM degradation because the trigger word line (of a different row from the v-cell) never stays high for more than a half clock cycle or gets asserted while the v-cell is accessed.

B. Trojan Type-1: Bridge

We have implemented Trojans causing bridge defects between SRAM circuit nodes, including Trojan Tb_QB_QB_x ($x \in \{2,3,4\}$) that creates bridges between the complementary nodes of two cells. Table IV provides the v-cell faulty behavior incurred by Trojan Tb_QB_QB_x ($x \in \{2,3,4\}$) in their on-states. As the number of PTs in the bridging path increases, the Trojan impact is reduced. Eventually, Tb_QB_QB_4 causes

978-1-5386-3775-3/18 $31.00 © 2018 IEEE

TABLE III
IMPACT OF TWO OTHER TYPE-1 TROJANS ON A 32X64 SRAM ARRAY.

Parameters	Golden	Ts_QB_Vss_4_WL		Ts_BL_Vss_2	
		Troj. Untrig.	Troj. trig.	Troj. untrig.	Troj. trig.
SNM-hold (V)	0.42	0.42	0.20	0.42	0.42
SNM-read (V)	0.24	0.24	-	0.24	0.24
Read access time (ns)	0.26	0.26	-	0.26	0.27
Write access time (ns)	0.84	0.86	-	0.84	0.85
Standby power (nW)	1.43	1.43	-	1.43	-
Read Energy (fJ)	118.29	118.29	-	118.30	-
Write Energy (fJ)	110.95	111.22	-	111.38	-

TABLE IV
PAYLOAD OF TROJAN Tb_QB_QB_X (X∈{2,3,4}).

Operation	x=2	x=3	x=4
QB1=0, QB2=1, read QB2	RDF	Normal	Normal
QB1=1, QB2=0, read QB2	RDF	Normal	Normal
QB1=0, QB2=0, write-1 QB2	Cell2 write failure	Cell2 write failure	Normal
QB1=1, QB2=1, write-0 QB2	CFid/RDF in Cell1	Normal	Normal
QB1=1, QB2=0, hold QB2	Normal	Normal	Normal

TABLE V
IMPACT OF UNTRIGGERED Tb_QB_QB_X ON A 32X64 SRAM ARRAY.

Parameters	Golden	Trojan Untriggered		
		x=2	x=3	x=4
SNM-hold (V)	0.42	0.42	0.42	0.42
SNM-read (V)	0.24	0.24	0.24	0.24
Read access time (ns)	0.26	0.26	0.26	0.26
Write access time (ns)	0.84	0.85	0.86	0.86
Standby power (nW)	1.43	1.45	1.45	1.45
Read Energy (fJ)	118.29	118.35	118.35	118.35
Write Energy (fJ)	110.95	110.99	111.07	111.19

TABLE VI
IMPACT OF UNTRIGGERED To_QB_VDD ON A 32X64 SRAM ARRAY.

Parameters	Golden	Trojan Trigger Lines			
		WL	WL, Q1	Q1, Q2	BL1, BLB1
SNM-hold (V)	0.42	0.42	0.41/0.42	0.41/0.42	0.41/0.42
SNM-read (V)	0.24	0.24	0.23	0.23	0.23
Read access time (ns)	0.26	0.26	0.26	0.26	0.26
Write access time (ns)	0.84	0.85	0.85/0.86	0.85/0.86	0.85/0.86
Standby power (nW)	1.43	1.43	1.43	1.44	1.42
Read Energy (fJ)	118.29	118.35	118.30	118.35	118.37
Write Energy (fJ)	110.95	110.71	110.69	110.69	109.39

merely a weak fault that compromises the v-cell SNMs. Conversely, Tb_QB_QB_2 leads to faulty behavior in each operation that involves different values at the two bridged nodes. In particular, read destructive faults occur when one of the v-cells is being read if the other one holds a different value. When both v-cell nodes hold value 0, the pulling-down effect of the bridging path prevents a successful write-1 operation on either of them. However, while both the victim nodes hold value 1, a write-0 performed on one of them will pull down both v-nodes. Depending on clock frequencies, the actual faulty behavior can be an instantaneous idempotent coupling fault or an RDF in the standby v-cell following the write, as shown in Fig. 5(f and g). Table V demonstrates the negligible impact on SRAM performance, power, and SNM caused by the Trojans untriggered.

C. Trojan Type-2: Open

Among the four Type-2 Trojans, the ones with trigger condition from a word line cause weak faults by temporarily reducing the hold-SNM of the v-cell to a negative value when the control word line is asserted (Fig. 5(j)). Since a word line is only on for a short time during a clock cycle, it is not adequate to cause a data flip in the v-cell. However, transient noise due to supply fluctuation, neighborhood signal coupling, or radiation, when paired with the negative noise margin, can easily modify the stored value. On the other hand, the two implemented Trojans that use data pattern or bit line pairs as the trigger condition cause explicit faulty behaviors, as the Trojan can be on during both standby or access mode of the v-cell. As shown in Fig. 5(h and i), the on-state Trojan incurs data retention fault after 175 μs in a standby v-cell, and the Trojan triggered by data pattern and bit line pair lead to $(w0)(r0)^2$ and $(w0)(r0)^3$ dynamic faults, respectively. The difference is because one of the bit lines is pulled down slightly during the evaluation phase, causing a weakly conducting path at the open defect spot. If the word line is asserted for a long time during a clock cycle, the bit line swing could turn the Trojan pMOS PT on and eliminate the explicit payload effect. Table VI demonstrates the impact of the untriggered Trojans.

V. CONCLUSION

We have presented a fundamentally new class of hardware Trojan attacks, where embedded memory in a chip, e.g. cache, can be subject to malicious implantation in an untrusted design house or foundry. We have designed and implemented various forms of such Trojans in SRAM arrays and shown that they can evade industry-standard post-manufacturing testing. We have also shown these Trojans can have unique payloads, which can affect data integrity in embedded cache memories, leading to undesirable and potentially catastrophic consequences at the system level. Different structures of memory Trojans are implemented and analyzed, comprehensively exploring the design space of trigger/payload locations and patterns. The Trojans, while inactive, cause minimal impact on SRAM performance, power, and stability. Future research will include test chip fabrication and measurement, demonstration of complex and powerful system-level attacks leveraging such Trojans, and novel test solutions to detect these Trojans.

VI. ACKNOWLEDGEMENT

The work is funded in part by National Science Foundation (NSF) grants 1603483 and 1623310.

REFERENCES

[1] D. Rosso, "International Technology Roadmap for Semiconductors Explores Next 15 Years of Chip Technology," *International Technology Roadmap for Semiconductors*, 2014.
[2] T. Lovric, "On Design Vulnerability Analysis and Trust Benchmarks Development," *ICCD*, 2013.
[3] S. Bhunia, et al.,"Hardware Trojan Attacks: Threat Analysis and Countermeasures," *Proceedings of the IEEE*, 2014.
[4] T. Hoque, et al., "Golden-Free Hardware Trojan Detection with High Sensitivity Under Process Noise," *JETTA, Springer*, 2016.
[5] S. T. King, et al., "Designing and Implementing Malicious Hardware," *LEET*, 2008.
[6] R. Saeidi, et al., "SRAM Hardware Trojan," *IST*, 2016.
[7] S. Kan, et al., " Triggering Trojans in SRAM circuits with X-propagation," *DFTS*, 2014.
[8] A. Das, et al., "Detecting/Preventing Information Leakage on the Memory Bus due to Malicious Hardware," *DATE*, 2010.
[9] F. Stellari, et al., " Revealing SRAM memory content using spontaneous photon emission," *VTS*, 2016.
[10] K. Cheng, et al., "Memory Test Experiment: Industrial Results and Data," *IEEE TCAD*, 2002.
[11] S. Hamdioui, et al., "Testing Static and Dynamic Faults in Random Access Memories," *VTS*, 2002.
[12] S. Hamdioui, et al., "Linked Faults in Random Access Memories: Concept, Fault Models, Test Algorithms, and Industrial Results," *IEEE TCAD*, 2004.
[13] Van de Goor, Ad J, "Testing Semiconductor Memories: Theory and Practice," *John Wiley & Sons, Inc.*, 1991.
[14] S. Kan, et al., " Enhancing Embedded SRAM Security and Error Tolerance with Hardware CRC and Obfuscation," *DFTS*, 2015.

High efficient low cost EEPROM screening method in combination with an area optimized byte replacement strategy which enables high reliability EEPROMs

Gregor Schatzberger, Friedrich Peter Leisenberger, Peter Sarson, Andreas Wiesner
ams AG
Premstaetten, Austria
gregor.schatzberger@ams.com

Abstract— **Modern high reliability EEPROM technologies use advanced screening methods to screen out weak bit cells. In combination with standard ECC (Error Correction Code) methods EEPROMs can be produced to withstand more than 50k endurance cycles at 150°C and have a data retention of 10 years at 150°C. Both the number of endurance cycles and the data retention define the lifetime of an EEPROM memory. After successfully passing the standard EEPROM screening tests only randomly distributed intrinsic failures will be activated during the EEPROM's lifetime. This work will present a highly advanced screening procedure in combination with an area efficient byte replacement strategy that enables an increase in the number of endurance cycles by a factor of 2.5 while the increase in area of the EEPROM memory plane is only 5%.**

Keywords— *Electrically erasable programmable read only memory (EEPROM), Error Correction Code (ECC), advanced screening methods, data retention, endurance cycle, byte replacement strategy*

I. INTRODUCTION

Non-volatile memories for the automotive market require high reliability in order to fulfil the customer's expectations. Therefore, EEPROM screening methods must be in place to predict the number of endurance (program/erase) cycles the memory can withstand in the field. These prediction methods are used to filter out weak bit cells that will fail within the specified number of endurance cycles. If an ECC is used the number of endurance cycles can be guaranteed or even extended depending on the chosen ECC algorithm. For an automotive application, the ECC must be turned off during the EEPROM screening according to the Automotive Electronics Council (AEC) standard to prevent the ECC masking unwanted process related yield losses. ECC should only cover potential failures present in the field over the EEPROM's lifetime.

After successfully passing the EEPROM screening, only randomly distributed intrinsic failures will be present during the device lifetime [1]. It is hard to catch such failures. Therefore, many automotive applications implement an ECC to ensure coverage. To implement such algorithms, redundant EEPROM bit cells are added to compensate for such potential failures. As the exact locations for such failures are hard to predict, the ECC simply adds to all address locations of the memory redundant EEPROM bit cells. Depending on the used ECC algorithm additional memory area is needed.

Making an in-depth analysis of the intrinsic failures, it turns out that there are only a few failures within one memory plane that are activated before the EEPROM enters the wear out region and the bit cells starts to collapse at a high rate. The beginning of the wear out region indicates the end of the bathtub curve and the bit cells start to fail at a very high rate. These failures are randomly distributed and cannot be covered by any ECC. If there are only a few bit cells within one memory array that will fail during the life time and we use an ECC to cover those failures we use a lot of area to tackle that problem. It would be much more efficient to only replace the bit cells that will fail during the EEPROM's lifetime. The solution to this issue is a screening routine that can predict the number of endurance cycles of each bit cell in the memory array and keep the test time under control. This paper presents

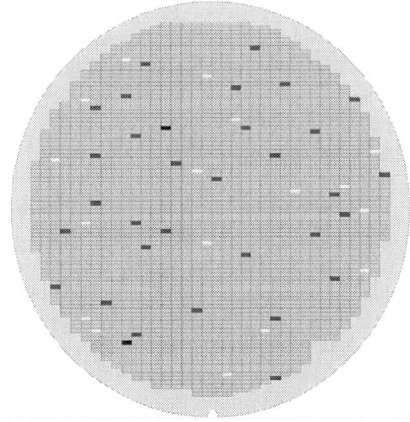

Fig. 1 Wafer map of an EEPROM (1kx8) after standard EEPROM screening, yield = 98.1%

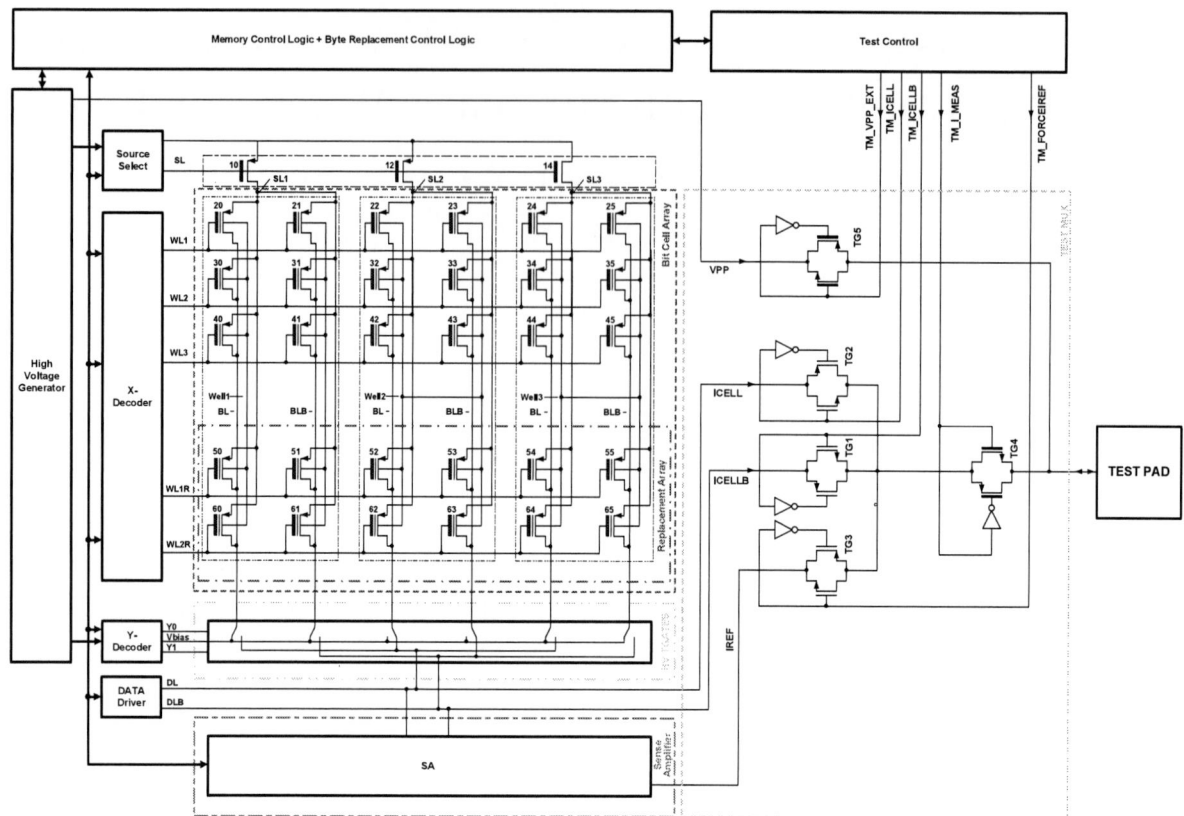

Fig. 2 EEPROM weak bit screening test concept

a screening methodology that is capable of predicting the lifetime of each single EEPROM bit cell in the memory array and finding the exact location of weak bit cells with an acceptable test time increase.

A standard EEPROM production test flow is split into 2 parts [2]. In the first part, all basic functional and weak bit screening tests are executed followed by a data retention pattern written to the memory. After the data retention bake step, part 2 of the production test flow is executed checking for the correct data retention pattern. The resulting wafer map is shown, Fig. 1. The yellow dots in the wafer map mark continuity fails, the red dots are functional failures, and the two blue dots are related to weak bit fails of the EEPROM.

II. TEST CONCEPT FOR WEAK BIT SCREENING

To enable an efficient screening algorithm in terms of sensitivity, area and test time the EEPROM needs to include special test modes. The program and erase state of EEPROMs based on Fowler-Nordheim tunneling can be measured in two ways. The threshold voltage or drain to source current can be used to judge if a bit cell is fully programmed or fully erased. In the following description of the screening algorithm the drain to source current (Icell) will be used.

The EEPROM needs to have a test mode that is capable of measuring the Icell current of each individual bit cell in the memory plane. The EEPROM architecture used in the example, Fig. 2, is fully differential. That means a pair of bit cells, e.g. bit cell 20 and 21, form one data bit of the EEPROM. To store a logic '0' bit 20 is erased and bit 21 is programmed. A fully differential memory array already includes redundancy in the memory and the sense amplifier (SA) readout margin is optimal. As shown in the block diagram, Fig. 2, special switches (TG1-TG5) are used in parallel to the signal path from the bit cells (20 to 65) to the sense amplifier (SA). The Test Control Logic closes the special t-gates TG1 or TG2 depending upon which bit should be connected to the TEST PAD via TG4. The SA inputs are switched to high impedance by the Memory Control Logic and by the Byte Replacement Logic. The selected bit cell defined by the address-bus of the memory is connected to the TEST PAD. For technologies based on p-channel bit cells the drain to source current of the bit cell can be measured against ground, whereas for n-channel based bit cell currents are measured against the positive supply. The actual applied address forced via the X-Decoder and Y-Decoder to the memory plane define which byte of the plane is measured. As only one TEST PAD is available, special test modes are used to select the individual bits of the selected byte that are measured. To ensure that both bit cells of the data bit can be measured, a separate test mode is used to switch from the left bit cell to the right bit cell. To measure bit cell 20 the appropriate address must be selected via the memory decoder and TG2, TG4 must be closed. Switching to bit cell 21 is done

978-1-5386-3775-3/18 $31.00 © 2018 IEEE   48

by opening TG2 and closing TG1. This test approach allows bit cell current distribution from the EEPROM memory plane to be measured in order to judge if a bit cell is a weak bit or not.

The drawback to this concept is the dramatic increase of test time as the memory size increases. Each measurement needs a defined setup time to measure the current flowing through the TEST PAD to the production tester either to ground or positive supply. The required setup time is limited by the tester.

To overcome that problem a test mode using an external reference current is implemented [3]. The external reference current is forced through the TEST PAD by closing TG4 and TG3 and is distributed to the sense amplifier (SA) of the memory, Fig. 2. Depending on the used memory concept (fully differential, pseudo differential or single ended) the external reference current is used to measure indirectly the value of the selected bit cell current. This is done by sweeping the external reference current. The sweep direction and step size depend on the chosen technology. The advantage of using this method, which we call digital margin test, is that all bits of the selected address can be tested at the same time. The output is the standard digital output of the EEPROM. Each applied current can be tested with the standard read access time. With the applied external current the whole address range can be checked. This enables fast testing because the time consuming setup of the external reference current is only needed once for each current step before the whole address range is checked with this reference level.

One important point for a successfully screening is that all tested bit cells are treated the same way especially during program and erase. This makes it necessary to have full control over the program and erase voltage and the program and erase timing. Most of the modern EEPROMs have an internal high voltage generator providing the necessary high voltage during program and erase steps. As these high voltage generators are based on internal references like a bandgap voltage generator we face the situation of having a variation of the output of the high voltage generator due to process and mismatch variations. The same is true for the internal oscillator, which is also a part of the high voltage generator. These variations make it impossible to setup a sensitive endurance screening procedure. Therefore, it is mandatory to have access to the internal high voltage node (VPP) of the EEPROM in a special test mode to have full control on the high voltage level and timing by closing TG5 and opening TG4, Fig. 2. The internal high voltage node VPP is connected to the TEST PAD. Now the production tester can generate the required high voltage and timing during program and erase by forcing the signals to the TEST PAD. This ensures that all bit cells see exactly the same program and erase pulse thus ensuring the appropriate precision for the high endurance screening.

III. WEAK BIT SCREENING

The screening procedure developed by Mielke [4] is based on the fact that weak EEPROM bit cells can be erased faster compared to standard bit cells. This effect can be tested in two possible ways. The first possibility is to have a fixed erase time with the erase operation being executed at different erase

voltages, and the second possibility is that the erase voltage is fixed and the erase time is varied. Both methods show the same results. Mazzali [5] uses a screening based on Mielkes [4] approach by testing the EEPROM with three different shorter erase pulses compared to the standard erase pulse. After each short test erase pulse the memory is read with the standard sense amplifier. If a bit is already erased a bit replacement algorithm is used to overcome the early endurance fails. Mazzali [5] stated in his invention that the exact number where the bit cell will fail cannot be predicted with his method but by replacing the weak bit cells the number of endurance cycle stated in the datasheet can be guaranteed. The drawback of both inventions is that the number of endurance cycles cannot be predicted accurately.

A closer look at the endurance behavior of EEPROMs based on Fowler-Nordheim tunneling shows that after a correct infant mortality screening, only a few bit cells inside the memory have a potential risk of failing before the EEPROM reaches the wear out region, Fig. 3. To extend the number of endurance cycles close to the wear out region, weak bit cells must be identified and screened out or replaced by bit cells which are capable of withstanding a high number of endurance cycles. A programmed bit cell delivers a drain to source current in the range of Icell1max and Icell1min and an erased bit cell a current between Icell0max and Icell0min. As an area efficient EEPROM is required, the focus is on the bit cell size because this is the main area contributor. The size of the bit cells is pushed to the process limits to ensure the most efficient bit cells in terms of area. A bit cell current distribution of the Icell current for program and erase across the memory plane can be observed, the range of the cell current being caused by process variations which change from wafer to wafer and lot to lot. Therefore an adaptive method is required to take these variations into account rather than using a fixed current value.

From Fig. 3 it can be seen that 50k endurance cycles can be safely screened and secured with standard screening procedures with an acceptable yield loss and test time. To extend the number of endurance cycles to 125k, the algorithm must be able to find all weak bits in the extended endurance range, for example B1, B2, B3 and B4, Fig. 3. Such weak bits have no hard failure mechanism like short to ground or supply,

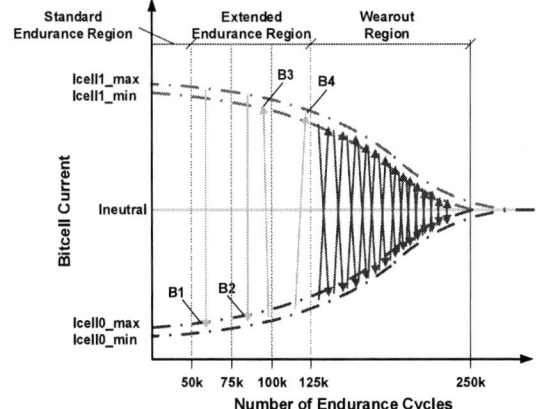

Fig. 3. Endurance behavior and intrinsic fails of an EEPROM

978-1-5386-3775-3/18 $31.00 © 2018 IEEE

instead the failure modes can be charge traps in the tunnel oxide or interpoly oxide caused by contamination or imperfections in the oxide structure, which become activated during the cell's lifetime. The failures close to the maximum allowed number of endurance cycles in the extended endurance range are the most difficult to find. As discussed earlier, the program and erase bit cells have a bit cell current distribution caused by mismatch and process variation. To safely push out the maximum number of endurance cycles the screening procedure must be able to catch all infant mortality and weak bits in the extended endurance range. An activated weak bit can lead to a stuck at high (B3, B4) or stuck at low (B1, B2). That means depending on the technology and failure mode a fast erase bit is not necessarily a fast programing bit cell or vice versus. Therefore, it is essential to prove and screen both the program and the erase state.

To implement an efficient screening method, it is necessary to know when the EEPROM bit cell will fail due to endurance cycling and the exact location of the bit cell. This can be achieved by using a byte replacement strategy which can be used to significantly increase the yield of the technology. It is useful to define so called cycle classes, which will give us a prediction on how many endurance cycles an EEPROM bit cell can withstand. As stated at the beginning, modern EEPROM's with standard screening procedures are capable of 50k endurance cycles at 150°C. To extend this number for example to 125k, we define three cycle classes 75k, 100k and 125k. As shown in Fig. 3 most of the EEPROM bit cells can handle the 125k write cycles. Only a few bits are not able to achieve this high number. These bit cells are of interest. Depending on the defect, those bit cells can fail from 50k to 125k endurance cycles. In Fig. 3 four examples are shown; failure B1 is activated close to 60k endurance cycles and produces a stuck at "low". B2 shows the same failure mode and is activated in the range of 85k endurance cycles. B3 generates a stuck at "high" failure close to 100k endurance cycles and B4 shows the same failure near 125k. All four failures must be safely detected by the screening procedure to enable a correct replacement of the weak bit cells so that the EEPROM can fulfill 125k write cycles.

Cycle class statistics of the endurance performance must be generated to get a better understanding of the failure modes of the technology and how they are activated. Generating such statistics is a huge effort because endurance cycling of EEPROMs is very time consuming. It has been found that the endurance behavior of EEPROMs follows a Gaussian distribution. Most of the bit cells of a mature EEPROM process are able to withstand a high number of endurance cycles, the number varies depending on the technology used, Fig. 3. The exact numbers strongly depend on the technology and the defect density from the production line. After a successful infant mortality screening, the likelihood of failure B1 (< 75k endurance cycles) is very low. When the number of endurance cycles increases the likelihood of getting a fail (B2, B3 and B4) also increases. From the Gaussian distribution and a given sigma value, the maximum number of endurance cycles and the cycle classes of an EEPROM technology can be extracted from the characterization data. The number of cycle classes depends on the sensitivity of the screening algorithm and the chosen

replacement strategy. To minimize the area overhead for implementing the replacement strategy, the number of cycle classes should be kept to a minimum.

IV. EXPERIMENTING WITH WEAK BIT SCREENING

For an efficient screening of fast bits that will fail during the lifetime, determined by cycling the EEPROM, the memory is setup in external VPP mode as explained in Fig. 2. Therefore, TG4 is opened, TG5 is closed and TG1, TG2 and TG3 are opened. All t-gates are controlled using associated control signals driven from Test Control. The production tester has now full control on the high voltage pin of the EEPROM and as a first step of the screening a standard program pulse is applied to the EEPROM. The standard program pulse uses the same level and timing on VPP (VPROG) as the internal high voltage generator would provide for typical program condition. Depending on the chosen technology we have to take care that all bit cells in the area are programmed. As some technologies are data sensitive, for programming a special test mode must be used which is capable of programming both EEPROM bit cells forming one data bit. The standard readout with the sense amplifier (SA) would lead to an unreliable result because both bit cells have the same status (programmed). As there is no difference between the two cells the offset and noise of the sense amplifier (SA) defines the output result.

To judge the programmed state of each bit cell, we switch in a second step of the screening to the external reference current mode, by opening TG5 and closing TG3 and TG4. To check the programming status of each bit cell we have to know the exact drain to source current (Icell) of each bit cell. Therefore, the external reference current that is internal distributed to all sense amplifiers is swept in defined steps either from a low to a high value or vice versa depending on the used technology. For simplification the positive ramp is explained. As a starting point a current with a level which is 50% of a good programmed bit cell current is applied from the tester to the EEPROM via the TEST PAD. All addresses are checked with this current level and the result should be equal to "1" because all bit cells show a higher current compared to the external reference current applied to the memory. In the case of a fully differential bit cell array, the second bit cell must be

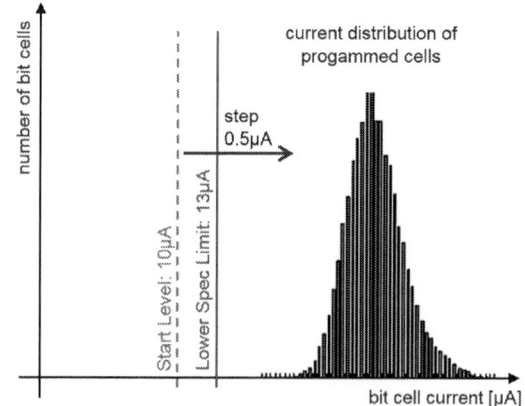

Fig. 4. Graphical presentation of the bit cell current distribution

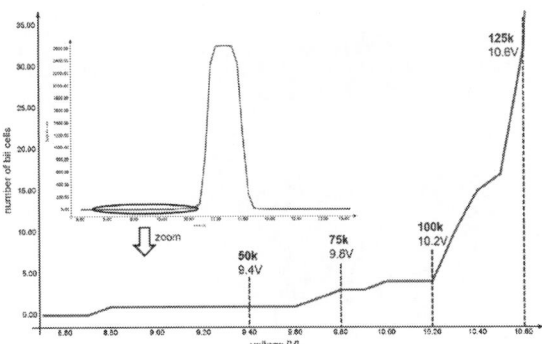

Fig. 5. Erase cycle class voltage distribution

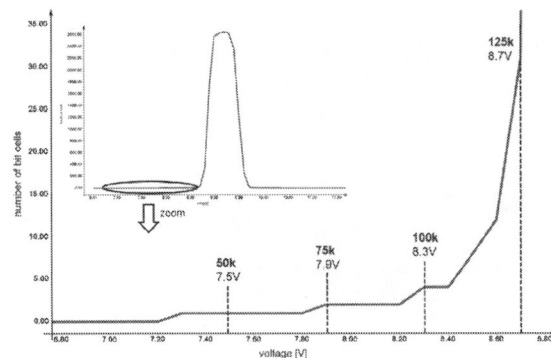

Fig. 6. Program cycle class voltage distribution

checked as well by using a special test mode which allows comparing the second bit cell. All second bit cells should be "1" as well. After the two readouts of the whole memory with the standard read out speed, the external reference current is increased by a defined step and all bit cells are checked again as described above. The external reference current is increased with the defined step size until all bit cells are changed from "1" to "0". What happens when a bit cell changes its state from "1" to "0"? The external reference current becomes larger compared to the Icell current of the bit cell. That means we indirectly measure with this method the value of each bit cell current in the EEPROM array and we can draw a bit cell current distribution of the programmed bit cells, Fig. 4.

The accuracy of this method depends on the step size used to increase the external reference current. The value of the step size depends on the production tester used (resolution of the current sink) and the noise level of the sense amplifier (SA) and of course on the test time. This value will vary for each technology and must be found during characterization. From the distribution, the median value is calculated and divided by two to get the external reference current needed for the weak bit testing. Extracting the reference current value out of the Icell distribution for each memory on the wafer ensures that the screening procedure tracks the process and wafer variations needed to enable a weak bit screening which is sensitive enough.

To screen for weak bits the memory is switched back to external VPP mode and a soft erase pulse is applied to the memory. Care must be taken for fully differential EEPROMs as stated above to erase both bit cells of one data bit. The VPP used for the soft erase pulse is lower than the standard erase voltage level but the timing used during erase stays the same as for a standard erase pulse. So, only VPP is changed and the timing is fixed. Different VPP levels can be assigned to different cycle classes. After the soft erase operation has taken place the EEPROM is switched to external reference current mode. The derived reference current from the digital margin test is applied and the whole address range of the memory is tested. If no bit cell is erased, we can state that we will have no endurance failure for that cycle class. The next step of the algorithm requires reprogramming the bit cell array again to avoid charge accumulation effects on the EEPROM bit cells and then repeating the test procedure with the next cycle class

erase voltage, until all cycle classes are tested. If we find a failure during the erase cycle class testing, the test program either marks the memory as fail part or stores the failing address if a byte replacement strategy memory is used to improve the endurance performance.

The exact values of the soft erase pulses used to judge the cycle class of each individual bit cell in the memory array depends on the technology used and can be found during characterization, Fig. 5. The small plot in the graph shows the number of fails for each soft erase pulse. Of interest is the left tail of the distribution where the weak erase bit cells are located. In the zoom of the left tail we can obtain the voltages of the different cycle classes and the number of fails during qualification.

As stated earlier, a fast erasable bit cell is not necessarily a fast programmable bit cell. Therefore, the screening procedure must be repeated to check for weak programed bits in the memory. Thus, the whole memory is switched back in special external VPP test mode allowing both bit cells of the memory to be erased at the same time. Then soft programming pulses are applied to the memory to check for the programming cycle classes, Fig. 6. To judge if bit cells are weak programmed bits, the external reference current mode is used in the same way as for the erase cycle class test, but the polarity is switched. Therefore, the expected value of the sense amplifier is now '0' not '1', which is complementary to the fast erase bit testing.

An interesting observation can be made by comparing the voltage levels for weak programming and weak erasing bits, Fig. 5 and Fig. 6. The required values for the cycle classes program and erase differ a lot. The reason for this behavior is the asymmetric program/erase characteristic of the EEPROM bit cell itself, resulting in the fact that the bit cell can be more easily programmed than erased.

It is possible to deliver the above described screening method for higher endurance performance EEPROM in two ways. The first solution is to accept the yield loss and only send the passing parts to the customer. The second possibility is to add additional circuitry to the memory to replace weak bit cells as described in [7]. A big advantage of this screening method is that the exact failure location is known and how many endurance cycles the weak bit cell can endure in the field.

978-1-5386-3775-3/18 $31.00 © 2018 IEEE

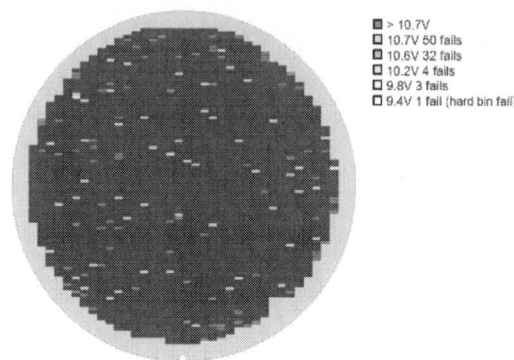

Fig. 7. Weak bit erase wafer map

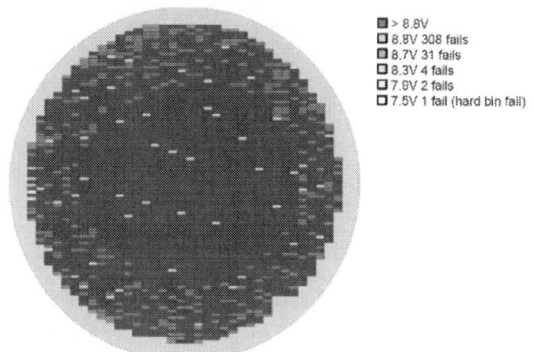

Fig. 8. Weak bit program wafer map

Depending on the program/erase failure rates of the process, additional replacement rows are added to the memory. If the failure rate is low, no column replacement strategy is needed. This saves area avoiding the additional high voltage well needed to realize this function. These additional EEPROM bit cells must also be screened with the screening procedure to define the reliability of the replacement cells. After testing all EEPROM bit cells in the memory, the replacement logic replaces the weak bit cells in the memory array with replacement bit cells which are able to fulfil the endurance specification of the EEPROM. The replacement strategy can be even expanded further as described in [7]. This dynamic replacement approach uses EEPROM counters to safely enable the correct switching from the weak bit cell to the replacement bit cell, because only the sum of the endurance performance of the weak bit cell and the replacement bit cell must fulfill the endurance specification. More logic circuitry is needed to enable this feature.

A very interesting finding was made during the process qualification of the new screening procedure. While the fast bit erase distribution was randomly distributed across the wafer as expected (Fig. 7), we found that the fast bit program distribution showed a higher concentration of fast program bits on the edge of the wafer, Fig. 8. That points clearly to a manufacturing weakness. Currently the production line has started several programs to improve this parameter.

V. CONCLUSION

This paper showed a weak bit screening method which increases the endurance performance of an EEPROM process close to the wear out region. The number of endurance cycles is improved by the factor of 2.5 with the new screening algorithm. The increase in test time depends on the number of cycle classes to be tested. Standard testing only includes one voltage for fast erase bit testing and one voltage for fast program bit testing including reading out the memory with the external reference current mode. For the weak bit screening with four cycle classes shown in this paper, three additional tests for both program and erase operation are added which increases the overall test time of the EEPROM block by ~12%.

The yield loss realized by using the advanced screening method strongly depends on the defect density achieved by the production line and is in the range of 15%. This can be significantly improved by using a simple byte replacement strategy which brings up the yield again into the high 90% region. Using a dynamic bit replacement strategy is even more efficient, especially if bigger memories (small mega bit range) are taken into account. Depending on the defect density of the process, the increase in area for the replacement bit cells is in the range of ~5% of the memory plane area. The additional increase in area of the overall memory depends on the chosen replacement strategy and is in the range of ~8% for the simple replacement to ~17% by using the dynamic approach.

Recent process optimizations have been triggered in the production line to improve the fast program yield issue and even further increase the endurance performance of the technology. Further optimization of the endurance performance will require further improvements to the screening algorithm and are currently under investigation.

REFERENCES

[1] Muroke, P. "Flash Memory Field Failure Mechanisms", 44th Annual Int. Reliability Physics Symposium, San Jose 2006.

[2] G. Schatzberger, F. P Leisenberger, P. Sarson, "Yield Improvement of an EEPROM for Automotive Applications While Maintaining High Reliability", IEEE 34th VLSI Test Symposium, pp. 1-6, Las Vegas, NV (April 2016).

[3] Peter Sarson, G. Schatzberger, F. Leisenberger, "Fast Bit Screening of Automotive Grade EEPROMs - Continuous Improvement Exercise", IEEE Transactions on VLSI (2017), DOI 10.1109/TVLSI.2016.2634589

[4] Neal R. Mielke Patent US 4,963,825 „Method of Screening EPROM Related Device for Endurance Failure".

[5] Stefano Mazzali Patent EP 0,686,978, B1 „A method for in-factory testing flash EEPROM devices".

[6] Johannes Fellner, Gregor Schatzberger Patent US 8,189,409 B2 "Readout Circuit for Rewriteable Memories and Readout Method for Same".

[7] Gregor Schatzberger, Fritz Leisenberger, Peter Sarson Patent Registration EP 17,158,340.4 "Memory Arrangement and Method for Operating or Testing a Memory Arrangement".R. Nicole, "Title of paper with only first word capitalized," J. Name Stand. Abbrev., in press.

[8] Y. Yorozu, M. Hirano, K. Oka, and Y. Tagawa, "Electron spectroscopy studies on magneto-optical media and plastic substrate interface," IEEE Transl. J. Magn. Japan, vol. 2, pp. 740-741, August 1987 [Digests 9th Annual Conf. Magnetics Japan, p. 301, 1982].

[9] M. Young, The Technical Writer's Handbook. Mill Valley, CA: University Science, 1989.

2018 IEEE 36th VLSI Test Symposium (VTS)

Test Challenges and Solutions for Emerging Non-Volatile Memories

Mohammad Nasim Imtiaz Khan and Swaroop Ghosh

School of Electrical Engineering and Computer Science, The Pennsylvania State University, University Park, PA, 16802

Email: muk392@psu.edu and szg212@psu.edu

Abstract—At the end of Silicon roadmap, keeping the leakage power in tolerable limit has become one of the biggest challenges. Several promising non-volatile memories (NVMs) are being investigated by the scientific community to address the issue. Some of the NVMs such as Spin-Transfer Torque RAM, Magnetic RAM, Resistive RAM, Phase Change Memory and Ferroelectric RAM have already entered the mainstream computing. However, the unique characteristics of these NVMs bring new fault models such as statistical and stochastic retention failures, magnetic and thermal tolerance failures, voltage droop and ground bounce induced read and write failures and long latency failures. In this work, we summarize new test failure mechanisms in NVMs and associated test challenges. We also propose new test methodologies, test patterns and Design-for-Test (DFT) techniques to characterize new failure models and compress test time.

Index Terms—Emerging NVM Test Challenges, Retention Test, Latency Test, Voltage Droop Test, Tolerance Test.

I. INTRODUCTION

Several emerging Non-Volatile Memory (NVM) technologies e.g., Spin-Transfer Torque RAM (STTRAM), Magnetic RAM (MRAM), Resistive RAM (RRAM), Phase Change Memory (PCM) and Ferroelectric RAM (FRAM) have drawn significant attention due to low power operation, high density/speed and the inherent non-volatility [1]–[5]. These memories have already penetrated the semiconductor market as discrete chips. Examples include MRAM/STTRAM by Everspin [6], CBRAM (a variant of RRAM) [7] by Adesto Tech, PCM by Intel [8] and FRAM by Cypress [9]. Although these memories are promising, their unique characteristics introduce new test challenges that call for designing new test methods and/or repurposing existing Static RAM (SRAM), Dynamic RAM (DRAM) and Embedded DRAM (eDRAM) test flows. A well-defined test methodology will facilitate broad adoption of these promising memory technologies in variety of systems and applications. Fig. 1a presents test flow for conventional memories. Various March tests are performed on conventional memories to characterize read/write operations, address decoding and identify failures (e.g. coupling between neighboring cells). Different tests are also performed to identify optimal settings of different read/write assist techniques (e.g. power gating, word-line over-drive and under drive, supply voltage collapse during write [10] and negative bit-line during write [11]). For DRAM and eDRAM, data retention is also characterized [12].

Table I summarizes the features of a subset of NVM technologies and compares them with the established memories

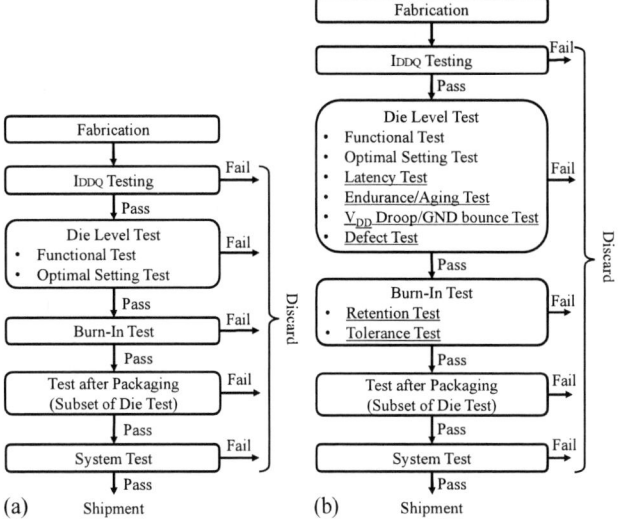

Fig. 1: Test flow: (a) conventional memory; (b) proposed for emerging NVMs.

TABLE I: Comparison between Various Memory Technologies

Feature	SRAM	DRAM	STTRAM	MRAM	DRAM	PCM
Read/Write	✓	✓	✓	✓	✓	✓
Capacitive coupling	✓	✓	✗	✗	✗	✗
Retention	✗	✓	✓	✓	✓	✓
Refresh	✗	✓	✗	✗	✗	✗
Magnetic susceptibility	✗	✗	✓	✓	✗	✗
Thermal susceptibility	✗	✗	✓	✓	✓	✓
Humidity susceptibility	✗	✗	✗	✗	✓	✗
Latency variation	✗	✗	✓	✓	✓	✓
Asymmetric read/write latency	✗	✗	✓	✓	✓	✓
High write current/voltage	✗	✗	✓	✓	✓	✓
Voltage droop	✗	✗	✓	✓	✓	✓
Asymmetric read/write current	✗	✗	✓	✓	✓	✓
Ground bounce	✗	✗	✓	✓	✓	✓
Endurance	✗	✗	✓	✓	✓	✓
Backhopping	✗	✗	✓	✓	✗	✗

e.g., SRAM/DRAM. It is evident that existing test infrastructure is not sufficient to validate emerging NVM specific features. Rationale is provided below:

Retention time: NVMs can have retention time of several years. DRAM/eDRAM undergoes special retention tests to certify the refresh rate which could range between 0.2ms to tens of ms [12], and does not affect total test time significantly. The same test flow will increase NVM retention test time

978-1-5386-3775-3/18 $31.00 © 2018 IEEE   53

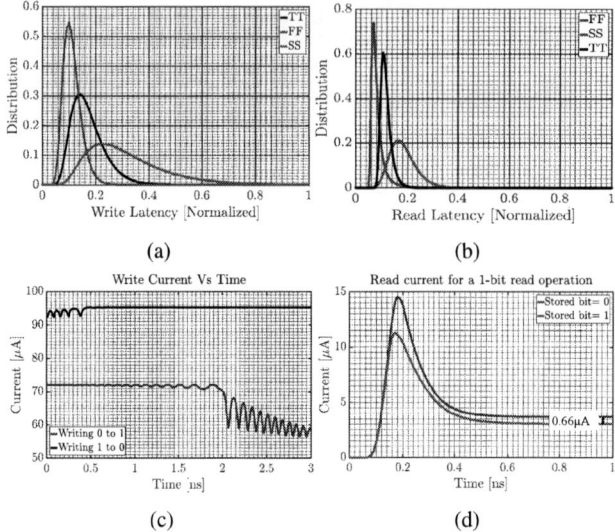

Fig. 2: STTRAM latency distribution (a) write and (b) read; high/asymmetric (c) write and (d) read current.

drastically. The stochastic nature of retention of STTRAM and MRAM [13] makes retention characterization complex and time consuming. Therefore, techniques for retention test time compression is desirable for NVMs.

Long latency: Conventional SRAM latency is in the order of fraction of a nanosecond [14] and typically not a strong function of process variation (PV)/temperature (PVT). DRAM/eDRAM latency could be high but not as sensitive to PV. However, emerging NVMs suffer from high write/read latency (Fig. 2a-2b, ~1-3ns) [15] and the latency varies significantly due to stochasticity and PVT which leads to read/write failure [15]. Therefore, it is necessary to test read/write latency within the operating temperature under PV. Design features are also required to allocate few extra cycles to read and write if needed, instead of slowing the system frequency.

High current: NVMs also suffer from high write/read current (Fig. 2c-2d). The amount of current needed to switch the state is high (~100μA/bit for STTRAM and MRAM). For a full cache line (512-bit) write, this high current can result in supply voltage droop and ground bounce. Write/read operation can fail due to this, especially when the farthest memory bank is accessed. Therefore, read/write failure due to voltage droop and ground bounce should also be tested.

Sensitivities/Tolerances: Spintronic memories e.g., STTRAM and MRAM are susceptible to external magnetic field [16]. It is shown in [16] that STTRAM bits can be corrupted by external magnetic and/or thermal field. The reason is, data in STTRAM/MRAM is stored in the form of free layer magnetization of Magnetic Tunnel Junction (MTJ) and an external magnetic field can flip its orientation. Everspin lists the maximum magnetic tolerance for their MRAM chips during write/read/standby to be 8000A/m (~100Oe) [6]. Therefore, magnetic tolerance for spintronic memory should be tested. All NVMs are susceptible to temperature [2], [17]–[19]. Temperature variation can change

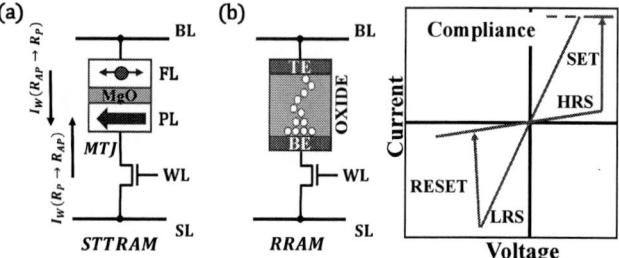

Fig. 3: (a) STTRAM bitcell; and (b) RRAM bitcell and I-V curve.

the energy barrier between memory states (in turn affecting retention) which leads to read/write and data reliability failures. Therefore, temperature tolerance for NVMs should also be tested. RRAM is also sensitive to environment (N_2/O_2/air) which requires similar attention [20].

In this work, we describe the above test challenges in NVMs, repurpose some of the conventional tests, and propose new test methods and test patterns with associated Design-for-Test (DFT) circuits to solve these challenges. Fig. 1b presents a representative test flow for NVMs. Proposed new test methods for NVMs are underlined (Fig. 1(b)). For the sake of brevity, we restrict the discussion to two flavors of NVMs namely, STTRAM and RRAM and introduce other NVMs as necessary. *To the best of our knowledge, this is the first attempt to study emerging NVM test challenges and corresponding solutions.*

In particular, we make following contributions in this paper: we (a) discuss new fault models specific to NVMs; (b) propose new DFT circuit for latency test; (c) present new test challenges introduced by NVMs; (d) describe new test methods for electrical characterization; and, (e) review existing NVM test methodologies.

In this paper, Section II presents basics of NVMs. Sections III-VI describes test for read/write latency, voltage droop/ground bounce, magnetic/thermal tolerance and retention time respectively. Section VII presents other fault models and test methods. Section VIII draws conclusions.

II. BASICS OF NVMs

In this section, we present basics of NVMs. For the sake of brevity, we restrict our discussion to two flavors of NVMs namely, STTRAM and RRAM.

A. Basics of STTRAM

STTRAM cell (Fig. 3a) contains MTJ as storage element which contains a free (FL) and a pinned (PL) magnetic layer. The resistance of the MTJ stack is high (low) if FL magnetic orientation is anti-parallel (parallel) compared to the PL. MTJ can be toggled from parallel (P) (data '0') to anti-parallel (AP) (data '1') (or vice versa) using current induced Spin-Transfer Torque by passing the appropriate write current (> critical current) from source-line to bit-line (or vice versa).

B. Basics of RRAM

RRAM contains an oxide material between Top/Bottom Electrode (TE/BE) (Fig. 3b). RRAM resistive switching is

978-1-5386-3775-3/18 $31.00 © 2018 IEEE

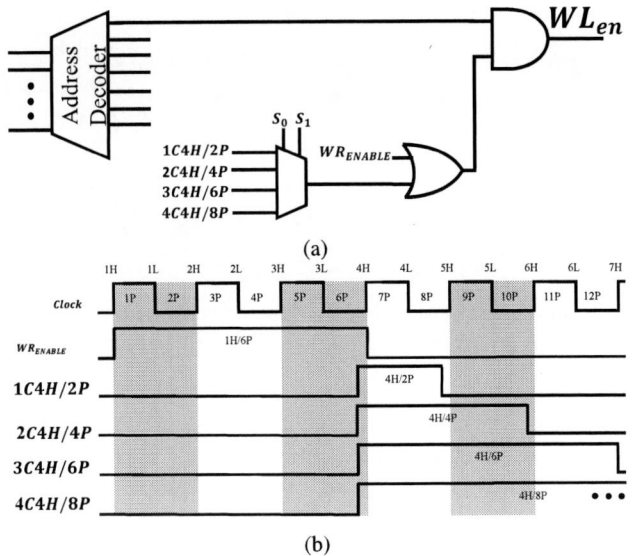

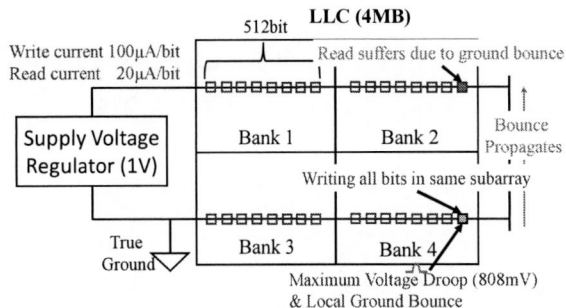

Fig. 5: STTRAM LLC write showing worst-case voltage droop and ground bounce. Ground bounce propagates to nearest banks and can cause read failure if read is performed in parallel from those banks.

chip fails for even x+5 cycle of write, discard the chip or slow down clock.

3. Repeat step 1 and 2 for other data patterns like all '1' and checkerboard pattern i.e. $\{w1 \Uparrow r1 \Downarrow\}$, $\{wCB \Uparrow rCB \Downarrow\}$ and $\{w\overline{CB} \Uparrow r\overline{CB} \Downarrow\}$. If the memory is not written with maximum cycles for all patterns, discard the chip or slow the clock.

4. Repeat step 1, 2 and 3 with certified write cycle and decrease read cycle (each time by 1) starting from (y+4) till read failure is captured. Certify the read latency with minimum cycles that ensure read passes for all patterns. If that value is more than maximum target read cycles, discard the chip.

5. If stochastic accuracy needs to be captured, repeat testing write/read for certified latency for N times (N could be 10-100). If desired accuracy is not met, discard the chip.

For test time compaction, only the worst-case write latency could be checked. For example, in case of STTRAM, writing $0 \rightarrow 1$ requires more write time than $1 \rightarrow 0$ [21].

B. Consideration to other NVMs

All NVMs suffer from wide latency distribution due to PVT. Therefore, latency test should be done for all NVMs. It should be noted that the proposed test steps for STTRAM can be extended to any other NVMs as the test is not memory specific.

IV. VOLTAGE DROOP/GROUND BOUNCE TEST

In this section, we discuss voltage droop/ground bounce due to high write current of NVMs, and propose test flow to capture corresponding failures. The discussion is focused on STTRAM. However, other NVMs are also considered.

A. Voltage Droop in NVMs and Write Failure

The write current ($\sim 100 \mu$A/bit for STTRAM/MRAM) of NVMs is high. For a full cache line (512-1024 bit) write, extremely high current (50-100mA) is drawn from the power supply. This high write current can result in droop due to the presence of interconnect resistance between the power supply source and destination which can cause write failure. This is especially true for the farthest bank of the cache [22]. Fig. 5 shows the cache organization for 4MB LLC and simulation shows that supply voltage can droop to 0.808V from 1V when writing all the bits to $Bank_4$ [22] (Fig. 6a). This voltage might not be sufficient to write the cell in desired write time and

---

Fig. 4: (a) Proposed DFT circuit for latency test; (b) timing waveform of different input signals of Fig. 4(a).

due to oxide breakdown and re-oxidation which modifies a Conduction Filament (CF). Conduction through the CF is primarily due to transportation of electrons in the oxygen vacancies. These vacancies are created under the influence of electric field due to applied voltage. The two states of the RRAM are termed as Low Resistance State (LRS) and High Resistance State (HRS). The process of switching the state to LRS (HRS) is known as SET (RESET). Fig. 3b also shows the I-V characteristics during SET/RESET process.

III. READ/WRITE LATENCY TEST

In this section, we propose test flow and DFT circuit for latency test under process and stochastic variation.

A. Test for read/write latency

Note that conventional March test can be run at multiple clock frequencies to search the frequency at which the memory array passes. However, it will increase test time as well as degrade the system performance if the clock is slowed down. We propose a DFT circuit (Fig. 4a) for latency test. Fig. 4b shows the timing waveform of different input signals of Fig. 4a. The basic idea is to increase the number of cycles allocated for write if the latency test fails. This approach will improve the test time and can also be used to mitigate write/read failures. $Write_{Enable}$ can be OR'ed with phase extension signals (1C4H/2P, 2C4H/4P, 3C4H/6P and 4C4H/8P) to increase the width of the pulse. For latency test we assume that the memory writes (reads) in x (y) clock cycle @fGHz. The steps are:

1. Write '0' to entire memory with x cycles. Then read with (y+5) cycles to verify write i.e., $\{w0 \Uparrow r0 \Downarrow\}$. Read uses more cycles to avoid read failure and capture only write failure.

2. If write fails, repeat step 1, with x+1 cycles using proposed DFT and keep increasing in case of write failure till maximum targeted clock cycle of write (let's say, x+5). If

978-1-5386-3775-3/18 $31.00 © 2018 IEEE   55

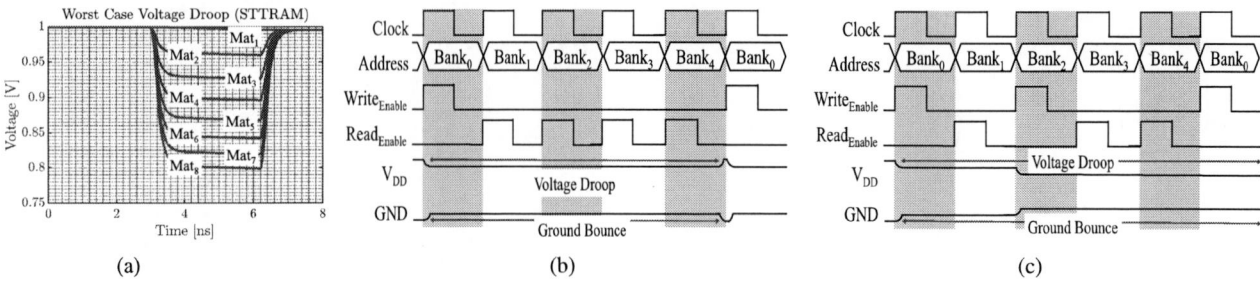

(a)   (b)   (c)

Fig. 6: (a) Worst-case voltage droop [22] when writing all the bits to the different MATs of the farthest bank ($Bank_4$) of Fig. 5; (b) four reads are initiated between two writes. We call it 1X mode; (b) three reads and one write are initiated between two writes. We call it 2X mode.

thereby write failure may occur. We need to trigger worst-case voltage droop during write to capture write failures.

B. Ground Bounce in NVMs and Write Failure

For the example shown in Fig. 5, the total write current will be dumped into the local ground of that particular sub-array. Thus, local ground voltage will bounce as there is a parasitic resistance between the local ground of that sub-array and the true ground of the chip. Due to ground bounce along with voltage droop, bitcell are going to get lower voltage headroom which might not be sufficient to write at target latency.

C. Read Failure due to Voltage Droop and Ground Bounce

NVM write latency is typically a few nanoseconds and therefore requires multiple clock cycle. For e.g., with a clock frequency of 2GHz and write (read) latency of 2.5ns (1ns), required number of clock cycles are 5 (2). However, the throughput will degrade if memory access is restricted for 5/2 cycles. Therefore, parallelism is used to perform write/read in successive cycles. This can cause read failure due to supply droop/ground bounce triggered by a parallel write to a different bank. For example, when $Bank_4$ is being written:

i) Read operations can be initiated in the next four clock cycles in other banks (Fig. 6b). These data are processed in pipeline to maintain high throughput. Due to ground bounce propagations to those banks, read operations initiated in clock cycles 2, 3, 4 and 5 will experience failure owing to, (a) poor sense margin at lower voltage; (b) $CELL_{TMR}$ loss due to higher access transistor resistance at lower word-line voltage. We call this write/read scheme 1X write.

ii) Multiple writes can be initiated with read. In Fig. 6b, there will be more current drawn from the supply which might add to the existing supply droop or ground bounce due to the second write and lead to write/read failure. We call this write/read scheme 2X write.

D. Test for Voltage Droop and Ground Bounce

Voltage droop/ground bounce test needs to validate write failure due to lower write voltage and read failure due to a parallel write to a different bank. Since write current is maximized when stored and new write data are both '0', we propose a data test pattern that causes maximum droop using $0{\rightarrow}0$ write for $(N_{col}-1)$ bits in a row and $0{\rightarrow}1$ write for

the last bit where N_{col} is the number of bits in a row. Further, when write is initiated in one bank, read needs to be performed in other bank to check for read failure. Therefore, we propose following test steps for voltage droop/ground bounce test considering 8-bit write data:

1. Identify memory banks that are farthest from the voltage regulator (potential candidates for highest droop/ground bounce). Physical design can be used to find these banks.

2. First write 0x00 to all addresses of all banks of the memory and perform read to verify write success {w0⇑r0⇓}. If write fails then discard the chip.

3. Write 0x01 (old data has all '0' and new data has a single '1', so the supply current is close to the maximum) to all addresses of the bank, identified at step-1 and verify write success by reading i.e. {wN⇑rN⇓}, where N=0x01. After initiating each write in the bank, initiate reads in other banks. If only write operation fails decrease the clock frequency or increase V_{dd} by small steps and repeat until 100% write success is achieved. If read fails along with/without write, increase V_{dd} approach is the only solution. Certify the chip with the maximum clock frequency/minimum V_{dd}. If these value does not meet target specification, discard the chip.

4. Repeat step 3 with other data patterns that have one data '1' (i.e. 0x02, 0x04, 0x08, 0x10, 0x20, 0x40 and 0x80) after performing step-2 to flush all 1's.

5. Above steps certify 1X write. If validation of 2X/3X write are needed, step 1-4 could be followed with modification of step-3 to initiate multiple writes to different banks during writing to the farthest bank (identified in step 1).

E. Consideration to other NVMs

All NVMs possess high write/read current, and are designed for high density leading to more parasitic R and C. Therefore, other NVMs will also suffer from read/write failure due to voltage droop/ground bounce. It should be noted the proposed test for STTRAM can also be extended to them.

V. MAGNETIC AND THERMAL TOLERANCE TEST

A. Magnetic Tolerance Test

FL of MTJ in STTRAM/MRAM can be flipped by external magnetic field [16]. Therefore, we propose methods to test magnetic field tolerance of spintronic memory during write, read and retention considering STTRAM as test case.

978-1-5386-3775-3/18 $31.00 © 2018 IEEE

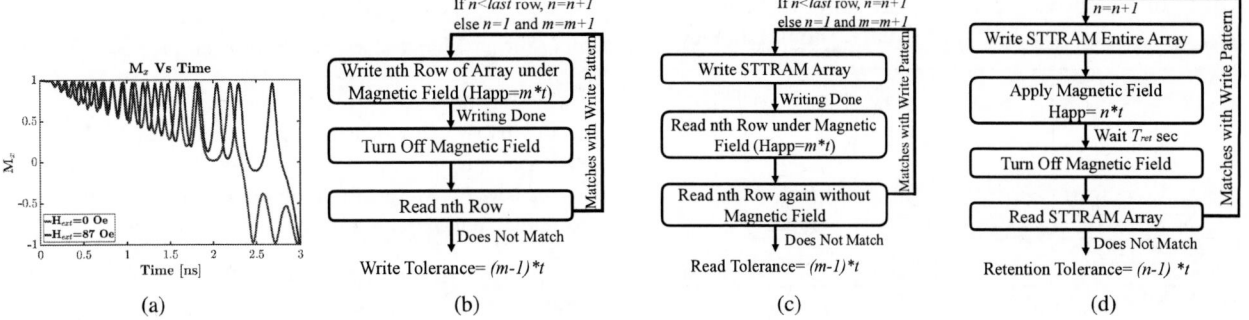

Fig. 7: (a) FL magnetic orientation (M_x) vs time for write tolerance; (b)-(d) Algorithms for write, read and retention tolerance.

Write Tolerance: This is the maximum magnetic field under which STTRAM can be written successfully at specified write current with specified write latency. Write tolerance is found to be 86Oe for write current/latency of $357\mu A$/3ns (Fig. 7a) for the STTRAM employed in this work. With an external magnetic field of 87Oe, MTJ may or may not go back to its previous state which indicates that the bit is not written successfully (write error).

Read Tolerance: This is the maximum external magnetic field under which STTRAM can be read successfully without disturbing the bits for a specific read current within a specified read latency. From simulations, we find a read tolerance of 370Oe for read current/latency of $20\mu A$/1ns.

Retention Tolerance: This is the maximum magnetic field STTRAM array can sustain without any bit-flip for a specified period of time under retention. Simulation shows a retention tolerance of 319Oe (320Oe flips the bit) for a 3ns period.

The procedure to identify the write/read/retention tolerances are shown in Fig. 7b, 7c and 7d respectively. In the above tests, the choice of step size, t, should be made judiciously to keep test time low. Magnetic tolerance of a bit for write/read/retention are dependent on the data that is being written (for write) or stored (for read/retention). For write tolerance, the worst-case is writing $0 \rightarrow 1$ (requires higher write time [21]). For both read/retention tolerance, the worst-case is when stored bit is '1'. This is because '0' (parallel) is the preferred state i.e. $1 \rightarrow 0$ transition requires less effort.

B. Thermal Tolerance Test

Tests must certify the chip in typical target temperature range (-10C to 90C):

(a) At high temperature, energy barrier between two states of the memory reduces which leads to reduction in data retention. This can also cause read failure as the resistance difference between two states reduces. Therefore, manufacturer needs to test retention and read failure at high temperature.

(b) At low temperature, the energy barrier between two states increases and therefore the write/read time increases causing write/read failures.

Test methodology: The proposed magnetic tolerance test method (Section V.A) can be used for thermal tolerance as well by applying external temperature instead of magnetic field (can be combined with standard hot/cold test). The highest

temperature at which read does not fail and data retention meets min target specification, can be used to certify the chip for the upper limit of thermal tolerance.

C. Consideration to other NVMs

Proposed magnetic tolerance test can be extended to all spintronic memories and temperature tolerance test can be extended to all NVMs.

VI. RETENTION TIME TEST

In this section, we present challenges with NVM retention test and summarize solutions proposed in prior works (with STTRAM as test case. However, the methodologies could be extended to RRAM and PCM as well).

A. Challenges in Retention Test

Conventional retention tests will incur high test time if used for emerging NVMs. Our estimate suggests that the test time of STTRAM with 1yr retention will be 1.15yr with eDRAM test approach (highest retention to total test time ratio of 86.5% is reported in [12]). However, if the retention is compressed to $\sim 3\mu s$, the test time could be improved by $\sim 1.3 \times 10^8 X$. STTRAM/MRAM retention is stochastic i.e. the same bit shows different retention when measured multiple times [13] which makes retention test even more challenging.

B. Test Time Compression

Using Temperature (for all NVMs): STTRAM cell with 1yrs of retention shows a reduced retention of 24.8s at 200C [23]. Therefore, it can be inferred that at room temperature a cell should have 1yrs of retention if it passes for the rated retention (24.8s) at 200C. Thereby, achieving $\sim 1.3 \times 10^6 X$ reduction in test time.

Using Weak-Write (for all NVMs): A disturb current passing through a NVM bitcell, lowers its retention time especially for STTRAM. Therefore, weak-write test can be used for reducing STTRAM retention test time [23], [24]. High temperature [23] and efficient DFT [23], [24] can be added for further reduction in retention time and test time.

Using External Magnetic Field (for spintronic memories): In [25], a novel Magnetic Burn-in (MBI) and MBI with high temperature (MBI+BI) are proposed to reduce retention test time for STTRAM. The idea is to apply external magnetic field antiparallel to the magnetic orientation of MTJ FL. This will reduce retention (test time) time of the cell (chip).

978-1-5386-3775-3/18 $31.00 © 2018 IEEE

TABLE II: Comparison between MBI, MBI+BI and Weak-Write based Retention Test Time Compression Methodology

	MBI (220Oe @ 25C) [25]	MBI+BI (220Oe @ 125C) [25]	Weak-write (w/ 184μA) [23]	Weak-write (w/ 184μA) [24]
Reduced ret. (/bit)	~41.7μs	~3.03μs	~162.8μs	~162.8μs
Test time w /LS	~3.42s	~0.25s	~13.33s	-
Test time w /EMACS	-	-	-	~1min
DFT overhead	Minimal	Minimal	Yes	Yes
Extra write power	No	No	Yes	Yes

C. Comparison of Different Proposed Schemes

We compare MBI/MBI+BI [25] with weak-write based retention test time compression methods proposed in [23], [24] for an 64MB memory with a word size of 512-bits and original retention of 10yrs (Δ=60KbT). For calculating test time using Linear Search proposed in [23] for MBI, MBI+BI and [23], the considered resolution is 10ns, read/write latency is 1ns/1ns, number of rows written simultaneously is 128 and number of iteration is 10^3. For EMACS [24], number of rows written simultaneously is considered as 500. The results are summarized in Table I. MBI+BI produces lowest test time (0.25s). However, MBI/MBI+BI requires a magnetic chamber with read/write capability and incur minimal DFT changes. The weak-write based methods are power hungry (constant disturb current) and require major DFT changes.

D. Considerations for Other NVMs

Retention time of all NVMs depend on temperature [2], [17]–[19]. It should also be noted that retention of any NVM bitcell can be reduced by passing current in the direction that will flip the stored data. Therefore, both temperature-based technique proposed in [23] and weak-write based techniques proposed in [23], [24] for STTRAM can be extended to other NVMs as well. However, magnetic field-based technique [25] can be extended to only spintronic memories like MRAM.

VII. DISCUSSION

Sensitivity to environment: RRAM resistance of two states (R_{LRS} and R_{HRS}) and the forming voltage depend on gases e.g., N_2, O_2 and air [20]. If the chip is hermetically sealed, RRAM will not suffer from these. However, it can still be exposed to these elements (e.g., O_2 generated from SiO_2). Therefore O_2/N_2 chambers are required to test NVMs e.g., RRAM to certify such sensitivities.

Sneak-path testing: Sneak-path current is a major issue in high density crossbar memory which leads to leakage though unselected cells and read failure due to reduced sense margin. In [26], a sneak path test method and DFT circuit is proposed. The DFT circuit controls sneak path number that are concurrently enabled and compares sneak-path to ideal current to detect different faults (e.g. stuck at, coupling etc.).

Other fault models: Different fault models like stuck at faults, coupling faults, transition fault, incorrect read fault, read disturb fault etc. are investigated in prior works [26],

[27]. Sneak-path testing [26] and test pattern to activate the faults (during March test) [27] have been proposed.

VIII. CONCLUSION

In this paper, we analyzed test challenges of NVMs, repurposed conventional test techniques and proposed added steps to cover new NVM-specific faults. We also proposed new test algorithms and DFT circuits to facilitate the tests.

ACKNOWLEDGMENT

This work is supported by Semiconductor Research Corporation (#2727.001), National Science Foundation (#CNS-1722557, #CCF-1718474 and #DGE-1723687) and DARPA Young Faculty Award (#D15AP00089).

REFERENCES

[1] Nigam, A., et al. "Delivering on the promise of universal memory for spin-transfer torque RAM", ISLPED, 2011.
[2] Worledge, D. C., et al. "Switching distribution and write reliability of perpendicular spin torque MRAM", IEDM, 2010.
[3] Wu, Yi., et al. "Recent progress of resistive switching random access memory (RRAM)", SNW, 2012.
[4] Pirovano, A., et al. "Low-field amorphous state resistance and threshold voltage drift in chalcogenide materials", TED, 2004.
[5] Kang, Y. M., et al. "The challenges and directions for the mass-production of highly reliable, high density 1T1C FRAM", ISAF, 2008.
[6] https://www.everspin.com/file/882/download.
[7] https://www.everspin.com/file/882/download.
[8] https://ark.intel.com/products/97544/Intel-Optane-Memory-Series-16GB-M_2-80mm-PCIe-3_0-20nm-3D-Xpoint.
[9] http://www.cypress.com/file/140901/download.
[10] Kim, K., et al. "Transient Cell Supply Voltage Collapse Write Assist Using Charge Redistribution", TCAS II, 2016.
[11] Mukhopadhyay, S., et al, "SRAM Write-Ability Improvement with Transient Negative Bit-Line Voltage", TVLSI, 2011.
[12] Yang, H.-Y., et al. "Testing Methodology of Embedded DRAMs", TVLSI, 2012.
[13] Brown, W.F., "Thermal fluctuations of a single-domain physics", JAP, 1963.
[14] Karl, E., et al. "A 4.6 GHz 162Mb SRAM Design in 22nm Tri-gate CMOS Technology with Integrated VMIN-enhancing Assist Circuitry", ISSCC, 2012.
[15] Motaman, S. et al. "Impact of Process-Variations in STTRAM and Adaptive Boosting for Robustness", DATE, 2015.
[16] Jang, J. W. et al. "Self-correcting STTRAM under magnetic field attacks", DAC, 2015.
[17] Fang, Z., et al. "Temperature Instability of Resistive Switching on HfOx-based RRAM Devices", EDL, 2010.
[18] Russo, U., et al. "Analytical Modeling of Chalcogenide Crystallization for PCM Data-Retention Extrapolation", TED, 2007.
[19] Rodriguez, J. A., et al. "High Temperature Data Retention of Ferroelectric Memory on 130nm and 180nm CMOS", IMW, 2016.
[20] Ke, Jr-J., et al. "Surface effect on resistive switching behaviors of Zno", APL, 2011.
[21] Ghosh, S., et al., "Security and privacy threats to on-chip Non-Volatile Memories and countermeasures", ICCAD, 2016.
[22] Aluru, R. K., et al. "Droop Mitigating Last Level Cache Architecture for STTRAM", DATE, 2017.
[23] Iyengar, A., et al. "Retention Testing Methodology for STTRAM", in IEEE Design & Test, 2016.
[24] Yoon, Insik, et al. "EMACS: Efficient MBIST Architecture for Test and Characterization of STTRAM Arrays", ITC, 2016.
[25] Khan, M. N. I., et al. "Novel magnetic burn-in for retention testing of STTRAM", DATE, 2017.
[26] Kannan, S., et al. "Sneak-Path Testing of Crossbar-Based Nonvolatile Random Access Memories", IEEE TNano, 2013.
[27] Chintaluri, A., et al. "A Model Study of Defects and Faults in Embedded Spin Transfer Torque (STT) MRAM Arrays", ATS, 2015.

Special Session on Reliability and Vulnerability of Neuromorphic Computing Systems

Shimeng Yu
Electrical Engineering and
Computer Engineering
Arizona State University
Tempe, AZ
shimeng.yu@asu.edu

Chenchen Liu
Electrical and Computer
Engineering
Clarkson University
Potsdam, NY
chliu@clarkson.edu

Wujie Wen
Electrical and Computer
Engineering
Florida International University
Miami, FL
wwen@fiu.edu

Yiran Chen
Electrical and Computer
Engineering
Duke University
Durhma, NC
yiran.chen@duke.edu

Abstract—**This is the summary of the special session on reliability and vulnerability of neuromorphic computing systems.**

Keywords—neuromorphic, reliability, vulnerability

I. INTRODUCTION

This special session highlights some latest studies on the reliability of neuromorphic computing system (NCS). In specific, three invited speakers will respectively 1) summarize reliability effects of resistive memory (ReRAM) based NCS; 2) discuss enhancing schemes of the reliability of ReRAM based NCS; and 3) present vulnerabilities of deep learning systems to various security attacks and the relevant solutions.

II. INVITED TALKS

A. *(Talk 1) Reliability Effects of Resistive Synaptic Devices on Neuromorphic Computing System Performance.*

Presenter: Shimeng Yu, Arizona State University

Abstract: Resistive synaptic devices are attractive for the replacement of SRAM in the hardware implementation of artificial neural networks (ANNs). However, one of the challenges for this emerging technology is the reliability failures due to data retention and write endurance degradation. This work investigates the impact of these two failures in the multilayer perceptron (MLP) using the NeuroSim+ simulator. For the retention failure in offline classification, we consider various possible conductance drift scenarios. The results confirm that faster degradation on the classification accuracy is highly correlated with larger deviation in the weighted sum. For the endurance failure in online learning, the strength of conductance tuning is assumed to become weaker overwrite pulse cycles. The analysis suggests that the learning accuracy is less impacted because the network is able to adapt itself and activate more synapses to participate in the weight update when the tuning capability of synapses are weakened.

B. *(Talk 2) Reliability Analysis and Enhancement of the ReRAM based Neuromorphic Systems*

Presenter: Chenchen Liu, Clarkson University

Abstract: Deep Neural Networks (DNNs) has gained immense success in the cognitive application and greatly pushed today's artificial intelligence forward. The biggest challenge in executing DNNs is their extremely data-extensive computations. Computing efficiency of speed and energy in the traditional computing platform is hence constrained. Neuromorphic computing has emerged as a promising solution to solve this challenge own to their high computation and communication efficiency. The resistive memory (ReRAM) based synaptic network has been widely investigated and applied to the neuromorphic computing system with fast computation and low design cost. As the resistive device (or memristor) continue to mature and achieve higher density, the bit failure within crossbar array can become a critical issue. These can degrade the computation accuracy significantly. In our research, we explored practical solutions to rescue the neuromorphic hardware from device defects, and hence enhance the neuromorphic system reliability. The significance of synaptic weights was evaluated theoretically and statistically initially, and then a retraining and a remapping design in algorithm and hardware level were developed. The evaluations based on real-device testing data showed that our design can recover 99.3% classification accuracy on MNIST and 96.2% accuracy on CIFAR-10 when 20% random defects were considered.

C. *(Talk 3) Exploiting Deep Learning System-level Vulnerabilities from the Intelligent Supply Chain*

Presenter: Wujie Wen, Florida International University

Abstract: Deep Neural Network (DNN) has recently become the ``de facto" technique to drive the artificial intelligence industry. However, there also emerge many security issues along with the proliferation of DNN based intelligent systems. Existing DNN security studies, such as adversarial or poisoning attacks, usually narrowly conduct at the software algorithm level, with the misclassification as their primary goal. In this talk, we will discuss the DNN security problems from another angle—the more realistic system-level attacks introduced by the emerging intelligent supply chain across DNN hardware accelerators, computing frameworks and algorithmic models. A novel and practical neural Trojan attacking framework capable of various malfunction, as well as its ``Prototype" on the NVIDIA Jetson TX2 platform, will be presented. We will also discuss potential defense techniques to mitigate such system-level attacks.

IP Session on ISO26262 EDA

Art Schaldenbrand,Cadence, UK
Yervant Zorian, Synopsis, USA
Stephen Sunter, Mentor, USA
Peter Sarson, Dialog Semiconductor (organizer) , UK

I. INTRODUCTION

This IP session will primarily focus on how ISO26262 is being used in conjunction with fault injection to measure the fault coverage of analogue circuits; however, one talk will discuss how ISO26262 is being to determine the functional safety aspect of digital circuits on advanced process nodes.

II. THE CHALLENGE OF FUNCTIONAL SAFETY IN AMS DESIGN. (ART SCHALDENBRAND)

III. ISO 26262 FUNCTIONAL SAFETY SUBSYSTEM FOR AUTOMOTIVE SOCS (YERVANT ZORIAN)

In this talk, we will discuss the latest Synopsis toolset for dealing with Analog Fault Simulation and fault injection into A/MS devices.

IV. MEASURING ISO 26262 METRICS AND FAULT COVERAGE SIMULTANEOUSLY FOR A/MS CIRCUITS (STEPHEN SUNTER)

There are five key metrics specified by ISO 26262: single point fault metric (SPFM), probability metric for random hardware failures (PMHF), latent fault metric (LFM), and diagnostic coverage with respect to residual and latent faults (DC-Res, DC-Lat). These can all be measured automatically for a circuit design, effectively in a single step, while simultaneously measuring fault coverage of tests. Results are presented for a mixed-signal function comprising cells from the ITC'17 analog/mixed-signal benchmark circuits.

2018 IEEE 36th VLSI Test Symposium (VTS)

An Inter-Layer Interconnect BIST Solution for Monolithic 3D ICs

Abhishek Koneru and Krishnendu Chakrabarty
Department of Electrical and Computer Engineering, Duke University, Durham, NC, USA

Abstract—Monolithic three-dimensional (M3D) integration offers higher-density integration compared to 3D integration based on through-silicon vias. Advances in testing are however needed to screen defects in M3D integration. We propose a built-in self-test solution to target shorts and opens in ILVs. In the proposed solution, scan cells at the interface of two layers are stitched into a twisted-ring counter (TRC) using their functional outputs and ILVs. The interface-register cells launch and capture tests, and a test path consists of ILVs and a multiplexer. We map the problem of minimizing the length of the wires added to stitch the TRC to that of finding a minimum-cost Hamiltonian circuit in a weighted bipartite graph. Since the weighted Hamiltonian circuit problem is NP-Complete, we propose a heuristic algorithm for this problem. Simulation results show that the proposed test solution can detect all opens and shorts.

I. INTRODUCTION

It is becoming increasingly difficult to sustain device scaling in an economically viable manner due to challenges associated with interconnect scaling, lithography of small features, and process variations. The semiconductor industry is therefore exploring three-dimensional (3D) integration to reduce power consumption with smaller device footprint, and higher performance [1]. In today's 3D integration process, separately manufactured dies/wafers are integrated into the same package, and through-silicon-vias (TSVs) are used to connect dies to each other. However, the keep-out-zone (KOZ) required for TSVs and limitations on the die alignment precision impose limits on the device integration density. A minimum KOZ of 3 μm is required at the 20 nm technology node [2], and the die alignment precision is currently limited to 0.5 μm [3].

Monolithic 3D (M3D) integration is an emerging technology in which transistors are processed layer-by-layer on the same wafer. It is attractive for 3D integration because it has the potential to achieve higher device density compared to TSV-based 3D stacking [4]. Sequential integration of transistor layers requires high-density vertical interconnects, known as the inter-layer vias (ILVs). The size and pitch of an ILV are one to two orders of magnitude smaller than those of a TSV [5]. In order to realize such high-density vertical interconnects, the inter-layer dielectric (ILD) thickness is being aggressively scaled [6, 7].

In the M3D fabrication flow, the bottom-layer transistors and their associated interconnects are first processed using a standard high-temperature process, e.g., silicon-on-insulator (SOI), FinFET, or bulk CMOS technology [8]. Next, a thin silicon layer is created over the bottom layer [8, 9]. Fig. 1 presents an illustration of silicon layer deposition over a base layer using SmartCut [15]. The top-layer transistors are then processed under a strict thermal budget [8]. Finally, ILVs are processed to connect the two layers. The last three steps are repeated for additional layers.

Sequential integration is enabled by: (i) a low-temperature process to create a thin silicon film over the bottom layer, and (ii) a process to realize top-layer transistors without damaging the underlying interconnects and degrading the bottom-layer transistors. Several approaches have been proposed in the literature to create a thin silicon film at low temperatures, e.g., the modified Smart-Cut process [15] originally used to manufacture SOI wafers. The M3D technology developed by CEA LETI (CoolCube^{TM}) has attracted the attention of companies such as IBM, ST Microelectronics, and Qualcomm [8]. The number of layers (both functional and test) can be limited by the likelihood of performance degradation due to the high-temperature processing steps. However, techniques to prevent damage have been proposed in the literature [16]. For example, shielding layers can protect processed layers from damage due to the high-temperature steps.

Although ILVs are comparable in size and pitch to conventional vias, they differ significantly in terms of fabrication steps and yield. The yield for processing ILVs is speculated to be much lower compared to a conventional via. Yield learning is especially important for a new technology such as M3D integration. Due to aggressive scaling of ILD thickness, the ILVs are prone to shorts, opens, and delay defects [6, 10]. Incomplete metal filling or voids in an ILV increase resistance and impact circuit timing [6, 11]. Large voids in the ILV can result in an open path, thereby causing a catastrophic failure. Impurities may increase resistance and interconnect delay. ILVs are also affected by voids that arise in the ILD since they are processed after the wafer-bonding step. Such voids lead to an increase in ILV resistance. If the size of a void in the ILD is large enough to impact two neighboring ILVs, then it leads to a resistive short between the two ILVs. The resistance of the defect depends on the thickness of the void.

Conventional scan-based test pattern generation cannot be re-used to test ILVs since they do not provide the diagnostic resolution to diagnose and characterize ILV faults. We may not be able to pinpoint to a specific ILV using conventional scan-based diagnosis since it may produce large number of candidates for one failure. Pre-bond TSV testing methods such as [12] is infeasible for ILVs since bare ILVs cannot be be exposed, and the state-of-the-art wafer-probe technology cannot support the pitch requirement for ILVs (100 nm to 200 nm). Techniques proposed to carry out post-bond TSV testing can be extended to M3D for post-assembly testing. For example, the proposed IEEE Std. P1838 [13], which defines a standardized architecture to test the TSVs, can be extended to M3D and used to test the ILVs. However, a wrapper cell has to be inserted at both ends of a TSV in order to test that TSV using P1838. A similar memory element is needed at both ends of an ILV in order to test an ILV by extending P1838 to M3D ICs. However, the overhead due to the insertion of memory elements on all ILVs can be significant since the number of ILVs in M3D ICs is expected to be an order of magnitude higher compared to TSV-based 3D ICs.

Due to these differences and the difficulty in extending TSV test methods to ILVs, we propose a new post-assembly test solution. The proposed solution assumes a DfT architecture based on dedicated test layers. In the proposed ILV built-in self-test (BIST) solution, interface-register cells in a test

978-1-5386-3775-3/18 $31.00 © 2018 IEEE

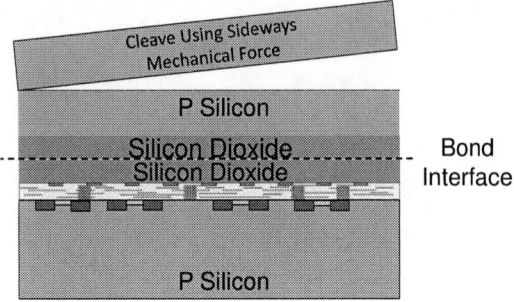

Fig. 1: Illustration of silicon layer deposition over a base layer.

layer are stitched into a twisted-ring counter (TRC) using their functional outputs and the ILVs. The interface-register cells are connected to form a long shift register, and the complement of the output of the last interface-register cell is connected to the input of the first interface-register cell.

A test is launched from one interface-register cell and captured on another interface-register cell. A metal wire is used to connect those two interface-register cells through a multiplexer. The multiplexer is inserted to switch between functional and ILV-test modes. Therefore, it is important to minimize the length of wires used to connect interface-register cells so that the impact on functional performance is minimized. The problem of minimizing the wire length can be mapped to the problem of finding a minimum-cost Hamiltonian circuit in a complete bipartite graph. As explained in Section III, the wire-length minimization problem is different from the scan-chain routing problem, which has traditionally been mapped to the traveling-salesman problem [14]. The weighted Hamiltonian circuit problem is NP-Complete on complete bipartite graphs. Therefore, we propose a heuristic technique for solving this problem efficiently.

We show that the proposed solution detects all hard opens and shorts in the ILVs. We validated the detection of all hard opens and shorts using HSpice simulations. The worst-case time complexity of the proposed wire-length minimization heuristic is $O(n^3)$, where n is the number of interface-register cells. We also highlight the effectiveness of the proposed heuristic by comparing the total wire length required to form a TRC using the proposed method with a baseline that connects the interface-register cells in a greedy manner. The total wire length required to form a TRC using the baseline is 2% to 14% higher compared to the proposed method.

The rest of the paper is organized as follows. Section II presents an overview of a recently proposed DfT solution based on dedicated test layers; the interconnect BIST method prsented in this paper is based on this DfT technique. Section III describes and evaluates the proposed BIST solution. Finally, Section IV concludes the paper.

II. DfT Architecture Overview

We consider M3D ICs that contain at least two functional layers. We refer to the layer that is processed first (last) as the bottom (top) layer. Without any loss of generality, we make the assumption that the external I/O pins are located on the top layer. The signals going from a particular layer to another layer in the direction of external I/Os is referred to as the *primary interface*, and the collection of signals going to a layer in the direction opposite to the external I/Os is referred to as

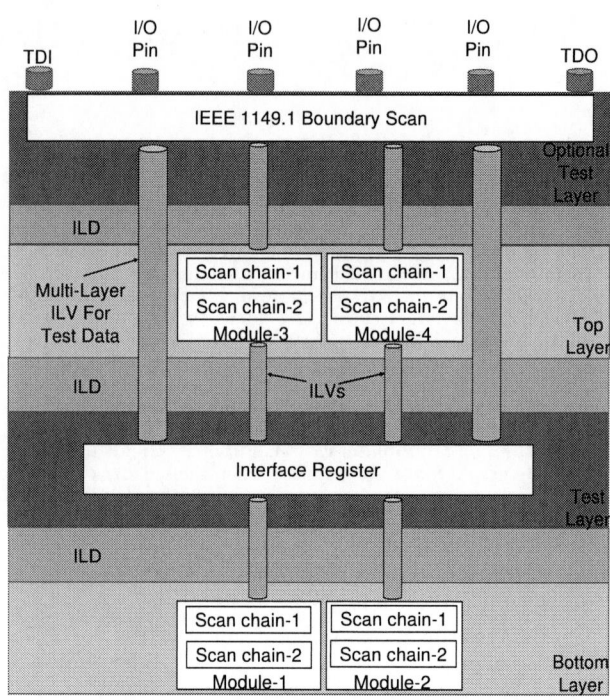

Fig. 2: Overview of the M3D DfT architecture [17].

a *secondary interface* corresponding to this layer.

The DfT architecture assumed in this paper is based on dedicated test layers, which provide controllability and observability to the signals at the interfaces of functional layers [17]. A register ("interface register") is added to the test layer to control and observe the signals at the interface of two functional layers. By placing the interface register on a dedicated test layer, additional any test structures are not added to the functional layer apart from conventional test structures.

An overview of this architecture is presented in Fig. 2 for two functional layers and two dedicated test layers. The external I/Os, located on the uppermost test layer, are wrapped by IEEE Std. 1149.1. The test layer for the uppermost functional layer is optional, and IEEE Std. 1149.1 boundary scan can be placed in the uppermost functional layer itself. A limited number of additional pins are required for boundary scan, of which two (TDI and TDO) are shown. The functional layers have intra-layer DfT such as scan-chains, test data CoDeCs, IEEE Std. wrappers, and test access mechanisms. However, all the layer-level DfT is located in the test layer, thus facilitating the reuse of test structures. In addition, the test layer can be common for multiple functional layers if they are processed using the same technology and do not require isolation.

A register is added to the test layer to control and observe the signals at the interface of two functional layers. We refer to this register as the interface register. By placing the interface register on a dedicated test layer, we are not adding any test structures to the functional layer apart from conventional test structures. Note that the proposed solution will not replace scan-based test. It will just aid in pinpointing faulty ILVs. Moreover, the impact on functional performance due to the interface register is not significant since the structure of an interface register is similar to that of a IEEE Std. 1500 wrapper. The operating modes supported by a test layer are:

978-1-5386-3775-3/18 $31.00 © 2018 IEEE

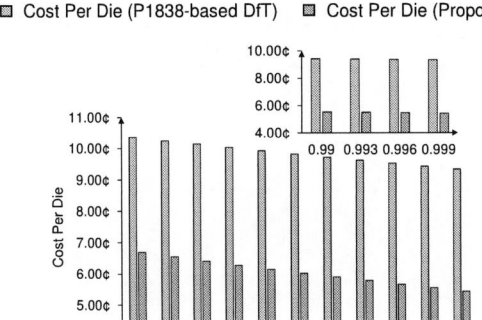

Fig. 3: Cost per die of an M3D IC using data for several ILV process yield values for b19.

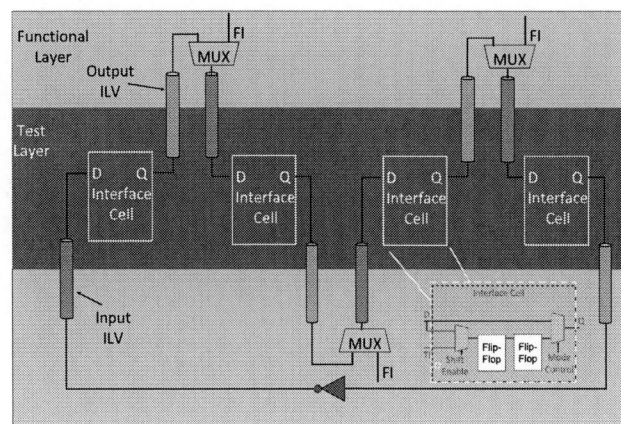

Fig. 4: Illustration of a four-bit TRC using interface cells.

1) *Partial-assembly test*: test access via dedicated probe pads;
2) *Post-assembly test*: test access via external I/O pins;
3) *Primary test*: test the functional layer for which the test layer forms the primary interface;
4) *Secondary test*: test the functional layer for which the test layer forms the secondary interface;
5) *Bypass*: only the bypass register is included in the test access chain;
6) *Interconnect test*: test the interface register;
7) *Turn*: the test-access mechanism (TAM) turns upwards from the test layer;
8) *Elevate*: the TAM goes downwards towards the next layer.

In [17], we compare the cost of this DfT architecture with the cost of a potential solution based on P1838. For P1838, we assume that a wrapper cell, which requires at least three gates [13], is added at the two ends of each ILV. We also assume a TAP interface for the external I/Os, located on the uppermost layer. A wrapper cell based on 1149.1 requires at least four gates [18].

We performed the cost comparison analysis for an IWLS benchmark. In [19], block-level partitioning of this benchmark into two layers was performed, and the physical design details presented in that paper are used for our analysis. We carried out the cost comparison analysis using data obtained for an advanced foundry process. Fig. 3 presents the cost-per-die of the b19 benchmark as function of the ILV yield. For b19, we observe that the savings in cost-per-die is as high as 40% for this DfT solution compared to the P1838-based solution.

More details about this DfT architecture and its cost effectiveness can be found in [17].

III. PROPOSED ILV BIST

At-speed transition-based delay-fault testing facilitates the detection of hard opens/shorts as well as resistive defects. A new test solution is required to enable at-speed testing of ILVs. Test solutions for TSV-based 3D ICs cannot be extended to M3D because of the following reasons. First, inserting a boundary register in every layer to test the ILVs as stand-alone objects is not efficient for M3D since the number of ILVs in M3D ICs is expected to be an order of magnitude higher. Second, testing an ILV as a part of shore logic (e.g., in [20]) by launching a transistion along an inter-layer path is not desirable since it requires significant changes to test-generation flows.

The size of the Extest chain will be prohibitively large since scan flip-flops from the functional layers will be part of that chain, in addition to interface-register cells from the dedicated test layers. Moreover, test pattern generation for the ILVs may become harder since a higher number of care bits are required to justify a value to an ILV; these additional constraints will make it more difficult to obtain high fault coverage, and also result in an increase in the number of test patterns.

We propose an ILV BIST solution that enables stand-alone testing of the ILVs without incurring significant area overhead. In the proposed solution, interface-register cells are stitched into a TRC using their functional outputs and ILVs. Fig. 4 shows a four-bit TRC formed using interface-register cells. ILVs connected to the input and output of an interface-register cell are defined as the input and output ILVs, respectively. In a TRC, the inverse of the output of the last cell is connected to the input of the first cell. Such a counter circulates a stream of ones followed by zeros around the ring. For example, in a counter with an initial register value of 0000, the repeating pattern is: 0000, 1000, 1100, 1110, 1111, 0111, 0011, 0001, and 0000. Once a TRC is initialized to a specific pattern, the pattern in the counter repeats every $2N$ clock cycles, where N is the number of cells in the counter. We show later that we can detect all opens and shorts in the ILVs by using these patterns. We also evaluate the size of the delay defect that can be detected by using a TRC.

In order to implement a TRC using interface-register cells, a multiplexer is inserted along with an output ILV to switch between functional and interconnect-test modes. The control signal for the multiplexers is generated in the dedicated test layer and fed to the functional layers. Therefore, two additional ILVs per test layer are required to enable ILV BIST. Since a test is launched from one interface-register cell and captured on another interface-register cell, interface-register cells have to equipped with at-speed launch and capture capability.

For a TRC, we define two types of operations–shift and functional. A shift operation is defined as a sequence of shifts controlled by the clock until the contents of the TRC are read out from the TDO pin. In a functional cycle, the counter is placed functional mode and clocked for one cycle. The TRC is initialized to a known value by shifting in data through the interface register. The counter is clocked at functional frequency for some clock cycles and the pattern at the end of functional cycles is shifted out through the interface register. In

978-1-5386-3775-3/18 $31.00 © 2018 IEEE   63

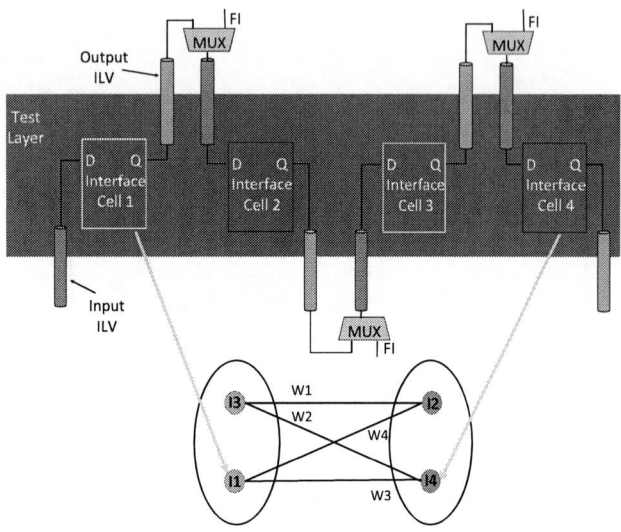

Fig. 5: Illustration of mapping a set of input and output ILVs to a complete bipartite graph.

order to minimize the impact of test circuitry on performance, it is essential to minimize the length of wires used to connect interface-register cells into a TRC. This reduces capacitive-loading effects on the nets connected to the ILVs. Subsequently, we refer to the problem of minimizing the total length of the wires used to form the TRC as the wire-length minimization problem. We next map the wire-length minimization problem to finding a minimum cost Hamiltonian circuit in a complete bipartite graph.

A. Problem Description

In order to implement a TRC using interface-register cells, we need to connect an output ILV to an input ILV. If the ILVs are treated as nodes in a graph and an edge between two nodes is used to denote a valid connection between two ILVs (input ILV and output ILV), then the resultant graph is bipartite. A *bipartite graph* is a graph whose vertices can be divided into two disjoint and independent sets such that every edge connects a vertex in one set to one in another. Fig. 5 shows an example of a complete bipartite graph. The Manhattan distance between an input ILV and an output ILV is used to represent the weight of an edge between the corresponding ILVs. The wire-length minimization problem then maps to finding a minimum cost Hamiltonian circuit in the weighted bipartite graph. The wire-length minimization problem is different from scan-chain routing problem since we can only visit an input (output) ILV from an output (input) ILV. There is no such constraint for the scan-chain routing problem, which has traditionally been mapped to the traveling-salesman problem [14]. We have shown that the weighted Hamiltonian circuit problem is NP-Complete for complete bipartite graphs [21].

B. Detectability of Opens and Shorts

A short between two wires is commonly referred to as a bridging fault. It can be detected by driving opposite values on the two interconnects that make up the bridge. Let us consider an n-bit TRC. We assume that the TRC is initialized to $b_1, b_2,, b_n$. We know that an n-bit TRC has a period of $2n$ functional cycles, i.e., the pattern repeats every $2n$ functional cycles. We can stop the functional clock after a pre-determined number of cycles and use shift operations to read out the contents of the TRC. The shifted-out data can then be compared

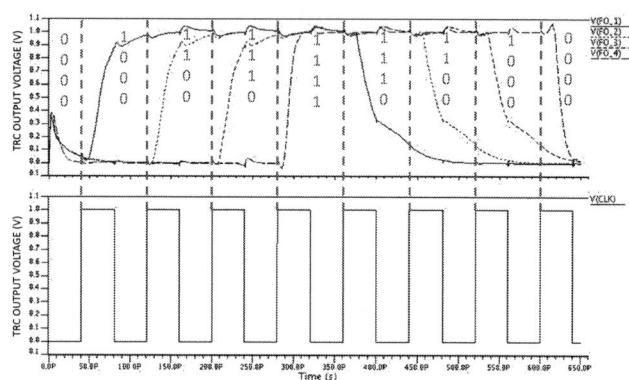

Fig. 6: Simulation result for a fault-free four-bit TRC. with the expected response in order to detect a fault.

We first show using an example that not all shorts can be detected if a shift operation is performed only once after $2n$ functional cycles. Assume that a short arises between the ILV connected to the input of the first interface cell, and the ILV connecting the nth interface cell and the inverter. This fault shorts the inverter, i.e., the value from the nth interface cell is directly fed to the first cell instead of being inverted. Since the inverter is shorted because of the fault, the value from the nth interface is fed to the first interface in each functional cycle without being inverted. The period of the counter becomes n, i.e., the pattern repeats after every n cycles. Therefore, the pattern after $2n$ functional cycles is the same as the initial pattern. As a result, the fault is not detected if a shift operation is carried out after $2n$ cycles.

The following theorem shows that all shorts are detected if the shift operation is performed twice in a specific manner. The proof is given in [21].
Theorem 1: The sequence of patterns generated by the TRC detects all shorts between ILVs if the contents of the TRC are shifted-out after n and $2n$ cycles.

An open tends to behave like a stuck-at fault when the ILV is used to connect gates, which is true for gate-level M3D. We next prove that the patterns generated by a TRC can detect all opens. First, we show using a counter-example that not all opens are detected if a shift operation is performed only once after $2n$ functional cycles. Assume that an open arises on the ILV connected to the input of the first interface cell. This fault disconnects the inverter from the first interface cell. Since the inverter is open because of the fault, the value from the nth interface is not inverted and fed to the first interface in each functional cycle. The TRC is no longer periodic and the fault may not be detected if a shift operation is carried out after $2n$ cycles.

The next theorem shows that all opens are detected if the shift operation is performed twice. The proof is given in [21].
Theorem 2: The sequence of patterns generated by the TRC detects all opens on ILVs if the contents of the TRC are shifted-out after n and $2n$ cycles.

C. HSpice Simulation Results

We carry out HSpice simulations on a four-bit TRC designed using interface cells. The simulations were carried out using a 45 nm predictive technology model [22]. The schematic of a four-bit twisted-ring counter is shown in Fig. 4. The mode control signal in each interface-register cell is assigned a DC value such that the each cell operates in functional mode (FI is selected). The counter is initialized to the all-zero state. Such a TRC circulates a stream of ones followed by zeros around the ring (Fig. 6).

We next inject opens and shorts in the TRC. Fig. 7(a) presents simulation results for the TRC after an open is injected at the output of the first cell. We can see that the pattern after four and eight functional cycles is 1000, which is different from the expected patterns (1111 and 0000). Therefore, this fault is detected. Fig. 7(b) presents the simulation results after a short is injected between the outputs of the first and second cell. The pattern after four and eight functional cycles is 0000, which is different from the expected patterns (1111 and 0000). Therefore, this fault is detected as well.

D. Wire-Length Minimization Heuristic

Since the wire-length minimization problem is NP-Complete, we propose an efficient heuristic to solve this problem. The proposed heuristic can be used to carry out wire-length optimization during the physical design phase of an M3D IC. Fig. 8 describes the proposed heuristic. First, we balance the number of input and output ILVs by carrying out k-means clustering on the set that has higher number of nodes. We need to balance the number of input and output ILVs since a Hamiltonian circuit cannot be found in an unbalanced bipartite graph. We represent each cluster by a representative node at the center of the cluster. We then construct a balanced-complete-weighted bipartite graph using the resulting set of input and output ILVs. Next, we find a Hamiltonian circuit in the balanced bipartite graph by using a well-known heuristic proposed in [1]. In this heuristic, a minimum-cost matching consisting of edges that connect input and output ILVs is found using a heuristic for maximum-weighted bipartite matching (MWBM). A minimum-cost tour that visits all the input ILVs is also found using a heuristic for the traveling salesman problem (TSP). The resulting Hamiltonian circuit consists of visiting the input ILVs in the sequence specified by the tour and using the matching edges to make sure that an output ILV is visited after an input ILV.

The worst-case computational complexity of k-means clustering algorithm is $O(nki)$, where n is the number of vertices, k is the number of clusters, and i the number of iterations needed until convergence. The number of iterations until convergence is often small, and the results only improve slightly after the first 10 iterations. Therefore, the number of iterations until convergence is limited to 10 in most implementations of k-means clustering algorithm. As a result, we can consider i as a constant in our analysis. We then obtain the runtime of the k-means clustering algorithm as $O(n^2)$ as the number of clusters can be at most $n-1$. We use the Munkres algorithm [23] to find minimum cost matching for the balanced-complete-weighted bipartite graph. The run time of the Munkres algorithm is $O(n^3)$ in the worst case. We then use the Christofides algorithm to find a minimum cost tour that traverses through all the vertices in one set in the balanced bipartite graph. The worst-case run time of the Christofides algorithm is $O(n^3)$ [24]. Finally, we obtain the minimum-cost Hamiltonian circuit by visiting the ILVs in the sequence specified by the tour and using the matching edges to make sure that an output ILV is visited after an input ILV. The worst-case runtime of this step is $O(n)$. Therefore, the runtime of the proposed heuristic is $O(n^3)$ in the worst case.

We utilize four IWLS'05 benchmarks (Table I) to evaluate the effectiveness of the proposed algorithm. These benchmarks are assumed to constitute the bottom layer in an M3D IC; therefore, each I/O port corresponds to an ILV. Moreover, since the data distributed with these benchmarks do not contain the position information for I/Os (i.e., ILVs in this case), two types of placement configurations are considered. In Configuration 1 (random placement), all ILVs are randomly placed on each benchmark. In Configuration 2 (uniform placement), all ILVs are uniformly placed, where the distances from one input or output ILV to its adjacent output or input ILVs are the same. Therefore, we do not modify the IWLS benchmarks. We only utilize the die area and I/O information to generate different ILV placements for each benchmark.

We generate the distance matrices for each benchmark

TABLE I: Design details of IWLS benchmarks.

Benchmark	Input Ports	Output Ports	Die Footprint (mm^2)
vga_lcd	89	109	0.123
ethernet	96	115	0.079
DMA	686	262	0.017
wb_conmax	1130	1416	0.031

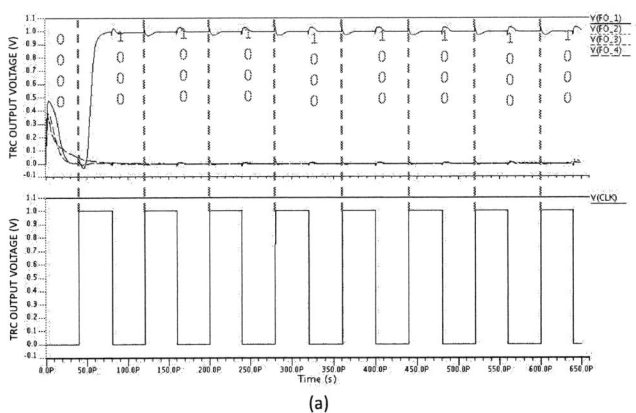 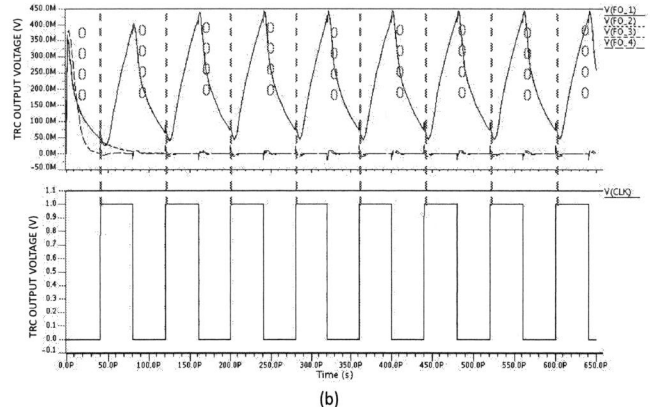

(a)   (b)

Fig. 7: Simulation results after fault injection: (a) open on output of first cell, and (b) short between outputs of first and second cell.

978-1-5386-3775-3/18 $31.00 © 2018 IEEE

Fig. 8: Algorithm for connecting interface-register cells into a twisted-ring counter

---

Input: Coordinates of input ILVs $V = \{v_1, v_2, ..., v_n\}$ and Coordinates of output ILVs $U = \{u_1, u_2, ..., u_m\}$

Output: A minimum cost Hamiltonian Circuit C that traverses through all ILVs, and every pair of adjacent ILVs in C contains an input ILV and an output ILV

1: **if** $(n > m)$ **then**
2: Construct a graph $G = (V, E)$: $v_i \in V$ represents input ILV i, and the distance between input ILV i and input ILV j denotes the weight of an edge between v_i and v_j
3: Reduce $V = \{v_1, v_2, ..., v_n\}$ to $\{v'_1, v'_2, ..., v'_m\}$ using k-means clustering, where v'_i is the centroid of cluster i
4: **else**
5: Construct a graph $G = (V, E)$: $v_i \in V$ represents output ILV i, and the distance between output ILV i and output ILV j denotes the weight of an edge between v_i and v_j
6: Reduce $U = \{u_1, u_2, ..., u_m\}$ to $U = \{u'_1, u'_2, ..., u'_n\}$ using k-means clustering, where u'_i is the centroid of cluster i
7: **end if**
8: $D \leftarrow$ Distance matrix between V and U
9: Construct a complete-weighted bipartite graph $B = (V \cup U, E)$, where the edge weights are given by D
10: $T \leftarrow TSP(G)$, where T is a tour that traverses through all vertices in G
11: $M \leftarrow MWBM(B)$, where M is a perfect matching for B
12: $C \leftarrow T \cup M$
13: **return** P

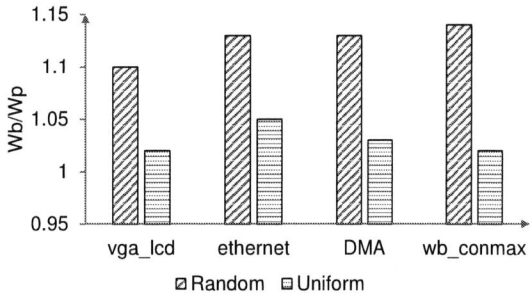

Fig. 9: Ratio of total wire length obtained using the proposed heuristic and baseline for random and uniform placements of ILVs.

based on the two placement configurations defined above. Once the distance matrices are generated, they are provided as input to the proposed heuristic implemented in a Python script, and the Hamiltonian circuit that corresponds to a tour of ILVs with minimum wire length is generated. We define a baseline method in which we start at a random ILV and pick the ILV that is visited next in a greedy manner, i.e., an output (input) ILV nearest to an input (output) ILV. The total wirelength required to form the TRC using the proposed method and baseline method are denoted by W_p and W_b, respectively. Fig. 9 shows the results in terms of the ratio W_b/W_p. It can be seen that W_b is 2% to 14% larger than W_p. Therefore, the proposed algorithm reduces the total wire length required to form the TRC.

IV. CONCLUSIONS

We have presented an ILV BIST solution for M3D that enables testing stand-alone testing of ILVs without incurring

significant area overhead. The proposed solution can detect all opens and shorts. We proposed a heuristic to solve the wire-length minimization problem for BIST and showed that the proposed algorithm is effective in reducing the total wire length required for test application.

ACKNOWLEDGEMENTS

The authors thank TM Mak for helpful comments and feedback on this paper.

REFERENCES

[1] Z. Or-Bach. *CHIPS 2020 VOL. 2: New Vistas in Nanoelectronics*, pages 51–91. Springer International Publishing, 2016.
[2] S. Kannan et al. Device performance analysis on 20nm technology thin wafers in a 3D package. In *IRPS*, 2015.
[3] A. B. Kahng. The ITRS design technology and system drivers roadmap: Process and status. In *DAC*, pages 1–6, May 2013.
[4] K. Arabi et al. 3D VLSI: A Scalable Integration Beyond 2D. In *ISPD*, 2015.
[5] P. Batude et al. 3-D Sequential Integration: A Key Enabling Technology for Heterogeneous Co-Integration of New Function With CMOS. *IEEE JETCAS*, 2(4):714–722, 2012.
[6] A. Koneru et al. Impact of Electrostatic Coupling and Wafer-Bonding Defects on Delay Testing of Monolithic 3D Integrated Circuits. *J. Emerg. Technol. Comput. Syst.*, 2017.
[7] A. Koneru and K. Chakrabarty. Analysis of Electrostatic Coupling in Monolithic 3D Integrated Circuits and its Impact on Delay Testing. In *ETS*, 2016.
[8] P. Batude et al. 3DVLSI with CoolCube process: An alternative path to scaling. In *VLSIT*, pages T48–T49, June 2015.
[9] L. Brunet et al. (invited) direct bonding: A key enabler for 3d monolithic integration. *ECS Transactions*, 64(5):381–390, 2014.
[10] A. Koneru et al. Impact of Wafer-Bonding Defects on Monolithic 3D Integrated Circuits. In *EPEPS*, 2016.
[11] O. D. Patterson et al. Detection of resistive shorts and opens using voltage contrast inspection. In *Advanced Semiconductor Manufacturing Conference*, pages 327–333, May 2006.
[12] B. Noia and K. Chakrabarty. Pre-bond probing of TSVs in 3D stacked ICs. In *ITC*, Sept 2011.
[13] E. J. Marinissen et al. IEEE Std P1838: DfT standard-under-development for 2.5D-, 3D-, and 5.5D-SICs. In *ETS*, May 2016.
[14] S. Makar. A layout-based approach for ordering scan chain flip-flops. In *ITC*, Oct 1998.
[15] I. Radu et al. Novel low temperature 3D wafer stacking technology for high density device integration. In *ESSDERC*, 2013.
[16] B. Rajendran et al. Pulsed laser annealing: A scalable and practical technology for monolithic 3D IC. In *3DIC*, pages 1–5, Oct 2013.
[17] A. Koneru et al. A Design-for-Test Solution for Monolithic 3D Integrated Circuits. https://goo.gl/prqXD5, In *ICCD*, 2017.
[18] E. J. Marinissen and Y. Zorian. Testing 3D chips containing through-silicon vias. In *ITC*, pages 1–11, 2009.
[19] S. Panth et al. Power-performance study of block-level monolithic 3D-ICs considering inter-tier performance variations. In *DAC*, 2014.
[20] K. Shibin et al. At-Speed Testing of Inter-Die Connections of 3D-SICs in the Presence of Shore Logic. In *ATS*, pages 79–84, Nov 2015.
[21] Supplementary document. https://goo.gl/MZGpmJ.
[22] Wei Zhao and Yu Cao. New generation of predictive technology model for sub-45nm design exploration. In *ISQED*, pages 6 pp.–590, March 2006.
[23] J. Munkres. Algorithms for the Assignment and Transportation Problems. *J. of the Society for Industrial and Applied Mathematics*, 1957.
[24] N. Christofides. Worst-case analysis of a new heuristic for the travelling salesman problem. In *Symposium on New Directions and Recent Results in Algorithms and Complexity*, 1976.

978-1-5386-3775-3/18 $31.00 © 2018 IEEE

A Built-In Self-Test Technique for Transmitter-Only Systems

Maryam Shafiee, Jennifer N. Kitchen, and Sule Ozev
School of Electrical, Computer, and Energy Engineering
Arizona State University
Tempe, AZ, USA

Abstract—Internet of Things (IoT) nodes used in environmental monitoring and smart city applications are becoming increasingly prevalent with over $200B projected market potential. These nodes typically employ one-way communications using a high-end transmitter without a corresponding receiver. Testing of such transmitter-only systems poses an additional challenge. Due to the lack of a receiver, low-cost test techniques, such as loop-back, cannot be used. In this paper, we present a low overhead built-in self-test (BIST) technique to characterize imbalances of IQ transmitters without a receiver, both for post-production and in-field test purposes. The proposed BIST uses simple circuitry and a single test setup. The target parameters are analytically computed independent from internal BIST parameters which eliminates the need for initial calibration phase. All measurements are in DC and no external RF signal generation is required. The overall measurement time, including the computation time, is less than 2ms. Simulation and measurement results show that the proposed method provides adequate estimation accuracy for digital calibration.

Keywords—Built-in self-test (BIST), IQ Transmitter, RF test

I. INTRODUCTION

Traditionally, majority of RF communications are based on two-way connection, requiring a transmitter and a receiver. However recently, with the advent of the Internet-of-Things (IoT), new applications for environmental monitoring, smart healthcare, and smart cities have emerged where large numbers of inexpensive sensor/information nodes are distributed to collect/broadcast information [1-3]. Such applications only require one-way transmission and include only a transmitter. For mid-range communications, a high-end transmitter is needed which can transmit information reliably with short burst times. This is especially true for a smart city management where the receiving counterpart may be moving with high speed, leaving no time for retransmission [4]. The simultaneous constraints of low cost and high-performance places a burden on the design, manufacturing, and test of these components. The RF transmitters for these applications are highly integrated which makes them increasingly susceptible to process, voltage, and temperature (PVT) variations. Accounting for these variations during the design phase requires tremendous amount of time for prediction of RF performance and optimizing it accordingly. Thus, there is an increasing gap between the need to relax the RF performance requirements at the design phase for rapid development and the need to provide high performance RF circuits that function with PVT variations. No matter how carefully designed, RF integrated circuits (ICs) manufactured with advanced technology nodes necessitate lengthy post-production calibration and test cycles with expensive RF instruments [5,6]. Hence, there is a growing interest in on-chip measurement of performance parameters for both post-production and in-field calibration purposes. Built-in self-test (BIST) and calibration of RF circuits can potentially enable production of low cost and robust electronics on rapidly evolving digital IC processes. BIST can replace the expensive RF test instrumentation and can be used for post-production and in-field testing of the device under test (DUT) to improve performance via digital calibration [7].

Cartesian transmitters have several important performance parameters, including I/Q gain and phase imbalance and baseband DC offsets. If these impairments can be measured at production time or in the field, they can be digitally calibrated in the baseband with minimum computational overhead [7]. Researchers have presented several techniques in the literature for the characterization of RF transceivers which target the entire transmitter-receiver chain. In [7-11], loop-back mode testing is proposed for specification test of RF transceivers. The analytical model for the entire path is extracted and analytical/numerical techniques are used to simultaneously solve transmitter and receiver parameters. In [7], a self-test method for zero-IF transceivers using loop-back and a small BIST circuitry is proposed to determine critical parameters, such as I/Q imbalance. However, techniques that rely on the presence of a full I/Q receiver are not applicable to transmitter-only systems.

By measuring the output power of the transmitter using RF amplitude measurement techniques [12-14], it is possible to determine important parameters of the transmitter. RF amplitude detection methods use additional circuitry to generate a DC or low frequency signal which is highly correlated to the DUT performance parameters. Conventional detectors, such as root mean square (RMS) detectors or envelope detectors, are subject to similar process variations as the DUT, which affects the measurement accuracy. To address this problem, other techniques for on-chip amplitude measurement that are independent of process variations are introduced recently in the literature [14].

978-1-5386-3775-3/18 $31.00 © 2018 IEEE

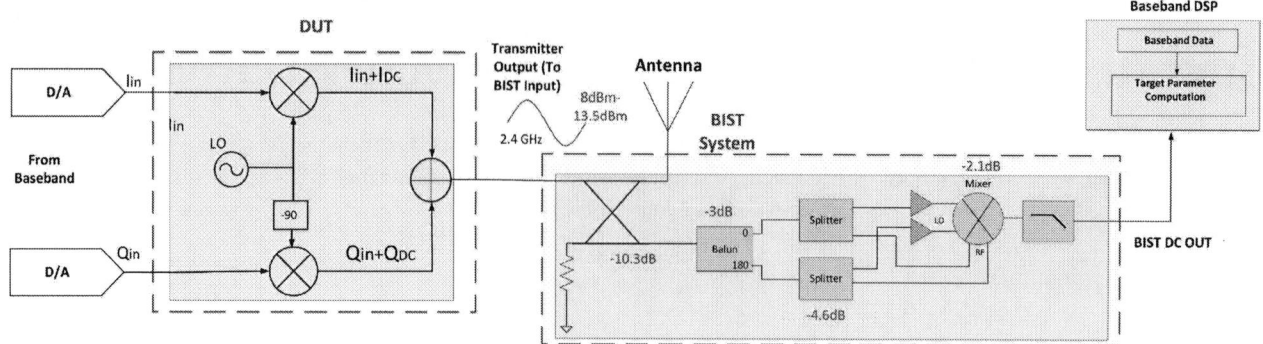

Fig.1. Simplified Transmitter and BIST model

However, since the I and Q signals are combined, amplitude measurement only does not provide adequate information for calibration. Moreover, in order to account for BIST variations, majority of amplitude measurement techniques would require a calibration phase that involves an external source.

In this work, we present a BIST method intended for the characterization of I/Q offset, gain and phase mismatch of IQ transmitters without relying on external equipment. The BIST circuit can be used for post-production and in-field calibration. The proposed BIST method is based on on-chip gain measurement as in prior works. However, in the proposed technique, variations in the BIST circuit do not affect the target parameter estimation accuracy since measurements are designed to be relative. The proposed BIST method uses full DC excitation in the baseband and DC measurements at the BIST output. This technique, unlike the previous approaches in literature [12-14], does not need any initial calibration for the BIST. Since performance characterization is independent of the internal BIST parameters, no knowledge of the BIST parameters is required.

II. PROPOSED METHODOLOGY

We propose a BIST method with simple circuitry that uses a single test setup to characterize the I/Q gain and phase mismatch as well as the DC offsets of transmitter-only systems. The transmitter can be characterized after production or periodically in the field.

A. BIST Model

Fig.1 shows the overview of the BIST system block diagram. The transmitter output is sensed via a directional coupler. The

majority of the signal power is conducted to the antenna and a trivial amount is fed to the self-mixing circuit to generate a DC signal at the BIST output. The generated baseband signal (BIST DC output), which is proportional to the input RF stimulus, is then further processed to compute the transmitter imbalances.

B. Analytical Derivations

Having the baseband input set at DC values of I_{in} and Q_{in}, the BIST DC output is given as in Equation (1), where the unknown parameters include gain (g) and phase ($\Delta\varphi$) mismatches, DC offsets (I_{DC}, Q_{DC}, V_{DC}) and BIST path gain (G_p). These 6 unknown variables are tabulated in Table I.

$$V_{o_{DC}} = V_{DC} + G_p \times \qquad (1)$$

$$\sqrt{\left((I_{in} + I_{DC}) + (1+g)Q\sin(\Delta\varphi)\right)^2 + (1+g)^2(Q_{in} + Q_{DC})^2(\cos\Delta\varphi)^2}$$

To determine all six unknown parameters, we need six linearly independent equations. The transmitter baseband inputs (I_{in} and Q_{in}) are the only test parameters we can set. The required linearly independent equations are constructed based on varying the input baseband levels by a known offset, Δ. Since the baseband inputs are set digitally, adding a pre-determined offset to the inputs I_{in} and Q_{in} will also generate an equivalent offset in the effective baseband inputs ($I_{in} + I_{DC}$, $Q_{in} + Q_{DC}$) without knowing the I_{DC} and Q_{DC} values. Since equation (1) has a quadratic dependency on the baseband inputs, adding offset to both or either input will generate linearly independent equations. To obtain the six equations, the baseband inputs are set as follows for each measurement:

$\{(I_{in}, Q_{in}) \rightarrow M_1, \quad (I_{in} + \Delta, \quad Q_{in}) \rightarrow M_2, \quad (I_{in}, Q_{in} + \Delta) \rightarrow M_3, (I_{in}, Q_{in} + 2\Delta) \rightarrow M_4, (I_{in} + \Delta, Q_{in} + \Delta) \rightarrow M_5, (I_{in} + 2\Delta, Q_{in} + 2\Delta) \rightarrow M_6 \}$

where M_i indicates the DC measurement for the ith step. Note that the BIST circuit DC offset is independent of the input and the path gain, G_p, is a scalar that multiplies all signals in identical fashion. In order to remove the unknown BIST DC offset, we only process

TABLE I. LIST OF TARGET PARAMETERS

Parameter	Description
$\Delta\varphi$	I/Q phase mismatch
g	I/Q gain mismatch
G_p	BIST path gain
I_{DC}	In-phase DC offset
Q_{DC}	Quadrature DC offset
V_{DC}	DC offset at the BIST output

978-1-5386-3775-3/18 $31.00 © 2018 IEEE

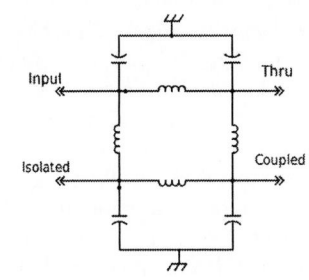

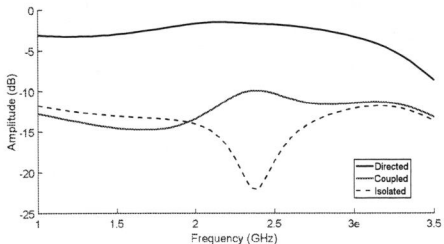

Fig.2. Lumped-element unequal-split coupler and its S-parameters

difference between two measurements. Equations (2-6) show the intermediate variables, E_1 through E_5, where the BIST DC offset is removed.

$$E_1 = (M_2 - M_1)^2 \qquad (2)$$

$$E_2 = (M_4 - M_1 - 2(M_3 - M_1))^2 \qquad (3)$$

$$E_3 = (4(M_3 - M_1) - (M_4 - M_1))^2 \qquad (4)$$

$$E_4 = (M_6 - M_4 - 2(M_2 - M_1))^2 \qquad (5)$$

$$E_5 = (M_5 - M_2 - (M_3 - M_1))^2 \qquad (6)$$

Furthermore, to remove the unknown path gain, G_p, we only process the ratios of the intermediate variables.

$$g = \sqrt{\frac{E_2}{(E_4 - 4E_5)}} - 1 \qquad (7)$$

$$\Delta\varphi = \sin^{-1}\left(\frac{E_5}{E_2}(1 + g)\right) \qquad (8)$$

$$I_{DC} = \Delta \frac{\frac{E_3}{E_2} - \frac{2E_1}{E_5} + \frac{E_4}{E_5} - 4}{\frac{2E_5}{E_2} - \frac{2E_4}{E_5} + 8} - I_{in} \qquad (9)$$

$$Q_{DC} = \frac{\Delta}{E_5}\left(E_1 - \frac{(E_4 - 4E_5)I_{in}}{\Delta} - \frac{E_4}{2} + 2E_5\right) - Q_{in} \qquad (10)$$

By solving Equations (2-6), we can analytically determine the target parameters. Equations (7-10) show the results of the analytical solution. The solution for the target parameters is independent of the circuit architecture since it can be used for any Cartesian transmitter.

III. BIST CIRCUIT DESIGN

The analytical derivation assumes that the BIST circuit works linearly and there are no additional unknowns due to the BIST circuit. The design of the BIST circuit is challenging due to these

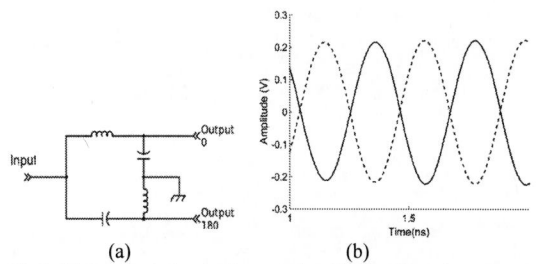

(a)   (b)

Fig.3. (a) Passive balun implementation (b) Signals at the outputs of the balun

constraints. The presented BIST circuit (shown in Fig.1) is implemented in 0.13μm CMOS technology. The supply voltage is 1.2V. Results are obtained at 2.4GHz although the method is extendable to any frequency. Since the input amplitude varies in a relatively wide range during test phase due to Δ step variations, the challenges in designing this BIST circuit are keeping the entire system in its linear region and keeping the voltage offset, gain, and phase offset of the BIST circuit independent of its input amplitude.

A. Unequally Split Directional Coupler

The directional coupler has asymmetric coupling factor between

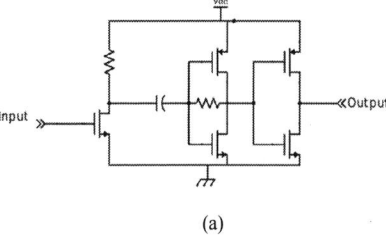

(a)

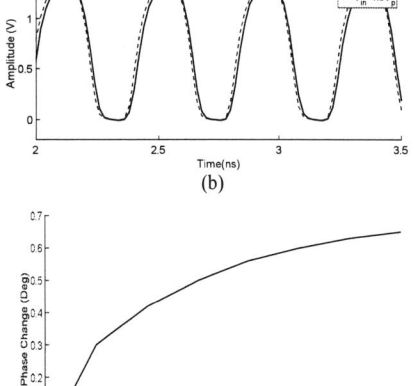

(b)

(c)

Fig.4. (a) Rail to rail amplifier circuit topology (b) Output for minimum and maximum test input (c) Phase shift with regards to BIST input signal

978-1-5386-3775-3/18 $31.00 © 2018 IEEE

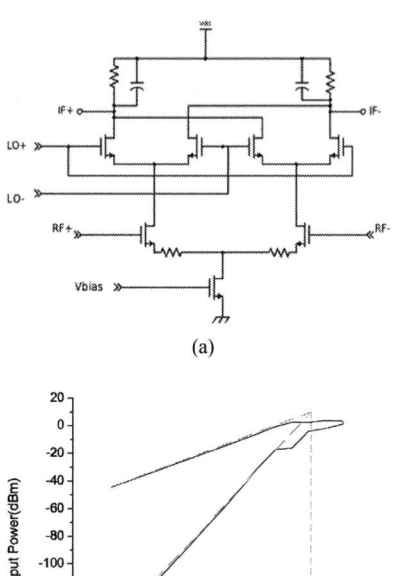

(a)

(b)

Fig.5. Gilbert cell mixer and its linearity evaluation. The projected input range is well below the IIP3 of the mixer.

the primary path and the BIST path for two purposes. First, in the primary path, it imposes a small insertion loss with minimum impact on the transmitter nominal operation. Second, in the coupling path, it attenuates the strong transmit signal to prevent saturating the BIST circuit. The narrowband unequal-split coupler is designed using lumped components [15]. The circuit realization and the S-parameters are shown in Fig.2. The coupler imposes a 1.2dB insertion loss with a coupling gain of -10.3dB. The insertion loss of the coupler is mainly due to inductor's finite Q-factor.

B. Balun

Since we do not need to provide gain in the BIST path, a passive structure can be used for balun to ensure the linearity of the BIST system over a wide range of input levels. The outputs of the balun are 180° out of phase, as shown in Fig.3, which also shows the circuit implementation.

C. Splitter and Rail to Rail Amplifier

A source follower topology is used for the splitter to branch out each of balun's output signals. It is then followed by amplifiers in the LO path to provide rail to rail signals at switching transistors. These amplifiers are required to make the mixer gain independent of the input signal power. Fig.4 (a) shows the circuit topology for the rail to rail amplifier. The rail to rail amplifier block uses a common source amplifier followed by an inverter chain. A known aspect of the inverter chain is varying output rise/fall time with respect to the input amplitude due to change in the slew rate. This

causes a delay between inverter chain's output signals, which reflects as a phase offset in the signals. If this phase offset depends on the input amplitude, it will not be canceled by relative measurements. Hence, keeping the phase offset constant while the input amplitudes vary is essential. Thus, the common source amplifier is employed to amplify smaller amplitude inputs closer to supply margins such that the delay difference between outputs at different input levels reduces to minimum. The output of the amplifier and the change in the phase shift due to varying input voltages for minimum and maximum input limits are depicted in Fig.4(b) and Fig.4(c) respectively. It is observed that the maximum phase change for the test input range is within 0.6° which is one of the major error source for the proposed method.

D. Mixer:
A conventional Gilbert cell mixer is designed to perform the self-mixing task. A resistive degeneration is employed to increase the linearity of the mixer. Since gain is not a concern for BIST, we sacrificed gain for more linearity. The mixer circuit realization is shown in Fig.5(a). Linearity of the mixer is evaluated by calculating the third order intercept point (IIP3) as illustrated in Fig.5(b). The IIP3 of the mixer is equal to 12dBm which guarantees a linear operation for the given test input range, which is depicted with the red arrow in Figure 5(b).

E. Low-Pass Filter:
A passive off-chip filter is placed at the chain end to filter out the small high frequency components of the output signal.

F. Link budget

The proposed technique is only valid within the linear operation of the BIST hence it is very important to assure a wide dynamic range for the BIST circuit while testing. The BIST input (transmitter

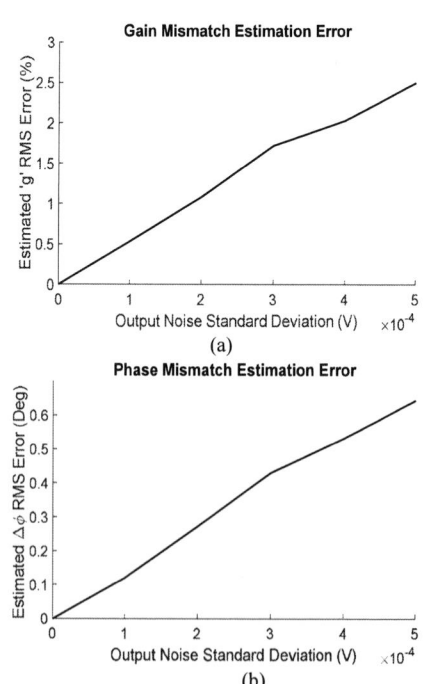

(a)

(b)

Fig.6. (a) Phase mismatch estimation error (b) Gain mismatch estimation error vs. input noise standard deviation

978-1-5386-3775-3/18 $31.00 © 2018 IEEE

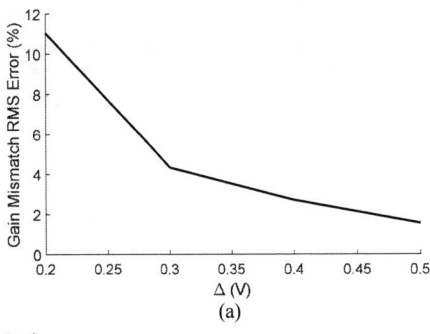

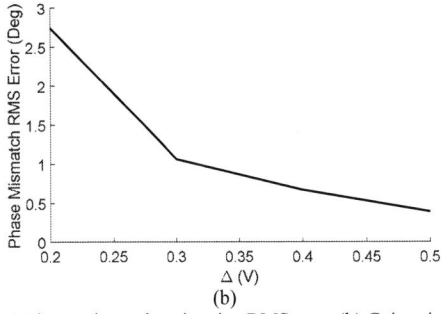

Fig.7. (a) Phase mismatch estimation RMS error (b) Gain mismatch estimation RMS error vs. Δ variable

output) range is required to place between 8dBm and13dBm to cover the necessary span for maximum Δ=0.4V. The gains and losses of the BIST building blocks are designed such that the entire chain works in the linear region. The corresponding gains of each block are shown in Fig.1. The signal at RF port of the mixer must be below the IIP3 and P1dB (≈IIP3-9.8dB) of the mixer. The RF signal span is highlighted in the graph which shows it is well below IIP3.

IV. EXPERIMENTAL RESULTS

In order to evaluate the proposed BIST technique, we use three

TABLE II. EXTRACTED TARGET PARAMETERS VS. ACTUAL USING MEASUREMENT

Parameter	Actual	Estimated
I	968 mV	958 mV
Q	881 mV	896 mV
1+g	1.001	0.984
Δφ	16°	15.8°
G_p	0.688	0.692

Fig.8. Lab Measurement Setup

experimental set-ups: (a) Matlab model based experimental evaluation for a large range of impairment and environmental conditions, (b) Hardware demonstration using off-the-shelf components, and (c) post-layout circuit simulations[1].

A. Matlab Model Simulation

The system model and proposed BIST technique are implemented in MATLAB and accuracy analysis is conducted

by simulations. First, we investigate the effect of output noise on estimation accuracy. The output measurement noise is varied and the phase and gain mismatch estimation RMS error is calculated accordingly. Fig.6 shows the characterization error in gain and phase mismatch with increasing environmental noise. The baseband step variation is set to Δ=0.4V. The injected gain mismatch is 5% and the injected phase mismatch is 3°. These values are selected based on EVM limit for WLAN standard which is less than 5.6%. The EVM for injected gain and phase mismatch is obtained using (11).

$$EVM \approx \sqrt{1 - \cos(\Delta\varphi) + \frac{g^2}{4}} = 4.5\% \qquad (11)$$

Adding estimation error from Fig.6 for 0.5 mV measurement error, to the existing gain and phase mismatches, the new EVM is equal to 5.1%. Thus, the error for the proposed technique adds only 0.6% to EVM, which is within the measurement uncertainty of EVM [16].

Next, we investigate the effect of Δ on the accuracy of the proposed technique. The gain mismatch is set to %5, the phase mismatch is set to 3° and the measurement error standard deviation is set to 0.5 mV. As we increase the Δ value in our measurements, the accuracy of the proposed technique increases. Fig.7 shows the accuracy of the BIST technique with varying input offset (Δ) value. The Δ level however, cannot be raised to any arbitrary value. The limiting factor here is keeping the transmitter and the BIST circuits in their linear operating region and to avoid saturation of the transmitter/BIST path.

B. Hardware Demonstration Using Discrete Components

The BIST method is also verified using lab equipment and off-the-shelf components as shown in Fig.8. The measurement setup includes, a signal generator to produce I and Q signals at 1GHz, a discrete combiner ZF-2-4+ to add the two signals and a discrete mixer, ZFM-150+, to down convert two signals to DC. The resulting signal is then down converted to DC using discrete mixer ZFM-150+. The Δ value is set to 0.4V. The measurement data is tabulated in Table II. It is observed that the gain mismatch estimation error is less than 1.7% and phase mismatch estimation error is 0.2°.

C. Post-Layout Simulation

Chip layout of the proposed BIST circuit is shown in Fig.9. The total area overhead is 0.247 mm² which is less than 4.2% of a Cartesian transmitter manufactured in the same process and in the

---

[1] The BIST IC is scheduled to tape out in November.

978-1-5386-3775-3/18 $31.00 © 2018 IEEE

same frequency band [17]. Fig.10 shows the BIST output for varying input amplitude levels. The DC offset is taken out from the output. The BIST circuit behavior is adequately linear within the test signal range. The BIST circuit adds no significant additional error to the I/Q imbalance computation. To prove the concept, the transmitter is emulated in Matlab and transmitter output voltages for M_1-M_6 is applied to the BIST circuit. The target parameters then are retrieved and compared with actual values. Table III shows the results. The total power consumption of the BIST is equal to 12.1mW.

D. Test Time

The proposed technique requires six DC measurements over six baseband input frames and simple mathematical calculations. Each frame is 200µs which is sufficient time for the filter settling and sampling. Including computation time, the overall test time is estimated to be less than 2ms.

V. CONCLUSION

In this paper, we propose a fast and process-robust BIST technique to characterize Cartesian transmitters. The mathematical model is simulated in Matlab for accuracy analysis of the method. Hardware measurements are performed to validate the BIST methodology. The measurements show that the gain mismatch estimation error is less than 1.7% and phase mismatch estimation error is 0.2°. The BIST circuit is designed in 0.13µm process. The circuit implementation of each block is presented. The post extraction results show that the design works in the linear region for the desired test input span and the BIST circuit poses only a slight degradation in the accuracy compared with MATLAB simulations.

ACKNOWLEDGEMENTS

This work is supported by the Semiconductor Research Corporation and the National Science Foundation.

REFERENCES

[1] S.Rmnath, et al., "IOT based localization and tracking", *IEEE International Conf. IOT and Applications*, 2017

[2] Y.Li, et al., "A novel fully synthesizable all-digital RF transmitter for IOT applications", *IEEE Trans. Computer-Aided Design of Integrated Circuits and Systems*, 2017

[3] L.Mainetti, et al, "An IoT-aware AAL System for Elderly People", *International Multidisciplinary Conference on Computer and*

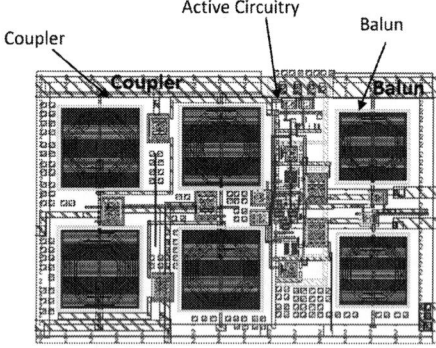

Fig.9. Chip layout of the proposed BIST

TABLE III. ESTIMATED TARGET PARAMETERS VS. ACTUAL USING CIRCUIT POST-EXTRACTION SIMULATION

Parameter	Actual	Estimated
I	1.1 V	1.088 V
Q	1.1 V	1.118 V
1+g	0.95	0.931
$\Delta\varphi$	3°	3.5°
G_p	0.091	0.0911

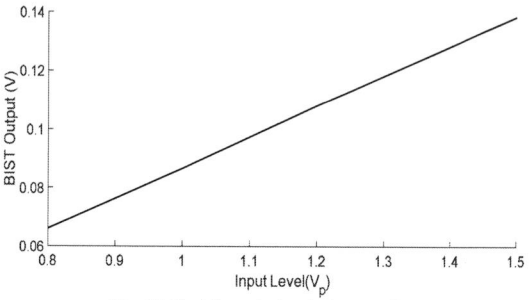

Fig.10. Post-layout simulation results

Energy Science (SpliTech) , 2016

[4] D.Kyriazis, et al., "Sustainable smart city IOT applications: Heat an electricity management & echo-conscious cruise control for public transportation", *IEEE 14th International Symposium and Workshops on World of Wireless, Mobile and Multimedia Networks (WoWMoM)*, 2013

[5] C.F.Cheang, et al, " A Combinatorial impairment-compensation digital predistorter for sub-GHZ IEEE 802.11 af-WLAN CMOs transmitter covering a 10x-wide RF bandwidth", *IEEE Trans. Circuits and Systems 1* ,vol.62, no.4,2015

[6] V.Natarajan, et al, " Yield recovery of RF transceiver systems using iterative tuning-driven power-conscious performance optimization", *IEEE Design and Test*, vol32, no.1 ,2015

[7] J.W.Jeong, A.Nassery, J.Kitchen, S.Ozev, " Built-in self-test and digital calibration of zero-IF RF transceivers", *IEEE Trans. on VLSI Systems*, vol.24, no.6, June 2016

[8] A. Halder, S. Bhattacharya, G. Srinivasan, and A. Chatterjee, "A system level alternate test approach for specification test of RF transceivers in loopback mode," in *Proc. 18th Int. Conf. VLSI Design*, Jan. 2005, pp. 289–294.

[9] E. S. Erdogan and S. Ozev, "Detailed characterization of transceiver parameters through loop-back-based BiST," *IEEE Trans. Very Large Scale Integr. (VLSI) Syst.*, vol. 18, no. 6, pp. 901–911, Jun. 2010

[10] M. Valkama, M. Renfors, and V. Koivunen, "Advanced methods for I/Q imbalance compensation in communication receivers," *IEEE Trans. Signal Process.*, vol. 49, no. 10,2001

[11] I.-H. Sohn, E.-R. Jeong, and Y. H. Lee, "Data-aided approach to I/Q mismatch and DC offset compensation in communication receivers," *IEEE Commun. Lett.*, vol. 6, no. 12, 2002

[12] S. Bhattacharya and A. Chatterjee, "Use of embedded sensors for built-in-test of RF circuits," in *Proc. International Test Conference* ,2004

[13] D.Han and A.Chatterjee ," Robust built-in test of RF ICs using envelope detectors ", in *proc. 14th Asian Test Symposium*, 2005

[14] J.W.Jeong, J.Kitchen and S.Ozev, "Robust amplitude measurement for RF BIST applications", in *Proc. 20th European Test Symposium*, 2015

[15] Inder Bahl, "Lumped elements for RF and microwave circuits", *Artech House*, ch.12, 2003

[16] E.Akar, G.Srinivasan, F.Taenzler and S.Ozev, "Optimized EVM testing for IEEE 802.11a/n RFICs", International Test Conference, 2008

[17] S.Ock, H.Song and R.Gharpurey, "A Cartesian feedback-feedforward transmitter IC in 130nm CMOS, in *Proc. Custom Integrated Circuits Conference*, 2015

Exploiting Built-In Delay Lines for Applying Launch-on-Capture At-Speed Testing on Self-Timed Circuits

Omar Al-Terkawi Hasib*, Daniel Crépeau†, Thomas Awad†, Andrei Dulipovici‡, Yvon Savaria* and Claude Thibeault‡

*École Polytechnique of Montréal, Montréal, Canada

Email: omar.al-terkawi-hasib@polymtl.ca, yvon.savaria@polymtl.ca

†Octasic, Montréal, Canada

Email: daniel.crepeau@octasic.com, thomas.awad@octasic.com

‡École de Technologie Supérieure, Montréal, Canada

Email: andrei.dulipovici@etsmtl.ca, claude.thibeault@etsmtl.ca

Abstract—The application of scan-based at-speed delay testing on asynchronous circuits is not trivial. Their unorthodox design leaves them generally incompatible with traditional synchronous design and test tools, as well as standard automatic test equipment. The correct generation of at-speed test clocks and the use of conventional automatic test patterns generation (ATPG) tools are some of the problems that face the application of at-speed testing on asynchronous circuits. This paper presents a method of applying scan-based at-speed testing on single-rail bundled-data handshake-free (self-timed) asynchronous circuits by taking advantage of built-in delay lines. The proposed test method uses launch-on-capture scan-based testing with endpoint masking and generates the test patterns using conventional ATPG tools. The proposed test is applied on circuits in a self-timed microprocessor fabricated in 28nm FD-SOI CMOS technology. This method is validated by the reported test coverage and simulation results, along with post-silicon test results on a Teradyne FLEX tester.

I. INTRODUCTION

At-speed delay testing is important to guarantee the correct behavior of circuits when operating at the rated speed. In the case of asynchronous circuits, there are many different ways for data propagation through the system [1], and whether a delay test is needed or not depends on the type of asynchronous circuit being tested [2]. For instance, most high-speed asynchronous designs need to be delay tested, since they have to satisfy some timing conditions in order to operate correctly [3].

The diverse asynchronous circuit design styles make it difficult to standardize the design and test process of such circuits using conventional design tools that are normally used for synchronous circuits [1]. Part of the difficulty stems from the fact that several asynchronous design styles are based on custom C-elements. In addition to the hassle of using custom standard cells for C-elements, testing circuits that adopted such design styles for structural and delay faults using conventional methods becomes a challenge. Several papers have proposed methods for testing circuits designed according to such asynchronous design styles [2]–[5] by modifying C-elements and latches to apply the required test.

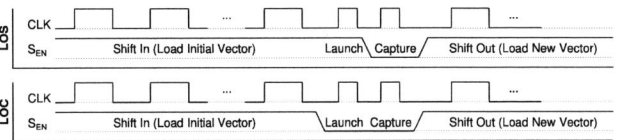

Fig. 1. Timing diagram of the classical LOS and LOC.

When applying at-speed testing on asynchronous circuits, it is essential to use a test clock speed equal to the one used during operation. This requires a careful study and effective reuse of the timing mechanisms of the asynchronous structure during at-speed testing. In general, the global asynchronous system speed can be inferred in several ways. However, the exact speed of operation of internal asynchronous clocks after fabrication is normally not known due to the effects of process variations on circuit delays. Thus, the use of external test clocks with predefined frequencies by the automatic test equipment (ATE) is not adequate in this case.

At-speed testing is normally applied by one of two well-known scan-based delay testing techniques: launch-on-shift (LOS) or launch-on-capture (LOC) [6], [7]. Both techniques use synchronous test clocks to shift in test vectors and an asynchronous scan enable signal to switch between shifting vectors and applying the test. The frequency of the synchronous test clock is normally controlled to apply the at-speed test after shifting in the test vectors. Fig. 1 shows the timing diagram of standard synchronous LOS and LOC. The main difference between LOS and LOC is when the scan enable signal (S_{EN}) is deactivated. This changes the mechanism by which the test vector is generated and launched into the circuit under test (CUT). This also creates a difference in the timing constraints related to the scan enable signal. LOS has a higher test coverage in general, however, satisfying the timing constraints on the S_{EN} can be challenging. This makes LOC an attractive alternative.

To apply at-speed testing on asynchronous designs, the

978-1-5386-3775-3/18 $31.00 © 2018 IEEE

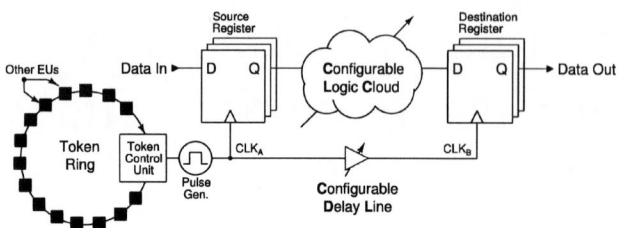

Fig. 2. Simplified schematic of the execution unit (EU).

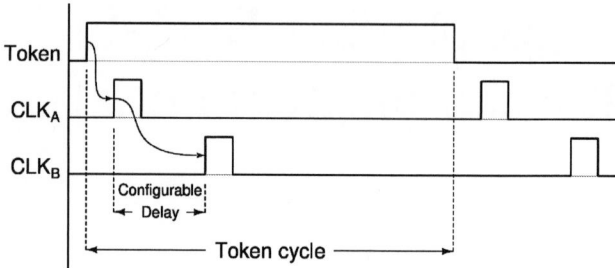

Fig. 3. Timing diagram of the EU operation.

synchronous test clock has to be adapted to the CUT in order to capture the test results at the same speed as the asynchronous system clock. Since asynchronous design styles are diverse, the method of triggering the asynchronous at-speed test clock can be unique to the asynchronous CUT.

In [8], Octasic has presented a unique asynchronous process architecture. This processor is built using a set of self-timed core processing units, named execution units (EUs), that are asynchronously managed using control structures, called token rings. There is a set of timing conditions that needs to be met for correct operation of this processor, and although the correct operation can still be achieved by reducing the speed of some parts of the system, at-speed testing is required to ensure that the design is truly tested for delay defects.

In this paper we look at the timing conditions that govern the operation of the asynchronous processor architecture proposed by Octasic in [8] and propose a method of applying scan-based at-speed testing using conventional ATPG tools. This architecture has only been test functionally, but no scan-based at-speed test has been applied before. The proposed scan-based test strategy takes advantage of the built-in delay line in each EU to apply LOC at-speed delay testing.

The rest of this paper is organized as follows. First, section II explains the functionality of the targeted self-timed structure and presents the timing constraints associated with it. Section III then presents the proposed test strategy including the test structures, test modes, proposed LOC strategy and test pattern generation considerations. This section also briefly discusses how to exploit the delay line in the targeted structure to achieve more than just at-speed delay testing. The coverage results, simulations, as well as post-silicon test results are reported in section IV. Finally this paper concludes in section V by summarizing our main findings.

II. TARGET ARCHITECTURE

In [8], Octasic presented a single-rail bundled-data handshake-free (self-timed) asynchronous processor architecture. This architecture uses mainly flip-flops rather than latches and it was successfully used in several generations of commercialized ICs. The architecture of [8] uses 16 EUs to build an ad-hoc like processing pipeline. Every EU can be configured to process any instruction in the processor. The processing steps, resources and timing are managed by token rings. In this system, tokens are signals that circulate in token rings between all EUs. There is one token ring for each processing

instruction. Once a token is acquired by a EU, the EU holds the token and starts processing the related instruction.

A simplified schematic of an EU is presented in Fig. 2. When an EU acquires a token, a pulse is generated at node CLK_A. The pulse clocks the input flip-flops (source register) sending the data through the preconfigured combinational logic cloud. At the same time, the pulse travels through the delay line to the output flip-flops (destination register) and triggers the capture of the result of the processed data. Since the EU is a group of multiple different processing clouds, the delay of the delay line is configured to match the delay of the logic cloud used. This depends on the instruction that is being processed (i.e. the type of the token acquired). Fig. 3 shows the timing diagram of this process.

For the EU to work properly, there are two timing constraints that need to be satisfied [9], [10]: a setup time constraint and a hold time constraint. This is similar to the synchronous design conditions. However, since the system is self-timed and uses a configurable delay line (CDL), the setup and hold time constraints govern mainly the delay of the CDL with respect to the delays of the configurable logic cloud (CLC). The following is the setup timing constraints:

$$t_{CK-Q} + t_{CLC} \le t_{CDL} - t_{setup} \qquad (1)$$

Where t_{CK-Q} is the delay for the data to appear at the output of the flip-flop after the clock edge arrival, t_{CLC} is the combinational cloud delay, t_{CDL} is the delay of the CDL and t_{setup} is the setup time for the output flip-flops. As can be seen from this equation, the delay of the CDL needs to be appropriately selected to ensure proper operation. For the hold time constraint, two conditions need to be met:

$$T_{token} \ge t_{CDL} \qquad (2)$$

$$t_{CK-Q} + t_{CLC} \ge t_{hold} + (t_{CDL} - T_{token}) \qquad (3)$$

Where T_{token} is the time for the token to circle around the token ring and come back to the same EU, t_{hold} is the hold time for the output flip-flops and the rest of the terms are as previously described in equation (1). The first condition in the hold constraint governs the relationship between the token ring delay and the CDL delay. This relationship is usually satisfied by the fact that the token ring is much longer than the CDL. However, this puts an upper limit on the delay of

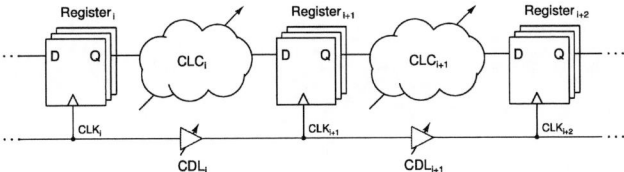

Fig. 4. Generalized multi-stage self-timed structure using the targeted asynchronous design block.

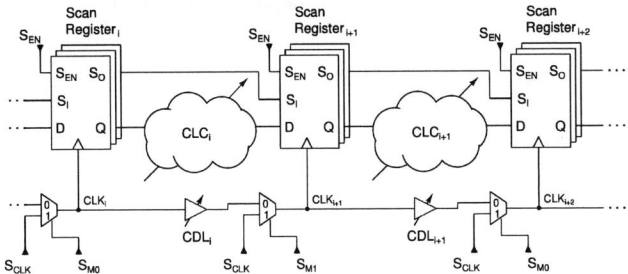

Fig. 5. Test structures inserted into the targeted multi-stage self-timed structure.

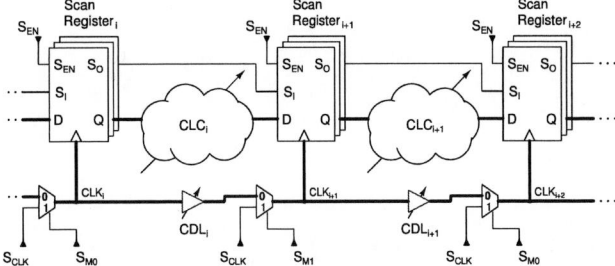

Fig. 6. Multi-stage self-timed structure configured for regular operation.

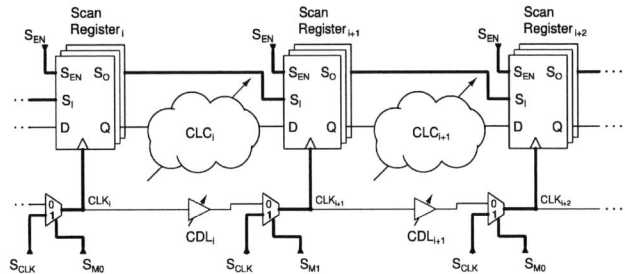

Fig. 7. Multi-stage self-timed structure configured for shifting the test vectors synchronously.

the CDL. The second constraint is similar to the synchronous hold time constraint, however, since a single pulse launches and captures the data, a hold violation is very unlikely to occur in this architecture if the first condition of the hold constraint is satisfied. Hence, the delay testing strategy will focus solely on checking the setup time constraint by testing the delays in the logic cloud against the delay of the CDL.

III. PROPOSED AT-SPEED TEST STRATEGY

In this paper, the proposed at-speed strategy will focus on applying at-speed test clocks by taking advantage of the built-in CDLs while using conventional scan-based testing to apply the test. The idea is to preconfigure the logic cloud and the delay line, scan-in the test vectors in using a synchronous test clock, then use the same clock to trigger at-speed launch and capture of the test vector using the CDL, and lastly scan-out the result.

A. Test Structures

A generalized view of the targeted self-timed system is shown in Fig. 4. Depending on the number of stages, this generic structure can be used to represent multiple parts inside the targeted processor. In this structure, the data would travel from one stage to another, and a clock pulse, traveling at a speed slightly slower than the slowest data in the cloud, controls the launch and capture of the data from one stage to another. Notice that only one pulse would travel down the multi-stage path at a time. This should not be confused with how the processing pipeline is constructed in such an architecture with multiple EUs.

In order to apply at-speed testing for this architecture, two types of test structures need to be put in place. The first test structure is scan chains. One of the advantages of this

asynchronous architecture is that it does not depend on C-elements nor latches. Hence, there is no need to design special standard cells for C-elements, nor there is a need to develop new methods of scanning the data or use extra clocks (as in the L1L2* methods [4], [5]). This makes the scanning of data exactly like synchronous architectures where scan flip-flops have scan-enable (S_{EN}) and scan-in (S_I) signals and where the scanning operation is driven by a synchronous test clock.

The second test structure is a circuit that manages the insertion of the synchronous test clock into the system and the process of switching between different clocking schemes. This is done here with a simple multiplexer. Fig. 5 shows the simplified multi-stage self-timed structure after the test structures have been inserted. The two main changes are (i) the replacement of normal registers by chained scan registers, (ii) the insertion of a multiplexer just after each CDL to allow synchronous clocking (with S_{CLK}) with interleaved scan mode signals (S_{M0} and S_{M1}).

B. Test Modes

The proposed testable structure (seen in Fig. 5) allows for one of three different modes. Firstly, the regular operation mode (mission mode) is configured by setting all the test control signals (S_{EN}, S_{M0} and S_{M1}) to 0. Fig. 6 highlights the activated paths during mission mode. Notice that the added test logic are only multiplexers that do not add much delay to the system. Moreover, both the data path and the clock path have a 2-input multiplexer added to them. The one on the clock path is obvious, while the one in the data path is built into the scan flip-flops and is used to chose between the data input (D) or the scan input (S_I), as in regular synchronous designs [11].

978-1-5386-3775-3/18 $31.00 © 2018 IEEE   75

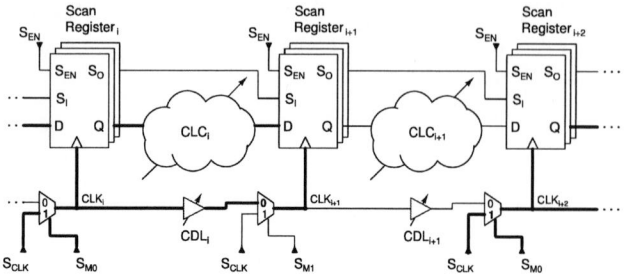

Fig. 8. Multi-stage self-timed structure configured for at-speed launch and capture.

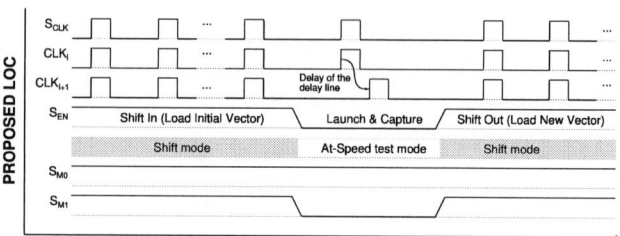

Fig. 9. Timing diagram of the proposed LOC strategy

TABLE I
SUMMARY OF TEST MODES

Mode	S_{EN}	S_{M0}	S_{M1}
Mission	0	0	0
Shift	1	1	1
At-Speed test	0	1 (0)	0 (1)

Thus, no major timing impacts are expected with the insertion of those multiplexers. Nonetheless, adjustments can be easily made to the delay line to negate the effects of any delay added by the multiplexers.

The second mode is the shift mode. This mode is used to scan-in or scan-out the test vectors. All the test control signals are set to 1, and the shifting of the test vectors is done with a synchronous scan clock (S_{CLK}). The paths that are activated during the shift operation are highlighted in Fig. 7.

The last mode is the at-speed test mode. To apply at-speed testing on this multi-stage self-timed structure, all the logic clouds and delay lines are preconfigured. The multi-stage self-timed structure is at-speed tested in an interleaved fashion, where one out of two consecutive stages will be tested at a time, thus requiring two steps to at-speed test the whole structure. This interleaved testing fashion simplifies programming the ATPG tool and is designed to work with the proposed LOC strategy (section III-C).

To apply the at-speed test, the S_{EN} signal is set to 0, while the scan mode signals are set to opposite values. This setting would activate the paths highlighted in Fig. 8, where half of the registers would be launching the test vectors and the other half would be capturing the results. When the test vectors are ready, the S_{CLK} is pulsed only once. This pulse will launch the test vector into the preconfigured logic cloud, travel through the preconfigured delay line and capture the test result. When the scan mode signals are inverted, the register roles are switched and the second half of the structure is tested. This test is at-speed because, with respect to the logic clouds, the delay lines are configured to have the same speed as in the regular operation mode. Thus the data is captured at the speed of regular operation. Table I summarizes the different modes of operation and the settings of the test control signals.

C. Proposed LOC Strategy

To apply the above at-speed test method, a LOC based strategy is adopted. Not only does LOC remove the tight timing constraints from the test control signals, but it also simplifies programming the synchronous ATPG tool for the asynchronous structure (as will be explained in section III-D). Applying LOC on the targeted asynchronous architecture is done in three stages: shift-in, launch and capture, shift-out. As mentioned earlier, this procedure is repeated twice to test the entire multi-stage self-timed structure where the only difference between each repetition is the value of the scan mode signals.

Fig. 9 shows the timing diagram of the proposed LOC strategy. The test starts by setting the system to shift mode (as described in Table I). The initial test vector is shifted into the scan chains by pulsing the scan clock (S_{CLK}). Once the test vector is completely loaded, the system is switched to the at-speed test mode. To ensure correct operation, enough time must be given for the test control signals to stabilize.

There is an important difference between a conventional synchronous LOC and the proposed LOC. The fact that the capture pulse is internally generated in the targeted asynchronous system means that, in the proposed LOC, the S_{CLK} will only be pulsed once. This pulse travels to CLK_i, where it launches the data into the logic cloud and then travels through the delay line. When it arrives at CLK_{i+1}, it triggers the capture of the test results. Since the delay line is preconfigured to match the logic under test, this delay test happens at-speed. Next, the system is set again to the shift mode to scan-out the test results and scan-in the next test vector.

D. Pattern Generation Considerations

In order to correctly generate the patterns with conventional ATPG tools that target synchronous designs for this asynchronous structure, while avoiding invalid test results, the following two issues must be addressed with regards to the use of a delay line to apply at-speed testing. The first issue arises from the fact that the capturing registers are considered to be in a different time-domain as compared to the launching registers, due to the presence of delay lines. This leads to the potential of having unexpected values in the launching registers at the end of the test. This issue is solved by masking the launching registers so that the results are only read on the capturing registers. The second issue arises

978-1-5386-3775-3/18 $31.00 © 2018 IEEE

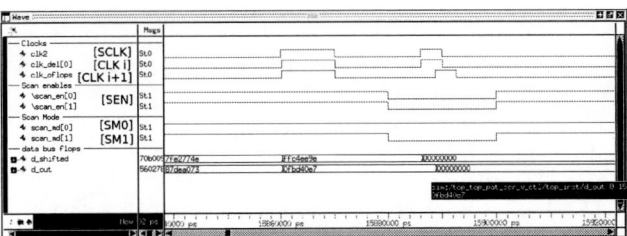

Fig. 10. Post-synthesis simulation of the proposed LOC on the picoALU showing the transition from the shift mode to the at-speed test mode. The signal names with reference to Fig. 9 are shown next to the original design names.

TABLE II
COVERAGE RESULTS

Design Unit Name	Number of Testable Faults	Number of Undetected Faults	Number of Patterns	Test Coverage (%)
PicoALU00	1711	16	213	99.06
PicoALU01	1723	28	213	98.37
PicoALU02	1723	28	213	98.37
PicoALU03	1735	40	213	97.69
Overall: PicoALUs	6892	112	213	98.37
AnARM Mul. Unit00	14041	430	202	96.94
AnARM Mul. Unit01	14044	433	202	96.92
AnARM Mul. Unit02	14040	432	202	96.92
AnARM Mul. Unit03	14044	433	202	96.92
Overall: AnARM Mul. Units	56169	1728	202	96.92

from the need to write time-plates for the synchronous pattern generation tool in order to correctly describe the LOC timing. As seen in Fig. 9, the S_{CLK} does not pulse during the capture cycle, moreover, none of the test control signals change. Thus, for LOC, the time-plate for describing the capture cycle is a dummy time-plate that is used to stall the input signals until the pulse internally captures the data. Notice that since none of the signals in LOC changes between the launch and capture cycles, the use of synchronous ATPG tools for pattern generation is simplified. If a LOS approach was to be used, an additional scan enable signal would be necessary for a separate control of the launching and capturing registers. Thanks to the interleaved scheme, no flip-flop would have to switch from a launching ($S_{EN} = 1$) to a capturing ($S_{EN} = 0$) mode between the two clock pulses. Thus, the S_{EN} signals would not be critical, as they are for regular synchronous designs. In addition to the fact that it does not require an extra S_{EN} signal, the LOC scheme benefits from a greater ATPG tool flexibility, provided by the "named capture procedure" [12]. The development of a LOS scheme is part of our future work.

E. More Than At-Speed Testing

The CDL in the targeted self-timed structure opens the possibility to use the same LOC technique to do more than just at-speed testing. The CDL can be designed to accommodate faster-than-at-speed testing, reliability testing and/or speed binning. Moreover, by accommodating shorter delay lines in the CDL, we are able to generate faster-that-at-speed tests that fail the CUT intentionally. By doing so, we can validate that the applied test strategy indeed works when testing fabricated chips. In terms of the CDL structure, there are different methods to build delay lines, however, in all cases, the added adjustments to the CDL might incur an area overhead on the existing design. Thus, there is a trade-off between the number of delay stages that can be added to the CDL to accommodate additional test features and the area of the design.

IV. RESULTS

In this section, the proposed at-speed test method is validated on two circuits that use the targeted Octasic self-timed structure. The first circuit is a processor that contains 4 arithmetic logic units (called picoALUs). The proposed test concept and implementation flow are validated on the picoALUs with

pattern generation, coverage reports and simulation results. The picoALUs serve as a proof of concept and were not fabricated. The second circuit is a fabricated fully operational self-timed ARM processor (called AnARM). The same test structure and test flow are applied on parts of 4 instances of a 32-bit multiplier unit in the AnARM. The AnARM was tested with a Teradyne FLEX tester, and the post-silicon test results will be reported for the proposed test strategy.

The technology that is used for all steps of design, test and fabrication is the STMicroelectronics 28nm FD-SOI CMOS technology. Scan chain creation and pattern generation were done using Mentor Graphics Tessent. Tessent was also used to apply the proposed LOC scheme using the "named capture procedures". The LOC timing is validated with post synthesis gate simulation using the test bench that was automatically generated from Tessent. The simulation of the LOC on the picoALU shows (in Fig. 10) the LOC signals behavior for the transition from the shift mode to the at-speed test mode. Notice that this behavior is identical to the one described in Fig. 9. The moment of the launch and capture are indicated in the figure by the pulses that occur on the clock signals (CLK_i and CLK_{i+1}) when the S_{EN} is low.

The test patterns were generated under the transition delay fault model. The coverage results for the 4 picoALUs and 4 circuits inside the AnARM processors are shown in Table II. The table reports the number of testable faults, the number of faults undetected by the pattern, the number of applied patterns and the test coverage percentage. As mentioned earlier, in the case of the AnARM, the test was applied on parts of 4 instances of a 32-bit multiplier unit. To be specific, one stage in each of the 4 multi-stage multiplier units was at-speed tested. As reported in the table, an overall of 98.37% test coverage was obtained on the picoALU circuits and an overall of 96.92% test coverage for the 4 targeted stages in the multiplier units of the AnARM. Notice that the number of patterns in all cases is the same for all units because the patterns are applied simultaneously.

For the post silicon results, the at-speed patterns were applied on 25 chips of the AnARM using a Teradyne FLEX tester. The CDL in the tested circuits in the AnARM was modified to get more than just at-speed testing. Two fault-injection signals (FI_0 and FI_1) were used to configure the CDL delay. Depending on those two signals, we are able to

978-1-5386-3775-3/18 $31.00 © 2018 IEEE

TABLE III
POST-SILICON TEST RESULTS OF 25 ANARM CHIPS

Test Speed	$FI_1 FI_0$	Pass (%)	Fail (%)
At-speed	00	100	0
Faster-than-at-speed level 1	01	100	0
Faster-than-at-speed level 2	10	24	76
Faster-than-at-speed level 3	11	0	100

TABLE IV
NUMBER OF FAILING ENDPOINTS WHEN USING FASTER-THAN-AT-SPEED
LEVEL 2 TEST SPEED

Chip ID	Average Number of Failing Endpoints	Chip ID	Average Number of Failing Endpoints
2	0	20	56.5
3	0	21	61
4	113	22	1.7
5	14.5	23	77.6
6	14.5	29	1
8	25.3	30	45
10	0	34	8
11	15	37	16
12	0	41	8.9
14	36.4	42	19.3
16	0	43	5.3
18	2.1	44	0
19	18.7	-	-

apply delay testing with an at-speed clock and 3 speed levels of faster-than-at-speed clocks. Using a timing analysis tool, a timing report of the delays of the 200,000 longest paths in the circuit was generated. The CDL was designed to apply faster-than-at-speed clocks that are faster than the reported paths by 25%, 50% and 75% when FI_0 and FI_1 are 01, 10 and 11, respectively. Normal at-speed testing is activated by setting FI_0 and FI_1 to 00.

Table III shows the percentage of passing and failing chips in the post silicon test for the four different cases. When the chips start to fail the test, In the case of $FI_0 FI_1 = 10$ (faster-than-at-speed level 2), each chip fails with a different number of failing endpoints. Table IV shows the average number of failing endpoints for each chip under the faster-than-at-speed level 2 test. This average was obtained by repeating the at-speed test 11 times. The results in this table can be used to bin the circuit under test by speed grades where the slower the chip, the larger the number of failing endpoints becomes. Furthermore, in future work, the same test can be applied after putting the chips under stress and seeing how those numbers change to study the reliability of the design and the aging effects. Some might wonder why the chips started to fail at level 2 of faster-than-at-speed and not at level 1. This is possibly because the patterns used are regular transition-fault-model patterns that are not generated by a timing aware engine. This implies that the ATPG algorithm selects some short paths for testing the circuits. Lastly, the fact that increasing the speed

of the capture clock (CDL delay) failed the chips confirms that the applied LOC is working properly.

V. CONCLUSION

This paper has proposed a launch-on-capture based at-speed test strategy that works with the single-rail bundled-data handshake-free (self-timed) asynchronous structure, such as the one used by Octasic in [8]. This test strategy is completely compatible with conventional test and ATPG tools. By using multiplexers in the data and clock paths, the proposed strategy allowed for conventional test vector scanning and took advantage of the built-in delay lines to apply the at-speed testing. The proposed test strategy have been validated with test coverage reports, simulation results and post-silicon test results for circuits fabricated in the 28nm FD-SOI CMOS technology.

ACKNOWLEDGMENTS

The authors would like to thank CMC Microsystems and CMP for providing the CAD tools and access to the 28nm FD-SOI CMOS technology, NSERC for providing partial funding, Octasic for providing financial support and scientific guidance, and Prof. Gordon Roberts at McGill University for access to the Teradyne tester.

REFERENCES

[1] A. Yakovlev, P. Vivet, and M. Renaudin, "Advances in asynchronous logic: From principles to gals noc, recent industry applications, and commercial cad tools," in *2013 Design, Automation Test in Europe Conference Exhibition (DATE)*, March 2013, pp. 1715–1724.

[2] F. Shi and Y. Makris, "Testing delay faults in asynchronous handshake circuits," in *2006 IEEE/ACM International Conference on Computer Aided Design*, Nov 2006, pp. 193–197.

[3] G. Gill, A. Agiwal, M. Singh, F. Shi, and Y. Makris, "Low-overhead testing of delay faults in high-speed asynchronous pipelines," in *12th IEEE International Symposium on Asynchronous Circuits and Systems (ASYNC'06)*, March 2006, pp. 11 pp.–56.

[4] F. te Beest and A. Peeters, "A multiplexer based test method for self-timed circuits," in *11th IEEE International Symposium on Asynchronous Circuits and Systems*. IEEE, March 2005, pp. 166–175.

[5] F. T. Beest, A. Peeters, M. Verra, K. van Berkel, and H. Kerkhoff, "Automatic scan insertion and test generation for asynchronous circuits," in *Proceedings. International Test Conference*. IEEE, 2002, pp. 804–813.

[6] J. Savir, "Skewed-load transition test: Part i, calculus," in *Test Conference, 1992. Proceedings., International*, 1992.

[7] J. Savir and S. Patil, "Broad-side delay test," *IEEE Trans. Comput.-Aided Design Integr. Circuits Syst.*, vol. 13, no. 8, pp. 1057–1064, 1994.

[8] M. Laurence, "Introduction to octasic asynchronous processor technology," in *Asynchronous Circuits and Systems (ASYNC), 2012 18th IEEE International Symposium on*, 2012, pp. 113–117.

[9] T. Awad, M. Laurence, M. Filteau, P. Gervais, and D. Morrissey, "Clock signal propagation method for integrated circuits (ics) and integrated circuit making use of same," U.S. Patent 8 130 019, Mar. 6, 2012.

[10] M. Fiorentino, O. Al-Terkawi, Y. Savaria, and C. Thibeault, "Self-timed circuits fpga implementation flow," in *New Circuits and Systems Conference (NEWCAS), 2015 IEEE 13th International*, June 2015, pp. 1–4.

[11] M. Bushnell and V. Agrawal, *Essentials of Electronic Testing for Digital, Memory and Mixed-Signal VLSI Circuits (Frontiers in Electronic Testing)*. Springer, 2005.

[12] X. Lin, R. Press, J. Rajski, P. Reuter, T. Rinderknecht, B. Swanson, and N. Tamarapalli, "High-frequency, at-speed scan testing," *IEEE Design Test of Computers*, vol. 20, no. 5, pp. 17–25, Sept 2003.

978-1-5386-3775-3/18 $31.00 © 2018 IEEE

Special Session on Bringing Cores Closer Together: The Wireless Revolution in On-Chip Communication

Terrence Mak, University of Southampton, UK
Hiroki Matsutani, Keio University, Japan
Partha Pratim Pande, Washington State University, USA (Organizer)

I. Introduction

The emerging field of NoC with Wireless interconnects is actively being pursued by a number of researchers worldwide, from a variety of different perspectives, ranging from very high levels of abstraction (e.g., system architecture) to very low levels (physical layer and transceiver design). Successful solutions will likely adopt and encompass elements from all or at least several levels of abstraction and rely on interdisciplinary concepts from multi-core architectures, integrated circuits, 3D ICs, digital communications, complex networks, and optimization techniques. This special session will provide a timely and insightful journey into various challenges and emerging solutions regarding the design of future NoC architectures. By scope and contents, this special session represents an engaging proposition to attendees belonging to both academia and industry.

II. Surfing on the Chip: Wired and Surface-Wave Integration for Network-on-Chip Architectures (Terrence Mak)

Network-on-chip (NoC) has been introduced as the backbone for many-core on-chip communication. Yet, merely metal-based NoC implementation offers limited performance and power scalability in terms of multicast and broadcast traffics. To meet up the proper scalability and huge on-chip communication demands in modern applications, we proposed a hybrid interconnects approach to addresses the challenges for on-chip multicast communication. This new NoC architecture combines and utilizes both regular metal wires and new type of wireless-NoC, surface-wave, achieving an unprecedented performance and reliability. More than that, new schemes such as multicast routing and arbitration will be discussed, in order to address the system-level multicast-challenges in the proposed architecture. We also discuss the hybrid integration, which was used to avoid overloading the network, to alleviate the formation of traffic hotspots and to avoid deadlocks, and these challenges are typically associated with multicast communications. Results demonstrated the effectiveness of the proposed approach in terms of power consumption (up to 10x) and performance (more than 22x) comparing to the metal-based design. This study explored promising potential of the proposed architecture for high-performance and reliable future NoC-based many-core processors.

III. A Building Block 3D System with Inductive-Coupling Through Chip Interfaces (Hiroki Matsutani)

In this talk, we will introduce a building block computing system that consists of multiple functional chips connected with inductive coupling wireless through chip interfaces. As the functional chips, we implemented four chip types: 1) Low-power microprocessor chip, 2) Coarse-grained reconfigurable accelerator chip, 3) key-value store memory chip, and 4) Convolutional neural network accelerator chip. These chips equip wireless transceiver circuits for vertical communication and they form a wireless 3D Network-on-Chip (NoC). We will illustrate the wireless 3D NoC design and the prototype system fabricated in a 65nm process.

IV. Designing Energy Efficient and Reliable Manycore Chip Enabled by Millimeter-Wave Wireless Links (Partha Pratim Pande)

The continuing progress and integration levels in silicon technologies make complete end-user systems on a single chip possible. This massive level of integration makes modern manycore chips all pervasive in domains ranging from weather forecasting, astronomical data analysis, and biological applications to consumer electronics and smart phones. Network-on-Chips (NoCs) have emerged as communication backbones to enable a high degree of integration in manycore platforms. Despite their advantages, an important performance limitation in traditional NoCs arises from planar metal interconnect-based multi-hop communications, wherein the data transfer between far-apart blocks causes high latency and power consumption. The latency, power consumption, and interconnect routing problems of NoCs can be simultaneously addressed by replacing multi-hop wired paths with high-bandwidth single-hop long-range wireless links. In this talk, we will present design of the millimeter (mm)-wave wireless NoC architectures. We will present detailed performance evaluation and necessary design trade-offs for the small-world network-enabled wireless NoCs with respect to their conventional wireline counterparts in presence of both conventional CMP and emerging big data workloads. We will discuss how Machine Learning can be exploited to design energy efficient Wireless NoC architectures. We will finish this presentation by discussing how the wireless NoC paradigm can enable realization of datacenter-on-chip using heterogeneous processing cores.

Innovative Practices on Functional Testing and Fault Simulation for FuSa

Anandh Krishnan, Intel Corporation
John van Gelder, Xilinx Inc.
Mayukh Bhattacharya, Synopsys Inc.
Sreejit Chakravarty, Intel Corp. (Moderator)
Prashant Goteti, Intel Corp. (Organizer)

I. INTRODUCTION

In this IP session, there will be 3 presentations focusing on functional testing and fault injection for automotive functional safety applications as well as a discussion on fault simulation and modeling for relevant analog test content. The 1st presentation will discuss verification solutions that accelerate fault injection for diagnostic coverage to meet ASIL requirements. The 2nd presentation discusses the use of focused random testing to achieve better functional coverage for automotive products. The 3rd presentation will discuss the various challenges associated with analog fault coverage in the absence of standards and discusses approaches to address them.

II. ACCELERATING FAULT INJECTION AND DIAGNOSTIC COVERAGE FOR FUSA (ANANDH KRISHNAN)

Safety is an integral part of the development process for automotive IPs which are steadily growing in modern automotive vehicles. ISO 26262 provides guidelines on meeting safety requirements for automotive products. Fault injection is an essential component of functional verification of safety mechanisms and to evaluate the diagnostic coverage capability. A diagnostic coverage assessment needs to be provided to justify that the IP meets certain ASIL requirements – ranging from A (least strict) to D (most strict). This presentation explores ideas on the leading edge of verification solutions to accelerate fault injection for obtaining diagnostic coverage in the functional safety flow. The design model and the designer work flow were kept unperturbed in the proposed approach.

III. FUNCTIONAL RANDOM TESTING ON THE ATE TARGETING CPU SOC FOR AUTOMOTIVE GRADE PRODUCTS (JOHN VAN GELDER)

With automotive grade products becoming more complex, new innovative techniques are encouraged to achieve better functional coverage. Structural test patterns (ATPG scan, MBIST, LBIST etc.) still provide the majority of test coverage. However, focused random testing on the ATE can provide additional coverage and find structural test coverage holes prior to production. Starting with a post silicon verification focused random tool, a smaller version was created that can be run on the device on the ATE without using external devices. The random tool provides additional coverage compared to targeted directed tests. In the unlikely event of a RMA, this tool provides a platform to develop additional tests to cover unforeseen test escapes.

IV. FAULT SIMULATION AND MODELING FOR ANALOG TEST (MAYUKH BHATTACHARYA)

We present various challenges associated with the problem of analog fault coverage in the absence of clear standards. We begin with the practical issues surrounding the catastrophic fault coverage problem - starting from fault models. We put special emphasis on the problem of deriving coverage metrics under money, time, and compute resource budgets by considering simulation speed, fault-equivalence, dynamic simulation control, fault-likelihood estimation, etc. Next, we present the challenges of parametric fault coverage estimation and possible approaches towards addressing them.

2018 IEEE 36th VLSI Test Symposium (VTS)

Broadcast-Based Minimization of the Overall Access Time for the IEEE 1687 Network

Zhanwei Zhong*, Guoliang Li[†], Qinfu Yang[‡], Jun Qian[‡], and Krishnendu Chakrabarty*
*ECE Dept., Duke University, Durham, NC, USA   [†]AMD Inc. Beijing, China   [‡]AMD Inc. Shanghai, China

Abstract—The IEEE Std. 1687 enables flexible access to on-chip instruments through the JTAG test-access port. This flexibility enables the minimization of the overall access time (OAT), and a number of techniques have been proposed to achieve this goal. However, these techniques do not utilize the broadcast feature in the 1687 network. In order to further reduce the OAT, we present an efficient test-scheduling method that exploits the broadcast feature for instrument access. Two optimization solutions are then proposed—the first solution minimizes the OAT without the retargeting time, while the second one reorders the configurations so as to minimize the overall retargeting time among configurations. Three industry test cases are used to evaluate the effectiveness of the proposed method.

I. INTRODUCTION

The IEEE Std. 1687 (IJTAG) [1] enables flexible access to the on-chip instruments through the JTAG test access port (TAP) [2], and it is now being increasingly advocated for post-silicon validation, production test, fault diagnosis, and fault monitoring [3]. To ensure flexibility of access, a hardware component called the Segment Insertion Bit (SIB) has been introduced [4]. The test data register (TDR) for the 1687 network is composed of either one or several SIBs. Besides the serial-data-in (SDI) port and the serial-data-out (SDO) port, the SIB has a hierarchical interface port (HIP), which is connected to a 1687 network segment [5]. A SIB has two states: 1) if it is opened, it includes the segments on the HIP in the scan path; 2) if it is closed, it shifts the data directly from its SDI port to its SDO port, excluding the segments on the HIP. The state of the SIB is set by first scanning in a control bit into its register and then updating its state register on a capture, shift, and update cycle (CSU) [5].

The overall access time (OAT) is a major concern in the use of the 1687 network, especially for production test using automatic test equipment (ATE). Prior work related to the minimization of the OAT includes: test-time analysis [6], design automation [4], network retargeting techniques [7], [8], and test scheduling under resource and power constraints [5], [9]. A drawback of these methods is that they are not able to reduce the OAT significantly when the same data is written to multiple instruments. In order to explain this problem, we consider the following example. As shown in Fig. 1(a) and Fig. 1(b), SIB-1 is a "doorway" SIB [4] that determines whether the path from SIB-2 to SIB-4 is included in the scan path. In addition, I-2, I-3, and I-4 are the instruments corresponding to SIB-2, SIB-3, and SIB-4, respectively. Suppose that the same data with k bits is required to be written into I-2, I-3, and I-4. If the 1687 network writes the data serially, $3k$ bits are needed. However, if we broadcast the data to the instruments, only k bits are needed.

In this paper, we present an efficient test-scheduling strategy that exploits the broadcast feature available in the 1687 network to minimize the OAT. In the proposed strategy, the process of writing a test pattern into the register of an instrument is referred to as a *task*. The tasks are placed into two different types of configurations (the supercast configuration and the hybridcast configuration), based on the computed values of cost per bit (CPB). In addition, to minimize the OAT, we develop two ILP optimization problems and obtain optimal solutions. The first solution minimizes the OAT without the retargeting time, while the second solution minimizes the retargeting time by reordering the supercast configurations and hybridcast configurations. We present

scheduling and optimization results for three industry test cases to highlight the effectiveness of the proposed method.

The remainder of this paper is organized as follows, Section II provides an overview of a broadcast-enabled 1687 network, and formulates the problem of minimizing the OAT using broadcast feature. Section III presents the proposed test-scheduling strategy that exploits the broadcast feature. In Section IV, two optimization problems are described and solved. Section V presents experimental results. Finally, Section VI concludes the paper.

II. TEST ARCHITECTURE AND PROBLEM FORMULATION

A. The BETA Architecture

The Broadcast Enabled Test Access (BETA) architecture is an implementation of the 1687 network for SOC designs; see Fig. 2. It is a hierarchial network and it consists of the following four types of building blocks:

1) *Instrument (I)*. Each instrument has multiple registers that are specified by the opcode stored in the "I_OPCODE" register in the Instrument. In an instrument group, we can only access the same type of registers, because "I_OPCODE" is broadcast to all instruments in a group. Hence, if we want to access another type of register in an instrument group, a new "I_OPCODE" should be broadcast to the instruments.

2) *Instrument Manager Controller (IMC)* is a 1149.1 TAP controller that manages a group of instruments (i.e., an instrument group). This block also has multiple instances located at the bottom of the BETA architecture. The instrument group is organized to operate in two modes: the daisy-chain mode and the broadcast mode. In the daisy-chain mode, the instruments are serially connected and a scan path that goes through all instruments is created. In the broadcast mode, the IMC directly connects to each instrument so that it can broadcast the same test data to all instruments simultaneously. The "BROADCAST_EN" register in the IMC determines whether the daisy-chain mode or the broadcast mode is used for the instrument group. Note that if some of the instruments are not designated to receive the test data, we only

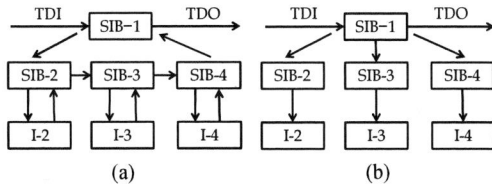

(a)   (b)

Fig. 1. Data written in (a) the daisy-chain mode, and (b) the broadcast mode.

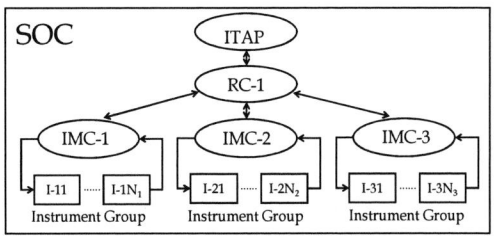

Fig. 2. An overview of the BETA architecture.

978-1-5386-3775-3/18 $31.00 © 2018 IEEE   81

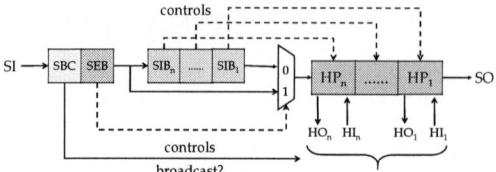

Fig. 3. Simplified view of the SIB design in the BETA architecture.

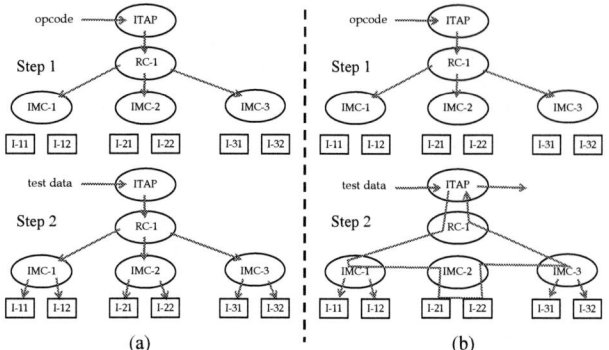

Fig. 4. Illustration of the (a) supercast, and (b) hybridcast configuration.

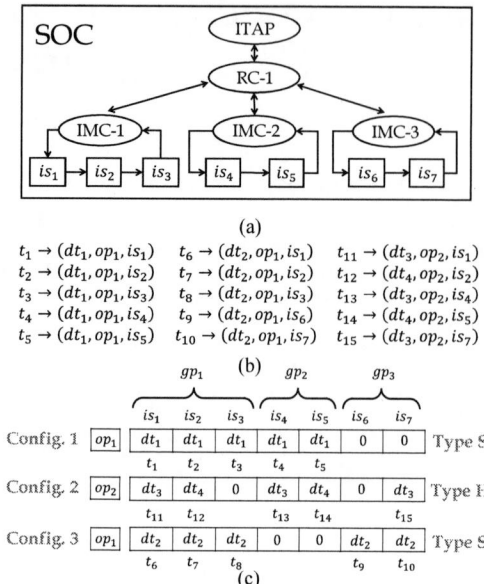

(a)

$t_1 \rightarrow (dt_1, op_1, is_1)$ $t_6 \rightarrow (dt_2, op_1, is_1)$ $t_{11} \rightarrow (dt_3, op_2, is_1)$
$t_2 \rightarrow (dt_1, op_1, is_2)$ $t_7 \rightarrow (dt_2, op_1, is_2)$ $t_{12} \rightarrow (dt_4, op_2, is_2)$
$t_3 \rightarrow (dt_1, op_1, is_3)$ $t_8 \rightarrow (dt_2, op_1, is_3)$ $t_{13} \rightarrow (dt_3, op_2, is_4)$
$t_4 \rightarrow (dt_1, op_1, is_4)$ $t_9 \rightarrow (dt_2, op_1, is_6)$ $t_{14} \rightarrow (dt_4, op_2, is_5)$
$t_5 \rightarrow (dt_1, op_1, is_5)$ $t_{10} \rightarrow (dt_2, op_1, is_7)$ $t_{15} \rightarrow (dt_3, op_2, is_7)$

(b)

		gp_1			gp_2		gp_3		
		is_1	is_2	is_3	is_4	is_5	is_6	is_7	
Config. 1	op_1	dt_1	dt_1	dt_1	dt_1	dt_1	0	0	Type S
		t_1	t_2	t_3	t_4	t_5			
Config. 2	op_2	dt_3	dt_4	0	dt_3	dt_4	0	dt_3	Type H
		t_{11}	t_{12}		t_{13}	t_{14}		t_{15}	
Config. 3	op_1	dt_2	dt_2	dt_2	0	0	dt_2	dt_2	Type S
		t_6	t_7	t_8			t_9	t_{10}	

(c)

Fig. 5. Illustration of (a) the BETA architecture, (b) an example task list with 15 tasks, and (c) the network configuration result.

need to set the "BYPASS_EN" bits in those instruments to "1". In this way, the corresponding instruments will be bypassed.

3) *Router Controller (RC)* is a 1149.1 TAP controller that only has a SIB register. This block has multiple instances located in the middle layer of the BETA architecture. The functionality of an RC is the same as that of a "doorway" SIB, i.e., determine whether the building blocks below in the hierarchy are included in the scan path.

4) *Interface Tap Controller (ITAP)* is a 1149.1 TAP controller that is located at the top of the BETA architecture. It is used to: 1) send opcode and test data to the blocks below in the hierarchy (i.e., RCs, IMCs and instruments); 2) receive test data from the blocks below it. In addition, since it is the only I/O port of the BETA architecture linked to the outside world, all the test data is fed through the serial input port of this block.

B. SIB Design

To facilitate reconfigurability in a 1687 network, a SIB is introduced to enable variable-length scan paths. Compared to traditional SIB design, the SIB in the BETA architecture with broadcast feature is more complex; see Fig. 3. The control bits are defined as follows:

1) *Selective Broadcast Enable Bit (SBC)*: When $SBC = 1$, broadcast is enabled, and the test data fed in the serial input (SI) port broadcast to all Hierarchial Ports (HPs) that are selected by the corresponding SIBs. When $SBC = 0$, broadcast is disabled and the Hierarchial Ports from HP_n to HP_1 are connected serially.

2) *SIB Exclusion Bits (SEB)*: When $SEB = 1$, the SIBs are not included into the scan path (they are bypassed). This feature can be use to reduce the SIB overhead reported in [6]. When $SEB = 0$, the SIBs are included in the scan path.

3) *SIB*: When $SIB = 1$, the segment on the corresponding HP is included into the scan path. When $SIB = 0$, the segment on the corresponding HP is bypassed.

C. Network Configurations

A network configuration specifies how ITAP, RCs, IMCs and instruments are connected, and it determines how data is sent to the registers in the instruments. In the proposed access method, two types of network configurations are used: the supercast configuration and the hybridcast configuration.

In the supercast configuration shown in Fig. 4(a), the SBC bit in ITAP, RCs and IMCs are all set to "1" in Step 1 and Step 2. This

implies that the same opcode is broadcast to all IMCs, and the same test data is broadcast to all non-bypassed instruments. Because broadcast is used everywhere in BETA, we refer to it as *supercast* (short for super-broadcast) to emphasize its importance.

In the hybridcast configuration shown in Fig. 4(b), the same opcode is first broadcast to all IMCs in Step 1 . However, in Step 2, the 1687 network is configured in a way that ITAP, RCs, and IMCs are serially connected (daisy-chain). For the instruments below each IMC, they can be either in the daisy-chain mode or in the broadcast mode. For example, the instruments below IMC-2 are serially connected and a scan path from I-21 to I-22 is formed (i.e., in the daisy-chain mode). On the other hand, IMC-1 and IMC-3 directly connect to instruments below and broadcast the test data to them (i.e., in the broadcast mode).

D. Problem Formulation

Based on the above description of the BETA architecture, we consider the following problem formulation in this work.

Input: (1) The BETA architecture for a specific SOC; (2) The task list, in which each task specifies: (i) the test data to be written, (ii) the opcode of the target data register, and (iii) the name of the target instrument.

Output: A sequence that contains the supercast configurations and the hybridcast configurations.

Constraints: (1) Only write operations are contained in the tasks (read operations are not involved); (2) The tasks are unordered (no dependency among tasks);

Objective: Minimize the OAT with respect to the given task list and the give BETA architecture.

Note that, in this paper, two specific BETA architectures are used: they are referred to as *BETA A* for industry *Design A* and *BETA B* for industry *Design B*. Based on these two BETA architectures, three test cases (i.e., task lists) are used to test the industry designs. The specifications for the industry designs, the BETA architectures and the test cases are given in Section V.

In order to illustrate the problem formulation, we use the following example. Consider the BETA architecture and the task list shown in Fig. 5(a) and Fig. 5(b), respectively. With these inputs, we obtain a sequence of three network configurations shown in Fig. 5(c), which minimize the OAT. Note that "S" refers to a supercast configuration,

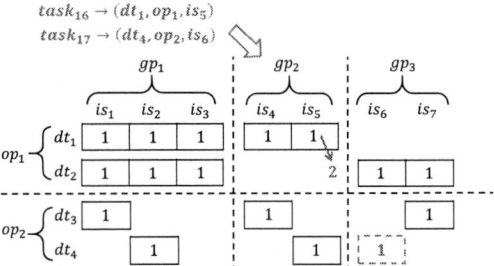

$task_{16} \rightarrow (dt_1, op_1, is_5)$
$task_{17} \rightarrow (dt_4, op_2, is_6)$

Fig. 6. Illustration of creating the lookup table.

Algorithm 1 $create\_lookup\_table(task\_list)$

Input: The tasks list $task\_list$;
Output: The lookup table $lookup\_table$;
1: initialize $lookup\_table$;
2: **for** $task$ in $task\_list$ **do**
3:    $dt, op, is := task.dt, task.op, task.i$; $gp := get\_group(is)$;
4:    **if** $lookup\_table[op][gp].find(dt) = null$ **then**
5:       $lookup\_table[op][gp][dt] = new\_hash(\ )$
6:    **end if**
7:    $count := lookup\_table[op][gp][dt]$;
8:    **if** $count.find(is) = null$ **then**
9:       $count[is] = 1$;
10:   **else**
11:      $count[is]++$;
12:   **end if**
13: **end for**
14: **return** $lookup\_table$;

Fig. 7. Pseudo-code for creating the lookup table.

---

while "H" refers to a hybridcast configuration. For each configuration: (1) op is the opcode used in a configuration; (2) gp is the name of an instrument group; (3) is is the name of an instrument; (4) dt is the data to be written ("0" means no data will be written). For example, in configuration "1" (supercast), op_1 will be broadcast to all IMCs and data dt_1 will be broadcast from is_1 to is_5 (except is_6 and is_7). On the other hand, in configuration "2" (hybridcast), op_2 will be broadcasted to all IMCs, but data dt_3 and data dt_4 will be written into the corresponding instruments serially (daisy-chain). It is important to note that the order of these configurations minimizes the retargeting cost (i.e., transition cost between configurations). The numbers (in red) on the left side indicates the order of the three configurations.

III. PROPOSED SCHEDULING STRATEGY

The broadcast feature is useful when multiple instruments share the same opcode and the same test data. However, based on the task table, it is hard to determine whether a task should be completed using the broadcast mode or the daisy-chain mode. Therefore, we develop a lookup table data structure that gathers the tasks with the same opcode and test data together. Next, based on the created lookup table, we design two algorithms to place all tasks in either supercast configurations or hybridcast configurations.

A. Creating the Lookup Table

We next describe the four-dimensional hash-table based $lookup\_table$ data structure. The first and the second dimensions of the $lookup\_table$ are indexed by the value of op (the opcode) and the value of gp (the name of the instrument group), respectively. Each element $lookup\_table[op][gp]$ is a hash table indexed by the value of dt (the test data). Finally, each element $lookup\_table[op][gp][dt]$ is a hash table indexed by the value of is (the name of the instrument). In summary, in the lookup table, the value of $lookup\_table[op][gp][dt][is]$ is equal to the number of times test data dt is written into the register specified by opcode op in the instrument named is, which belongs to instrument group gp.

Consider the example shown in Fig. 6. The lookup table (in black) is created by 15 tasks (the numbers in black add up to 15). Based on this lookup table, it is clear that: (1) is_1 to is_3 are in group gp_1, is_4 and is_5 are in group gp_2, and is_6 and is_7 are in gp_3; (2) only two opcodes op_1 and op_2 are used in the 15 tasks, and (3) four data patterns (from dt_1 to dt_4) are written into the instruments. Besides, the numbers also convey additional information. For example, we need to write the three instruments in group gp_1 once with opcode op_1 and data dt_1. On the other hand, in group gp_2, we only need to write instrument is_4 with opcode op_2 and data dt_3.

Next suppose that two additional tasks ($task_{16}$ and $task_{17}$) are added. Then we have to update the lookup table as follows:

1) For $task_{16}$, because there is already a value "1" indexed by dt_1, op_1 and is_5, we only need to increase this value by 1.

2) For $task_{17}$, there is no such a value in the lookup table indexed by dt_4, op_2 and is_6. Therefore, we create a value of "1" instead.

The pseudo-code for creating the lookup table is shown in Fig. 7. The computational complexity of this algorithm is $\mathcal{O}(n)$, where n is the total number of tasks.

B. Definition of Cost Per Bit (CPB)

In this subsection, we define two parameters that are used in the next subsection to construct the supercast configurations. We first introduce some notations: (1) N_{shift} is the total number of bits that are shifted in group gp; (2) N_{eff} is the effective number of bits written into the registers in group gp; (3) N_{is} is the number of the instruments in group gp; (4) L_{dt} is the length (in bits) of test data dt; (5) L_{op} is the length (in bits) of opcode op; (6) N_{val} is the number of values in $lookup\_table[op][gp][dt]$ (some values might not exist).

The first CPB parameter is defined as the data transmission cost for instrument group gp in the broadcast mode:

$$CPB_{brc}(gp) = \frac{N_{shift}}{N_{eff}} = \frac{L_{op} + N_{is} + L_{op} + L_{dt}}{L_{dt} \times N_{val}} \quad (1)$$

In order to broadcast data dt to N_{val} non-bypassed instruments, $L_{op} + N_{is}$ bits are first shifted in to bypass some of the instruments. Next, $L_{op} + L_{dt}$ bits are needed to select the target registers and broadcast data dt to all non-bypassed instruments. Because there are N_{val} non-bypassed instruments, the number of effective bits written into the registers is $L_{dt} \times N_{val}$.

The second CPB parameter is defined as the data transmission cost for instrument group gp in the daisy-chain mode:

$$CPB_{dc}(gp) = \frac{N_{shift}}{N_{eff}} = \frac{L_{op} + N_{is} + L_{op} + L_{dt} \times N_{val}}{L_{dt} \times N_{val}} \quad (2)$$

In order to serially shift data dt to N_{val} non-bypassed instruments, $L_{op} + N_{is}$ bits are again shifted in to bypass some of the instruments. Next, a total of $(L_{op} + L_{dt} \times N_{val})$ bits are needed to shift data dt to all non-bypassed instruments. Because there are N_{val} non-bypassed instruments, the number of effective bits we have written into the registers is again $L_{dt} \times N_{val}$.

C. Constructing the Supercast Configurations

The data structure of a supercast configuration is shown in Fig. 8. It consists of two data fields: (1) opcode op that specifies the target data register, and (2) an array that contains the data values to be written. The "0" values in the array indicate that we do not need to write data to those instruments. To construct supercast configurations, retrieval operations are needed. Each retrieval operation is specified by opcode op and data dt, because it constructs a supercast configuration that broadcasts data dt to the data registers specified by op.

An example of a retrieval operation specified by op_1 and dt_1 is shown in Fig. 8. In this operation, the information in the first line of the lookup table will be retrieved. For group gp_1, we first compute the values of CPB_{brc} and CPB_{dc}. Next, we compare these two values.

978-1-5386-3775-3/18 $31.00 © 2018 IEEE

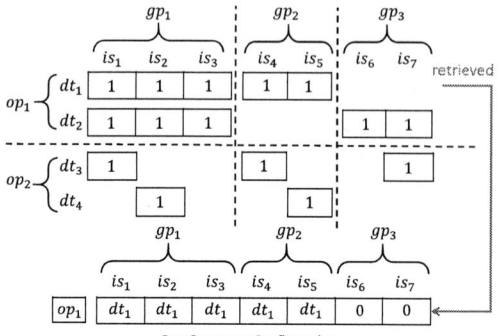

Fig. 8. Illustration of a retrieval operation for a supercast configuration.

Algorithm 2 $get\_supercast\_configs(lookup\_table, param)$

Input: The lookup table $lookup\_table$ and the user-defined parameter $param$;
Output: The supercast configurations $supercast\_configs$ and the updated lookup table $lookup\_table$;
  1: initialize $supercast\_configs$;
  2: **for** (op, dt) in $lookup\_table$ **do**
  3:     **while** true **do**
  4:         $record := retrieve(op, dt, lookup\_table)$;
  5:         **if** $not\_empty(record)$ **then**
  6:             $supercast\_configs.push\_back(op, record)$;
  7:         **else**
  8:             break;
  9:         **end if**
10:     **end while**
11: **end for**
12: **return** $supercast\_configs$, $lookup\_table$;

Fig. 9. Pseudo-code for constructing the supercast configurations.

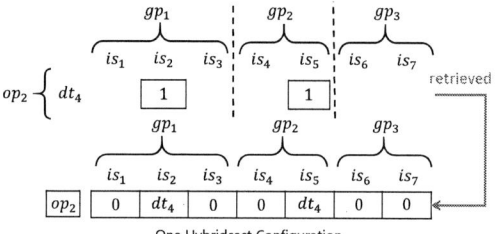

Fig. 10. Illustration of a retrieval operation for a hybridcast configuration.

Suppose that $param$ ($param > 1$) is a user-defined parameter. If the value of CPB_{brc} is equal to or smaller than the value of $param \times CPB_{dc}$, the three values in the lookup table indexed by gp_1 will be decreased by 1 and then be removed (because they are now equal to 0), and the first three elements of the array in the supercast configuration will set to "dt_1"s. However, if the value of CPB_{brc} is larger than the value of $param \times CPB_{dc}$, nothing will be changed in the lookup table and the first three element of the array in the supercast configuration will remain "0"s.

Similar procedures are also carried out for the values in group gp_2 and group gp_3. Suppose the computed values of CPB_{brc} is smaller than the values of CPB_{dc} for both group gp_1 and group gp_2. The result of this particular retrieval operation is shown in Fig. 8.

The pseudo-code for constructing the supercast configurations is shown in Fig. 9. Function $retrieve$ carries out the retrieval operation (line 4). It returns the test data to be written for the selected instrument. The computational complexity of this algorithm is $\mathcal{O}(p \times m)$, where p is the total number of (op, dt) pairs in the lookup table, and m is the total number of instruments.

D. Constructing the Hybridcast Configurations

The data structure of a hybridcast configuration is nearly the same as that of a supercast configuration. The only difference is that in a supercast configuration, the data values (except "0" values) should

---
Algorithm 3 $get\_hybridcast\_configs(lookup\_table)$

Input: The lookup table $lookup\_table$;
Output: The hybridcast configurations $hybridcast\_configs$;
  1: initialize $hybridcast\_configs$;
  2: **for** op in $lookup\_table$ **do**
  3:     **while** true **do**
  4:         $record := retrieve(op, lookup\_table)$;
  5:         **if** $not\_empty(record)$ **then**
  6:             $hybridcast\_configs.push\_back(op, record)$;
  7:         **else**
  8:             break;
  9:         **end if**
10:     **end while**
11: **end for**
12: **return** $hybridcast\_configs$;

Fig. 11. Pseudo-code for constructing the hybridcast configurations.

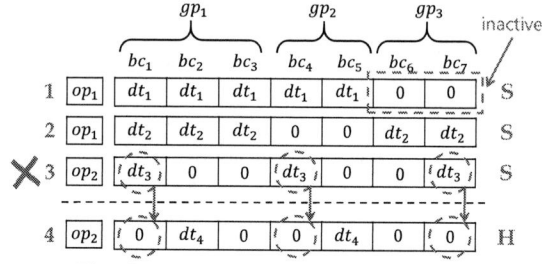

Fig. 12. Illustration of configuration integration.

be the same; however, in a hybridcast configuration, the data values (except "0" values) can be different.

An example of a retrieval operation specified by op_2 is shown in Fig. 10. The lookup table shown here is the one after we construct the supercast configurations. Therefore, most of the values are retrieved and removed (when equal to 0). In this example, only one line of values exist. For group gp_1, only the value indexed by is_2 exists, we retrieve this value to the hybridcast configuration and remove this value in the lookup table. Next, for group gp_2 and group gp_3, similar procedures are carried out. Finally, the hybridcast configuration after this retrieval operation is shown in Fig. 10.

The pseudo-code for constructing the hybridcast configurations is shown in Fig. 11. The computational complexity of this algorithm is $\mathcal{O}(k \times m)$, where k is the total number of different op values, and m is the total number of instruments.

IV. SCHEDULE OPTIMIZATION

Recall that we have obtained the supercast configurations and the hybridcast configurations based on the steps described in Section III. In this section, we introduce two integer linear programming (ILP) problem formulations [10] and utilize their solutions to further reduce the OAT. The first solution minimizes the OAT without the configuration retargeting (transition) time, and the second solution minimizes the overall retargeting time.

A. Motivational Example

Fig. 12 shows three supercast configurations (the upper part) and one hybridcast configuration (the lower part), which are retrieved from the lookup table shown in Fig. 6. However, the OAT for these configurations can be further decreased. For example, if we *integrate* the third supercast configuration with the first hybridcast configuration, no additional OAT cost is added. However, the OAT is reduced by about 100 clock cycles (the average retargeting cost for the BETA architecture), because there are only 3 configurations now instead of 4 configurations. Based on this observation, we need to find out a set of supercast configurations to be integrated with the hybridcast configurations, so that the OAT can be further reduced. To achieve this goal, we have developed an ILP model as described next.

978-1-5386-3775-3/18 $31.00 © 2018 IEEE

1: $\mathbb{GP}$: the set of names for all instrument groups in the BETA architecture.
2: $\mathbb{OP}$: the set of all opcodes in the BETA architecture;
3: $NGH_{op,gp}$: the total number of active instrument groups named gp in the hybridcast configurations specified by opcode op;
4: C_{re}: the average cost (bits) for retargeting among configurations;
5: $C_{op,i}$: the length of data dt in the supercast configuration specified by $d_{op,i}$.
6: $N_{op,i}$: the number of non-zero values in the supercast configuration specified by $d_{op,i}$.
7: NS_{op}: the total number of configurations in $supercast\_configs$ specified by opcode op;
8: NH_{op}: the total number of configurations in $hybridcast\_configs$ specified by opcode op;
9: $DGS_{gp,op,i}$: equal to 1 if instrument group gp in the supercast configuration specified by $d_{op,i}$ is active; otherwise 0.

Fig. 13. Definition of the parameters in the ILP model of Section IV.A.

B. Integration of Configurations

The set of parameters considered for the ILP model is defined in Fig. 13. Note that an *inactive* instrument group refers to an instrument group that contains all "0"s in a configuration; see Fig. 12. Otherwise, it is an *active* instrument group. The binary variables, $d_{op,i}, op \in \mathbb{OP}, 1 \le i \le NS_{op}$, are defined such that $d_{op,i}$ is equal to 1 if the ith supercast configuration specified by opcode op should be integrated with the hybridcast configurations specified by the opcode op.

Prior to optimization, the number of hybridcast configurations NH_{op} specified by the opcode op (refered to as NH_{op}) is given by:

$$NH_{op} = \max_{\forall gp \in \mathbb{GP}} \{NGH_{op,gp}\} \tag{3}$$

After optimization, the number of the hybridcast configurations nh_{op} specified by the opcode op (referred to as nh_{op}) is given by:

$$nh_{op} = \max_{\forall gp \in \mathbb{GP}} \left\{ NGH_{op,gp} + \sum_{i=1}^{NS_{op}} d_{op,i} \times DGS_{gp,op,i} \right\} \tag{4}$$

Note that Equation (4) is non-linear. However, we can linearize it as shown below:

$$nh_{op} \ge NH_{op,gp} + \sum_{i=1}^{NS_{op}} d_{op,i} \times DGS_{gp,op,i}, \forall gp \in \mathbb{GP} \tag{5}$$

Based on the variables and constraints defined above, our objective is to minimize the cost given by:

$$\Delta C = C_{re} \times \sum_{op \in \mathbb{OP}} \left(nh_{op} - NH_{op} - \sum_{i=1}^{NS_{op}} d_{op,i} \right)$$
$$+ \sum_{op \in \mathbb{OP}} \sum_{i=1}^{NS_{op}} C_{op,i} \times d_{op,i} \times (N_{op,i} - 1) \tag{6}$$

The value of ΔC is equal to the change in OAT after we apply the integration based on the ILP solution. Because our goal is to minimize the OAT, ΔC is always non-positive. In the first term, the value of $nh_{op} - NH_{op}$ is equal to the number of newly added hybridcast configurations specified by opcode op after integration. On the other hand, the value of $\sum_{i=1}^{NS_{op}} d_{op,i}$ represents the number of deleted supercast configurations specified by opcode op because of integration. Therefore, the contribution of the first term to ΔC indicates the cost change incurred by the change in the number of configurations. In a supercast configuration, we only need to broadcast the test data once. However, if a supercast configuration is integrated into a hybridcast configuration, the data should be sent to each instrument individually. The second term computes the cost increase induced by this case.

The pseudo-code for configuration integration is shown in Fig. 14. The computational complexity of this algorithm is $\mathcal{O}(d \times m)$, where d is the total number of "$d_{op,i}$"s that are equal to 1, and m is the total number of instruments.

Algorithm 4 $integrate\_configs(supercast\_configs, hybridcast\_configs, d)$

Input: The supercast configurations $supercast\_configs$, the hybridcast configurations $hybridcast\_configs$ and the binary decision variables d;
Output: The updated supercast configurations $supercast\_configs$, and the updated hybridcast configurations $hybridcast\_configs$;
1: **for** (op, i) in d **do**
2:    $idx := get\_config\_index(op, i)$;
3:    $target\_configs := get\_hybridcast\_configs(op)$;
4:    **for** $config$ in $target\_configs$ **do**
5:       $integrate\_configs(supercast\_configs[idx], config)$;
6:       **if** $is\_empty(supercast\_configs[idx])$ **then** break;
7:       **end if**
8:    **end for**
9: **end for**
10: **return** $supercast\_configs$, $hybridcast\_configs$;

Fig. 14. The pseodu-code for integrating configurations.

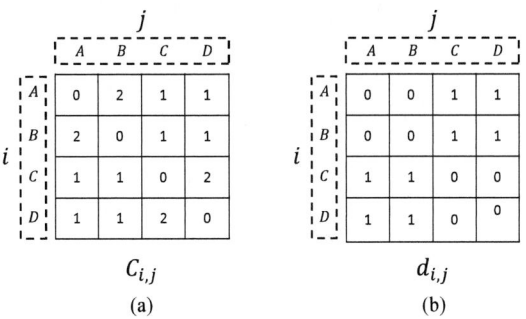

Fig. 15. The values of (a) the retargeting costs, and (b) the decision variables.

C. Minimization of Retargeting Time

Recall that we have already obtained a optimized set of the supercast configurations and the hybridcast configurations. Next, we need to reorder these configurations so that the retargeting time is minimized. In order to illustrate this problem, we consider the following example. Suppose that we are given four different network configurations A, B, C and D, and the retargeting time between each pair of configurations is shown in Fig. 15(a). The parameter $C_{i,j}$ represents the retargeting time from configuration i to configuration j. Note that the retargeting time from configuration i to configuration j is the same as that from j to i. Our objective is to find a sequence that begins from O (the initial configuration) and goes through all other configurations exactly once such that we can minimize the overall retargeting time (e.g., the sequence $O \rightarrow B \rightarrow C \rightarrow A \rightarrow D$ is the solution for this example). However, with this objective, it is hard to formulate the problem using ILP. Rather than finding a sequence that starts from O, we can instead find a cyclic sequence that minimizes the overall retargeting time (e.g., the cyclic sequence $D \rightarrow B \rightarrow C \rightarrow A \rightarrow D$). The overall retargeting time of the sequence $O \rightarrow B \rightarrow C \rightarrow A \rightarrow D$ and of the cyclic sequence $D \rightarrow B \rightarrow C \rightarrow A \rightarrow D$ are nearly the same. Hence, our objective now is to find a cyclic sequence that minimizes the overall retargeting time.

Based on the above discussion, we now develop the ILP model. The binary decision variables $d_{i,j}$ are defined as follows: $d_{i,j}$ and $d_{j,i}$ are both equal to 1 if there is a retargeting from configuration i to configuration j. Suppose that, in this example, the cyclic sequence $D \rightarrow B \rightarrow C \rightarrow A \rightarrow D$ minimizes the overall retargeting time. According to the parameters in Fig. 15(a), the minimized retargeting time can be expressed as:

$$C_{min} = C_{A,D} + C_{D,B} + C_{B,C} + C_{C,A}$$
$$= (C_{A,D} + C_{D,A} + C_{B,D} + C_{D,B} \tag{7}$$
$$+ C_{B,C} + C_{C,B} + C_{A,C} + C_{C,A})/2$$

Using the binary decision variables, the minimized retargeting time can now be collectively expressed as:

TABLE I
ILP Model for Configuration Ordering.

Objective:
$\text{Minimize} \sum_{i=0}^{N-1} \sum_{j=0}^{N-1} (C_{i,j} \times d_{i,j})$
Subject to:
$d_{i,j} = d_{j,i}, \forall\ 0 \leq i,j \leq N-1$
$d_{i,j} = 0, \forall\ i = j,\ 0 \leq i,j \leq N-1$
$\sum_{j=0}^{N-1} C_{i,j} = 2, \forall\ 0 \leq i \leq N-1$
$\sum_{i=0}^{N-1} C_{i,j} = 2, \forall\ 0 \leq j \leq N-1$

Algorithm 5 $get\_optimized\_configs(d, supercast\_configs, hybridcast\_configs)$

Input: The binary decision variables d, the supercast configurations $supercast\_configs$, and the hybridcast configurations $hybridcast\_configs$.
Output: The optimized configurations $optimized\_configs$;
1: $unordered\_configs := join(supercast\_configs, hybridcast\_configs)$;
2: $N := unordered\_configs.size(\ )$;
3: $order := new\_array(N, 0);\ order[0] := random(N-1)$;
4: **for** $1 \leq i \leq N-1$ **do**
5:      $idx = find\_index(d[order[i-1]], 1)$;
6:      $order[i] := idx$;
7:      $d[i-1][idx] := 0$;
8:      $d[idx][i-1] := 0$;
9: **end for**
10: initialize $ordered\_configs$;
11: **for** $0 \leq i \leq N-1$ **do**
12:      $ordered\_configs.push\_back(unordered\_configs[order[i]])$;
13: **end for**
14: **return** $ordered\_configs$;

Fig. 16. Pseudo-code to determine the order of the configurations.

$$C_{min} = \sum_{i \in \mathbb{C}} \sum_{j \in \mathbb{C}} (C_{i,j} \times d_{i,j}) \qquad (8)$$

where $\mathbb{C}$ is the set of all configurations (except configuration O).

Based on an inspection of the $[d_{i,j}]$ matrix shown in Fig. 15(b), we can easily derive the following two constraints:

$$\sum_{j \in \mathbb{C}} d_{i,j} = 2, \forall\ i \in \mathbb{C} \qquad (9)$$

$$\sum_{i \in \mathbb{C}} d_{i,j} = 2, \forall\ j \in \mathbb{C} \qquad (10)$$

After we obtain an optimal cyclic sequence, we need to transform it back to a sequence that begins with O (the initial configuration). We can randomly choose a configuration in the cyclic sequence, delete the transition to this configuration, and add a transition from O to it. For example, if the optimum cyclic sequence is $(D \rightarrow B \rightarrow C \rightarrow A \rightarrow D)$, then we randomly choose a configuration (e.g., A), delete the transition from C to A, and add the transition from O to A. Finally, we get the sequence $O \rightarrow A \rightarrow D \rightarrow B \rightarrow C$.

Based on the above example, we can now easily derive the ILP model for the minimization of the retargeting time among the supercast configurations and the hybridcast configurations; see Table I. Here, the value of N is the number of all configurations.

The pseudo-code for updating the order of configurations is shown in Fig. 16. The computational complexity of this algorithm is $\mathcal{O}(c^2)$, where c is the total number of all configurations.

V. Experimental Results

In this section, we evaluate the proposed test scheduling strategy using state-of-the-art industry designs. The specifications of two SOCs and their BETA architectures are shown in Table II. *Design A* is a relatively low-end SOC with about 5.5 billion transistors; however, *Design B* is a "monster" SOC product that has about 12.5 billion transistors. Note that "# of levels" in Table II refers to the network depth of the BETA architecture.

In Table III, we make a comparison between four methods. Test-time reduction is defined as the percentage reduction compared to the

TABLE II
Specifications for Industry Designs and Their BETA Architectures

Specifications	Design A	Design B
# of Transistors	~5.5 billion	~12.5 billion
BETA Name	BETA A	BETA B
# of RCs	6	13
# of IMCs	56	63
# of instruments	294	659
# of levels	5	6

TABLE III
Scheduling and Optimization Results

M-I. Not using the broadcast feature
M-II. Proposed method without ILP-based optimization
M-III. Proposed method with the first ILP-based optimization (no retargeting time optimization)
M-IV. Proposed method with both ILP-based optimizations

Design	Test Case	OAT (clock cycles) / Test Time Reduction			
		M-I	M-II	M-III	M-IV
A	1	31053	12671 / 60%	12353 / 61%	11816 / 62%
B	2	56188	17065 / 70%	16569 / 71%	15935 / 72%
B	3	76145	43592 / 43%	42786 / 44%	41967 / 45%

baseline method (M-I) used for production testing. For example, for Test Case 2, the OAT for M-II (17065 clock cycles) is 70% smaller than the OAT for M-I (56188 clock cycles). It is clear that the broadcast feature significantly reduces the OAT. In addition, additional OAT reduction is obtained using the two ILP-based optimizations. Even a 1-2% reduction is significant because the ILP models are applied to already optimized schedules. Note that, in the manufacture test of each chip (i.e., Design A or Design B), Test Case (i.e., 1, 2, or 3) will be written in for hundreds of times. Therefore, the overall test time reduction will be approximately 2 seconds for each chip, which represents a significant amount of time saving for manufacture test.

VI. Conclusion

We have presented an efficient test-scheduling approach that exploits the broadcast feature in IEEE 1687 to reduce the OAT. Two ILP models and their solutions have been presented for further optimization. The first solution minimizes the OAT without the retargeting time, while the second solution minimizes the retargeting time. Three industry test cases have been used to evaluate the efficiency of the proposed method. Note that, in this work, the test tasks have only write operations. In future work, we will consider read operations in the optimization solution.

References

[1] IEEE Standard Committee *et al.*, "IEEE standard for access and control of instrumentation embedded within a semiconductor device," *IEEE Std. 1687-2014*, 2014.
[2] IEEE Standard Committee *et al.*, "IEEE standard test access port and boundary scan architecture," *IEEE Std. 1149.1-2001*, 2001.
[3] R. Cantoro *et al.*, "Automatic generation of stimuli for fault diagnosis in IEEE 1687 networks," in *IOLTS'16*, pp. 167–172, 2016.
[4] F. G. Zadegan *et al.*, "Design automation for IEEE P1687," in *DATE'11*, pp. 1–6, 2011.
[5] S. S. Nuthakki *et al.*, "Optimization of the IEEE 1687 access network for hybrid access schedules," in *VTS'16*, pp. 1–6, 2016.
[6] F. G. Zadegan *et al.*, "Test time analysis for IEEE P1687," in *ATS'10*, pp. 455–460, 2010.
[7] R. Baranowski *et al.*, "Modeling, verification and pattern generation for reconfigurable scan networks," in *ITC'12*, pp. 1–9, 2012.
[8] R. Krenz-Baath *et al.*, "Access time minimization in IEEE 1687 networks," in *ITC'15*, pp. 1–10, 2015.
[9] F. G. Zadegan *et al.*, "Test scheduling in an IEEE P1687 environment with resource and power constraints," in *ATS'11*, pp. 525–531, 2011.
[10] A. Schrijver, *Theory of linear and integer programming*. John Wiley & Sons, 1998.

2018 IEEE 36th VLSI Test Symposium (VTS)

Efficient Parallel Testing: A Configurable and Scalable Broadcast Network Design Using IJTAG

Saurabh Gupta[†,‡], Jae Wu[†], and Jennifer Dworak[‡]
[†]NVIDIA Corporation, Santa Clara, CA   [‡]Southern Methodist University, Dallas, TX
[‡]{sgupta, jdworak}@smu.edu, [†]jwu@nvidia.com

Abstract—To meet high performance requirements, System-on-Chips (SoCs) may include multiple replicated copies of functional embedded cores. To reduce the time required to apply identical test data to these replicated cores, we designed a novel broadcast network architecture that harnesses IEEE Std 1687 (IJTAG). Our architecture provides highly configurable broadcast/multicast and daisy modes, allowing one to selectively apply test data to any combination of embedded modules. The broadcast network is also scalable, supports hierarchical network architectures, and can be easily interfaced with other IJTAG-compliant test architectures. The new broadcast network provides a trade-off between network reconfigurability and the programming overhead of the network reconfiguration bits. It saves up to 70-80% of the test time in a sample test data broadcast scenario compared to serial and prior broadcast IJTAG networks. Compared to a serial network, our broadcast network requires only one extra reconfiguration bit.

I. INTRODUCTION

Modern System-on-Chips (SoCs) include several embedded cores, such as DSP engines, memory cores, communication modules, and intellectual property (IP) cores. Depending on the intended application, SoCs may also include multiple replicated copies of these embedded cores and IP modules. Moreover, increasingly complex SoC design necessitates an increasing number of embedded instruments to be included in the chip for test, debug, monitoring, characterization and configuration of the chip and the board on which it is placed.

IEEE Std 1687 (IJTAG) [1] provides a standard interface for the reconfigurable access and control of on-chip embedded instruments. It was also designed to provide interface compatibility with the IEEE 1149.1 (JTAG) [2] test access port (TAP) and TAP controller. However, IJTAG does not use the instructions in the IEEE 1149.1 Instruction Register (IR) to configure the IJTAG portion of the test network. Instead, IJTAG uses scan data within the shift path itself to dynamically reconfigure the active shift path by opening (or closing) access to new shift path segments.

Replicated copies of functional embedded cores and their associated test/debug embedded instruments require identical test data. Thus, the same test data needs to be shifted several times through the network for each instance of the replicated embedded cores. This results in long test data write times in a standard IJTAG scan network—because the data is shifted through the network serially from the Test Data In (TDI) port to the Test Data Out (TDO) port.

To solve this problem, we propose a configurable and scalable broadcast network architecture using IJTAG. Some of the major features of this architecture are the following:

- Reconfigurable broadcast and daisy chain modes.
- Common test data can be broadcasted to any combination of instruments in the IJTAG network while the remaining instruments stay bypassed. SIBs do not change their open/close states during the broadcast mode.
- Switching the network from broadcast mode to daisy mode results in a fixed length daisy mode shift path configuration, each time. The resulting daisy chain configuration after exiting the broadcast mode does not depend on the prior broadcast mode configuration or the test data shifted during the last broadcast cycle.
- When this network is switched from broadcast to daisy mode, the resulting shift path is of the shortest length possible in the network. All the SIBs in the network reset to their closed state when switching from broadcast to daisy mode. This reduces the amount of time required to reconfigure the shift path for the next daisy mode operation that follows the previous broadcast mode operation.
- The broadcast network architecture is scalable. During the design phase, the designer does not have to identify the groups of clients that would receive broadcast data.
- The new broadcast network architecture supports hierarchy while maintaining the above mentioned feature set.
- Similar to the traditional IJTAG network, the new broadcast network architecture can interface with the TAP controller. Thus it can be used in any IEEE 1149.1-compliant test architecture.

In the remainder of this paper, Section II presents an overview of prior work in test time reduction through test data broadcast. Next, Section III presents some of the relevant architectural features of an IJTAG test network. Section IV discusses prior IJTAG based broadcast network designs. Section V discusses the new broadcast network architecture using IJTAG and the different features of this architecture. Finally, Section VI concludes the paper.

II. PRIOR WORK

In VLSI testing, the concept of broadcasting test data has been traditionally associated with applying the same test data to multiple internal scan chains. This was first proposed in [3] and [4]. Other scan architectures, such as Illinois scan have also used broadcast scan modes [5], [6]. Several compression based test architectures, such as [7], [8], and [9] have used broadcast scan schemes. The conceptual modular architecture using a Test Access Mechanism (TAM) and a core test wrapper

978-1-5386-3775-3/18 $31.00 © 2018 IEEE

was published in [10]. A TAM and wrapper architecture for parallel testing of identical cores was discussed in [11]. Parallel testing of multiple identical cores using a broadcast-based TAM has also been discussed in [12], [13], [14], [15].

The IEEE 1500 Standard [16] for embedded core-based integrated circuits defines a wrapper parallel port (WPP) as a parallel interface to the IEEE 1500 wrapper. This parallel access mechanism provides increased data bandwidth to the wrapper core. However, the IEEE Std 1149.1 [2] for board testing and the IEEE Std 1500 [16] do not provide broadcast scan mechanisms to apply identical test data to the replicated copies of embedded cores in an SoC. In contrast, broadcast network designs to apply identical data to multiple test instruments were briefly discussed in the IEEE Std 1687 [1, pp. 230-232]. In addition, a Parallel-IJTAG network architecture and some optimizations were discussed in [17], but a data broadcast mode was not included in that discussion.

III. IJTAG Overview

Fig. 1 shows an overview of the JTAG architecture and the interfacing of the IEEE 1687 IJTAG network to JTAG. Note that additional test data registers (TDRs) may also be present and selected through placing an appropriate instruction in the instruction register (IR).

An IEEE 1687 test network provides a scalable plug-and-play interface and access mechanism for on-chip embedded instruments. An important advantage of IEEE 1687 is that the active scan path is reconfigured using a distributed control mechanism based on the contents of the scanned data bits. Thus, adding new embedded instruments to the IJTAG network at design time and other IJTAG network design modifications do not require redesigning the IEEE 1149.1 IR decoder. IJTAG also allows access to any combination of hundreds of instruments without requiring new instructions.

Distributed control of an IJTAG network is provided through special data-side scan registers with shift and update stages that generate the local control signals. A scan register that generates this type of local control signal is referred to as a *Network Instruction Bit* (NIB) [17], [18]. Examples of local control signals that could be generated by a NIB include *"local reset"* and *"deny"*. Such signals can be asserted by placing a correct value in the update cell of the NIB during the UpdateDR (Update Data Register) state of the JTAG TAP controller, as shown in Fig. 2. Such local control is valuable because it allows operations to be applied to only a specific portion of the IJTAG network or even to a specific instrument.

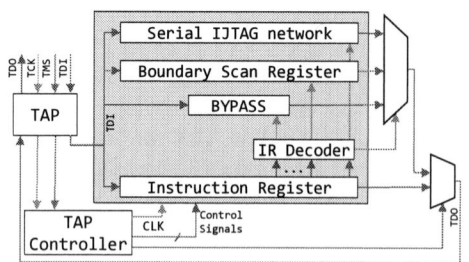

Fig. 1: Interfacing of serial IJTAG network to JTAG [17]

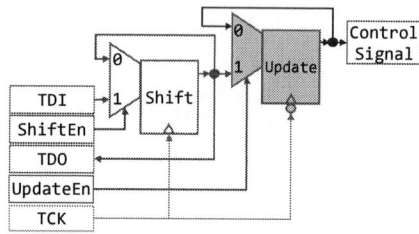

Fig. 2: Network Instruction Bit [17]

A Segment Insertion Bit (SIB) is a special type of NIB that is used in an IJTAG network to dynamically reconfigure the active shift path by adding (or removing) a scan segment to (or from) the active TDI-TDO shift path, as shown in Fig. 3. The *SIB_Select** signal in Fig. 3 can be asserted or de-asserted by placing an appropriate value in the update cell of the SIB. When the *SIB_Select** signal is asserted, it causes the ScanMux to select the input connected to the TDO2 signal, which is then fed to the SIB shift cell. Thus the "Extra scan segment" between TDI2 and TDO2 gets added to the the active shift path. The SIB is considered to be in "open" state during this mode. The *SIB_Select** signal also allows the Capture, Shift, and Update control signals from the TAP controller to propagate to the "Extra scan segment". When the *SIB_Select** signal is de-asserted, the active shift path shown consists of only the SIB cell between TDI and TDO. Thus, the "Extra scan segment" is bypassed. The SIB is considered to be in the "closed" state during this mode.

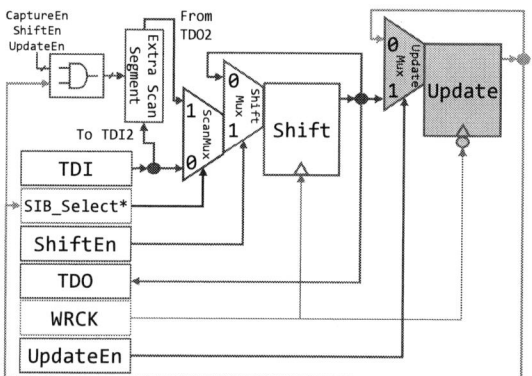

Fig. 3: SIB design with extra scan segment also shown

IV. Previous Broadcast Network Designs

An example broadcast network design from the IJTAG standard is shown in Fig. 4. The network design shown in Fig. 4 supports both broadcast and daisy chain modes. Here, the network can be operated in broadcast mode by asserting the broadcast control signal of the Network Instruction Bit (NIB). During the broadcast mode, all three instruments can be loaded simultaneously. However, there is no way to configure the broadcast mode such that some instruments remain bypassed during broadcast and only a subset of these instruments receive the broadcasted data. Also, there is no way to reconfigure the active shift path in daisy scan mode either.

Another example broadcast network design from the IJTAG standard is shown in Fig. 5. This design can be used to

978-1-5386-3775-3/18 $31.00 © 2018 IEEE

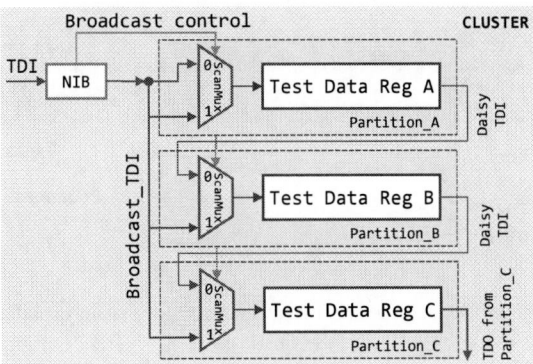

Fig. 4: Broadcast and daisy scan example from the IEEE Std 1687, [1, p. 232]

broadcast test data to all three instruments in the network. It can also access and control exactly one instrument at a time by asserting Sel1, Sel2, or Sel3. However, it does not provide a daisy mode so that all the instruments can be read out concurrently in series. Similar to the design in Fig. 4, with this design there is no way to configure the broadcast mode such that some instruments are bypassed during broadcast and only a subset of these instruments receive the broadcast data.

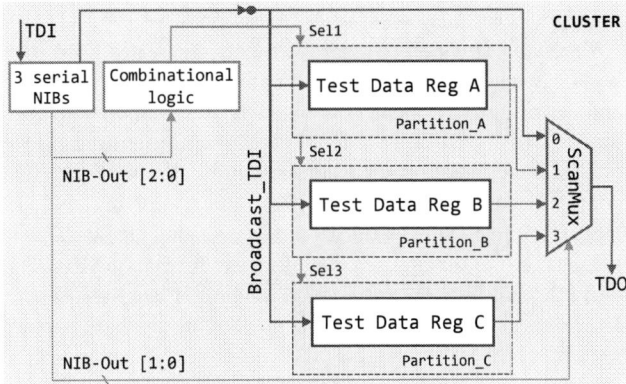

Fig. 5: Broadcast design example with exclusive access from the IEEE Std 1687, [1, p. 231]

In contrast, our new broadcast network architecture provides configurable broadcast and daisy modes, and improves test data write times with very low additional area overhead.

V. PROPOSED BROADCAST NETWORK ARCHITECTURE

As discussed in Section II, the IEEE Std 1687 describes two example broadcast networks. However, those networks broadcast the test data to all the instruments in the network. Thus, a careful decision has to be made during design regarding which instruments should be collected together in a broadcast group. A reconfigurable broadcast mode would give more flexibility during the design phase as well as during the actual testing phase. A network that provides reconfigurable broadcast (and daisy) modes and supports hierarchical broadcast networks and that has a low increase in the area overhead compared to a serial IJTAG network is desirable. The new broadcast network was designed keeping these design features in mind; using IJTAG also makes the architecture scalable. The new IJTAG

broadcast network can be interfaced to JTAG circuitry similar to the interfacing of the IJTAG network shown in Fig. 1.

In the subsequent subsection we begin by describing our broadcast network design with a single level of hierarchy. This is followed by design improvements required to support hierarchical broadcast networks.

A. Broadcast Network - Without Hierarchy

Fig. 6 shows a conceptual view of the new IJTAG broadcast network. The example network is part of a cluster that is divided into three partitions—A, B, and C. Each partition consists of a TDR and its corresponding SIB that is designed to support broadcast modes. Compared to a regular serial IJTAG network with a flat-architecture [17], [19], the design consists of an extra ScanMux per TDR. The broadcast_TDI signal can be routed along with other control signals.

1) NIB as Partition Broadcast Control Bit (BCB): The partition Broadcast Control Bit (BCB) at the cluster level is implemented using a NIB. The 1-bit partition BCB-Out signal from the partition BCB feeds the select signal of the scan-muxes in each partition. When the partition BCB asserts the BCB-Out signal, the network switches to broadcast scan mode. When the BCB-Out signal is de-asserted, the network switches back to daisy (or serial) scan mode. The BCB-Out signal from the partition BCB is also applied to the modified SIBs in the network. With reference to the SIBs, the BCB-Out signal is referred to as the *SIB-Freeze* signal in this paper.

2) Broadcast Segment Insertion Bit: The modified SIB design used in the broadcast network of Fig. 6 is shown in Fig. 7. Compared to the SIB design in Fig. 3, the broadcast SIB has some additional features—a frozen SIB state, a shift cell with a feedback path forced to zero during broadcast mode, and a pipelined UpdateEn. In terms of area overhead, the broadcast SIB requires two extra logic gates, and a pipeline flip-flop when compared to the SIB design in Fig. 3. These SIB design features that enable efficient broadcast mode operations are described subsequently.

Frozen SIB State: As mentioned earlier, the partition BCB-Out signal also serves as the SIB-Freeze signal for the broadcast SIB shown in Fig. 7. Based on value of the SIB-Freeze

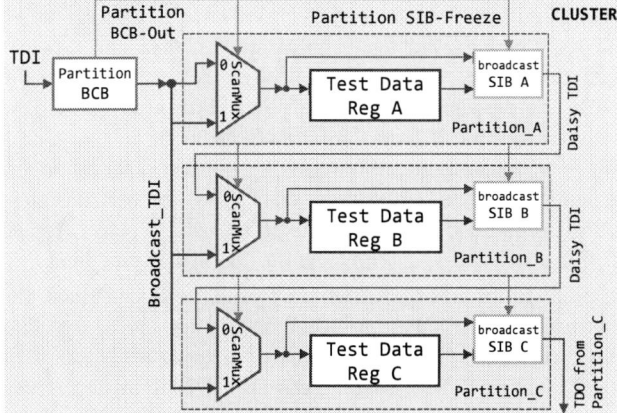

Fig. 6: Proposed IJTAG reconfigurable broadcast network - Without hierarchy

978-1-5386-3775-3/18 $31.00 © 2018 IEEE

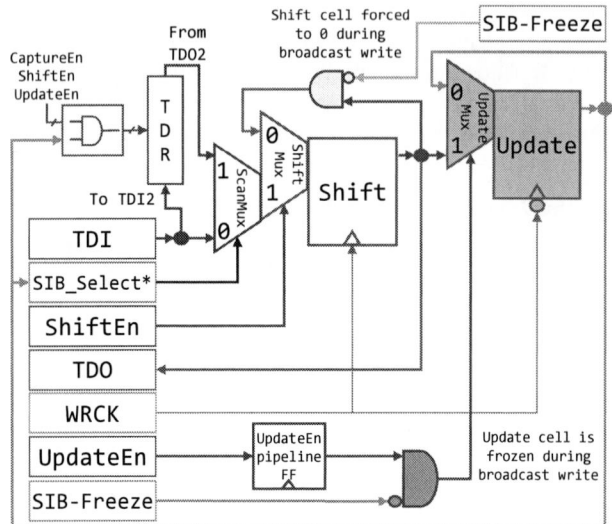

Fig. 7: Modified SIB design with SIB-Freeze, pipelined UpdateEn, and forced shift cell reset for the new configurable broadcast network

signal and the value in the update cell, the broadcast SIB shown in Fig. 7 can be in one of four different states. These four SIB states and the corresponding SIB-Freeze and update cell values are given in Table I.

During daisy (i.e. serial) mode, the SIB-Freeze remains deasserted and the broadcast SIB works similarly to the SIB shown in Fig. 3. When the SIB is closed (update cell = 0), the corresponding TDR is bypassed. When the SIB is open (update cell = 1), the corresponding TDR is active and connected to the active daisy chain. Before switching to broadcast mode, in the last daisy shift operation each SIB is placed in an appropriate open or closed state, depending on the TDRs that need to receive the identical test data during broadcast.

During broadcast mode the SIB-Freeze signal remains asserted. This prevents the UpdateEn signal from reaching the update mux of the SIB, and thus update operations cannot be performed on the SIB. In other words, during the broadcast mode, the SIB does not change its open/closed state. Without a freeze, during the broadcast mode, all the broadcast SIBs will switch between their open/closed states. Once the target TDRs have received the broadcasted test data, the network can be switched back to daisy scan mode to reconfigure the SIBs. To reduce the number of clock cycles that would be required to switch from broadcast scan mode to daisy scan mode and reconfigure all the SIBs for the next broadcast/daisy operation, we made two more changes to the SIB design—during broadcast mode the shift cell's feedback path is forced to zero, and the UpdateEn signal to the SIB is pipelined.

Shift cell's feedback path is forced to zero: When the network is switched from broadcast to daisy mode, the SIBs that were open during the broadcast mode will add their corresponding TDRs to the active shift path in daisy scan mode. Thus, the resulting shift path will consist of all the TDRs that were receiving the broadcasted test data along with all of the broadcast SIBs in the network. Unfortunately,

TABLE I: Broadcast SIB States

SIB State	SIB-Freeze	Update cell	TDR	Scan mode
Closed and unfrozen	0	0	Bypass	Daisy_Byp
Open and unfrozen	0	1	Active	Daisy
Closed and frozen	1	0	Bypass	Broadcast_Byp
Open and frozen	1	1	Active	Broadcast

any network reconfiguration, after switching from broadcast to daisy mode, would require shifting the new SIB configuration data through the TDRs connected to the active shift path. This would result in longer network reconfiguration overhead after each broadcast to daisy mode switch (depending on the number and length of the TDRs that were receiving the broadcasted data.)

To avoid this, the SIBs are designed such that the feedback path of the shift cell is forced to a zero during broadcast mode. Thus, when the broadcast mode ends, and the SIBs are unfrozen, an UpdateDR operation would close all the SIBs in the network resulting in minimal active shift path length. The network would now require fewer clock cycles to reconfigure it for the next daisy scan mode operation.

Pipelined UpdateEn signal: The network is switched from broadcast mode to daisy mode by shifting a zero in the BCB and performing an UpdateDR. This UpdateDR operation resets the BCB-Out/SIB-Freeze signal, as shown by the red arrows in Fig. 8. As mentioned earlier, the shift cells of all the SIBs contain a zero at this point. Now, after applying the first UpdateDR to end the broadcast mode, a second UpdateDR will place a zero in the update cells of all the broadcast SIBs. This will close all the SIBs to produce a minimal length shift path. However, a second consecutive UpdateDR operation would require cycling through the TAP controller's Capture-Shift-Update states, denoted as *TAP C* in the top half of Fig. 8.

To avoid these additional cycles, we have inserted a single flip-flop in the path of the UpdateEn signal of all the broadcast

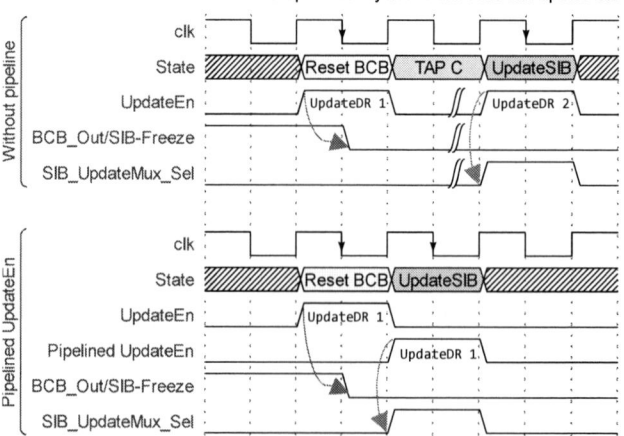

Fig. 8: SIB unfreeze-update operations, with and without pipelined UpdateEn

SIBs in the broadcast network. The first UpdateDR operation resets the BCB and deasserts the SIB-Freeze signal. The corresponding timing diagram is shown in the bottom half of Fig. 8. Now, when the delayed UpdateEn signal from the UpdateDR operation reaches the broadcast SIBs in the next clock cycle, the SIB-Freeze signals for the broadcast SIBs are already zero—unfreezing those SIBs. Thus, the delayed UpdateEn signal propagates to the update cell and allows the zero currently in the shift cell to be placed in the update cell on the next falling clock edge. This closes all the broadcast SIBs. As a result, a single UpdateDR operation resets the BCB and closes all the SIBs in two consecutive clock cycles. This, avoids the need to perform two consecutive UpdateDRs as required in a design without pipelined UpdateEn signals to the broadcast SIBs.

Since the TAP controller has to cycle through its state machine for the Capture-Shift-Update (CSU) signals, the delayed UpdateEn signal to the broadcast SIBs should be able to update the broadcast SIBs before any other CSU signals arrive at the broadcast SIB or the corresponding TDR.

B. Broadcast Network - With Hierarchy

Fig. 9 shows a conceptual view of the new IJTAG broadcast network with two levels of hierarchy. The example network is a part of the chip level test network, which is divided into three clusters—I, II, III. Each cluster may further consist of the cluster level broadcast network (partition BCB and partitions A, B, C) shown in Fig. 6 along with the additional broadcast SIB and scan mux per cluster as shown in Fig. 9. The broadcast SIBs (I, II, and III) in Fig. 9 operate in a similar manner as the broadcast SIBs shown in Fig. 6.

1) Partition Broadcast Control Bit (BCB) with frozen state: For the hierarchical broadcast network to operate as intended, it is important to add frozen states to the partition BCB of every cluster in the network. This is similar to the frozen state of the broadcast SIBs we discussed earlier. Further optimizations similar to the broadcast SIB design, such as delayed UpdateEn and forced shift cell reset, can also be used.

2) Cluster Broadcast Control Bit (BCB): The cluster BCB at the chip level is implemented using a NIB similar to the partition BCB that we showed earlier in Fig. 6. The control signal from the cluster BCB generates cluster BCB-Out and cluster SIB-Freeze. The control signal also functions as the partition BCB-Freeze for the partition BCBs in the clusters.

C. Test Time Improvements

Our proposed broadcast network (shown in Fig. 6 without hierarchy and in Fig. 9 with hierarchy), provides a trade-off between network reconfigurability and SIB (and NIB) programming overhead. To evaluate this trade-off and test time improvements, we compare four IJTAG networks using 8 32-bit TDRs shown in Fig. 10—a serial IJTAG network in (a), prior broadcast networks in (b) and (c), and finally (d) based on our proposed broadcast network (from Fig. 6). These 8 TDRs are part of two embedded cores. Because our proposed network provides reconfigurable broadcast/daisy

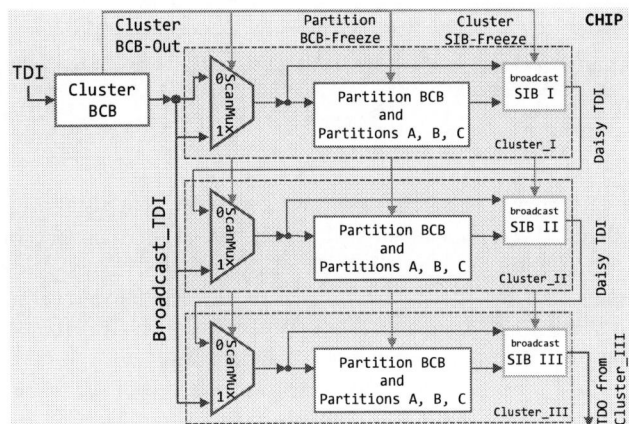

Fig. 9: Proposed IJTAG reconfigurable broadcast network - With hierarchy

scan modes, these TDRs from the two embedded cores do not need to be grouped in clusters in Fig. 10d as compared to Fig. 10 (b), and (c). The total number of (SIBs+NIBs) in the networks shown in Fig. 10 (a), (b), (c), and (d) are 8, 4, 10, and 9 respectively.

The number of clock cycles required to perform read/write operations on the instrument TDRs are reported in Table II. The reported values are calculated using the same method that was used in [17], [19]. For each of the test cases the number of clock cycles are reported assuming that the network was in reset state before the read/write operation [20]. Thus, the initial state of all the SIBs and NIBs was assumed to be closed. The test cases were chosen to evaluate the networks in four different scan configurations—all 8 TDRs are in daisy scan mode (*Daisy_all*), some TDRs are in daisy scan mode and remaining are bypassed (*Daisy_byp*), all TDRs receive identical data (*Broadcast_all*), and finally some TDRs receive identical data and remaining are bypassed (*Broadcast_byp*).

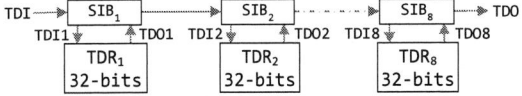

(a) Serial IJTAG network (flat-structure) with 8 TDRs of size 32-bits each

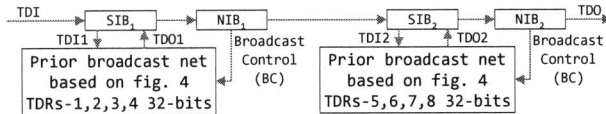

(b) IJTAG network with two clusters of broadcast networks based on Fig. 4

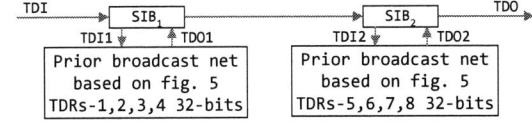

(c) IJTAG network with two clusters of broadcast networks based on Fig. 5

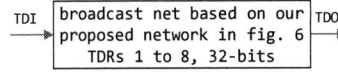

(d) IJTAG network based on the proposed broadcast network in Fig. 6

Fig. 10: Sample IJTAG networks used in Table II

TABLE II: No. of clock cycles required for read/write of test data using four different IJTAG network designs.

Test Case	Fig. 10a	Fig. 10b	Fig. 10c	Fig. 10d
Daisy_all	282	274	338	284
Daisy_byp	154	274	180	156
Broadcast_all	282	82	101	53
Broadcast_byp	154	274	180	53

Daisy_all: This test case involves reading out data from all the TDRs. As mentioned earlier, the IJTAG network in Fig. 10b has the lowest number of (SIBs+NIBs). Thus, it requires the lowest test time for a *Daisy_all* operation. However, it is important to note that the network clusters in Fig. 10b do not provide any reconfigurable daisy modes.

Daisy_byp: This test case involves reading out data from TDRs 1, 2, 5, and 6. The networks in Fig. 10 (a) and (d) can reconfigure the network in daisy mode such that only these four TDRs are active and the remaining TDRs are bypassed. The network in Fig. 10d contains one extra shift bit (BCB) in the chain compared to Fig. 10a. This extra bit needs to be shifted in the initial network reconfiguration step and while reading the TDRs 1, 2, 5, and 6. Thus, the proposed network requires two extra clock cycles compared to the serial network. The network in Fig. 10b, does not contain any SIBs within the broadcast clusters and thus it cannot reconfigure the active shift path during daisy scan mode. Also, the network in Fig. 10c can only read one TDR per embedded core at a time (daisy scan is not supported). Thus, these two prior broadcast networks require 43% and 14% more clock cycles, respectively.

Broadcast_all: This test case involves broadcasting data to all the TDRs. Now, even though the network in Fig. 10b has low SIB programming overhead, having two separate broadcast clusters results in 35% more test clock cycles compared to our proposed network in Fig. 10d. Also, in Fig. 10b, without clusters, spanning multiple TDRs from different embedded cores would result in very long daisy scan mode operations.

Broadcast_byp: This test case involves broadcasting test data to TDRs 1, 2, 5, and 6. In this scenario, our proposed brodcast network can broadcast the identical test data to TDRs 1, 2, 5, and 6 by using its reconfigurable broadcast mode while keeping the remaining TDRs bypassed. However, other networks do not provide this feature and will need to shift data in daisy scan mode. Thus, in this case our proposed network reduces test time by 70-80% of the clock cycles.

VI. CONCLUSION

In this paper we introduced a new IJTAG based broadcast network architecture to improve the test data access times for multiple identical embedded cores. We have shown that the new broadcast network is scalable, and it does not require careful design time considerations regarding the instruments that should be collected together in a broadcast group. Our Brodcast Control Bit (BCB) and modified broadcast SIB designs enable efficient hierarchical broadcast networks. We made further optimizations to our broadcast SIB design to reduce the overhead required to program the SIBs and reconfigure the network when switching from broadcast to daisy modes. Compared to a serial IJTAG network, our design requires only one additional scan-mux, two extra logic gates, and an extra delay flip-flop per TDR in the network. In our sample test scenario we showed that by making use of its reconfigurable broadcast mode our proposed broadcast network dramatically reduced test time up to 70-80% of the clock cycles compared to IJTAG based serial and prior broadcast networks.

REFERENCES

[1] "IEEE Standard for Access and Control of Instrumentation Embedded within a Semiconductor Device," *IEEE Std 1687-2014*, pp. 1–283, Dec. 2014.

[2] "IEEE Standard for Test Access Port and Boundary-Scan Architecture - Redline," *IEEE Std 1149.1-2013 (Revision of IEEE Std 1149.1-2001) - Redline*, pp. 1–899, May 2013.

[3] K.-J. Lee, J.-J. Chen, and C.-H. Huang, "Using a single input to support multiple scan chains," in *1998 IEEE/ACM International Conference on Computer-Aided Design. Digest of Technical Papers (IEEE Cat. No.98CB36287)*, Nov. 1998, pp. 74–78.

[4] ——, "Broadcasting test patterns to multiple circuits," *IEEE Transactions on Computer-Aided Design of Integrated Circuits and Systems*, vol. 18, no. 12, pp. 1793–1802, Dec. 1999.

[5] I. Hamzaoglu and J. H. Patel, "Reducing test application time for full scan embedded cores," in *Digest of Papers. Twenty-Ninth Annual International Symposium on Fault-Tolerant Computing (Cat. No.99CB36352)*, Jun. 1999, pp. 260–267.

[6] F. F. Hsu, K. M. Butler, and J. H. Patel, "A case study on the implementation of the Illinois Scan Architecture," in *Proceedings International Test Conference 2001 (Cat. No.01CH37260)*, 2001, pp. 538–547.

[7] P. Wohl, J. A. Waicukauski, J. E. Colburn, and M. Sonawane, "Achieving extreme scan compression for SoC Designs," in *2014 International Test Conference*, Oct. 2014, pp. 1–8.

[8] C. Barnhart, V. Brunkhorst, F. Distler, O. Farnsworth, A. Ferko, B. Keller, D. Scott, B. Koenemann, and T. Onodera, "Extending OP-MISR beyond 10 times; scan test efficiency," *IEEE Design Test of Computers*, vol. 19, no. 5, pp. 65–73, Sep. 2002.

[9] L.-T. Wang, K. S. Abdel-Hafez, X. Wen, B. Sheu, S. Wu, S.-H. Lin, and M.-T. Chang, "UltraScan: using time-division demultiplexing/multiplexing (TDDM/TDM) with VirtualScan for test cost reduction," in *IEEE International Conference on Test, 2005.*, Nov. 2005, pp. 8 pp.–953.

[10] Y. Zorian, E. J. Marinissen, and S. Dey, "Testing embedded-core based system chips," in *Proceedings International Test Conference 1998 (IEEE Cat. No.98CH36270)*, Oct. 1998, pp. 130–143.

[11] S. K. Goel and E. J. Marinissen, "Effective and efficient test architecture design for SOCs," in *Proceedings. International Test Conference*, 2002, pp. 529–538.

[12] S. Makar, T. Altinis, N. Patkar, and J. Wu, "Testing of Vega2, a chip multi-processor with spare processors." in *2007 IEEE International Test Conference*, Oct. 2007, pp. 1–10.

[13] G. Giles, J. Wang, A. Sehgal, K. J. Balakrishnan, and J. Wingfield, "Test access mechanism for multiple identical cores," in *2009 International Test Conference*, Nov. 2009, pp. 1–10.

[14] M. Sharma, A. Dutta, W. T. Cheng, B. Benware, and M. Kassab, "A novel Test Access Mechanism for failure diagnosis of multiple isolated identical cores," in *2011 IEEE International Test Conference*, Sep. 2011, pp. 1–9.

[15] F. Zhang, Y. Sun, X. Shen, K. Nepal, J. Dworak, T. Manikas, P. Gui, R. I. Bahar, A. Crouch, and J. Potter, "Using Existing Reconfigurable Logic in 3d Die Stacks for Test," in *2016 IEEE 25th North Atlantic Test Workshop (NATW)*, May 2016, pp. 46–52.

[16] "IEEE Standard Testability Method for Embedded Core-based Integrated Circuits," *IEEE Std 1500-2005*, pp. 1–136, Aug. 2005.

[17] S. Gupta, A. Crouch, J. Dworak, and D. Engels, "Increasing IJTAG Bandwidth and Managing Security through Parallel Locking-SIBs," in *2017 International Test Conference*, Nov. 2017.

[18] A. Crouch, "Tutorial IEEE 1687 IJTAG | Embedded Instruments ICL PDL | ASSET InterTech." [Online]. Available: https://www.asset-intertech.com/eresources/tutorial-ieee-1687-ijtag-embedded-instruments-icl-pdl

[19] F. G. Zadegan, U. Ingelsson, G. Carlsson, and E. Larsson, "Design automation for IEEE P1687," in *2011 Design, Automation Test in Europe*, Mar. 2011, pp. 1–6.

[20] S. N. Mozaffari, S. Tragoudas, and T. Haniotakis, "More Efficient Testing of Metal-Oxide Memristor-Based Memory," *IEEE Transactions on Computer-Aided Design of Integrated Circuits and Systems*, vol. 36, no. 6, pp. 1018–1029, Jun. 2017.

Securing IJTAG Against Data-Integrity Attacks

Rana Elnaggar[†], Ramesh Karri[‡], and Krishnendu Chakrabarty[†]

[†]Department of Electrical and Computer Engineering, Duke University, Durham, NC 27708
[‡]Department of Electrical and Computer Engineering, New York University, Brooklyn, NY 11201

Abstract—**The IEEE Std. 1687 (IJTAG) facilitates access to on-chip instruments in complex system-on-chip designs. However, a major security vulnerability in IJTAG has yet to be addressed. IJTAG supports the integration of tapped and wrapped instruments at the IP provider with hidden test-data registers (TDRs). The instruments with hidden TDRs can manipulate the data that is shifted through them. We propose the addition of shadow test-data registers by the trusted IJTAG integrator to protect the shifted data from illegitimate manipulation by malicious third-party IPs. In addition, we use information-flow tracking to identify the modified bits during the attack and the attacking instruments in an IJTAG network. We present security proofs, simulation results and the overheads associated with these countermeasures for various benchmarks.**

I. INTRODUCTION

With the relentless increase in the complexity of integrated circuits, there is a growing trend to integrate a large number of on-chip instruments to facilitate test, debug, and diagnosis. Instruments can be delivered as bare, wrapped or tapped [1]. Bare instruments are delivered without a test wrapper added at the IP provider; the wrapper is added later by the trusted network integrator. In contrast, wrapped instruments come with test wrappers integrated with the IP. Similarly, the tapped instruments come with an embedded tap controller. The IEEE Std. 1687 (also known as IJTAG) has been introduced to facilitate efficient access to these embedded instruments. IJTAG can be used to reconfigure scan chains using shifted-in data to include/exclude different embedded instruments [1].

A. Overview of IJTAG and Security Vulnerabilities

In a typical design, the IEEE Std. 1149.1 TAP controller is used as the master tap controller (MTAP) to orchestrate the IJTAG network [1]. The MTAP generates signals that determine the state of the MTAP finite state machine (FSM). The operations performed in the different FSM states are as follows. During the capture state, the output from an embedded instrument is loaded into its test data register (TDR). The TDR is a shift register that captures the output test response from the embedded instrument in parallel in the capture state and shifts it out serially. Similarly, the TDR applies the input test vectors into the embedded instruments in parallel in the update state. In the shift state, data is shifted throughout the IJTAG network. The shifted data is either the: (1) test responses captured from the instruments, or (2) test vectors generated by the MTAP to test embedded instruments, or (3) the configuration bits generated by the MTAP controller to reconfigure the IJTAG network. In the update state, test vectors are loaded into the TDR of the instruments and the configuration bits are loaded into the segment insertion bits (SIBs) in the IJTAG network.

Fig. 1 shows an IJTAG benchmark network (Mingle) with eight instruments $W_1, W_2.., W_8$ [2]. The SIB is a flip-flop with an update cell. When a value '1' is written into the update cell of the SIB in the update state, the TDRs behind the SIB are included in the scan chain. The scanMux control bit (SCB) is also used to reconfigure IJTAG networks.

Initially, the MTAP sees a scan chain with two flip-flops (SI→SIB1→SIB2→SO). To access the next level of the IJTAG network hierarchy, "01" is shifted into SIB1 and SIB2 during the shift cycle*. This opens SIB1 after the update state and keeps SIB2 closed. Now, the MTAP sees SI→SIB5→SCB3→SIB1→SIB2→SO that is 4 flip-flops long. To access W_5 and W_6, SIB5 and SIB6 need to be opened while keeping SIB2 closed and setting the select signal of SMUX3 (SCB3)[†] to select the output from SIB5 and block the output from SIB8. Therefore, when we shift in the sequence "0101" in the update state, SIB5 is opened and SIB6 and the TDR of W_5 are added to the scan chain. If the length of the TDR of W_5 is 32, the MTAP sees scan chain SI→W_5→SIB6→SIB5→SCB3→SIB1→SIB2→SO of length 37. To access W_6, SIB6 needs to be opened. The MTAP shifts in "01011" + 32 zeros. After the update state, SIB6 is opened, SIB2 remains closed and SCB3 selects the output from SIB5.

The IJTAG network interfaces with the wrapped/tapped instruments through control signals, and scan-in and scan-out ports. A wrapped/tapped instrument's TDR is hidden inside the test wrapper. Hence, the operations performed on the data as it is shifted through these black-box (i.e., wrapped and tapped) instruments are hidden from the IJTAG integrator. Consequently, IJTAG networks are vulnerable to data-integrity attacks launched by the hidden logic in the instrument that may

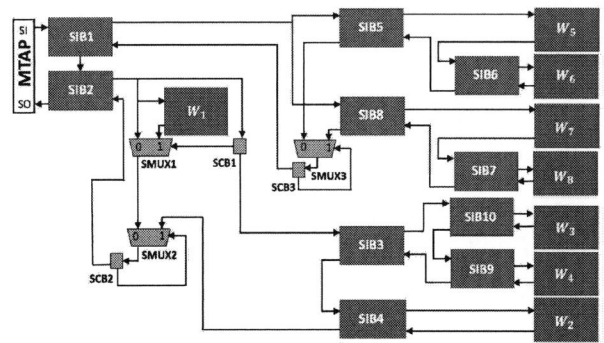

Fig. 1: Schematic of the Mingle benchmark network with 10 Segment Insertion Bits (SIBs), 3 Scan Control Bits (SCBs), 3 Scan Muxes (SMUXs), and 8 wrapped instruments [2].

---

*The most significant bit of the sequence is shifted in first
[†]SCB3 initially has a '0'

compromise the accuracy, consistency, and trustworthiness of the data.

B. Paper Contributions

The major contributions of this paper are as follows:

- We highlight data-integrity risks posed to the IJTAG network by wrapped and tapped third-party IPs.
- We describe a countermeasure based on shadow TDRs to ensure that: (1) the bits that are used to configure the scan chains are genuine; (2) the test data that is applied to the embedded instruments is authentic; (3) the test-data output that is captured from the embedded instruments reaches the master TAP controller unaltered.
- We prove that the IJTAG network with shadow TDR countermeasure operates correctly even in the presence of malicious instruments.
- We adapt information flow tracking by monitoring the taint information to identify: (1) the instruments that are the target of a data-integrity attack; (2) the modified bits during a data-integrity attack; and (3) the instruments that attempted to launch data-integrity attack.

The remainder of the paper is organized as follows. Section II describes related prior work. Section III describes the target threat model. Section IV and V present the proposed countermeasures, which are validated using formal proofs and simulation results. We conclude the paper in Section VI.

II. RELATED PRIOR WORK

Since IJTAG networks are reconfigured based on the shifted-in data, the hidden instruments are vulnerable to unauthorized access. One can restrict unauthorized access to segments of the IJTAG network by adding locks to the SIBs [3]. SIBs are unlocked only if a specific sequence of bits (i.e., the secret key) is shifted in. One can also use trap bits to permanently lock a SIB when the wrong key is shifted in. The SIB is then unlocked only on a global reset. Alternatively, honey traps can be used to return incorrect information about the length of the scan chain when an incorrect key is shifted in to obfuscate the IJTAG network structure [4]. To increase the difficulty of unlocking the SIBs, one can use an LFSR to generate secret keys [5]. One can also use physically unclonable functions to generate instrument-specific secret keys [6]. We can also insert a sequence filter at the MTAP to lock it when an unauthorized bit sequence is detected [7]. Model checking can be used to verify that IJTAG network cannot be accessed without a secret key [8]. However, the secret keys may be susceptible to power analysis [9]. Hence, data encryption can be used to secure the data that is shifted through the network. In addition, stub chains of varying lengths can be used to obfuscate the IJTAG network structure. Access to the IJTAG network can be blocked if the number of issued shift cycles exceeds a pre-specified range [10].

Prior work protects IJTAG networks from unauthorized user access; however, it does not secure the networks against data-integrity attacks launched by potentially malicious embedded instruments in the network. This paper advances IJTAG security by tackling the data-integrity vulnerabilities posed by

black-box third-party IPs. The proposed defenses complement prior countermeasures.

III. DATA-INTEGRITY ATTACKS ON IJTAG NETWORKS

Some of the wrapped and tapped instruments are potentially malicious because the network integrator has no control over the data as it is shifted through these black-box wrapped and tapped instruments. The attacks that an untrusted instrument can launch are as follows:

1) **Modify the configuration bits of the network:** This attack can open access to the instruments that the MTAP controller intended to keep closed or block access to the instruments that MTAP controller needs to test. Fig. 2 shows that the malicious instrument W_5 of Fig. 1 modifies the IJTAG configuration bits so that SIB5 and SIB6 are closed. Consequently, W_5 and W_6 are not accessed.

2) **Alter the test data used to test downstream instruments:** In Fig. 1, the malicious instrument W_5 can modify the test data applied to instrument W_6. As a result, W_6 responds with a different response. W_6 is deemed to be faulty, even though, it is fault-free.

3) **Change the test data output from the upstream instruments:** In Fig. 1, the malicious instrument W_6 can modify the test output of W_5. Instrument W_6 can mask the effect of the failure, and no corrective action is taken.

4) **Modify the configuration bits of the network and the returned test data:** This is a combination of the first and third attacks and it proceeds as follows. Assume that the MTAP controller in Fig. 1 shifts in data to include the TDRs of W_5 and W_6 in the scan chain. W_5 alters the network configuration to block the access to W_6, and opens an internal dummy scan chain instead, to obfuscate the change in the scan chain length and then shifts out a malicious test response. The MTAP controller still sees this as having been shifted by W_6.

5) **Sniff sensitive data sent to the downstream instruments or to the MTAP controller:** Malicious instruments can sniff the data that is shifted through it. This data can include the secret keys used to unlock SIBs. As a result, encryption of the data [10] is needed to protect the data sniffed by the malicious wrapped instruments. However, encryption of all the network data is not a practical solution as many embedded instruments are simple and cannot decrypt the received data.

The position of a suspicious instrument in the network impacts the types of the attacks that it can launch. Table I summarizes these attacks.

IV. SECURE INTEGRATION USING SHADOW REGISTERS

We assume that the IJTAG network integrator is trusted. This is a reasonable assumption as it is essential to have a trusted foothold to build trust into the IJTAG network in the

Position in network hierarchy	Modify or sniff data from	
	MTAP → instruments	Instruments → MTAP
First instrument in first-level	✓	✗
Last instrument in first-level	✗	✓
Any position in the middle	✓	✓

Table I: Position of the malicious instrument and attack capabilities.

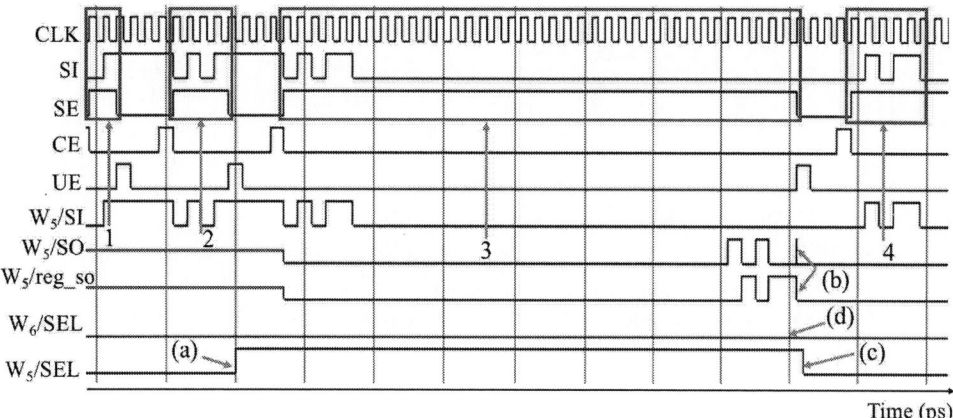

Fig. 2: Simulation of Mingle with malicious instruments W_5 and W_6; (a) W_5 is accessed; (b) output of W_5/SO is the complement of W_5/reg_so; (c)-(d) W_5 and W_6 are not accessed after the IJTAG configuration bits are modified.

presence of potentially untrustworthy instruments. The IJTAG integrator can secure the instruments in IJTAG networks by using a shadow TDR placed in parallel to a wrapped instrument to ensure that the shifted data in the network is not altered; see Fig. 3(a). We first introduce some terminology:

- **Shadow TDR** ($STDR$) is an n-bit shift register (the length of the corresponding wrapped instrument TDR). It is placed in parallel to wrapped instrument W as shown in Fig. 3(a). It ensures that genuine data is propagated through the network.

- **Controlling Multiplexer** (M) selects the output of W to be shifted out during the first n shift cycles after capture, and it selects the output of $STDR$ starting from shift cycle $n+1$ after capture.

- **Trusted Data** is generated by trusted sources and is not modified as it is shifted through the network. The following are trusted data: (i) The test response data that is captured from any instrument in the network; (ii) The network configuration and test data vectors that are generated by the MTAP controller.

- **Untrusted Data** passes through wrapped/tapped instrument W_i and it is not generated by W_i. This data is untrusted since it can be modified as it is shifted through the wrapped/tapped instrument.

A. Details of Shadow TDR-based Secure Integeration

The key idea here is that the trusted network integrator adds $STDR$ and M in parallel to the wrapped/tapped instrument, where $STDR_i$ shifts the same patterns that are shifted through W_i. For the first n cycles after capture, M_i shifts out the output from the wrapped/tapped instrument. M_i shifts out the output from $STDR_i$ starting from shift cycle $n+1$ after capture. If W_i changes the shifted out data, this modification will not impact the output, as M_i shifts data out from $STDR_i$ starting from shift cycle $n+1$ after capture. Therefore, STDR ensures that data passes untampered through the network.

B. Analysis of Security Properties

To prove that the shadow TDR secures the IJTAG network from data-integrity attacks, we introduce lemmas to specify the conditions necessary to ensure that the data is not tampered.

Lemma 1. *Consider a malicious wrapped instrument W_i with an n-bit TDR_i which is loaded during capture. The data shifted out of W_i can be trusted only during the first n shift cycles after capture.*

Proof. Since the length of the TDR of the wrapped/tapped instrument is n bits, n shift cycles are needed after the capture to shift out the TDR contents. Any data shifted out subsequently is either generated by an instrument upstream or the MTAP controller, and it can be modified by W_i. □

Lemma 2. *Consider an n-bit $STDR_i$ associated with a malicious wrapped instrument W_i. The data in $STDR_i$ is up-to-date only starting from $n+1$ shift cycle after the capture.*

Proof. $STDR_i$ is n bits long and it cannot load the data captured from W_i in parallel. Therefore, during the first n shift cycles after capture, $STDR_i$ contains outdated data from the MTAP (e.g., test sequence applied to W_i). This outdated data is replaced n shift cycles after capture with trusted data from upstream instruments or the MTAP controller. □

Theorem 1. *Consider a wrapped/tapped instrument W_i with an n-bit TDR. The data that is shifted out of the controlling multiplexer M_i is trusted if and only if M_i selects the output of W_i during the first n cycles after capture and selects the output of $STDR_i$ starting from shift cycle $n+1$ after capture.*

Proof. Suppose the output of M_i is untrusted. This condition is true if and only if the following conditions are satisfied: (1) the output of W_i is untrusted when it is selected by M_i during the first n shift cycles after capture; (2) the output of $STDR_i$ is malicious after the first n shift cycles. However, according to Lemmas 1-2, these conditions cannot be satisfied. As a result, the data generated by M_i is always trusted. □

Corollary 1. *If all wrapped/tapped instruments are malicious, the output of the IJTAG network will be trusted if and only if every instrument W_i is wrapped with a shadow TDR.*

C. Extension to instruments with variable-length TDRs

Test wrappers of embedded instruments may use IEEE Std. 1149.1 TDRs, IEEE Std. 1500 core wrappers or other custom designs [1]. As a result, the TDR may be of variable length. We propose a reconfigurable shadow TDR ($RSTDR$) that can

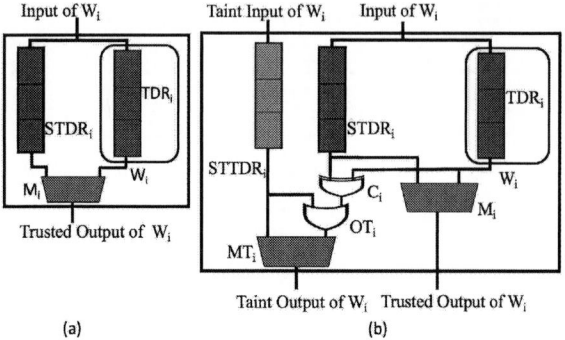

Fig. 3: (a) Design of the shadow TDR (STDR) countermeasure; (b) Design of the STTDR (STDR with IFT) countermeasure.

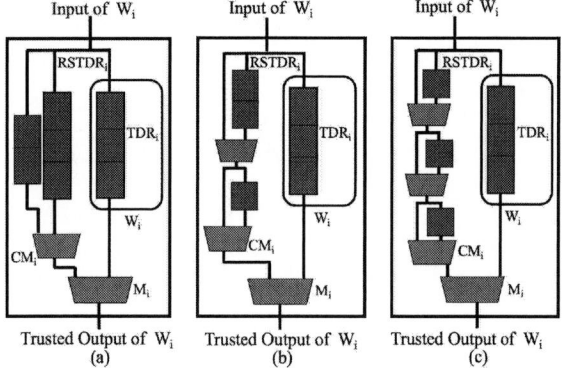

Fig. 4: Design (a), (b) and (c) of the RSTDR.

match the size of the TDRs inside the wrapped instrument. According to IJTAG, it is the responsibility of the tester (i.e., the MTAP) to know the length of the TDR of each instrument to shift the appropriate test vectors [1]. Three possible designs of $RSTDR$ are shown in Fig. 4. Fig. 4(a) shows shadow TDRs of all possible length choices inside a wrapped instrument placed in parallel with the selection done by a multiplexer. An alternative with lower overhead is shown in Fig. 4(b); here a TDR of length L (equal to the length of the longest wrapped-instrument TDR configuration) is added and divided into segments that are joined using multiplexers. A generalized version of this design connects the output of each flip-flop to a 2-1 multiplexer; as shown in Fig. 4(c). The control signals that determine the length of $RSTDR$ and the reconfiguration strategy depend on the network structure and choices made by the the network integrator.

D. Simulation and Synthesis Results

1) Benchmarks: We used the BasicSCB, Mingle, and TreeFlat benchmark networks described in [2]. We use these three IJTAG networks because they consist of wrapped instruments. For simulation, we use Mentor ModelSim [11]. For synthesis, we use Synopsys Design Compiler [12] with Synopsys 90 nm generic libraries.

2) Inserted malicious instruments: We designed a malicious wrapped instrument that does not alter its captured data, but complements any other data shifted out of it.

3) Simulation Results: We verified that the baseline IJTAG networks operate normally when only trusted instruments are integrated. Next, we inserted malicious instruments into the baseline networks and checked that these baseline networks without the shadow TDR countermeasure are susceptible to data-integrity attacks as shown in Fig. 2. The applied test sequences in Fig. 2 (shown as 1-4) are:

1) Shift in "01" to open SIB1 and keep SIB2 closed.
2) Shift in "0101" to open SIB5 and include W_5 and SIB6.
3) Shift in "01011" + 32 zeroes to access W_6.
4) Shift in "01011" + 64 zeroes to test the network configuration. The sequence "01011" ensures that the network is configured as SI $\rightarrow W_5 \rightarrow W_6 \rightarrow$ SIB6 $\rightarrow$ SIB5 $\rightarrow$ SCB3 $\rightarrow$ SIB1 $\rightarrow$ SIB2 $\rightarrow$ SO.

Fig. 5 shows that the protected Mingle network operates normally in the presence of the malicious instruments W_5 and W_6. In a trusted instrument, the scan output (W_5/SO) of the instrument receives data from the shift TDR. However, in Fig. 5, W_5/SO receives the complement of the data in the shift TDR (W_5/reg_so) except when W_5 is shifting out its captured contents. $STDR_5$ and the controlling Mux M_5 make the IJTAG network resilient to data-integrity attacks. Therefore, the output of M_5 is assigned the output of $STDR_5$ (SRO_W_5/SO) after the first n shift cycles after capture. Thus, the IJTAG network operates normally.

4) Area and Power-Consumption Overhead: To investigate the overhead for large, realistic IJTAG networks, we extend the BasicSCB, Mingle and TreeFlat benchmarks to include 1000 wrapped instruments. We add these extra instruments in series with the last instrument in the original network and investigate area and power-consumption overhead if the network integrator wraps only 2%, 10%, and 30% of the instruments with STDR. We assume that these instruments come wrapped/tapped from third-party providers, while the rest of the instruments are either wrapped by the trusted network integrator or are provided by trustworthy vendors. The results are shown in Table II assuming a 32-bit TDR in instrument. Overhead results for IFT are also included in the table—details of IFT are presented in the next section.

V. INFORMATION-FLOW TRACKING (IFT)

The shadow TDR ensures that the IJTAG network is resilient to data-integrity attacks launched by untrusted instruments. However, it does not identify the modified bits during the attack. We therefore propose the use of information-flow tracking to detect the bits modified by the attacker; this knowledge can be further used to identify the attempted attack and its target instrument. IFT relies on labeling insecure data and tracking their propagation in the system. It has been used to protect software applications against data confidentiality and integrity attacks [13]–[16]. In addition, it has been applied at the gate-level to detect hardware Trojans [17] and timing channels [18]. In this work, we use IFT to mark and track the modified bits as they are shifted through the IJTAG network. We first introduce IFT-related terminology.

- **A taint bit** of value '1' shows that the data shifted out from a suspicious instrument is modified during shifting.
- **A shadow Taint TDR** ($STTDR$) is a shift register of length n, where n is the length of the associated wrapped instrument

978-1-5386-3775-3/18 $31.00 © 2018 IEEE

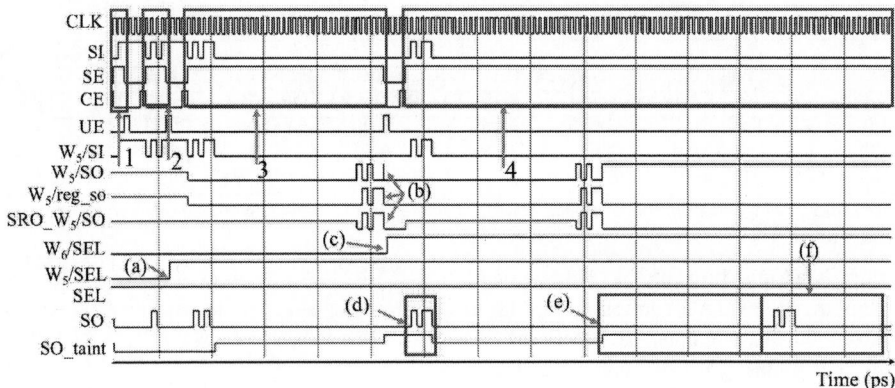

Fig. 5: Mingle IJTAG network with malicious instruments W_5 and W_6 secured using STDR + IFT countermeasures. (a) Indicates that W_5 is accessed; (b) Indicates that W_5 attempted to change data; (c) Indicates that W_6 is accessed; (d)-(f) Ones in SO_taint indicate that an attempt was made to change the corresponding bits.

TDR. It is placed in parallel to W; see Fig. 3(b). By using $STTDR$, taint bits can be shifted through the network to label the bits that are modified by malicious instruments.

- **A comparator** (C) is used to compare the shifted output bits of W with that of $STDR$.

- **Tainting OR-gate** (OT) outputs '1' if either the output of C or $STTDR$ is '1', and '0' otherwise.

- **A Taint Mux** (MT) selects the value of the taint bit from $STTDR$ to be shifted out during the first n shift cycles after capture, and after n shift cycles it shifts out the value of OT.

A. Details of IFT-based Security Solution

We replicate the IJTAG network's basic structures (multiplexers, shift registers and SIBs) to propagate the taint bit throughout the network. Fig. 3(b) shows the hardware structures added to a wrapped instrument by the network integrator to implement IFT-based countermeasure. When a malicious instrument W_i attempts to modify the shifted data, this attempt is captured by C_i as the output of W_i is different from $STDR_i$. During the first n shift cycles after capture, the output of MT_i is assigned the taint bit shifted out by $STTDR_i$ to propagate the taint information that represents data modifications by upstream instruments, otherwise, the output of MT_i is assigned the output of OT_i to propagate the taint information that shows the data modification by upstream instruments or W_i. The shifted-out taint bits are analyzed by the MTAP controller to identify the attempted attack.

We next prove that any attempt to modify the data as it passes through tapped/wrapped instruments is detected and observed by the MTAP controller.

Lemma 3. *Consider instrument W_i with an n-bit TDR. During the first n shift cycles after capture, the output of OT_i incorrectly taints the genuine captured data with value '1' (i.e., it generates false positives).*

Proof. Consider an embedded instrument W_i that has an n-bit TDR. During the first n shift cycles after capture, the bits that are shifted out of W_i are trusted (Lemma 1). However, the data that is shifted out of $STDR_i$ is outdated (Lemma 2), i.e., TDR_i and $STDR_i$ may contain different values. In this case, the output bits of OT_i have value '1', i.e., OT_i generates false positives although the data in the TDR of W_i is trusted. $\square$

Theorem 3: Suppose the taint mux MT_i selects the output of $STTDR_i$ during the first n shift cycles after capture and selects the output of OT_i starting from shift cycle $n + 1$ after capture. Any attempt to modify a shifted-out bit by a malicious wrapped/tapped instrument is captured and reported to the MTAP with no false positives.

Proof. We present a proof by contradiction.

(a) **Data modification by upstream instruments is detected:** Suppose the output of MT_i does not detect the modification performed by upstream instruments. This condition can be met if and only if the taint bits in $STTDR_i$ are not considered when evaluating the output of MT_i, since these bits record any modification by upstream instruments. However, this contradicts the configuration of MT_i; when the value of a taint bit in $STTDR_i$ is '1', this value is always propagated to all the input ports of MT_i. Therefore, any modification by any upstream instrument is detected at MT_i.

(b) **Data modification by W_i is detected:** Suppose that the output of MT_i does not capture data modification that is performed by W_i. This condition can be met if and only if the output of C_i is not taken into consideration *after* the first n shift cycles after capture. However, based on the configuration discussed earlier, *after* the first n shift cycles, MT_i selects the output of OT_i which propagates the output of C_i if it is equal to '1'. As a result, any modification applied by W_i after the first n shift cycles is detected at MT_i.

(c) **No false positives:** The output of MT_i has false positives if and only if the output of OT_i is selected by MT_i during the first n shift cycles after capture (Lemma 3). However, the output of OT_i is selected only after the first n shift cycles after capture. As a result, the output of MT_i has no false positives.

Based on (a), (b) and (c), we conclude the proof. $\square$

B. Simulation Results For IFT

Fig. 5 shows results for the Mingle benchmark when the IFT countermeasure supplements the shadow TDRs. The test sequences shown by arrows 1-4 in Fig. 5 are similar to the sequences in Fig. 2. However, the test sequence that has a label of '4' shifts a sequence of "01011" followed by 138 zeroes to observe SO_taint and the corresponding values of SO. These tests are taken from the testbench available with Mingle [2].

Bench-marks	Percentage of Instruments Secured by Countermeasures																	
	2						10						30					
	Area (%)		P_{dyn} (%)		P_{leak} (%)		Area (%)		P_{dyn} (%)		P_{leak} (%)		Area (%)		P_{dyn} (%)		P_{leak} (%)	
	STDR	IFT	STDR	IFT	STDR	IFT	STDR	IFT	STDR	IFT	STDR	IFT	STDR	IFT	STDR	IFT	STDR	IFT
TreeFlat	0.84	1.5	1.42	1.8	0.93	1.93	3.8	6.88	3.72	5.90	4.60	8.7	11.30	20.8	9.6	17.8	13.9	25.98
Mingle	1.07	1.72	2.18	2.55	0.91	1.78	4	7.07	4.37	6.53	4.6	8.69	11.54	21	10.95	18.62	13.86	25.99
BasicSCB	0.81	1.47	0.74	1.27	0.92	1.78	3.81	6.88	3.12	5.36	4.62	8.698	11.2	20.8	9.11	17.01	13.87	25.98

Table II: Percentage area and power-consumption overhead results. P_{dyn}: dynamic power overhead; P_{leak}: leakage power overhead; STDR : STDR only; IFT: STDR with IFT.

In Fig. 5(d), SO_taint is assigned a value of '1' to indicate that W_5 attempted to modify the IJTAG configuration bits. In Fig. 5(e), SO_taint shows that W_6 attempted to modify the bits that were captured from W_5 during the capture cycle. W_6 is identified as the attacker since there are no other activated instruments in the IJTAG network. In Fig. 5(f), the value assigned to SO_taint indicates that either W_5 or W_6 or both of them attempted to modify test sequence '4'.

C. Area and Power-Consumption Overhead

Table II shows the percentage increase in the area and power consumption if only 2%, 10%, and 30% of the instruments are wrapped with the proposed countermeasures and when the length of the TDR of all instruments is 32.

As expected, IFT increases area and power consumption overhead. In addition, it adds two extra pins to the MTAP controller (SI_taint and SO_taint). Nevertheless, the small overhead is acceptable as a tradeoff for securing IJTAG.

D. Extending IFT to Identify the Attacking Instruments

IFT identifies the bits that the malicious instruments attempted to modify. However, the number of times a bit was changed and the malicious instrument that lanuched the attack are unknown. For example, in Fig. 1, to keep SIB1, SIB5, and SIB6 open and SIB2 closed when testing W_5 and W_6, "01011" is shifted prior to the test sequences that are applied to W_5 and W_6. If W_5 modifies the first and the second bit, and W_6 modifies the second and the third bit, the overall taint information observed at the MTAP is "11100". However, the exact source of the attack and the number of times a bit is modified are unknown.

To identify the instrument(s) that launched the data-integrity attacks in an IJTAG network, one can associate a scan flip-flop with each instrument and connect these scan flip-flops to form an attacker-identification scan chain (AIS). Whenever an instrument attempts a data-integrity attack, a taint bit '1' is written into its associated scan flip-flop. AIS is placed in scan mode only to check if an attack occurred or to identify the attacker. This countermeasure can be integrated with the IFT-based countermeasure as follows; after MTAP observes SO_taint with a value '1', it enables the scan mode of AIS to identify the attacking instruments. Although all the attacking instruments can be identified using AIS, if there are multiple attackers in the IJTAG network, the source of each attack may still not be identified.

VI. CONCLUSION

We have presented security solutions for IJTAG networks that are vulnerable to data-integrity attacks. The shadow TDR-based countermeasure ensures the correct operation of IJTAG networks in the presence of data-integrity attacks. We have described the identification of the attacking instruments, the targets of the attack, and the modified bits by the attack.

Finally, we note that the proposed methods target attacks that are different from the attacks considered in prior work. A comparison in terms of the defeated attacks is shown in Table III. This comparison shows that a combination of proposed countermeasure and countermeasure in [10] improves the security of IJTAG networks against the three classes of attacks.

Attack type	Defeated?							
	[3]	[4]	[5] [6]	[7]	[9]	[10]	Proposed method
Data sniffing	✗	✗	✗ ✗	✗	✗	✓	✗
Unauthorized user	✓	✓	✓ ✓	✓	✓	✓	✗
Data modification	✗	✗	✗ ✗	✗	✗	✗	✓

Table III: Comparison of our approach to prior countermeasures.

REFERENCES

[1] "IEEE Standard for Access and Control of Instrumentation Embedded within a Semiconductor Device," IEEE Std. 1687-2014, 2014.

[2] A. Tšertov et al., "A Suite of IEEE 1687 Benchmark Networks," in *Proc. IEEE ITC*, 2016.

[3] J. Dworak et al., "Don't Forget to Lock your SIB: Hiding Instruments using P1687," in *Proc. IEEE ITC*, 2013.

[4] A. Zygmontowicz et al., "Making It Harder to Unlock an LSIB: Honeytraps and Misdirection in a P1687 Network," in *Proc. IEEE DATE*, 2014.

[5] H. Liu and V. D. Agrawal, "Securing IEEE 1687-2014 Standard Instrumentation Access by LFSR Key," in *Proc. IEEE ATS*, 2015, pp. 91–96.

[6] S. Kumar et al., "Securing IEEE 1687 Standard On-chip Instrumentation Access Using PUF," in *Proc. IEEE iNIS*, 2016, pp. 56–61.

[7] R. Baranowski et al., "Securing access to reconfigurable scan networks," in *Proc. IEEE ATS*, 2013, pp. 295–300.

[8] R. Baranowski et al., "Reconfigurable Scan Networks: Modeling, Verification, and Optimal Pattern Generation," *ACM TODAES*, 2015.

[9] I. S. Gupta et al., "Mitigating Simple Power Analysis Attacks on LSIB Key Logic," in *North Atlantic Test Workshop (NATW)*, 2017, pp. 1–6.

[10] S. I. Kan et al., "Echeloned IJTAG Data Protection," in *Proc. IEEE AsianHOST*, 2016, pp. 1–6.

[11] "Mentor ModelSim," https://www.mentor.com/products/fpga/verification-simulation/modelsim/.

[12] "Synopsys Design Compiler," https://www.synopsys.com/support/training/rtl-synthesis/design-compiler-rtl-synthesis.html.

[13] W. Enck et al., "Taintdroid: An information-flow tracking system for realtime privacy monitoring on smartphones," in *Proc. IEEE USENIX*, 2010, pp. 393–407.

[14] G. E. Suh et al., "Secure program execution via dynamic information flow tracking," in *Acm Sigplan Notices*, vol. 39, no. 11, 2004, pp. 85–96.

[15] F. Qin et al., "Lift: A low-overhead practical information flow tracking system for detecting security attacks," in *Proc. IEEE MICRO*, 2006.

[16] H. Yin et al., "Panorama: capturing system-wide information flow for malware detection and analysis," in *Proc.ACM CCS*, 2007, pp. 116–127.

[17] W. Hu et al., "Detecting hardware trojans with gate-level information-flow tracking," *IEEE Computer*, vol. 49, no. 8, pp. 44–52, 2016.

[18] J. Oberg et al., "Leveraging gate-level properties to identify hardware timing channels," *IEEE TCAD*, vol. 33, no. 9, pp. 1288–1301, 2014.

Special Session on Overcoming Reliability and Energy-Efficiency Challenges with Silicon Photonics for Future Manycore Computing

Sudeep Pasricha[*], Davide Bertozzi[†], Hui Li[§]
[*]Colorado State University, Fort Collins, Colorado, USA
[†]University of Ferrara, Italy
[§]Xidian University, China
sudeep@colostate.edu, brtdvd@unife.it, huili@xidian.edu.cn

Abstract – Silicon photonics has emerged in recent years as one of the most promising solutions to overcome the challenge of worsening chip-scale communication performance with technology scaling. Recent breakthroughs in silicon photonic device fabrication and CMOS integration have presented computer designers with an opportunity to devise on-chip optical networks that have significant advantages in bandwidth density, energy-efficiency, and propagation delay over traditional electrical solutions. Not surprisingly, the challenge of designing chip-scale silicon photonic communication fabrics is today actively being pursued by a number of researchers worldwide. Many semiconductor companies (e.g., Intel, IBM) have begun investing heavily into silicon photonics and are releasing functional prototypes. However, silicon photonic interconnects have high susceptibility to faults due to several factors such as homodyne and heterodyne crosstalk, process variations, and thermal fluctuations. Moreover, photonic devices can have a significant power dissipation footprint, which can increase further when compensating for potential faults. New network-centric circuits, architectures, tools, and protocols are required to overcome these challenges. This special session focuses on the overarching goals of enabling high fault resilience and energy-efficiency in emerging silicon photonic on-chip networks.

Exploring Power and Data Signaling Enhancements for Emerging Silicon Photonic Networks-on-Chip
Presenter: Sudeep Pasricha, Colorado State University

Light signals from a multi-wavelength laser source represent the fundamental modality used to activate photonic devices and propagate data in photonic networks-on-chip. Unfortunately, laser power overheads constitute a very high portion of overall power dissipation in photonic networks. Moreover, the multi-wavelength signals in photonic waveguides interfere with each other, resulting in crosstalk and sideband lobe truncation effects that can severely reduce the reliability during photonic data transmissions. This talk presents innovative solutions to enhance the signaling efficiency in photonic networks-on-chip. A novel low-overhead technique for run-time management of laser power in photonic networks will be discussed, which makes use of on-chip semiconductor amplifiers to achieve traffic-independent and loss-aware savings in laser power. To overcome the drawbacks of traditional on-off-keying based signaling, an innovative signaling approach with four-amplitude-level photonic signals will be presented, which doubles the aggregate bandwidth without increasing photonic hardware and incurred noise. The impact of these innovations is explored on various photonic network-on-chip architectures, to show their effectiveness in reducing bit-error-rate, area, and energy costs.

Toward a Cross-Layer Synthesis Methodology for Wavelength-Routed Optical Networks-on-Chip
Presenter: Davide Bertozzi, University of Ferrara

While the information and computing revolution is often credited to Moore's Law scaling, the complexity challenge has been actually addressed by electronic design automation, which is capable of transforming complex system-on-chip designs from high-level functional specifications into detailed geometric descriptions. Similarly, the uptake of emerging interconnect technologies depends not only on technology maturity, but also on the availability of tools and methodologies bridging the gap between system designers and technology developers. This talk illustrates an ongoing research effort to extend the paradigms and the methodologies of electronic design automation to the context of emerging silicon photonic networks. The talk goes into the details of an early-stage cross-layer synthesis methodology for wavelength-routed optical network-on-chip topologies. At the top layer of the design hierarchy, the goal is to capture all topology design points in a unified design framework. As the abstract view of a candidate topology design point is refined into an actual implementation, the intricacy of co-designing the electronic layer with a 3D-stacked optical layer is addressed. The emphasis here is on minimizing the design predictability gap when trying to fit tight spacing and place-and-route constraints. This work contributes towards evolving design automation beyond its electronic roots.

Thermal-Aware Design Methods in Optical Networks-on-Chip
Presenter: Hui Li, Xidian University

Silicon photonics is an emerging technology considered as one of the key solutions for future generation on-chip interconnects, providing several prospective advantages such as low transmission latency and high bandwidth. Optical networks-on-chip, utilizing silicon photonic interconnects at the chip scale, are a promising and "hot" research topic. However, they still encounter unresolved challenges in reliability, due to thermal variations, process variations, device defects, and so on. Thermal variations are inevitable, induced by the temperature sensitivity of silicon photonic devices. Under a given chip activity, this leads to a lower laser efficiency and a drift of wavelengths of optical devices (e.g., on-chip lasers and microring resonators), which in turn results in a higher Bit Error Ratio and consequently reduces the energy efficiency of optical interconnects. In general, this reliability issue could be alleviated or be tolerated from different aspects, such as with wavelength tuning, encoding technology, power management schemes, adaptive routing algorithms, and so on. This talk will focus on design methods from the perspective of thermal-induced reliability issues.

978-1-5386-3775-3/18 $31.00 © 2018 IEEE

Innovative Practices on Test in Japan

Koji Asami, Advantest, Japan, Yoshiro Tamura and Haruo Kobayashi, Gunma University, Japan
Jun Matsushima and Yoichi Maeda, Renesas Electronics, Japan
Kazumi Hatayama, Gunma University, Japan (Organizer)

I. Introduction

The IP session highlights three innovative practices in Asia, which include low cost testing of analog front-end circuits, evaluation of complex analog filter and power reduction of power-on self-test for automotive MCU.

II. Analysis and Evaluation Method of Complex Analog Filter
(Koji Asami, Yoshiro Tamura and Haruo Kobayashi)

An RC polyphase filter is well-known as a complex analog filter and has asymmetric frequency characteristics with respect to DC. It is composed of resistors and capacitors, and has complex analog input/output ports. Since this type of filter can partially realize Hilbert transform, it can be used for generating orthogonal signals. In the field of wireless communications, it can also be used for suppressing interference signals by combining the Low-IF receiver topologies.

In this presentation, the transfer function of the RC polyphase filter is analyzed and its relationship to the Hilbert transform is discussed. Furthermore, we introduce a technique to evaluate the mismatch characteristics between the real path and the imaginary path in the RC polyphase filter. These studies could lead the complex analog processing to apply for the wideband wireless communications.

III. A DFT based approach to functional safety for automotive MCU
(Jun Matsushima and Yoichi Maeda)

As required by ISO 26262 standard, it is essential to equip the automotive MCUs (Micro-Control Unit) with the functional safety mechanisms. Power-On-Self-Test (POST) is a well-known functional safety mechanism that has been widely deployed in various safety-critical systems. For automotive MCU, the POST is required to meet many constraints in terms of the specified fault coverage, the limited test application time and the low power consumption. We have developed these automotive MCUs in a limited development period and have further supplied them to the market. This presentation explain about reducing power consumption at POST and realizing POST with minimal change of design flow.

2018 IEEE 36th VLSI Test Symposium (VTS)

Efficient Generation of Parametric Test Conditions for AMS Chips with an Interval Constraint Solver

Felix Neubauer*, Jan Burchard*, Pascal Raiola*, Jochen Rivoir§, Bernd Becker* and Matthias Sauer*

*University of Freiburg, Germany
{neubauef | burchard | raiolap | becker | sauerm}
@informatik.uni-freiburg.de

§Advantest Europe GmbH, Germany
jochen.rivoir@advantest.com

Abstract—The characterization of analog-mixed signal (AMS) silicon requires a suitable pattern set able to exercise the parametric operational space to – among other tasks – validate the correct (specified) working behaviour of the device under test. As experience shows, most of the unexpected problems occur for very specific value combinations of a few test condition variables that were not expected to have an influence. Additionally, restrictions on the operational conditions have to be taken into account.

We present a method to efficiently create a set of test conditions to cover such a constrained search space with a user-defined density. First, an initial test condition set is generated using quasirandom Sobol sequences. Secondly, we analyse the test conditions to identify and fill uncovered areas in the parameter space using the in-house interval constraint solver iSAT3.

The applicability of the method is demonstrated by experimental results on a 19-dimensional search space using a realistic set of constraints.

Index Terms—Parametric Test Conditions, Analog, Mixed-Signal, Characterization, Constraints, Sobol, Formal Methods, SMT, Interval Constraint Solver, iSAT3

I. INTRODUCTION

Characterization of analog-mixed signal (AMS) circuits is a challenging problem requiring a well-designed set of input patterns [1], [2]. Among other tasks, characterization must validate that the device under test works as specified under all intended operating conditions. Operating conditions are defined by a large number of test condition variables, here called *dimensions*, that can be continuous, discrete or logical. Usually, operating conditions are constrained to not destroy the device (e. g. limit power consumption), avoid undesired behaviour (e. g. frequencies with too many spurs) or avoid unsupported or not implemented modes.

Experience shows that complex chips tend to have unexpected problems that occur for very specific value combinations of a few test condition variables that were not expected to have an influence. The goal of this work is to ensure that such problems will most likely be found, even in a black-box scenario, i. e. a scenario where a priori knowledge is (almost) not available.

As already indicated above experience of the industrial author shows that these unexpected problems depend on only a very few ($n \in \{2, \dots, 5\}$) test condition variables. Therefore it is not necessary to densely cover the high-dimensional space of all test condition variables, which would be practically impossible, rather it is sufficient to compute a characterization set, that densely covers all n-dimensional subspaces for small n, without leaving "large" untested areas between the (exercised) test conditions. Furthermore, because no a-priori knowledge is available for unexpected problems, all these subspaces should be covered as uniformly as possible. In the context of this work, a good coverage – and thus a good *density* of the characterization set

– will be indicated by a small *"Maximum Uncovered n-Cube (MUnC)"* in any n-dimensional subspace.

In the context of AMS circuits, typically constraints defining operational restrictions on the test condition variables exist (e. g. a power limit is provided). In the absence of such constraints, low discrepancy codes, like Sobol sequences [3], provide a fast and effective method to achieve a rather uniform distribution of test conditions in all subspaces [4], [5]. This is no longer guaranteed in the presence of constraints: test conditions violating the constraints have to be excluded because they are outside the given specification. This may cause a distribution bias and an increase in size for the MUnC (illustrated by Example 1 in Sec. II, Fig. 1). Hence more sophisticated methods are required to be able to take such constraints into account.

Contribution: This paper presents a method, that:

- generates a set of multi-dimensional test conditions (initial characterization set) using a Sobol sequence and corrects all test conditions violating the user-defined set of linear or non-linear constraints
- finds the maximum uncovered n-cube (MUnC) regarding a given number n of so-called active dimensions to estimate the density of the initial characterization set
- modifies the initial characterization set until a user-defined density is reached by specifically generating test conditions in uncovered regions of the input space.

The method utilizes a SAT Modulo Theory (SMT) encoding to find the MUnC and fill up the characterization set. The corresponding SMT formulas are solved by the interval constraint solver iSAT3 [6]–[9] which is able to reason directly over complex (non-)linear constraints in the real and integer domain. This results in a characterization set which provably holds a user-defined density.

The experimental results show that the method is able to fill up the initial characterization set in a feasible amount of time considering a large multi-dimensional test space with up to 19 dimensions and reasonable user-defined target densities as well as complex constraints. Furthermore, compared to characterization sets of the same size created by Sobol methods the achieved density is significantly higher, which shows the great potential of the proposed method.

Related Work: There are several approaches aiming at ordering or prioritizing a given parametric characterization set with respect to the significance of the test conditions, but without addition of further test conditions [10]. [11] presents an automated tool for sensitivity analysis and test condition generation, but this approach requires a detailed and accurate (SPICE) simulation model of the circuit and its defective behaviour. [12] aims at covering a set of circuit input parameters but does not support complex constraints.

978-1-5386-3775-3/18 $31.00 © 2018 IEEE

Structure of the paper: The remainder of the paper is structured as follows. Section II provides a description of the tackled problem as well as the fundamentals on Sobol sequences and the interval constraint solver iSAT3. The details of the proposed approach are covered in Section III, followed by the experimental results presented in Section IV. Finally, the paper is concluded in Section V.

II. GENERAL PROBLEM DESCRIPTION

AMS circuits usually have different types of input parameters including continuous (e. g. waveforms), discrete (e. g. register settings) and logic ones as well as static conditions (e. g. device number). Those input parameters – in the following referred to as *dimensions* – span a multi-dimensional test condition space.

Example 1. *Let's assume there are 4 dimensions: the voltage (continuous, value range: $[0,4]V$), the current (continuous, value range: $[0,5]A$), a mode (discrete; 3 different states) and a Boolean dimension which states whether an embedded component is activated or not.*

The main task of the approach presented in this paper is to get a high quality characterization set for such AMS chips. In the context of black-box characterization it is beneficial to use a characterization set well-distributed over the whole multi-dimensional test condition space in order to uniformly cover all fragments of that space. Thus as a quality measurement the *density* of such a set is used which is defined as the absence of test condition free spaces – so-called *Uncovered n-Cubes (UnC)* – larger than a given size. The size of such an n-cube is the *distance* from its central point to its side. For continuous dimensions the distance is measured in percent of the value range to be able to express this value in multiple dimensions with potentially different value ranges at once. For discrete dimensions it is sufficient to distinguish whether the distance of two values is zero or not. The largest UnC is called *Maximum Uncovered n-Cube (MUnC)*. Consequently a smaller MUnC implies a higher density of the characterization set and vice versa.

A good initial method for characterization set generation is using quasirandom low discrepancy sequences (e. g. Sobol [3], [13], as in this paper). Such sequences are often used for numerical integration, simulation, and optimization [5]. In contrast to pseudorandom numbers, they have the property to efficiently sample even a multi-dimensional space in a more uniform way [4], [14].

However, even this initial characterization set may leave UnCs of certain sizes (e. g. orange UnCs in Figure 1). Furthermore, usually there are additional dependencies between different input parameters expressed by so-called *parameter constraints*. In theory, those constraints can be arbitrarily complex, including linear and non-linear arithmetic or even transcendental functions. Figure 1 shows such a constraint for Example 1 including initial Sobol generated test conditions. The parameter constraint restricts the power of the chip by demanding *voltage · current < 9W* and thus excludes the blue area from the test condition space. This invalidates some of the generated test conditions as they violate the parameter constraint like the red points in the figure. Excluding such test conditions affects the density of the characterization set as existing UnCs are enlarged, lowering the quality of the set. An example can be seen in Figure 1 where the existing green UnC grows by the red part because of the removal of the red invalid test condition.

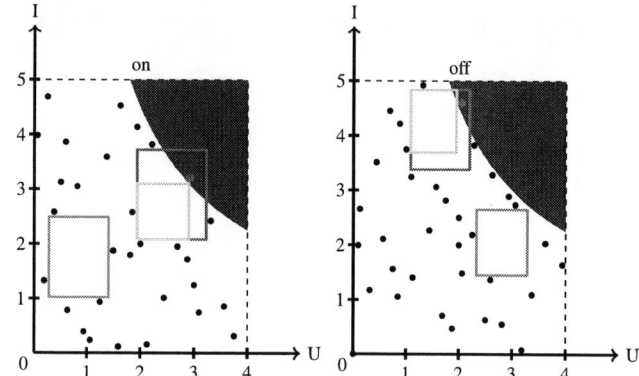

Fig. 1. Example 1 with the parameter constraint $U \cdot I < 9$ and an initial Sobol generated test condition set; active dimensions: U, I, on/off

Our approach tackles both issues by filling up the UnCs in the initial characterization set to improve the density and thus the quality.

As this involves handling continuous and discrete variables with a limited value range as well as parameter constraints of certain theories, it is natural to involve a SAT Modulo Theory (SMT) solver. SMT is an extension of the SAT problem. While SAT solvers search for a satisfying assignment of a propositional formula (or prove the absence thereof), SMT solver consider a Boolean combination of so-called theory atoms, which consist of arithmetic operations of the underlying theory and a relational operation, such as

$$(U \cdot I < 9) \wedge \cdots \wedge (((I < 1.25 - d) \wedge I_{act}) \vee \cdots \vee !on) \quad (1)$$

In this paper the interval constraint solver iSAT3 [7]–[9] is used to solve SMT formulas. iSAT3 is the third implementation of the iSAT algorithm [6], [15] and uses Interval Constraint Propagation (ICP [16]) to handle the theory part. iSAT3 is able to operate on the bounded real, integer and the Boolean domain supporting a wide range of operations including linear, non-linear and even transcendental functions (e. g. sine, cosine, exp).

As already mentioned, errors in AMS chips usually depend on only a few input parameters. Thus, it is sufficient to consider only relatively small subsets of dimensions (corresponding to the relevant input parameters). Those dimensions are referred to as *active*. In Figure 1 the active dimensions *voltage*, *current* and *on/off* are shown for Example 1. Generally speaking, all non-active dimensions are projected to the active ones. Following this thought it is sufficient to fill up the holes in the characterization set in relation to each possible subset of active dimensions.

III. APPROACH

In this section, a novel efficient approach for the generation of parametric test condition sets with a high density is described. Figure 2 provides the general work-flow. It consists of first generating a well-distributed initial characterization set taking into account the parameter constraints (Section III-A), then finding the MUnC to estimate the quality of the initial set (Section III-C) and finally generating test conditions to fill up the characterization set until a certain density is obtained (Section III-D). The latter two steps share the same basic encoding idea explained in Section III-B, while Section III-E describes some further optimizations.

A. Initial Characterization Set Generation

Finding a minimal set of test conditions that fulfils density conditions for all multi-dimensional subspaces of a given size at

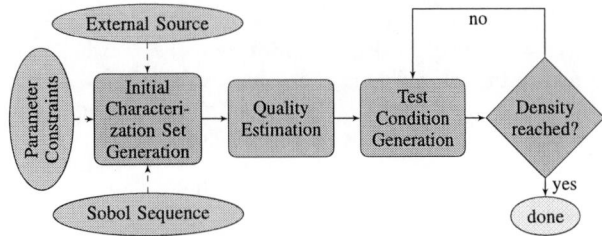

Fig. 2. Work-flow of proposed approach

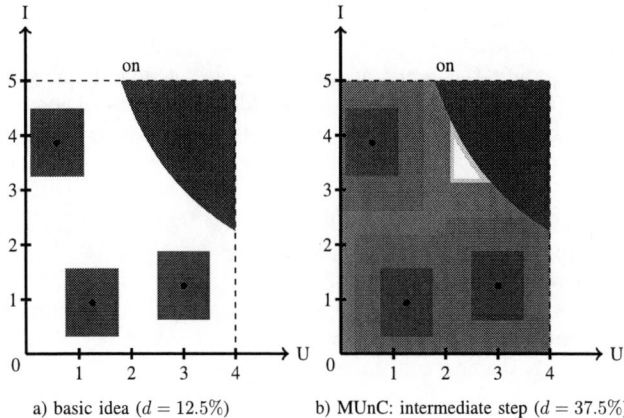

a) basic idea ($d = 12.5\%$)   b) MUnC: intermediate step ($d = 37.5\%$)

Fig. 3. Example 1: encoding idea & intermediate MUnC step both with distance d; active dimensions: U, I, on/off (only "on" is shown)

the same time is a non-trivial problem. The test conditions in such a set would have to be "perfectly uniformly distributed" for all such subspaces, i. e. test conditions projected to any relevant subspace should result in a uniform distribution in this subspace.

A Sobol sequence (Section II) is able to generate a well-distributed multi-dimensional test condition set efficiently, which is also well-distributed in lower-dimensional projections (however potentially creating large UnCs). We use such a sequence to generate the initial characterization set. To estimate the number of test conditions to be generated we calculate the minimal number of test conditions needed to cover a certain subspace of a given size with a given density and use this as a lower bound. As Sobol does not guarantee a "perfectly uniform distribution" the calculation is extended by an empirically identified (Sobol) factor (see Section IV-B), resulting in the following computation:

$$\#TC_{initial} = \left\lfloor f_{Sobol} \cdot \left(\frac{100}{|MUnC|} \right)^{\#actDim} \right\rfloor \quad (2)$$

with the targeted $size$ of the MUnC and the considered number of active dimensions $\#actDim$ in the subspace.

This Sobol generated initial set of test conditions forms the base of the following steps.

B. Basic Idea & Encoding

The next step is to analyse the initial test condition set and find the MUnC considering only a limited number n of active dimensions. Thereafter, the previously generated set is filled to close the uncovered spaces until a certain density is reached. Both of these tasks can be tackled by finding a new test condition which has at least a certain distance d from all other test conditions or proving no such point exists.

Figure 3a) shows the general encoding idea with the help of a graphical representation of Example 1 for two continuous and one discrete active dimensions including three candidates of the initial test set generated by a Sobol sequence. The red squares around each test condition indicate the area within the defined distance. These areas are already covered and considered to be "forbidden". Thus a new test condition is allowed to lie somewhere in the "white space" of the picture. For larger distances or a greater amount of test conditions more and more of the valid space is blocked until at some point it is not possible to find a new test condition with at least the considered distance to all other ones.

The task of finding a new test condition that does not violate any of the forbidden areas is encoded into an SMT formula which is solved by iSAT3. Formula 1 shows an excerpt of such an encoding for Example 1. The first part describes the parameter constraint (blue area) which always has to hold. The second part represents the "forbidden space" around one test condition. For the continuous dimensions the new test condition coordinates

must be outside of this space (e. g. $(I < 1.25 - d)$) while a new discrete value must not be equal to the existing one (e. g. $!on$).

While a satisfying assignment of the variables provides a new test condition, unsatisfiability of the formula proves its non-existence. As mentioned in Section II, not all dimensions are considered at the same time, but only a subset – the active dimensions. This subset has a fixed size but the members might vary. In the presented approach the solver can choose any subset of the dimensions when searching for the location of a new test condition to ensure that all possible sub-spaces are considered, filled up and hence covered. This information is also encoded into the formula (act variables, see Formula 1).

The problem space increases exponentially with the number of dimensions as each new dimension increases the possible value combinations by at least the factor 2. Considering not all dimensions at the same time decreases the complexity of the analysis tremendously even though all $\binom{\#dim}{\#actDim}$ possible subsets with a fixed number of dimensions have to be taken into account.

C. Quality Estimation

The MUnC can be found by applying the approach of Section III-B. Starting with a low distance the corresponding formula is created and solved. If a test condition was found the distance is increased by a given accuracy value and the formula is solved again. Figure 3b) shows the forth intermediate step for finding the MUnC in Example 1. In this example the accuracy of 12.5% was added to the distance (different red shades) in each step. The green triangle marks the last possible area to find a new test condition. This process is repeated iteratively until the result is "unsatisfiable". The last successfully checked distance marks the size of the MUnC regarding the chosen accuracy.

To improve the performance, the value of the target distance is handled as in a binary search.

D. Test Condition Generation

The goal of this step is to fill up the initial test condition set in order to reach a certain density. The principle of the idea described in Section III-B can also be applied for this task. Figure 3a) shows the space already covered (red) and the one still to cover (white). Solving the corresponding formula either provides a new test condition or proves that there is no new test condition as the given density is already reached. In the first case the new test condition is added to the set, the formula is updated

978-1-5386-3775-3/18 $31.00 © 2018 IEEE

and solved again. This is repeated until the characterization set density is reached.

Divide and Conquer: As mentioned in Section II, the complexity of this task increases with adding more dimensions. To tackle this challenge a Divide and Conquer approach is proposed. All dimensions restricted by one or more parameter constraint are handled separately from the dimensions without any restriction. This approach contains the following four steps:

(1) fill up the initial test condition set restricted to only the **constrained** dimensions
(2) fill up the initial test condition set restricted to only the **unconstrained** dimensions
(3) merge the new constrained and unconstrained test conditions and add them to the initial test condition set
(4) fill up the extended set considering all dimensions

Step 1 and 2 can be done by applying the method described above for the corresponding dimensions. It should be mentioned that as step 2 is only dealing with unconstrained dimensions, the SMT approach is not mandatory to be used. Other methods (e. g. a modified Sobol call) can also be applied.

In step 3, the newly generated test conditions from the first two steps are merged by putting the values of two test conditions together in the right order. The cross product of all test conditions (including the initial test conditions) is not feasible as it would produce a tremendously higher number of test conditions than necessary. Instead every test condition from step 1 is merged with exactly one test condition from step 2. If the two test condition sets do not have the same size, the smaller set has to be filled up, e. g. using (quasi)random values. This type of merging only considers cases in which the active dimensions are either all constrained or unconstrained but not mixed. This is why the next step is needed.

In step 4 the initial test conditions as well as the newly generated and merged test conditions from steps 1 to 3 are considered. This test condition set is filled up by using – again – the method described above but with the additional requirement that there has to be at least one constrained and one unconstrained dimension chosen to be active. This covers the shortcomings of the merge step. The resulting test condition set guarantees the desired density.

E. Optimizations

The dimensions considered in this scenario (see Section II) are either continuous with values from the real domain or discrete. This section describes a scaling method for the real domain to the integer domain and also backwards. The idea is to handle and generate test conditions including real values but only use integer values in the formula and during the solving process. Furthermore a test condition correction method is proposed which allows to continue to use test conditions which e. g. violate parameter constraints, instead of removing them completely.

Using integer scaling has two main advantages:

- As the (bounded) real domain is infinite it cannot be accurately modelled which can cause inconclusive solutions. However, the bounded integer domain is finite and thus there are no such modelling issues.
- Reasoning over the integer domain is much faster than reasoning over the real domain (see Section IV-C).

The accuracy of the integer scaling is defined by the so-called *intScale factor (isf)* given in percent. $isf = 0.5$ means for instance that all values of all dimensions should be mapped with an accuracy of 0.5% or in other words to the integer interval $[0, \frac{100}{isf}] = [0, 200]$. With this information the translations from the real to the integer domain and backwards can be done.

Let $x \in [x_{lb}, x_{ub}]$ be a real variable. The corresponding integer value $x_{int} \in [0, \frac{100}{isf}]$ is calculated as follows:

$$x_{int} = \left\lfloor (x - x_{lb}) \cdot \frac{100}{(x_{ub} - x_{lb}) \cdot isf} \right\rfloor \qquad (3)$$

The backwards calculation is similar:

$$x_{real} = x_{int} \cdot \frac{(x_{ub} - x_{lb}) \cdot isf}{100} + x_{lb} \qquad (4)$$

Using these calculations, all real values can be converted to the integer domain for the SMT formula and, after solving, the results can be translated back.

In general the translation from real to integer might result in accuracy losses. This can be seen in the following example:

Example 2. *Let* $x \in [0, 10]$ *be a real variable, scaled to* $x_{int} \in [0, 200]$ *and the constraint is* $x \geq 5.21$.

$$x_{int} \geq \lfloor 5.21 \cdot 20 \rfloor = \lfloor 104.2 \rfloor = 104$$
$$\textit{Valid solution: } x_{int} = 104$$
$$\Rightarrow x_{real} = 5.2 \not\geq 5.21 \; \lightning$$

In the SMT formula there are two different types of constraints to consider. The *distance* constraints defining the blocked area around a test condition (see Formula 1) can be transformed like in Example 2. The violation of such a distance constraint is not crucial because the corresponding test condition is still valid even if it is a little bit too close to another test condition. The parameter constraints on the other hand cannot be handled this way because they usually contain more than one dimension variable. The integer scaling would make those incomparable as the mapping target is always the same interval even for different real value ranges. Consequently the parameter constraints have to be handled separately, as shown by the following example.

Example 3. *Let* $x \in [-2, 4]$ *and* $y \in [0, 2.5]$ *be the real variables of two continuous dimensions. The constraint* $x - y > -2$ *is to be transferred to the integer domain* $(isf = 0.5)$ *by transforming each variable separately using Equation 4 and afterwards applying equivalence transformations to obtain only integer coefficients:*

$$x_{real} = x_{int} \cdot \frac{6}{200} - 2, \qquad y_{real} = y_{int} \cdot \frac{2.5}{200},$$
$$x_{int} \cdot \frac{6}{200} - 2 - y_{int} \cdot \frac{2.5}{200} > -2$$
$$\Leftrightarrow \qquad 12 \cdot x_{int} - 5 \cdot y_{int} > 0$$

In contrast to the distance constraints, parameter constraints must not be violated by test conditions. Consequently, both, all (quasi)randomly generated test conditions and – in case of integer scaling – all test conditions generated by iSAT3 have to be checked against the parameter constraints. If there is a violation the corresponding test condition is corrected by replacing all conflicting test condition values with the default value of the involved dimensions. Those default values are chosen in a way such that they fulfil all parameter constraints.

Note: For constraints where the integer scaling of the parameter constraints can be handled like in Example 3, no test condition generated by iSAT3 will violate the original constraint in the real domain, as only equivalence transformations are used.

	proposed approach			pure Sobol approach	
	MUnC (%)	#TC$_{initial}$	#TC$_{total}$	#TC	MUnC (%)
19 dimensions	10.0	1500	6560	6560	22.0
	9.0	2057	7716	7716	21.0
	8.0	2929	10219	10219	20.5
	7.0	4373	16211	16211	18.5
	6.0	6944	21554	21554	17.0
	5.0	12000	33632	33632	14.0
	4.0	23437	59041	59041	13.5
	3.0	55555	132382	132382	12.5
	2.0	187500	375911	375911	7.5

TABLE I
DENSITY COMPARISON BETWEEN SOBOL AND THE PROPOSED APPROACH
FOR THE SAME AMOUNT OF TEST CONDITIONS

IV. EXPERIMENTAL RESULTS

In this section the evaluation of the proposed method is presented. The results are based on data from an in-house AMS chip of Advantest. This chip has 9 input parameters containing 4 continuous and 5 discrete dimensions. One of these discrete dimensions is Boolean, the others consist of more than two states. Furthermore the continuous dimensions are restricted by linear parameter constraints, referred to as *basic* constraint set.

In order to obtain a second more complex example, 10 random dimensions (5 continuous, 5 discrete) are added as well as a more *complex* constraint set which contains quadratic and even cubic constraints over the continuous dimensions as well as one over a discrete dimension. In general, 3 dimensions are considered to be active. All experiments are performed on an Intel® Xeon® CPU E5-2643 @ 3.30GHz with 64GB RAM.

A. Evaluation for different Target Densities

First, different experiments for 9 and 19 dimensions using only the basic or in addition the complex parameter constraints are performed. The Sobol factor is chosen to be 1.5 (see Section IV-B). Figure 4 and 5 show the corresponding runtime and the size of the generated test condition set for different target densities represented by the MUnC size.

As there is – as far as we know – no other approach targeting the same task, the direct competitor is a pure Sobol approach which generates the whole test set using a Sobol sequence. As the Sobol approach cannot target for a certain test set density but only generates a given number of test conditions, the following experimental setup is chosen (see Table I): First the proposed approach is applied for the target MUnC sizes of 10.0% to 2.0% which generates test sets of certain magnitudes including #TC$_{initial}$ test conditions generated by Sobol. Afterwards a new test set with the same size is created by using the pure Sobol approach. The comparison of the MUnC sizes shows the superior quality of the (improved) test sets generated by the proposed approach compared to the ones generated using pure Sobol.

Another observation is that the additional use of the complex constraints does not seem to change the results significantly. This indicates that the overall complexity of the approach is not (only) the result of these parameter constraints. Furthermore, the figures show the exponential character of the task both in the target MUnC size and the number of dimensions not only for the runtime but also for the number of test conditions. Comparing different numbers of active dimensions (e. g. 2, 3 and 4) shows a similar picture. Nevertheless the proposed approach achieves very feasible runtimes, suitable for the application in the industrial context.

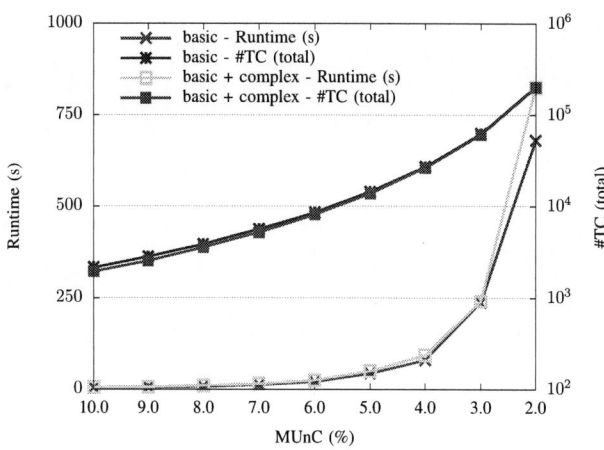

Fig. 4. Runtime (linear scale) and number of generated test conditions (logarithmic scale) for different target MUnC sizes (9 dimensions, 3 active, Sobol factor: 1.5)

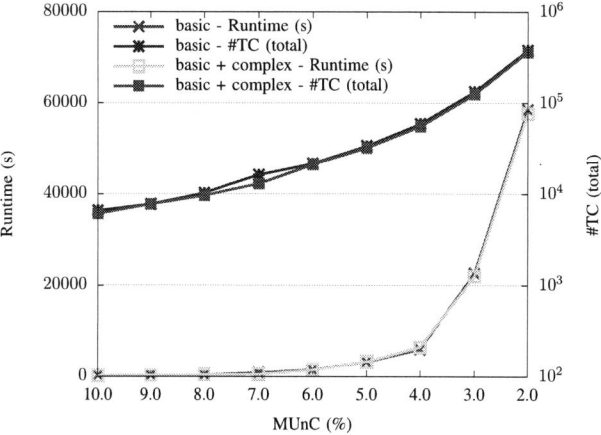

Fig. 5. Runtime (linear scale) and number of generated test conditions (logarithmic scale) for different target MUnC sizes (19 dimensions, 3 active, Sobol factor: 1.5)

B. Empirical Identification of the Sobol Factor

The performance of the approach as well as the number of test conditions to be generated until the user-defined test set density is reached highly depend on the quality and cardinality of the initial test set. As described in Section III-A, a Sobol sequence is used to get a well-distributed test set while the number of test conditions #TC$_{initial}$ is derived from the optimal distribution and the so-called *Sobol factor*. Figure 6 shows the connection between the runtime and the number of generated test conditions for different Sobol factors in the range $[0.5, 4.0]$. While the runtime is always decreasing when using a higher Sobol factor, the total number of test conditions first decreases and later increases again. This indicates that generating more initial test conditions with Sobol is beneficial to a certain limit. The runtime decreases significantly as less highly specialized but rather costly test conditions generated by iSAT3 have to be added to reach the target density. Passing that limit, the impact of a larger initial test set regarding the density decreases resulting in larger test condition sets for more or less the same runtime. Table II further confirms this, considering the MUnC of the initial test sets with different Sobol factors.

A Sobol factor of 1.5 seems to provide a good trade-off between runtime and number of test conditions.

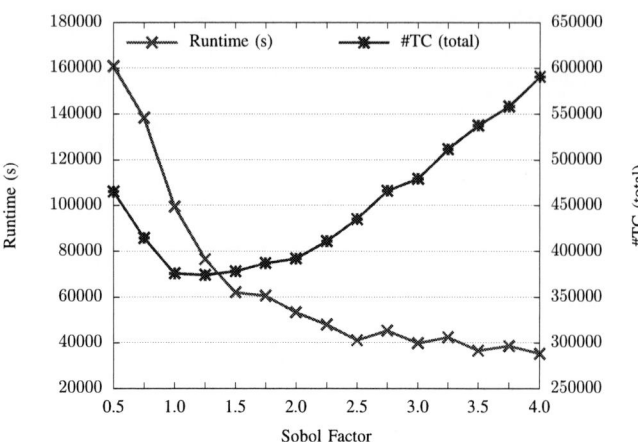

Fig. 6. Runtime and number of generated test conditions (both linear scales) for different Sobol factors (19 dimensions, 3 active, MUnC: 2.0%, basic constraints)

Sobol factor	0.5	0.75	1.0	1.25 - 1.75	2.0	2.25 - 3.0	3.25 - 4.0
MUnC	13.5	13.0	12.5	8.5	8.0	7.5	7.0

TABLE II
MUNC FOR DIFFERENT SOBOL FACTORS (3 ACT. DIM., MUNC: 2.0%)

C. Real Arithmetic vs. Integer Scaling

In Section III-E, the method, benefits and drawbacks of scaling the real domain to the integer domain were described. Table III shows the significant lower runtimes using integer scaling in average by the factor 50 for all target MUnC sizes, while the real computations even reach the timeout of 3 days in three cases. Furthermore an unneglectable amount of UNKNOWN results which can arise due to the infiniteness of the real domain (see Section III-E) appear while using real arithmetic which are avoided using integer scaling. However, in the performed testruns there were no test conditions generated with integer scaling that violate the parameter constraints. The only corrections were made during the random generation of test conditions where the parameter constraints are not considered at all.

To summarize, the experimental results show the efficiency of the proposed approach and its advantages over a pure (quasi)random test condition generation. The method generates test condition sets with a far better density than a purely Sobol generated set of the same size. Thus, to achieve the same density using Sobol, one needs more test conditions. Although the underlying problem has exponential complexity, the runtimes are very feasible for the practical usecase. Furthermore the last subsection shows the huge impact the use of integer scaling has on the performance.

	basic parameter constraints			
MUnC (%)	integer scaling		real arithmetic	
	#TC	Runtime (s)	#TC	Runtime (s)
10.0	6560	152.61	5642	3491.27
9.0	7716	226.11	7951	9720.65
8.0	10219	350.48	10591	21233.30
7.0	16211	800.17	14937	46032.80
6.0	21554	1458.12	19503	92468.10
5.0	33632	3117.49	29606	220001.00
4.0	59041	5900.25	-	TO
3.0	132382	22628.00	-	TO
2.0	375911	58479.30	-	TO

(19 dimensions)

TABLE III
COMPARISON OF THE NUMBER OF GENERATED TEST CONDITIONS AND THE RUNTIME FOR INTEGER SCALING AND REAL ARITHMETIC (19 DIMENSIONS)

V. CONCLUSION AND FUTURE WORK

In this paper, a method to generate and improve a parametric characterization set for AMS chips with various kinds of input parameters as well as complex constraints is presented. First, an initial test condition set is generated using a uniformly distributed Sobol sequence. Secondly, the maximum uncovered n-cube is calculated which is a quality indicator of the characterization set as it approximates the largest uncovered area in the constrained test space. At last, new test conditions are generated to fill up the characterization set until a user-defined density is reached. The latter two tasks are tackled by an SMT encoding of the problem utilizing the interval constraint solver iSAT3.

The experimental results show the benefits of the proposed method in contrast to a pure Sobol approach: we obtain a far better density for the same set size as well as a method for measuring and selectively improving the quality of the characterization set and all of that in a feasible amount of time.

Future Work: In this paper, linear and polynomial parameter constraints have been considered. A next step would be to support even more complex constraints including e. g. the modulo operation or transcendental functions such as *sine* or *exp*. Guiding the SMT solver into choosing more "clever" test conditions could improve the performance of the approach further.

REFERENCES

[1] L. S. Milor, "A tutorial introduction to research on analog and mixed-signal circuit testing," *IEEE Transactions on Circuits and Systems II: Analog and Digital Signal Processing*, vol. 45, no. 10, pp. 1389–1407, Oct 1998.

[2] R. Voorakaranam, S. S. Akbay, S. Bhattacharya, S. Cherubal, and A. Chatterjee, "Signature testing of analog and RF circuits: Algorithms and methodology," *IEEE Transactions on Circuits and Systems I: Regular Papers*, vol. 54, no. 5, pp. 1018–1031, May 2007.

[3] I. Sobol, "On the distribution of points in a cube and the approximate evaluation of integrals." *U.S.S.R. Comput. Math. Math. Phys.*, vol. 7, no. 4, pp. 86–112, 1969.

[4] I. Sobol and Y. L. Levitan, "The production of points uniformly distributed in a multidimensional cube," *Preprint IPM Akad. Nauk SSSR*, vol. 40, no. 3, 1976.

[5] I. Sobol, "Points which uniformly fill a multidimensional cube," *Mathematics, cybernetics series*, p. 32, 1985.

[6] M. Fränzle, C. Herde, T. Teige, S. Ratschan, and T. Schubert, "Efficient solving of large non-linear arithmetic constraint systems with complex boolean structure," *Journal on Satisfiability, Boolean Modeling, and Computation*, vol. 1, no. 3-4, pp. 209–236, 2007.

[7] K. Scheibler, S. Kupferschmid, and B. Becker, "Recent improvements in the SMT solver iSAT," in *MBMV 2013*.

[8] K. Scheibler and B. Becker, "Using interval constraint propagation for pseudo-boolean constraint solving," in *FMCAD 2014*.

[9] K. Scheibler, "Applying CDCL to verification and test: when laziness pays off," Ph.D. dissertation, University of Freiburg, Freiburg im Breisgau, Germany, 2017. [Online]. Available: https://freidok.uni-freiburg.de/data/12669

[10] P. N. Variyam and A. Chatterjee, "Enhancing test effectiveness for analog circuits using synthesized measurements," in *Proceedings. 16th IEEE VLSI Test Symposium*, Apr 1998, pp. 132–137.

[11] N. B. Hamida, K. Saab, D. Marche, B. Kaminska, and G. Quesnel, "LIMSoft: automated tool for design and test integration of analog circuits," in *Proceedings International Test Conference 1996. Test and Design Validity*, Oct 1996, pp. 571–580.

[12] A. Abderrahman, B. Kaminska, and E. Cerny, "Optimization-based multi-frequency test generation for analog circuits," *Journal of Electronic Testing*, vol. 9, no. 1, pp. 59–73, Aug 1996.

[13] P. Bratley and B. L. Fox, "Algorithm 659: Implementing sobol's quasirandom sequence generator," *ACM Trans. Math. Softw.*, vol. 14, no. 1, pp. 88–100, Mar. 1988.

[14] M. Bianchetti, S. Kucherenko, and S. Scoleri, "Pricing and risk management with high-dimensional quasi-monte carlo and global sensitivity analysis," *Wilmott*, vol. 2015, no. 78, pp. 46–70, 2015.

[15] C. Herde, "Efficient solving of large arithmetic constraint systems with complex boolean structure: proof engines for the analysis of hybrid discrete-continuous systems," Ph.D. dissertation, 2011.

[16] F. Benhamou and L. Granvilliers, "Continuous and Interval Constraints," in *Handbook of Constraint Programming*, ser. Foundations of Artificial Intelligence, 2006, pp. 571–603.

978-1-5386-3775-3/18 $31.00 © 2018 IEEE

Enhanced Hotspot Detection Through Synthetic Pattern Generation and Design of Experiments

Gaurav Rajavendra Reddy, Constantinos Xanthopoulos and Yiorgos Makris
gaurav.reddy@utdallas.edu, constantinos.xanthopoulos@utdallas.edu, yiorgos.makris@utdallas.edu
Department of Electrical and Computer Engineering, The University of Texas at Dallas, Richardson, TX 75080, USA

Abstract—Continuous technology scaling and the introduction of advanced technology nodes in Integrated Circuit (IC) fabrication is constantly exposing new manufacturability issues. Design hotspots are one of such problems, which are a result of complex design and process interactions. These hotspots are known to vary from design to design and foundries expect such hotspots to be predicted early and corrected in the design stage itself, as opposed to a process fix for every hotspot, which would be intractable. Various efforts have been made in the past to address this issue by using a known database of hotspots as a source of information. Most of those works use either Machine Learning (ML) or Pattern Matching (PM) techniques to identify and predict hotspots in new incoming designs. Almost all of those methods suffer from high false-alarm rates, mainly because (i) they are oblivious to the root causes of hotspots, and (ii) a large hotspot database to learn from is generally not available. In this work, we try to address these limitations by using novel hotspot Design of Experiments (DOEs) and synthetic pattern generation approaches. We analyze the effectiveness of the proposed method against the state-of-the-art on a 45nm process, using industry standard tools and designs.

I. INTRODUCTION

Continued technology scaling and the introduction of every advanced technology node in Integrated Circuit (IC) fabrication brings in new challenges for foundries. Lithography is a major challenge during technology development. As shown in Figure 1, in early technology nodes, the wavelength of light used in lithography was much smaller than the features being printed. In the latest nodes, this has reversed and lithography has become extremely challenging due to complex interactions between designs and sophisticated unit processes. To mitigate some of the lithography-related issues and ensure reliable manufacturing, various Resolution Enhancement Techniques (RETs) such as Optical Proximity Correction (OPC), Multi-patterning, Phase-shifted masks etc., are used. Despite employing RETs, certain areas in the design (layout), which are Design Rule Check (DRC) clean and Design For Manufacturability Guidelines (DFMGs) compliant, show abnormal and unexplained variation, causing parametric or hard defects. Such areas are termed as 'Hotspots' (popularly known as 'Lithographic hotspots' or 'Design weak-points'). The cause of hotspots is mostly attributed to their neighborhood (a set of polygons surrounding the hotspot area) which causes complex interactions of light during the lithography process. Since, hotspots vary from design to design, identifying their root causes and finding a fix for all such hotspots through process changes is extremely difficult, time consuming and

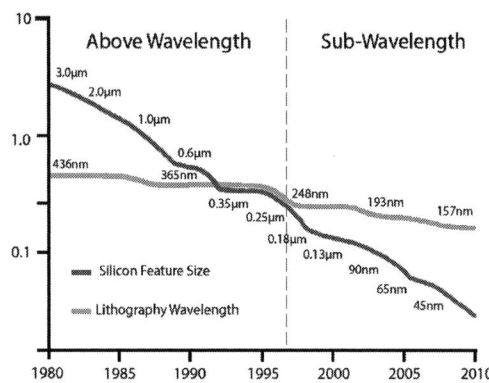

Fig. 1. Changes in lithography with silicon feature sizes [2]

expensive. Thus, in most cases, foundries create a database of known hotspots and restrict their presence in incoming customer designs. A hotspot database is usually populated through Failure Analysis (FA), inline inspections, lithographic simulations using well-calibrated lithographic models, etc. [1]. If a design pattern turns out to be a hotspot in later stages of fabrication, especially, after mask production, it may result in huge financial losses to the foundry. Hence, there is a great incentive to identify hotspots early and correct them in the design stage itself.

Many researchers have suggested pattern matching and machine learning-based techniques to identify and predict hotspots in new incoming designs. Unlike previous works [3], [4], where the focus has been on using more and more powerful machine learning tools, we take a novel approach to improving hotspot detection by increasing the information-theoretic content of the training data that these methods use. We call this process 'Database enhancement' and it involves two procedures: (i) 'Synthetic pattern generation' and, (ii) 'Design of experiments': Combined, these procedures enable a machine learning entity to effectively learn the 'root cause features' of hotspots. These procedures are also 'method agnostic', as they can be used with any of the previously proposed hotspot detection methods to improve their performance.

The rest of the paper is organized as the following: The State-Of-The-Art (SOTA) and its limitations are explained in detail in section II. The proposed methodology is presented in section III. Experimental results are discussed in section IV and conclusions are drawn in section V.

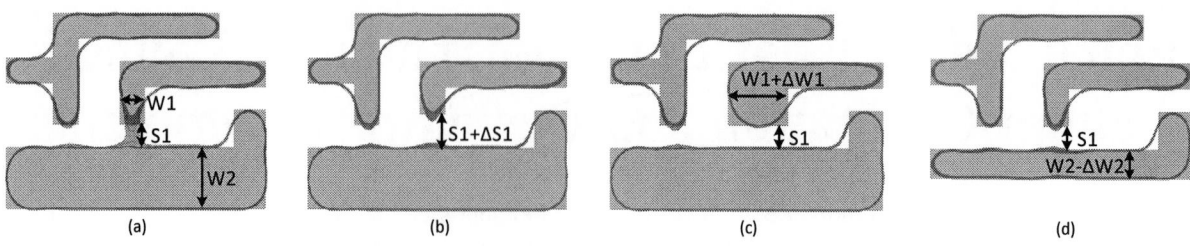

Fig. 2. (a) A hotspot pattern, (b-d) variants of pattern 'a' which are non-hotspots

II. THE STATE-OF-THE-ART AND ITS LIMITATIONS

Hotspot detection has been a topic of high interest in the past few years. Authors of [5], [6] have used pattern matching techniques, wherein a new design is compared to a database of previously seen hotspots and potential hotspot areas in the design are flagged. While these techniques are helpful in identifying known hotspots, they often fail in predicting unknown (never-seen-before) hotspots. To address this issue, machine learning techniques using Support Vector Machines (SVMs) [4], Artificial Neural Networks (ANNs) [7], multiple/meta classifiers [3] etc. were proposed. They essentially 'learn' (are trained) from a known database and use the trained model to make a prediction on new patterns. Over time, various flavors of these techniques have been proposed which have provided better accuracy, faster run-times etc., [8], [9]. However, most of them still suffer from high false alarm-rates, mainly because (i) these techniques are focused on learning the major differences between hotspots and non-hotspots, rather than their root cause features, and (ii) the lack of large hotspot databases limits their learning capabilities. Often, only a small number of known hotspots are used for learning [4], [5]. The 'Database enhancement' approach proposed in this work specifically addresses these issues.

A. False-Alarms In The State-Of-The-Art

The state-of-the-art machine learning-based hotspot detection techniques suffer from high false-alarm rates [3]–[5], [9]. The source of these false-alarms is illustrated using the following example. Figure 2 shows four patterns with their contours (Process Variability (PV) bands) obtained from litho simulations. Among them, pattern (a) is a hotspot due to a short between two of its polygons. Patterns (b-d) are very similar to pattern (a), but their subtle differences from pattern (a) makes them non-hotspots.

Case 1 - Let us assume that a machine learning based classifier is being trained to detect hotspots and among the patterns shown in Figure 2, only pattern (a) is part of its training dataset. During testing, if pattern (b) is presented to the classifier, it tends to classify it as a hotspot due to its close similarity to pattern (a). But, in reality, it is not a hotspot due to the increased space $S1 + \Delta S1$. The classifier made this error because it had failed to recognize $S1$ as a root cause feature of this pattern.

Case 2 - Let us assume that the classifier's training dataset includes both patterns (a) and (b). In this case, the classifier easily recognizes that the constrained space $S1$ makes this

pattern a hotspot and a relaxed space $S1 + \Delta S1$ would make it a non-hotspot. Then, if pattern (c), which is very similar to patterns (a) & (b) (also having a constrained space $S1$), is presented to the classifier, the classifier tends to call it a hotspot. But in reality, it is not a hotspot, because of the increased width $W1 + \Delta W1$. Here, the classifier predicted incorrectly because, during training, it had only recognized $S1$ as a root cause feature, but not $W1$. Similarly, the feature $W2$ is also a root cause feature.

From the above example, it becomes evident that, unless otherwise trained with many variants of a known hotspot, the ML entity assumes that all polygons in a pattern equally contribute towards making it a hotspot and fails to learn the root cause features. Without such learning, it remains oblivious to the subtle variations in similar-looking patterns and tends to misclassify them, creating large amounts of false-alarms. Hence, enhancing the database with sufficient variants of known hotspots becomes imperative towards empowering an ML entity to learn effectively.

III. PROPOSED METHODOLOGY

The proposed hotspot detection flow is shown in Figure 3. A high-level description is provided below and its major blocks are explained in detail in the next sub-sections. This flow is typically implemented at the foundry side and executed prior to mask fabrication; yet parts of it can be potentially incorporated into the Product Design Kits (PDKs) and transferred to the customer, in order to reduce design debug cycles.

A set of known hotspots and non-hotspots gathered from prior experience form the initial database. Design of Experiments (DOEs) is performed to increase the information-theoretic content of the initial database. As part of these experiments, synthetic variants (patterns) of known hotspots are generated and subjected to process simulations (litho/litho-etch) to determine which of the patterns are hotspots. Synthetic patterns, along with the initial database, form the enhanced database. Patterns in the enhanced database are converted into numerical feature vectors. Feature vectors are, then, subjected to dimensionality reduction and a machine learning-based classifier (i.e., an SVM) is trained using the dimensionality-reduced feature vectors. The trained model is, then, stored to evaluate future incoming designs. When a foundry receives a new design from its customers, the design is decomposed into smaller patterns and predictions are made on these patterns using the trained classifier. Patterns classified as hotspots are subjected to further investigation, flagged as areas of interest for inline inspections, and drive design fixes if warranted.

978-1-5386-3775-3/18 $31.00 © 2018 IEEE

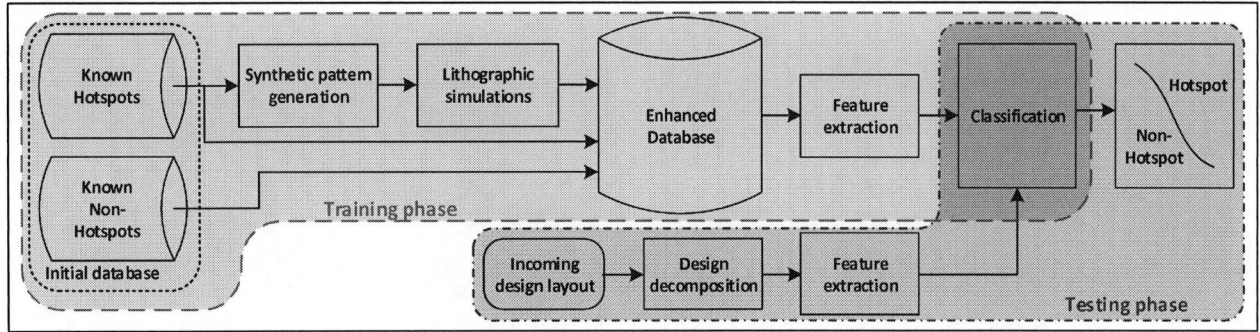

Fig. 3. The proposed machine learning-based Hotspot detection flow

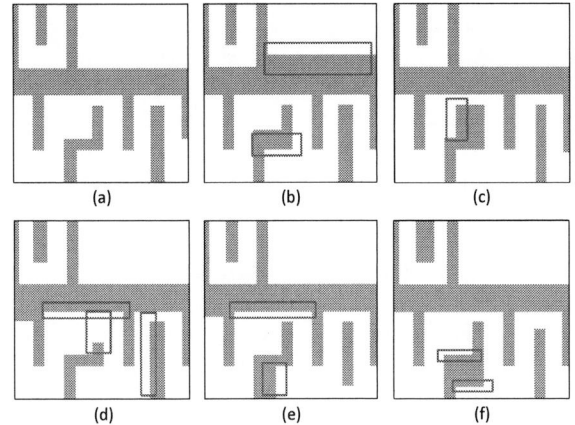

Fig. 4. (a) A hotspot pattern, (b-f) Synthetic patterns generated from pattern 'a'. Red markers indicate the subtle differences from pattern 'a'

A. Synthetic Pattern Generation and DOEs

For every hotspot in the initial database, multiple synthetic patterns are generated by changing one or more features at a time. Features such as corner to corner distances, jogs, line-end positions, layer spacing, layer area etc., are varied. Figure 4 (a) shows one such hotspot and Figures 4 (b-f) show some of its synthetic patterns. A time-efficient method for varying these features relies on perpendicularly moving the edges of one or more polygons in each snippet by a random distance. This approach allows to quickly generate multiple patterns whose variance can be easily controlled by two parameters. The first parameter p is the probability of any given edge to move or remain stationary. By increasing this probability, we effectively increase the number of polygons and their edges that are altered in the snippets. The second parameter d is associated with a distribution of distances, which is sampled for every polygon edge selected by the first parameter. The sampled value denotes the distance by which the edge will be displaced. These distance values follow a normal distribution centered at 0. In this way, most synthetic patterns are slight variants of the original pattern, thereby enabling us to learn the root causes effectively. However, the variation between generated patterns can be easily changed by varying the parameter d. Essentially, the parameter d can be thought of as the standard deviation of this distribution. Parameters p and d are varied based on domain knowledge and experimentation.

As expected, the above-mentioned procedure results in a plethora of patterns, many of which might not even pass the DRC. To ensure that valid layout topologies are generated and to make this process run-time efficient, we implemented a minimal DRC engine in Python, which we execute after every pattern is generated. This check ensures that most of the generated patterns are valid. However, since implementing complex design rule checks becomes complicated, all synthetic patterns which pass this minimal DRC check are also subjected to a full DRC using CalibreDRC. Through this approach, we can ensure that the vast majority of the generated patterns are DRC clean and usable. Synthetic patterns are, then, subjected to lithographic simulations to ascertain the ground truth about them. To this end, it is assumed that litho models are well-calibrated to the process, as is often the case in mature processes (with PDKs 1.0 and above). On the other hand, during early technology development, foundries may not have well-calibrated models readily available, but do have access to plenty of test-silicon. In those situations, simulation results from crude models can be used as a guide to direct actual silicon-based experiments.

The number of synthetic patterns necessary to significantly improve the information-theoretic content in the training set depends on the process node, design complexity, layer of interest etc. We have studied this dependency on a 45nm process and a detailed explanation can be found in the experimental results section. In general, these experiments are not run-time intensive, as they work with small layout snippets. Moreover, this is a one-time procedure, hence a large number of synthetic patterns could be generated. Synthetic patterns, along with their litho simulation results, are added into the initial database in order to create the enhanced database/dataset.

B. Feature Extraction

In all proposed machine learning-based hotspot detection schemes, hotspot and non-hotspot patterns are initially obtained in the form of layout snippets and then subjected to feature extraction, whereby the image snippet is transformed into a numerical feature vector which can be used to train/test a machine learning entity. Various feature extraction methods, such as bounded rectangle region-based [7], polygon fragment-based [4], concentric circle sampling-based [9], density transform [5] etc., have been proposed in the past, suited to

978-1-5386-3775-3/18 $31.00 © 2018 IEEE

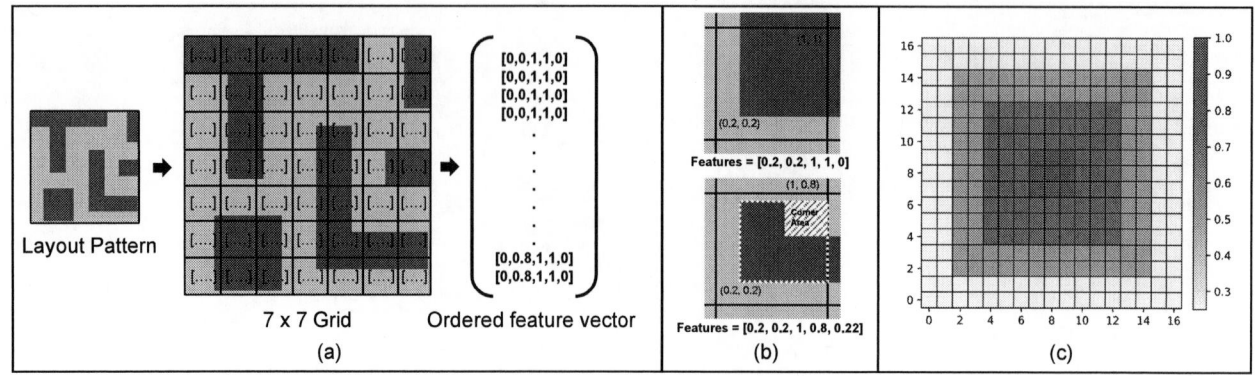

Fig. 5. Feature extraction (a) Grid based co-ordinate transform, (b) Features extracted from within a grid block. Top - features of a polygon without inner-corners, Bottom - features of a polygon with an inner-corner, (c) A plot showing the weights associated with a 17x17 grid used in this work

the detection flow they were used in. While every method has its own drawbacks and there is no clear winner among them, density transformation has been most widely used and works reasonably well. However, it fails to capture the minor variations which are crucial for effective learning.

In this work, we propose a novel feature extraction technique, which we call 'Co-ordinate Transform'. In co-ordinate transformation, as illustrated in Figure 5(a), an $n * n$ grid is overlapped on a pattern, where n is the number of blocks on each dimension of the grid. n is chosen such that each block covers only one, or a part of one polygon in the pattern. n is given by $(pattern\_size/layer\_half\_pitch)$. Polygons present within each block are transformed into five features: {bottom-left x co-ordinate, bottom-left y coordinate, top-right x co-ordinate, top-right y co-ordinate, corner info}. While the first four features capture most of the information, they fail to capture the inner corner information. Hence, a fifth feature 'corner info', shown in Figure 5(b), is necessary. This is given by the difference between the *polygon_bounding_box_area (area in white dotted line)* and the *layer_area (area in red)*. Features from every block of the grid are extracted and their ordered vector makes the final feature vector. While the five features capture the geometric information of polygons, their ordering captures the location of those polygons within the pattern.

Comparison to density transform: In density transform, different polygons may result in the same density value. For instance, if a polygon with area P_a is present inside a block with area B_a, the density of that block will remain P_a/B_a, irrespective of whether the polygon has a corner, is oriented vertically or horizontally, occupies the top, bottom, left or right portions of the block. To reduce the information loss due to these issues, density transform is often used with very fine grids (having large number of blocks). But such fine grids result in large number of features with very little variation within the features. This increases the ML model complexity and leads to over-fitting. On the other hand, coordinate transform is free from such issues as it uses coordinates to capture polygon information and uses coarse grids which, in turn, creates fewer features yet with more variation.

Weighted features: As an option, during co-ordinate transformation, weights can be assigned to various blocks of the grid, as shown in Figure 5(c). Typically, minor variations in the central area of patterns have high influence in causing hotspots, while this influence fades as we move towards the periphery. Given enough data, an effective machine learning entity can learn this variation by itself, but, adding domain knowledge like this helps to work with smaller datasets, faster training times etc.

C. Classification

Hotspot detection requires a robust two-class classifier which can learn a separation boundary between hotspots and non-hotspots with maximum margin. In this work, a non-linear SVM is used with a Radial Basis Function (RBF) kernel. As a pre-processing step, dimensionality reduction is performed using Principal Component Analysis (PCA). While retaining most of the variation in the dataset, PCA reduces the number of features into a smaller subset called 'principal components'. Working with principal components aids data visualization, reduces ML model complexity etc. Detailed discussion of SVMs and PCA is out of the scope of this work; For more information, the reader is referred to [10] and [11] respectively.

Class balancing: Typically, the number of known hotspots patterns available for training is smaller compared to known non-hotspot patterns. Training with such imbalanced datasets results in a skewed classifier and the classifier tends to favor the dominating class. To avoid this, the minority class is re-sampled (replicated) and the class sizes are equalized. We noted that, by doing so, we do not alter the information-theoretic content of the dataset; we only help evade the skewed learning problem. Handling imbalanced datasets is discussed in detail in [12].

D. Layout Decomposition

When a foundry receives a new design from its customers, in order to perform hotspot detection, the design needs to be decomposed into patterns. As shown in Figure 6(a), patterns have two important attributes: a) *Hotspot region:* The region where actual defects (shorts or opens) are anticipated. b)

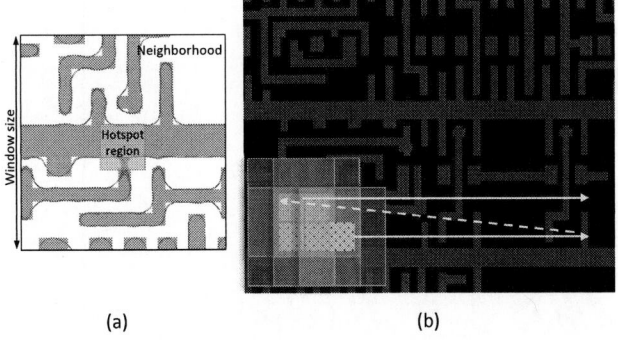

(a)   (b)

Fig. 6. Layout decomposition into patterns (a) Attributes of a pattern (b) Sliding window procedure

Neighborhood: The area surrounding the hotspot region, which could have possibly caused the defect. The extent (size) of the neighborhood varies from technology node to node and is often determined through experimentation. It is safe to use a larger neighborhood than suspected, because, if the polygons towards the periphery of the pattern are indeed benign, an effective machine learning entity would ignore them by assigning low weights to corresponding features. The bounding box encapsulating the neighborhood is popularly known as a 'Window'. To capture patterns from a layout, a moving window based approach is used, as shown in Figure 6(b). The step size for the window movement in both x and y directions should be carefully selected to ensure that the entire layout is covered and every part of the layout appears within the hotspot region of at least one pattern. Patterns obtained from this step form the test set and are tested for hotspots using the trained classifier.

IV. EXPERIMENTAL RESULTS

The objective of this work is to show that enhancing the training dataset using synthetic patterns indeed increases its information-theoretic content and, in turn, reduces false-alarms. To demonstrate this, we implemented the machine learning-based hotspot detection flow shown in Figure 3. The classifier in this flow is trained with and without an enhanced dataset and tested against a common testing dataset. In the rest of the paper, we refer to the classifier trained with an enhanced dataset as 'enhanced classifier' and to the one trained with the non-enhanced dataset as 'non-enhanced classifier'. The non-enhanced classifier is the State-Of-The-Art (SOTA). The difference between the prediction results of the two classifiers indicates the effectiveness of this approach.

For this analysis, we obtained a Register-Transfer-Level (RTL) code of an Advanced Encryption Standard (AES) encryption core from [13]. To create the design layout, the RTL code was synthesized, placed and routed using the Nangate open cell library [14] which is based on a 45nm PDK [15].

Generating the non-enhanced dataset: From a randomly chosen area of the AES layout, 40,000 patterns are captured on Metal1 (M1) layer using the Calibre Pattern Match tool. Litho simulations are performed using Calibre Litho-Friendly-Design (LFD) tool-kit [16], with the litho models provided in

TABLE I
EXPERIMENTAL SETUP

Parameter	Value
Patterns in the initial dataset	40,000
Patterns used for training	20,000 (50%)
Patterns used for testing	20,000 (50%)
Synthetic patterns used for enhancement (per hotspot in the training set)	200
Training set size after class balancing (non enhanced dataset)	38,426
Training set size after class balancing (enhanced dataset)	246,562
Synthetic patterns added to the test set	≈280 (per hotspot in the training set)
Test set size after class balancing and synthetic pattern addition	238,408
Window size (each side)	$8.5 * layer\_pitch$
Window step size (x and y direction)	$1.5 * layer\_pitch$
Grid size	$17 * 17$
Number of features	1445
Features used after PCA	300

the PDK. Simulation results ascertain the ground-truth for the captured patterns. 50% of this dataset is kept aside as the test set and never used in the training process. The other 50% is used for training and is referred to as 'non-enhanced dataset'[1].

Generating the enhanced dataset: The non-enhanced dataset generated in the previous step contains 787 hotspots. For every one of these hotspots, 500 synthetic variants are generated. Of them, approximately 480 passed DRC and, among them, 200 are used in training. Litho simulations are performed on all DRC-clean synthetic patterns and their ground truth (i.e. hotspot vs. non-hotspot) is obtained. The synthetic patterns along with the patterns in the non-enhanced dataset, together form the 'enhanced dataset'.

Both the enhanced and non-enhanced datasets are generated using co-ordinate transform and have weighted features as explained in section III-B.

About 280 synthetic patterns per hotspot, which are not used in training, are added to the common test set. Note that, these are DRC-clean patterns which could potentially occur in any future incoming designs. As explained in section II-A, these are the type of patterns which are often misclassified due to the subtle variations among them. Hence, it is imperative to have them in the test set, in order to validate whether the trained model is robust enough to classify them correctly.

As explained in section III-C, an SVM with an RBF kernel is used as a two-class classifier. While training a classifier, or any machine learning based entity, some of the model parameters require 'tuning' to learn effectively. In this case, to be fair, we have ensured that, both enhanced and non-enhanced

---
[1]Some of the prior works [5], [8], [9] have used the pattern database provided by ICCAD '12 CAD contest [17] as a benchmark. While we sought to use the same, that dataset is from 28nm and 32nm technologies for which Litho models are not publicly available. Few other works [3], [4] have used a set of designs whose source is not published. Hence, for this analysis, design layout was generated using open source RTL codes.

978-1-5386-3775-3/18 $31.00 © 2018 IEEE

TABLE II
EXPERIMENTAL RESULTS

Metric	SOTA	This work	Formula Used
hotspot hit rate	89.20%	95.70%	$\frac{predicted\_hotspots}{total\_hotspots}$
non-hotspot hit rate	41.50%	74.18%	$\frac{predicted\_non\_hotspots}{total\_non\_hotspots}$
Correct prediction rate	70.60%	87.31%	$\frac{hotspot_{hits}+non\_hotspot_{hits}}{total\_patterns\_tested}$
False positive rate	22.81%	10.06%	$\frac{false\_pos}{total\_patterns\_tested}$
False negative rate	6.59%	2.63%	$\frac{false\_neg}{total\_patterns\_tested}$
Total prediction error	29.40%	12.69%	$\frac{false\_pos+false\_neg}{total\_patterns\_tested}$
Total improvement from this work - reduction in prediction error by 56.8% (% change from 29.40% to 12.69%)			$\frac{tot\_err_{SOTA}-tot\_err_{this\_work}}{tot\_err_{SOTA}}$

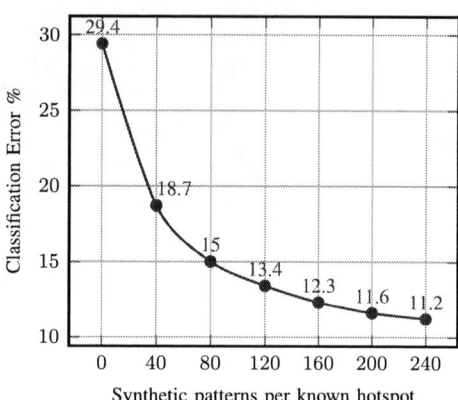

Fig. 7. Variation of prediction error w.r.t number of synthetic patterns used

classifiers are performing at their best by tuning their model parameters 'C' and 'gamma' using grid-search methods [18].

Training datasets are subjected to class balancing as explained in section III-C. Final training and testing set sizes, and other setup parameters are reported in Table I. The test set is tested by both the enhanced and non-enhanced classifiers. Results are reported in Table II. The results clearly indicate that the enhanced classifier performs better and has reduced the classification error by about 57%.

The number of synthetic patterns to be generated has to be decided by the user. To aid this process, we performed a study where, a non-enhanced dataset of 5,000 patterns was obtained and the number of synthetic patterns used to enhance the dataset was varied. As seen in Figure 7, increasing the number of patterns continuously reduces the classification error. When the error due to the addition of 0 synthetic patterns is taken as baseline, we can observe that, an addition of a mere 40 synthetic patterns (per known hotspot) reduces classification error by about 36% (% change from 29.40% to 18.7%). This testifies the effectiveness and the practicality of this approach.

V. CONCLUSION

We have discussed the problem of lithographic hotspots in advanced technology nodes, analyzed the state-of-the-art in this domain and highlighted that they suffer from high false-alarm rates. To address these issues, we have proposed a novel database enhancement approach which involves synthetic pattern generation and design of experiments. We have implemented the proposed flow using a 45nm PDK and have demonstrated about 57% reduction in classification error in comparison to the state-of-the-art.

ACKNOWLEDGMENT

This research has been partially supported by Semiconductor Research Corporation (SRC) through task 2709.001.

REFERENCES

[1] V. Dai, L. Capodieci, J. Yang and N. Rodriguez, "Developing DRC plus rules through 2D pattern extraction and clustering techniques," *SPIE Advanced Lithography*, pp. 727517 – 727517–10, 2009.

[2] "Synopsys," http://www.monolithic3d.com/blog/category/3d%20technology/2.

[3] Duo Ding, Bei Yu, J. Ghosh and D. Z. Pan, "EPIC: Efficient prediction of IC manufacturing hotspots with a unified meta-classification formulation," in *17th Asia and South Pacific Design Automation Conference*, 2012, pp. 263–270.

[4] D. Ding, J. A. Torres and D. Z. Pan, "High performance lithography hotspot detection with successively refined pattern identifications and machine learning," *IEEE Transactions on Computer-Aided Design of Integrated Circuits and Systems*, vol. 30, no. 11, pp. 1621–1634, 2011.

[5] W. Y. Wen, J. C. Li, S. Y. Lin, J. Y. Chen and S. C. Chang, "A fuzzy-matching model with grid reduction for lithography hotspot detection," *IEEE Transactions on Computer-Aided Design of Integrated Circuits and Systems*, vol. 33, no. 11, pp. 1671–1680, 2014.

[6] H. Yao, S. Sinha, C. Chiang, X. Hong and Y. Cai, "Efficient process-hotspot detection using range pattern matching," in *IEEE/ACM International Conference on Computer-aided Design*, 2006, pp. 625–632.

[7] D. Ding, X. Wu, J. Ghosh and D. Z. Pan, "Machine learning based lithographic hotspot detection with critical-feature extraction and classification," in *IEEE International Conference on IC Design and Technology*, 2009, pp. 219–222.

[8] K. Madkour, S. Mohamed, D. Tantawy and M. Anis, "Hotspot detection using machine learning," in *17th International Symposium on Quality Electronic Design*, 2016, pp. 405–409.

[9] H. Zhang, Bei Yu and E. F. Y. Young, "Enabling online learning in lithography hotspot detection with information-theoretic feature optimization," in *IEEE/ACM International Conference on Computer-Aided Design*, 2016, pp. 1–8.

[10] V. N. Vapnik, *The Nature of Statistical Learning Theory*, Springer-Verlag New York, Inc., 1995.

[11] S. Wold, K. Esbensen and P. Geladi, "Principal component analysis," *Chemometrics and Intelligent Laboratory Systems*, vol. 2, no. 1, pp. 37 – 52, 1987.

[12] H. He and E. A. Garcia, "Learning from imbalanced data," *IEEE Transactions on Knowledge and Data Engineering*, vol. 21, no. 9, pp. 1263–1284, 2009.

[13] "Opencores," http://opencores.org/.

[14] "Nangate OCL," https://www.nangate.com/?page_id=22.

[15] "FreePDK45," https://www.eda.ncsu.edu/wiki/FreePDK.

[16] "Calibre-LFD," https://www.mentor.com/products/ic_nanometer_design/design-for-manufacturing/calibre-lfd/.

[17] J. A. Torres, "ICCAD-2012 CAD contest in fuzzy pattern matching for physical verification and benchmark suite," in *IEEE/ACM International Conference on Computer-Aided Design*, 2012, pp. 349–350.

[18] "GridsearchCV," http://scikit-learn.org/stable/modules/generated/sklearn.model_selection.GridSearchCV.html.

2018 IEEE 36th VLSI Test Symposium (VTS)

Staggered ATPG with Capture-per-Cycle Observation Test Points

Yingdi Liu[1], Janusz Rajski[2], Sudhakar M. Reddy[1], Jędrzej Solecki[2], Jerzy Tyszer[3]

[1]University of Iowa
Iowa City, IA 52242, USA

[2]Mentor - A Siemens Business
Wilsonville, OR 97070, USA

[3]Poznań University of Technology
60-965 Poznań, Poland

Abstract—**This paper presents a new staggered test pattern generation scheme. It produces deterministic stimuli in the course of a test-per-clock-based process by using dedicated capture-per-cycle observation test points. These observation points, once inserted into a design, form dedicated scan chains with the capability of capturing test responses during shift cycles when other regular scan cells are loading test patterns. This new scan infrastructure enables one to generate more compact test patterns, reduce test pattern counts, systematically detect many additional faults, and keep the resultant silicon real-estate at the acceptable level. It appears that original scan cells of a design can provide good observability for staggered test patterns. Thus, capture-per-cycle observation test points are directly inserted at selected scan cells' inputs with a minimal impact on the design. Experimental results obtained for large industrial designs illustrate feasibility of the proposed ATPG and are reported herein.**

Keywords—*ATPG, deterministic test patterns, test point insertion, staggered test patterns, test-per-clock, test-per-scan.*

I. INTRODUCTION

Scan is one of the most influential and industry-proven structured design for test (DFT) technology. It allows direct access to memory elements of a circuit under test (CUT) by reusing flip-flops to form shift registers in test mode. The test-per-scan paradigm is then commonly used to feed serial inputs of the scan chains by means of ATE or another source of test patterns, with the same ATE or a test response compactor recording test responses captured earlier into individual scan cells and leaving the scan chains through their serial outputs. The resultant high controllability and observability of internal nodes makes it possible to generate high quality tests automatically, run a variety of diagnostic procedures, and, for example, debug the first silicon. Moreover, simple architecture of scan chains enables their automated stitching and insertion supported by EDA tools. Unintended consequences of the widespread adoption of scan-based testing are mainly related to the fact that all scan chains are filled with a test pattern before it is applied. As a result, the vast majority of test time is spent on shifting test data. The fact that the scan shift frequency is much lower than that of a capture (functional) mode remains a concern, too. In order to utilize the test application time in a more efficient manner, a tri-modal scan (TMS) has been proposed in [11].

TMS has scan cells partitioned dynamically to work in three modes acting as either mission memory elements, sources of test stimuli, or test response compactors. In the last two cases, scan cells form the actual scan chains. Scan chains in the stimuli mode resemble the conventional scan chains in the shift mode. However, test data is applied to the CUT every clock cycle, and these scan chains do not capture test responses. The latter functionality is assumed by scan chains in the compaction mode that accumulate test responses every clock cycle. At the same time, a single bit (per chain) of the resultant signature is always shifted-out. The remaining scan cells are kept in the mission mode. Test results propagating through the combinational part of the circuit can also reach the scan cells in the mission mode. These responses further circulate within the circuit and eventually reach the observation scan chains during the subsequent clock cycles. Since test patterns are applied every clock cycle, the scheme is time-efficient and allows one to complete a test within much shorter durations than done by conventional schemes.

Many test-per-clock schemes had been proposed prior to the introduction of TMS. Some of them were designed for logic built-in self-test (LBIST) or otherwise applicable pseudo-random test patterns. The schemes of [3], [4], [5], [6], [7], [8], [17], [18] may serve here as illustrative examples. Recently, an LBIST scheme has been proposed that aims to achieve high quality test offered by a conventional LBIST in much shorter time [9]. This is accomplished by applying test patterns every clock cycle, thus increasing the percentage of test time used for actual testing. Contrary, however, to TMS and other schemes using similar principles, this LBIST rests on hybrid test points [10]. In particular, observation test points capture fault effects every shift cycle into dedicated flip-flops that form separate scan chains. Their content is gradually shifted into a compactor shared with the remaining scan chains that deliver test responses captured once the entire test pattern has been shifted-in. The scheme has a low area overhead, is routing friendly, and allows flexible trade-offs between test time, test coverage, and area overhead.

Certain principles of TMS are used by the scheme of [19]. It employs shadow registers to capture test results during scan shift. The shadow flip-flops are directly associated with scan cells capable of observing the largest number of faults during successive shift cycles. The same flip-flops form a compactor that captures and accumulates results in a test-per-clock fashion. Although the scheme can enhance fortuitous detection of cell-aware faults, it may also significantly inflate the circuit sequential silicon real estate (even above 30% of the circuit's original scan cell count).

The work of J. Tyszer was supported in part by the Polish Ministry of Science and Higher Education under Grant DS-8133/18.

978-1-5386-3775-3/18 $31.00 © 2018 IEEE

Unlike LBIST or other methods deploying pseudo-random test patterns, a modern ATPG is capable of producing highly compact deterministic test patterns achieving a high fault coverage. The number of such patterns increases with every semiconductor technology node and the growing complexity of digital designs. Interestingly, the principle of test-per-clock schemes can also be used in the context of deterministic tests [1], [2], [12], [15]. As proposed in [12], such patterns can be generated from tests created for each individual fault. The test patterns are then applied to the CUT while test responses are observed by all primary outputs and scan cells in a test-per-clock manner. Another approach [15] generates tests by reusing overlapping bits between former test responses and current test vectors.

In this paper, we demonstrate how a state-of-the-art test-per-scan ATPG can be constrained to produce staggered test patterns. This is achieved through a test-per-cycle-based cube merging. Furthermore, the proposed staggered ATPG works synergistically with a limited number of capture-per-cycle observation test points inserted into designs in a manner similar to that of the logic BIST of [9]. Consequently, compatible test cubes can form highly compact test patterns. During fault simulation, the resultant patterns are also processed in a test-per-cycle fashion, and thus more faults can be dropped from a fault list. As a result, the proposed method makes it possible to further reduce ATPG pattern counts while maintaining the corresponding area overhead low. Since test patterns are applied every clock cycle, the scheme is extremely time-efficient and may enable a multiple-detection framework. Alternatively, given a target fault coverage, it is possible to reduce test application time drastically, thereby decreasing the overall manufacturing cost.

The rest of the paper is organized as follows. Section II describes a scan architecture used in conjunction with the proposed staggered ATPG. Section III presents the staggered test generation technique. A method to identify the most suitable locations for capture-per-cycle observation test points is proposed in Section IV. Experimental results obtained for large industrial designs are discussed in Section V, and the paper concludes with Section VI.

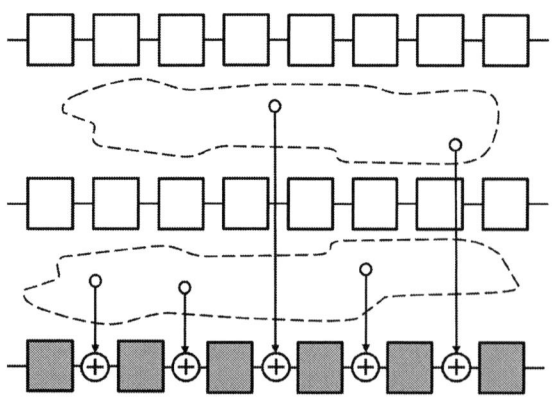

Fig. 1. Scan architecture.

II. SCAN ARCHITECTURE

In a conventional test-per-scan scheme, the operational rule is to feed serial inputs of the scan chains with test stimuli and capture test responses through scan chain serial outputs. Typically, all scan cells are controlled by a single scan enable signal. Thus, all scan chains are functionally indistinguishable, i.e., they all either shift data in and out or capture test results. In contrast to this paradigm, a test-per-clock architecture adapted in this work operates as shown in Fig. 1 [9]. The majority of memory elements form conventional scan chains (white blocks in the figure), i.e., they operate either in the shift mode (with the asserted scan enable) or in the capture mode. Scan cells serving the observation points (the grey boxes in Fig. 1) are arranged into separate scan chains operating exclusively in the compaction (capture) mode. These compaction chains accumulate test responses using XOR gates interspaced between their successive cells. A scan cell associated with a single observation point is shown in Fig. 2. The global test point enable (EN) signal activates observation points in the test mode, and disables them in the mission mode. Test results received from CUT (input D) are XOR-ed with data provided by another scan cell, thus incorporating shift and capture functionality within a single clock cycle. Although the capture-per-cycle observation test points may only work in the compaction mode, it is possible to modify them for the sake of scan chain integrity test, as shown in [11]. It is important to observe, however, that, contrary to the TMS scheme, this structure remains static, i.e., no scan cells can change their functionality dynamically.

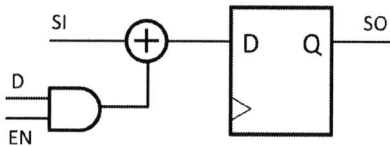

Fig. 2. Observation point.

Once a test is launched, test data are serially fed into the conventional scan chains through the scan serial inputs, while the chains' content drives the CUT. Note that after every *single shift cycle* there is a test pattern presented to the CUT. Test responses corresponding to these patterns are captured and accumulated by the observation test point chains every *single shift cycle* while regular scan cells are still operating in the shift mode (compare Fig. 1). Hence, it becomes possible to stagger ATPG patterns based on this functionality. Indeed, the key idea of our ATPG is to deploy the capture-per-cycle observation points to record data while regular scan cells are still being loaded, and to generate more compact test patterns by utilizing the subsequent vectors gradually filling the regular scan chains.

III. STAGGERED ATPG

Recall that a conventional ATPG framework typically comprises the actual test pattern generation and fault simulation. First, a test cube, i.e., a group of specified scan cells, is produced to detect a single fault. Successive test cubes are kept in a buffer, where, at some point, they become subject to a cube merging

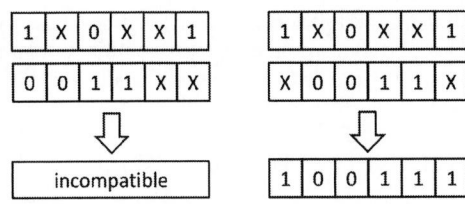

Fig. 3. Compatible test cubes.

process. Depending on specified bits, test cubes with no conflicting assignments are considered compatible and can be merged to form a single test pattern (Fig. 3). ATPG continues test generation and cube merging until a certain number of patterns is created. Usually, fault simulation follows to determine all faults detected by the newly formed tests. Clearly, ATPG iterates until all faults are covered. With the capture-per-cycle observation test points allowing tests every clock cycle, the above test-per-scan-based ATPG can be modified to generate staggered test patterns that fully leverage the test-per-clock scan design of Section II.

A. Test cube merging

As mentioned earlier, a test pattern can be obtained as a superposition of several compatible test cubes. With the test-per-scan paradigm in place, conflicting test cubes cannot be merged to form a single pattern. However, the test-per-clock approach provides an opportunity to merge even conflicting test cubes. Consider test cubes t_1 and t_2 shown in Fig. 4. It appears that loading vector t_2 suffices to apply test pattern t_1 (it occurs at shift cycle c_4), provided the corresponding specified values (grey boxes) in both vectors are the same. In this case, having observation points that capture test results every clock cycle, one can generate staggered patterns based on mutually shifted test cubes without compromising the fault coverage. Fig. 5 is another example where one can make incompatible test cubes mergable. This is achieved by shifting (adjusting) cubes along scan chains to form a more compact pattern, which activates all specified values of every initial test cube in a test-per-cycle manner. For example, a test pattern formed in Fig. 5 will apply successive specified bits as follows:

- bits A during the fifth shift cycle,

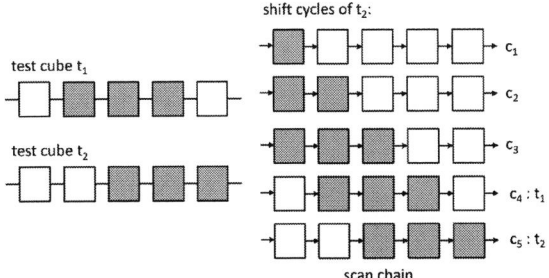

Fig. 4. Test cube specifications get covered by an intermediate shifting state of another test cube.

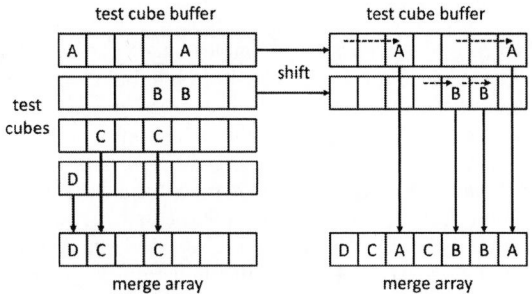

Fig. 5. Test cube merging.

- bits B during the sixth cycle,
- bits C during the seventh cycle, and
- bit D also during the seventh cycle.

If every potential fault effect can be recorded by at least one capture-per-cycle observation test point, one can perform this type of shift-and-merge process to further decrease the total number of required test patterns that need to be applied. Clearly, this approach is also well positioned to reduce the effective number of ATPG-produced test cubes. This is because many fortuitous detections are likely to be recorded by fault simulation of staggered patterns, as discussed below.

The test cube merging integrated with the staggered ATPG can be summarized as follows:

1. First, create an empty merge array M.
2. Merge test cube *t* with M provided it is compatible with the content of M.
3. If *t* is incompatible with M, check whether fault effects caused by *t* can be recorded by any observation point.
4. If so, inspect compatibility of *t* with M by shifting *t* towards the scan tail end until a shifted replica of *t* becomes compatible with M (then merge *t* with M); if all shift positions have been unsuccessfully tried, then skip *t*.
5. After traversing the entire test cube buffer, generate a pattern based on the values gathered in M.
6. Reset M and repeat the above steps until all relevant test cubes are examined.

Before the above procedure is carried out, one may presort the test cube buffer so that test cubes targeting faults that do not propagate to any observation points (faulty effects are exclusively captured by regular scan cells) are examined prior to the remaining test cubes. Note that these test cubes cannot be shifted along scan chains as their capture cycles occur only at particular time frames following completion of the scan upload phase. Test cubes that propagate errors to observation points having much more flexible capture principle (virtually every intervening scan shift cycle may record erroneous responses) can be positioned at any suitable time frame within the scan uploading sequence. As one may expect, experimental evidences indicate that indeed this approach provides better results in terms of test cube mergability. It is also worth noting that the test cube buffer is gradually refilled as ATPG keeps adding new test cubes for successive faults – here the presented procedure is fully compatible with the conventional ATPG flow.

978-1-5386-3775-3/18 $31.00 © 2018 IEEE

B. Fault simulation

Typically, test cubes target only a small number of faults while the remaining ones are detected by simulation. After test cube merging, all bits that remain unspecified are also filled with certain values, for example, pseudorandom ones. Unlike a conventional test-per-scan ATPG using a single fault simulation pass per pattern, a staggered ATPG process requires several fault simulation runs to analyze every intermediate stimulus generated during every shift cycle, and to drop all faults detected this way. In other words, besides a fault simulation pass corresponding to the original test pattern, additional (staggered) test patterns are formed and fault simulated with cell constraints disabling regular scan cells-based observation sites. This is done by shifting the current pattern towards the scan front end, i.e., in the opposite direction of the test cube compatibility check. The total number of staggered patterns depends on the scan size. Clearly, one may expect that in addition to faults targeted by the original pattern and its merged derivatives, more faults can be detected by simulation compared to a conventional test-per-scan ATPG.

In summary, the proposed staggered ATPG flow consists of the following steps:

1. Generate test cubes for selected faults.
2. Generate a derivate test pattern by first merging the test cubes targeting faults that propagate to regular scan cells only, and then the remaining test cubes; fill all unspecified bits.
3. Fault simulate the original test pattern.
4. Fault simulate staggered patterns by appropriately shifting the original test pattern.
5. Repeat steps 1 – 4 until no faults remain on a fault list.

IV. OBSERVATION TEST POINTS SELECTION

A proper selection of capture-per-cycle observation sites is essential for successful application of the staggered test generation scheme presented in the previous section. They can be either picked to improve the cube merging efficiency, or to increase the number of faults detected by fault simulation.

Recall that a given test cube excites a fault and propagates the fault effect to at least one observation point, including regular scan cells or primary outputs. Any circuit's internal node that has been assigned values such that the fault effect can propagate through is called a *detection gate* of the corresponding test cube. As shown in Section III (see also Fig. 5), not every test cube can be a subject to the shift-and-merge procedure. Only test cubes with detection gates being observable during shift cycles can be adjusted along scan chains for compatibility checks. In other words, only test cubes activating several dedicated capture-per-cycle observation points can have multiple (flexible) locations in the merge array. Clearly, the more test cubes with detection gates acting as dedicated observation points, the better test cube merging results one may expect during the staggered test generation.

Recall also that once the merging of compatible test cubes is completed, we assume that all the remaining unspecified bits are filled with pseudorandom values. The resultant fully specified patterns are fault simulated and all detected faults are dropped. This includes faults detected by staggered test patterns as well

as faults fortuitously covered by a fully specified pattern. It is worth noting that test patterns used in this process are formed by all shifted replicas of original patterns combined with captured test responses of former test patterns. Clearly, the capture-per-cycle observation test points need to be placed at proper locations to capture faults playing a significant role in determining the final pattern count of the staggered ATPG.

In the early stages of experiments, we have used two test point insertion techniques [10], [16], which have already proved to be successful in reducing test pattern counts and test application times. Although making those test points capture-per-cycle observation sites may allow one to detect a large number of faults, the very same observation points are seldom sensitized by a reasonable number of test cubes. Consequently, the method that has demonstrated its superiority in a convincing manner, which seemed to offer a practical and economical way to get sufficient coverage of randomly-detectable faults, employs regular scan cells. They were found to be good observation sites for both test cube generation and fault detection, and they can serve as the capture-per-cycle observation test points for the staggered test patterns.

We deploy an approach similar to that of [19] by inserting observation points directly at the inputs of certain regular scan cells to capture test results that otherwise would be lost due to the scan shift mode separating the scan cells from a combinational part of a design (see Fig. 6). The same method avoids the risk of having observation points inserted at internal nodes of design's functional paths. The candidate scan cells are selected based on fault detection profiles obtained by counting how many target faults are observed (and thus detected) by every individual scan cell during a baseline (reference) ATPG run. Although the staggered ATPG may behave differently, this information can still represent the likelihood of fault detection with respect to every individual scan cell. Moreover, additional faults may propagate through sensitized paths set by the test cubes towards scan cells that are likely to be reached. Note that linking observation points with scan cells having the highest fault detection

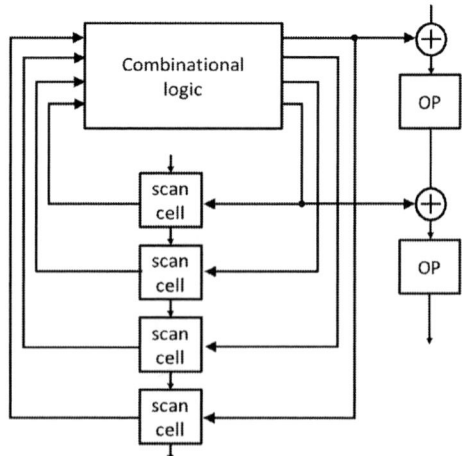

Fig. 6. Observation points insertion.

counts (targeted by test cubes), makes a large number of test cubes "flexible". Otherwise, we cannot shift and merge test cubes as the fault-effect targeted by the test cube cannot be captured by any observation point. Because of silicon area constraints, we only select a limited number of scan cells (approximately 1% – 2% of a design's total number of scan cells) with the largest fault detection counts, and observation test points described in Section II are inserted at these scan cells inputs.

V. EXPERIMENTAL RESULTS

Performance of the staggered ATPG working with the capture-per-cycle observation test points have been verified experimentally on 10 large industrial designs. They represent different design styles, different scan methodologies, and mirror the latest technology nodes. The basic data regarding these circuits, such as the number of gates, the number of scan cells, the length of the longest scan chain, and the fault population are listed in Table I. Although longer scan chains may help in getting more staggered test patterns, we only use test cases with a reasonable scan size in order to be able to accommodate on-chip test compression such as the EDT scheme [13].

TABLE I
CIRCUIT CHARACTERISTICS

	#Gates	#Scan cells	#Scan chains	Scan chain length	#Faults
D1	1.20M	75K	398	189	3.74M
D2	2.09M	145K	421	345	5.44M
D3	1.04M	60K	49	1220	1.29M
D4	1.38M	95K	172	555	2.99M
D5	4.83M	329K	604	545	9.19M
D6	2.53M	206K	370	557	3.25M
D7	7.43M	245K	426	576	19.34M
D8	15.49M	762K	1554	491	54.66M
D9	20.36M	894K	1769	505	76.03M
D10	3.22M	213K	501	425	8.80M

The experimental results for stuck-at faults are presented in Table II. Columns "Pattern count" gather the key data for both examined scenarios – the number of ATPG-produced test cubes in line with the conventional scenario and its staggered counterpart. In all experiments reported in this section, the staggered ATPG results were obtained after inserting observation test points, as discussed in Section IV, i.e., by using the method working with the test cube detection profiles. The total number of test points is kept below 2% of the entire scan cell population (see column "Observation points" of the table). This percentage is an industry-wide accepted standard. As can be seen, the staggered ATPG yields visible pattern count reduction in all examined cases. It varies from 1.4x to 2.7x for otherwise similar test coverage numbers provided by the conventional ATPG. For the sake of illustration, the respective pattern counts are visually compared in Fig. 7.

It is worth noting that fault simulation used in the experiments does not account for an unlikely event of aliasing that may occur when fault effects are masked within scan chains driven

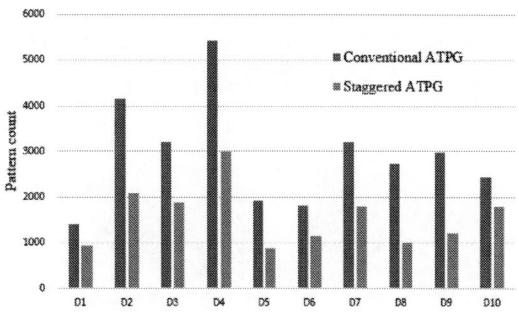

Fig. 7. Pattern count comparison.

by the observation test points. Fortunately, the likelihood of such an event is extremely small [14] as these scan chains form finite memory devices, where after several clock cycles (depending on a fault injection site) an error is shifted out. Moreover, missed faults remain ATPG targets.

VI. CONCLUSION

In this paper, we present a staggered ATPG working synergistically with the capture-per-cycle observation test points. This approach generates highly mergeable deterministic test patterns and detects many additional faults through staggered patterns fault simulation. The observation test points are inserted at the scan cell inputs based on the test cube detection profiles. This process has a minimum impact on a design compared to other test point insertion techniques. At the same time, test responses captured every clock cycle by means of the observation points visibly improve faults observability. Experimental results obtained for large industrial designs demonstrate – on average – 1.9x pattern count reduction for stuck-at patterns while preserving the original fault coverage.

REFERENCES

[1] H. Ando, "Testing VLSI with random access scan," *Proc. COMPCON*, 1980, pp. 50-52.

[2] D. Baik and K. K. Saluja, "Progressive random access scan: a simultaneous solution to test power, test data volume and test time," *Proc. ITC*, 2005, pp. 359-368.

[3] P. H. Bardell and W.H. McAnney, "Simultaneous self-testing system," US patent 4513418, Apr. 23, 1985.

[4] F. Corno, P. Prinetto, and M. Sonza Reorda, "Making the circular self-test path technique effective for real circuits," *Proc. ITC*, 1994, pp. 949-957.

[5] A. Jas, K. Mohanram, and N.A. Touba, "An embedded core DFT scheme to obtain highly compressed test sets," *Proc. ATS*, 1999, pp. 275-280.

[6] E. Kalligeros, X. Kavousianos, D. Bakalis, and D. Nikolos, "An efficient seeds selection method for LFSR-based test-per-clock BIST," *Proc. ISQED*, 2002, pp. 261-266.

[7] B. Konemann, J. Mucha, and G. Zwiehoff, "Built-in logic block observation techniques," *Proc. ITC*, 1979, pp. 37-41.

[8] A. Krasniewski and S. Pilarski, "Circular self-test path: a low cost BIST technique for VLSI circuits," *IEEE Trans. CAD*, vol. 8, no.1, pp. 46-55, Jan. 1989.

[9] S. Milewski, N. Mukherjee, J. Rajski, J. Solecki, J. Tyszer, and J. Zawada, "Full-scan LBIST with capture-per-cycle hybrid test

978-1-5386-3775-3/18 $31.00 © 2018 IEEE

TABLE II
EXPERIMENTAL RESULTS

	Conventional ATPG		Pattern reduction	Staggered ATPG with observation points		
	Test coverage [%]	Pattern count		Observation points	Test coverage [%]	Pattern count
D1	96.36	1400	1.47x	1,428	96.36	953
D2	97.26	4161	1.99x	2,613	97.22	2,089
D3	92.20	3204	1.71x	866	92.20	1,878
D4	97.68	5433	1.81x	1,913	97.66	3,000
D5	94.24	1914	2.15x	6,322	94.23	889
D6	97.63	1822	1.58x	3,389	97.62	1,151
D7	98.31	3195	1.78x	4,904	98.35	1,798
D8	99.04	2727	2.69x	12,970	99.04	1,015
D9	99.22	2967	2.43x	17,134	99.22	1,219
D10	98.05	2433	1.35x	3,723	98.04	1,802

points," *Proc. ITC*, 2017, paper 10.3.

[10] E. Moghaddam, N. Mukherjee, J. Rajski, J. Tyszer, and J. Zawada, "Test point insertion in hybrid test compression/LBIST architectures," *Proc. ITC*, 2016, paper 2.1.

[11] G. Mrugalski, J. Rajski, J. Solecki, J. Tyszer, and C. Wang, "Tri-modal scan-based test paradigm," *IEEE Trans. VLSI Systems*, vol. 25, no. 3, pp. 1112-1125, March 2017.

[12] O. Novak and J. Nosek, "Test-per-clock testing of the circuits with scan," *Proc. Int. On-Line Test Workshop*, 2001, pp. 90-92.

[13] J. Rajski, J. Tyszer, M. Kassab and N. Mukherjee, "Embedded deterministic test," *IEEE Trans. CAD,* vol. 23, No. 5, pp. 776-792, May 2004.

[14] J. Rajski, J. Tyszer, C. Wang, and S. Reddy, "Finite memory test response compactors for embedded test applications," *IEEE Trans. CAD*, vol. 24, no. 4, pp. 622-634, April 2005.

[15] W. Rao and A. Orailoglu, "Virtual compression through test vector stitching for scan based designs," *Proc. DATE,* 2003, pp. 104-109.

[16] S. Romersaro, J. Rajski, T. Rinderknecht, S.M. Reddy, and I. Pomeranz, "ATPG heuristics dependant observation point insertion for enhanced compaction and data volume reduction," *Proc. DFTVS*, 2008, pp. 385-393.

[17] Y. Son, J. Chong, and G. Russell, "E-BIST: Enhanced test-per-clock BIST architecture," *IEEE Proc. Comput. Digit. Techn.*, vol. 149, pp. 9–15, Jan. 2002.

[18] C. Stroud, "An automated BIST approach for general sequential logic synthesis," *Proc. DAC*, 1988, pp. 3-8.

[19] F. Zhang, D. Hwong, Y. Sun, A. Garcia, S. Alhelaly, G. Shofner, L. Winemberg, and J. Dworak, "Putting wasted clock cycles to use: Enhancing fortuitous cell-aware fault detection with scan shift capture," *Proc. ITC*, 2016, paper 2.3.

Special Session on Intelligent Sensor Nodes

Organizers: Kanad Basu (New York University) and Shreyas Sen (Purdue University)
Moderators: Kanad Basu (New York University) and Shreyas Sen (Purdue University)

I. INTRODUCTION

In a world connected by Internet of Things (IoT), sensor nodes play a vital role in monitoring, transmitting and processing useful data. In this session, we will discuss about four significant aspects of intelligent sensor nodes.

II. THE CHALLENGE OF LARGE-SCALE CONNECTIVITY: DESIGN, VALIDATION, AND DEBUG

Sandip Ray
University of Florida

Today we are living in an environment that includes millions of computing devices with diversity of attached sensors are monitoring, communicating, analyzing, and responding to highly sensitive, personalized information, often in real time. Designing, validating, and testing applications with such devices requires an end- to-end architecture that is resilient against functional bugs, security vulnerabilities, and fabrication defects that may manifest themselves in unanticipated use cases, together with corresponding paradigms for validation and testing. This talk will provide a general layout of the scope of this problem, focusing in particular on scale and complexity challenges, current state of the practice, and its limitations. In particular, trade-offs between validation, security, and intelligence will be discussed.

III. HIERARCHICAL CHECKING AND ADAPTATION STRATEGIES FOR ROBUST INTELLIGENT AUTONOMOUS SYSTEMS

Abhijit Chatterjee
Georgia Institute of Technology

Future robotic systems that operate without human intervention will be the extreme embodiment of intelligent autonomy. These systems will be able to sense the environment in which they work as well as the health of their own internal subsystems. Adaptation is achieved by changing the manner in which actuation is performed in response to sensor data. A hierarchical system of checks is proposed that trades off coverage with detection latency at different levels of the design: hardware, software and the application layer. It is seen that high coverage of failures that impact overall system function can be achieved through use of the proposed hierarchical checking methodology. At the same time, rapid reconfiguration of the autonomous system control mechanisms to accommodate different system health conditions, as indicated by the proposed checks, can be performed

with low latency. This leads to the development of real-time machine learning algorithms that allow an autonomous system to continuously learn from its operating environment and adapt in real time.

IV. SOLVING THE DRIFT PROBLEM OF BIOLOGICAL AND CHEMICAL SENSORS IN THE FIELD

Sule Ozev and Jennifer Blain Christen
Arizona State University

Low-cost biological and chemical sensor systems have many potentially transformative applications including personalized health care, environmental monitoring, and biological or chemical threat detection. In order to facilitate deployment outside the laboratory environment, the drift problem of these sensors must be resolved. We propose a hybrid analytical/statistical model for sensor drift and use this model to calibrate the sensor readings after they are transmitted to the host device. Our model is based on physical drift mechanisms that provide the functional form of the drift behavior. However, instead of relying purely on analytical formulations, we use an empirical approach to match the drift model parameters with data that is obtained at production time. Once the sensor system is deployed in the field, the IoT device transmits raw sensor readings to the host device, along with additional environmental information, such as temperature. The host device keeps track of elapsed time and environmental factors, and calibrates the sensor readings using the sensor drift model. Subsequent software calibration provides accurate measurements even if the sensors behavior drifts considerably.

V. ANALOG, MIXED-SIGNAL AND MEMS DFT AND ITS USE FOR INTELLIGENT SENSORS

Salvador Mir
Universite' Grenoble Alpes

Design-for-test techniques, in particular Built-in Self-test (BIST), are important for the validation of the functionality of integrated circuits during production testing and in-field monitoring. These techniques can be exploited for the control of the devices, enhancing device reliability and performance. This talk will briefly describe some advanced embedded techniques for analog devices (Analog-to-Digital converters, RF front-ends, MEMS) and will provide some insights on how they can be exploited in the domain of intelligent sensors for enhancing performance and optimizing power consumption while guaranteeing device specifications.

Innovative Practices on Silicon Photonics

Roy Meade, Ayar Labs, USA
Woosung Kim, Intel Corp., USA
Richard Otte, Promex Industries Inc., USA
Eugene Atwood, IBM Corp. (Organizer/Moderator)

I. Introduction

This IP session focuses on introduction of silicon photonics into first level packaging. Silicon photonics is driving heterogenous construction to bring III-V light sources into a silicon modulated and detected environment. Unique test challenges and solutions will be addressed.

II. Challenges and Opportunities in Integrated CMOS/Photonics Test (Roy Meade)

Driven by performance, cost, and time to market, semiconductor industry interest in more-than-Moore development has grown. The heterogeneous integration of silicon photonics is a prime example, and is being developed to address challenges in networking and Datacom. However, the adoption and integration of silicon photonics and III-V devices poses unique manufacturing and test challenges, and the ecosystem, while growing, is still nascent. To enable Known-Good-Die (KGD), new capabilities, such as optical vertical coupling, are needed for parametric test and wafer sort. After assembly, final test and burn-in have their own specific requirements. In this talk, the challenges of production silicon photonics test will be introduced, along with strategies to address the same.

Roy Meade – Roy is VP of Manufacturing for Ayar Labs, an early stage start-up commercializing DARPA funded silicon photonic technology. He has over 20 years of experience developing and managing semiconductor and fiber optic technology. Roy earned a Bachelor and Master of Science in Mechanical Engineering from the Georgia Institute of Technology, and an MBA from the Fuqua School of Business at Duke University. A Senior Member of the IEEE, he has more than 60 USPTO issued patents. Currently, he is a Principal Investigator for LytBit, a research project under ARPA-E's ENLITENED program.

III. Machine Learning Application For Silicon Photonics Transceiver Testing (Woosung Kim)

Multiple source agreement (MSA) optical transceiver specs often require same test result at multiple temperature conditions. It results in extensive test time and high manufacturing test cost. In this paper, we introduce machine learning based test methodology for silicon photonics transceiver manufacturing test time reduction. Machine learning technique estimates test output parameters at certain temperatures given prior test results. We complied more than 10,000 different input test parameters from various prerequisite tests to build a prediction model. For production implementation simplicity, we used supervised machine learning linear regression model with Tikhonov regularization and could reach $R2>0.97$ of predicted value correlation with physical measurement value.

Woosung Kim – Intel Corp. Woosung received the B.S. degree from Yonsei University, Seoul, Korea in 2006 and M.S., Ph.D. degrees from Washington University, St. Louis, USA, in 2012, 2013, respectively. He joined Intel Corporation, California, USA, in 2013. Since 2013, he has been working in silicon photonics, from photonics chip research to manufacturing test software development. His current main areas of interest are high volume manufacturing test software development and machine learning based manufacturing test improvement. Co-contributers also from Intel are Yeoh Hoe Seng, Yi-Shing Chang, and Suohai Mei

IV. The status, needs and potential solutions related to testing photonic devices and products including those that Incorporate Photonic Integrated Circuits (PIC) (Dick Otte)

The general status, needs and potential solutions for testing of optical devices, components and products, including those that incorporate Photonic Integrated Circuits (PIC), are presented. The test issues and needs at Manufacturing Readiness Levels and at wafer, die, subassembly and final assembly levels related to design for test, test cost minimization, optical signal access and optical parameters to be measured are addressed along with the difference between electrical and optical device testing. This material is drawn from work done by the AIM Academy Integrated Photonics Systems Roadmap Technical Working Groups (TWGs), the iNemi Optical Electronics Roadmap TWG and the AIM IP Rochester Test Assembly and Packaging team.

Richard Otte (Dick) has been President & CEO of Promex Industries Inc. since 1995. Promex is an ISO 9001:2015, ISO 13485:2016 and ITAR registered manufacturing services provider specializing in onshore heterogeneous assembly of medical, biotech and semiconductor products in its Santa Clara and San Diego, California facilities.

Systematic *b*-Adjacent Symbol Error Correcting Reed-Solomon Codes with Parallel Decoding

Abhishek Das and Nur A. Touba

Computer Engineering Research Center,
University of Texas at Austin, TX-78712
abhishekdas@utexas.edu, touba@ece.utexas.edu

Abstract—**With technology scaling, the probability of write disturbances affecting neighboring memory cells in nonvolatile memories is increasing. Multilevel cell (MLC) phase change memories (PCM) specifically suffer from such errors which affects multiple adjacent memory cells. Reed Solomon (RS) codes offer good error protection since they can correct multi-bit symbols at a time. But beyond single symbol error correction, the decoding complexity as well as the decoding latency is very high. This paper proposes a systematic *b*-adjacent symbol error correcting code based on Reed-Solomon codes with a low latency and low complexity parallel one step decoding scheme. A general code construction methodology is presented which can correct any errors within *b*-adjacent symbols. The proposed codes are compared to existing adjacent symbol error correcting Reed-Solomon codes, and it is shown that the proposed codes achieve better decoder latency. The proposed codes are also shown to achieve much better redundancy compared to symbol error correcting orthogonal Latin square (OLS) codes.**

Keywords—adjacent symbol error correction, Reed Solomon codes, parallel decoding, phase change memories, multilevel cell

I. INTRODUCTION

Multilevel Cell nonvolatile memories like phase change memories have recently been in focus due to their high density and lower costs [Papandreou 10]. Their ability to store multiple bits in a single cell as well as being byte-addressable makes them an attractive alternative to DRAM solutions. But with technology node scaling, the problem of write disturbance errors specifically in MLC PCM gets highly exacerbated [Jiang 14]. The read and write operations in MLC PCM are byte-addressable and are at the granularity of multi-bit symbols, thus making its performance sensitive to read and write access latencies. For lower technology nodes, the write disturbance error effectively manifests itself in the form of burst errors affecting multiple cells. As technology scales further, the cells become closer to each other. This worsens the write disturbance error problem thereby reducing the reliability of MLC PCMs. A similar problem is also found in spin-transfer torque magnetic random-access memory (STT-MRAM) using a magnetic tunnel junction (MTJ). [Yoon 17] showed that for dense memory bits and lower stored energy, a magnetic field- induced coupling between adjacent bits can cause significant change in the average retention time. As this technology matures enabling dense multilevel cells, magnetic field-induced coupling is expected to play a key role. Thus, this is also a form of burst error affecting neighboring cells in the memory with the performance being sensitive to memory access latencies.

Binary burst error correcting codes with fast decoding procedures [Klockmann 17] [Dutta 11] might not be suitable for such an application because of limited range of bursts that the codes can correct. Each memory cell can store multiple bits, and the write disturbance errors affects multiple cells thus affecting a large number of bits. Reed Solomon codes are highly suitable for these cases since the correction is on a symbol basis and they have very low redundancy. But Reed-Solomon codes suffer from complex decoding procedure which results in high decoding latency spanning multiple cycles. [Fujiwara 06] describes various methods proposed over the years to correct single byte errors with a parallel decoding methodology. But the main issue with single byte error correction is that the burst errors may affect multiple bytes at a time within the burst range and are not guaranteed to affect only a single byte.

An adjacent symbol error correcting Reed-Solomon code is also proposed in [Namba 15b]. But the codes have a high decoder latency and decoder area since they use $GF(2^m)$ operations where m is the number of bits per symbol. Also, the operations involve multiplication, squaring and inversion which increases the circuit complexity exponentially as the number of bits per symbol increases. Thus, it is unsuitable for low latency memory accesses. [Reviriego 13, 15] propose a method to correct double adjacent errors and triple adjacent errors respectively for binary bits using OLS codes. These methods can be extended [Namba 15a] to correct adjacent symbol errors as well. But, these codes have very high redundancy.

This paper proposes a methodology to correct *b*-adjacent symbol errors using Reed-Solomon codes. The codes are systematic by design and can have a maximum information symbol length $k = 2^m-1$, where m is the number of bits per symbol. This contrasts with the traditional Reed-Solomon codes which have a maximum block length $n = 2^m-1$. The codes have very low redundancy and have a parallel one-step decoding procedure. This makes the decoder latency and the decoder complexity very low. The rest of the paper is organized as follows. Section II reviews the Reed-Solomon codes and its extensions. Section III describes the proposed *b*-adjacent symbol error correcting Reed-Solomon codes in detail along with its construction and decoding procedures. Section IV evaluates the proposed codes and compares it with existing codes in terms of hardware complexity and redundancy. Section V presents the conclusion of this work.

II. REED-SOLOMON CODES

Reed-Solomon codes are a special case of non-binary BCH

978-1-5386-3775-3/18 $31.00 © 2018 IEEE

codes of length $n = q^m-1$ over $GF(q^m)$. The number of parity check symbols for RS codes is given by $n - k = 2t$, where t is the number of errors being corrected. For $q=2$, the RS code comprises of m-bit symbols which is used to construct the code. The general parity check matrix for a RS code is shown in equation (1). An extension for the SEC RS code exists and has the parity check matrix of the form shown in equation (2). It is a subclass of Hamming type codes over $GF(2^m)$ that has 2 check symbols [Bossen 70]. The codes consist of m-bit symbols unlike a binary Hamming code. Thus, a single error can result in a syndrome which is either equal to a column or is a multiple of a column in the parity check matrix. Any column in the parity check matrix can have (2^m-1) multiples i.e. all possible cases except the all-0 column. Thus, the number of possible syndromes for each single error is (2^m-1). As a result, for the case of extended RS codes, the number of possible syndromes equals $n(2^m-1)$. This can result in huge complexity, since the number of syndromes increases exponentially with m i.e. the number of bits per symbol. Thus, to make the decoding simpler, [Bossen 70] describes the use of companion matrix which transforms the H-matrix over $GF(2^m)$ to binary form.

$$H = \begin{pmatrix} 1 & 1 & 1 & \cdots & 1 \\ 1 & \alpha & \alpha^2 & \cdots & \alpha^{n-1} \\ 1 & \alpha^2 & \alpha^4 & \cdots & \alpha^{2(n-1)} \\ \vdots & \vdots & \vdots & \cdots & \vdots \\ 1 & \alpha^{2t} & \alpha^{4t} & \cdots & \alpha^{2t(n-1)} \end{pmatrix} \quad (1)$$

$$H = \begin{pmatrix} 1 & 1 & 1 & 1 & \cdots & 1 & 0 & 1 \\ 1 & \alpha & \alpha^2 & \alpha^3 & \cdots & \alpha^{n-1} & 1 & 0 \end{pmatrix} \quad (2)$$

All the 1's in the parity check matrix are replaced with identity sub-matrices I and α is replaced by the companion matrix T. The notation form of the new parity check matrix for a primitive polynomial $g(x) = x^2 + x + 1$, $k = 3$ and $n = 5$ is shown in equation (3). The binary form of the matrix is shown in equation (4).

$$H = \begin{pmatrix} I & I & I & 0 & I \\ I & T & T^2 & I & 0 \end{pmatrix} \quad (3)$$

$$H = \begin{pmatrix} 1 & 0 & 1 & 0 & 1 & 0 & 0 & 0 & 1 & 0 \\ 0 & 1 & 0 & 1 & 0 & 1 & 0 & 0 & 0 & 1 \\ 1 & 0 & 0 & 1 & 1 & 1 & 1 & 0 & 0 & 0 \\ 0 & 1 & 1 & 1 & 1 & 0 & 0 & 1 & 0 & 0 \end{pmatrix} \quad (4)$$

A companion matrix is a m x m non-singular binary matrix defined from a primitive polynomial $g(x)$ of degree m [Fujiwara 06]. The syndromes from the parity check matrix are constructed for each row in a general fashion by XORing all the bits for which the corresponding column is a 1. Since there are only 2 rows in the parity check matrix, the syndrome corresponding to any single symbol error will be of the form shown in equation (5).

$$\begin{pmatrix} S_1 \\ S_2 \end{pmatrix} = \begin{pmatrix} e_i \\ T^i e_i \end{pmatrix} \quad (5)$$

In (5), i is the location of the error and e_i is the error pattern, and S_1, S_2 are m-bit symbols. Thus, for column i, $T^i.S_1 \oplus S_2 = 0$, while for all other columns, $T^j.S_1 \oplus S_2 \neq 0 \; \forall \; j \neq i$. The decoder for the parity check matrix in equation (4) has been shown in

Fig. 1. This decoding procedure using companion matrix reduces the complexity of the circuit compared to syndrome comparison decoding method of Hamming codes by trading off decoder latency.

This decoding method is only useful for single symbol error correction. For $t > 1$, Reed-Solomon codes have a two-step decoding procedure which involves two polynomials, an error locating polynomial and an error magnitude polynomial. [Carrasco 08] describes various methods to compute both the error location and the error magnitude for t (> 1) errors.

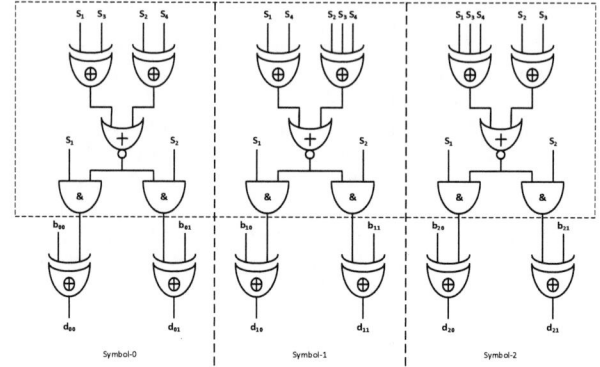

Fig. 1. Decoder Circuit for parity check matrix of equation (4).

III. PROPOSED CODES

This section describes the proposed b-adjacent symbol error correcting Reed-Solomon Codes. The key idea is to construct a new parity check matrix in such a manner that the first b-rows of the parity check matrix have at-most one 1 within b-adjacent columns in any given row. This can easily be done by interleaving 1's every b-columns. The main purpose of such a construction is to distinguish the error magnitude for the adjacent columns in error in a single step. This type of construction enables the error pattern of any erroneous symbol to be one of the syndrome symbols.

The lower (r-b) rows of the parity check matrix, where r is the total number of check symbols, is constructed by using rows from the original Reed-Solomon parity check matrix such that the following conditions are met.

1. All b-adjacent syndromes generated by XORing b-adjacent columns should be unique.

2. All syndromes for all possible combinations of columns within b-adjacent columns should be unique.

3. If multiples of a column is used to XOR instead of the original column in the above two cases, all syndromes thus generated should be unique.

Condition 1 ensures that no b-adjacent errors are miscorrected. Condition 2 ensures that any number of errors within the b-adjacent columns are not miscorrected. The unique syndromes exactly identify which b-adjacent columns contain the errors. Since Reed-Solomon codes correct non-binary m-bit symbols, each column can also have different multiples. These multiples can be identified from the upper b-rows of the parity check matrix. But to avoid mis-correction for different magnitudes of errors, condition 3 needs to be satisfied. Thus, we keep on adding rows to the bottom part of the parity check matrix until all the conditions above are satisfied. Finally, r parity check symbols are added to the parity check matrix by appending the columns from a $r \times r$ identity matrix, thus making it systematic.

Since the code is an extended form of the extended SEC RS code, the proposed codes are systematic by design. Thus, the proposed codes have the following parameters: maximum block length $n = k + r$, where the number of information symbols is $k = (2^m-1)$ and r is the number of check symbols needed to construct the parity check matrix. The systematic design of the proposed codes increases the speed of the encoding procedure, since now it only involves simple XOR operation of data symbols. Also, the parity check symbols sometimes need to be re-ordered or placed in such a manner that the conditions above are met without any additional rows. But this placement or re-ordering of the parity symbols in the codeword does not affect the encoding or decoding latency. An example of the parity check matrix for a double-adjacent symbol error correcting code for $k = 7$ using the companion matrix notation has been shown in equation (6). The binary form of the matrix has been shown in Fig. 2. Similarly, the parity check matrix for a triple adjacent symbol error correcting code for $k = 8$ using the companion matrix notation has been shown in equation (7).

$$H = \begin{pmatrix} I & 0 & I & 0 & I & 0 & I & 0 & I & 0 & 0 \\ 0 & I & 0 & I & 0 & I & 0 & 0 & 0 & I & 0 \\ I & T & T^2 & T^3 & T^4 & T^5 & T^6 & I & 0 & 0 & 0 \\ I & T^2 & T^4 & T^6 & T & T^3 & T^5 & 0 & 0 & 0 & I \end{pmatrix} \quad (6)$$

$$H = \begin{pmatrix} I & 0 & 0 & I & 0 & 0 & I & 0 & 0 & I & 0 & 0 & 0 & 0 \\ 0 & I & 0 & 0 & I & 0 & 0 & I & 0 & 0 & I & 0 & 0 & 0 \\ 0 & 0 & I & 0 & 0 & I & 0 & 0 & I & 0 & 0 & 0 & 0 & 0 \\ I & T & T^2 & T^3 & T^4 & T^5 & T^6 & T^7 & 0 & 0 & 0 & I & 0 & 0 \\ I & T^2 & T^4 & T^6 & T^8 & T^{10} & T^{12} & T^{14} & 0 & 0 & 0 & 0 & I & 0 \\ I & T^3 & T^6 & T^9 & T^{12} & I & T^3 & T^6 & 0 & 0 & 0 & 0 & 0 & I \end{pmatrix} \quad (7)$$

A. Decoding Procedure

The easiest decoding procedure is to simply compare all the syndromes and based on the computed syndrome, both the error location and magnitude can be found. But this method involves comparison of a very large number of syndromes. Also, the number of syndromes increases linearly with the number of symbols in a codeword, and exponentially with the number of bits per symbol. Thus, in order to reduce the complexity of the decoder we propose a decoding method based on the companion matrix. This method has less complexity, but the reduction in complexity comes at the cost of decoder latency.

As in a SEC RS code, the syndromes are computed by taking the XOR of all data bits whose corresponding column is 1 in the binary form of the parity check matrix. The syndromes themselves are made up of m-bit symbols. If b-adjacent symbols are in error starting from symbol i, the syndromes then are given by equation (8). If b-adjacent columns or any number of columns within b-adjacent columns are in error, then equation (9) is always true for the b-adjacent columns.

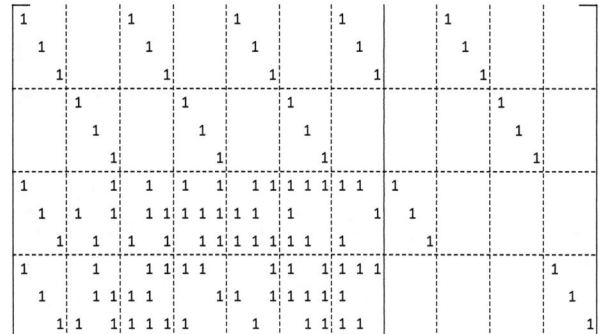

Fig. 2. Binary form of parity check matrix from equation (6).

Thus, if we extend the SEC RS code decoding using companion matrix so that b-adjacent columns are considered together, then the location can be computed using equation (9). The implementation of equation (9) is simply parallel XOR gates with the syndrome symbols as its inputs. If equation (9) is satisfied for all cases of $\beta \; \forall \; 1 \leq \beta \leq x$ and $x = (r-b)$, then b-adjacent errors have occurred from symbol i.

$$\begin{pmatrix} S_1 \\ \vdots \\ S_b \\ LS_1 \\ \vdots \\ LS_x \end{pmatrix} = \begin{pmatrix} e_i \\ \vdots \\ e_{i+b-1} \\ T^i e_i \oplus T^{i+1} \oplus \cdots \oplus T^{i+b-1} e_{i+b-1} \\ \vdots \\ T^{xi} e_i \oplus T^{x(i+1)} \oplus \cdots \oplus T^{x(i+b-1)} e_{i+b-1} \end{pmatrix} \quad (8)$$

$$T^{\beta i} S_1 \oplus T^{\beta(i+1)} S_2 \oplus \cdots \oplus T^{\beta(i+b-1)} S_b \oplus LS_\beta = 0 \quad (9)$$

Thus, the error location for each symbol is the OR of equation (9) for all β. Thus, a symbol d_i will have b different possibilities of adjacent columns that can be in error. The final indication of whether a symbol is in error or not is the AND of error location signals of all b adjacent columns it is part of. This is because if any one of the b different possibilities is in error, then it equates to 0 indicating that the symbol is in error. A symbol is error free if and only if all b possibilities equate to 1. The error pattern is obtained from S_α where α is the row number in the upper b-rows of the parity check matrix for which column i is a 1. The partial schematic diagram of the error pattern generator circuit for the parity check matrix in Fig. 2 has been shown in Fig. 3. The error pattern is then XORed with the received message symbols to get the decoded message symbols. Thus, the upper b rows of the syndromes are used to obtain the error magnitudes or patterns for adjacent errors. The lower $(r-b)$ rows of the syndromes are used to obtain the error location. The decoding procedure is thus a parallel one step procedure.

If any number of check symbols within b-adjacent symbols are in error, then there is at-least one case of β for which equation (9) is not satisfied. This simply indicates that none of the data symbols are in error. From Fig. 3, symbol d_6 has a different decoding procedure compared to other symbols since the parity check symbols have been re-ordered in the parity check matrix to satisfy the conditions discussed previously. In such a case, we consider the rows of the parity check matrix which have 1's in the parity check columns. These rows become the magnitude computation rows of the parity check matrix, while all the

978-1-5386-3775-3/18 $31.00 © 2018 IEEE

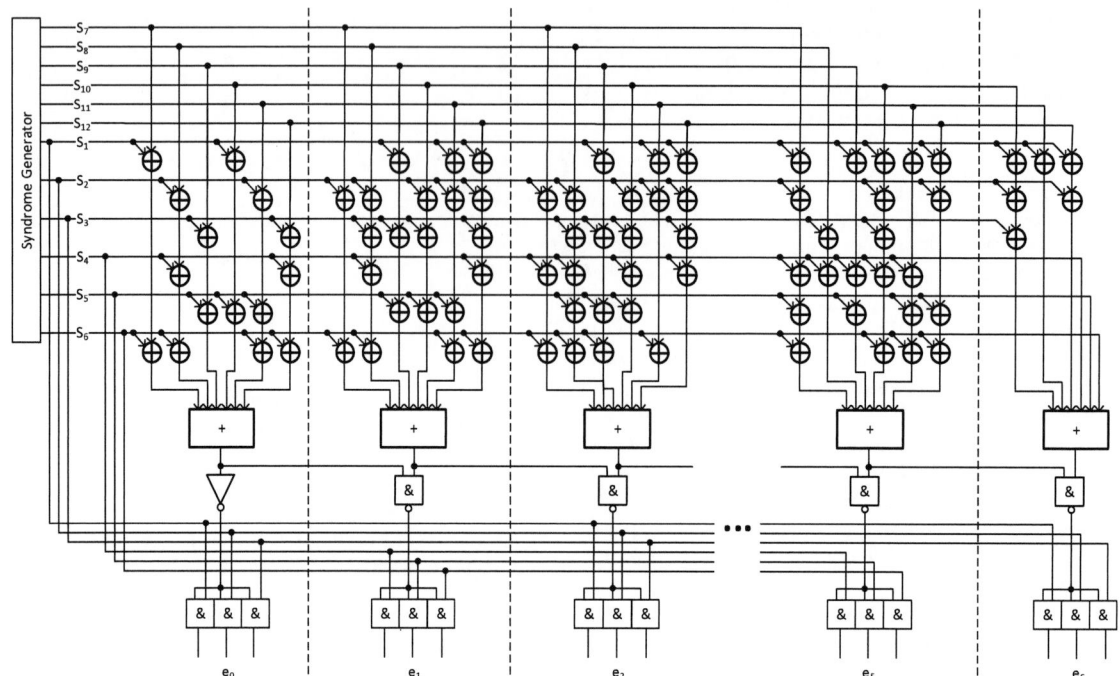

Fig. 3. Partial schematics of error pattern generator for double adjacent error correcting RS code for parity check matrix of equation (6).

remaining rows are used for error location.

Thus, the proposed codes can provide suitable error protection against burst symbol errors with low latency decoding. Consider $m = 6$ bits/symbol RS code as an example, three 2-bits/cell MLC memory cells can be concatenated to form a symbol. A double adjacent symbol error correction (DAsEC) scheme can protect at least 4 adjacent MLC memory cells and at most 6 adjacent MLC memory cells. Similarly, a triple adjacent symbol error correction (TAsEC) scheme can protect at least 7 adjacent MLC memory cells and at most 9 adjacent MLC memory cells. In general, if a symbol is made up of c MLC cells and a b-adjacent symbol error correcting proposed Reed-Solomon code is used then the minimum number of adjacent memory cells protected by the proposed code is $C_{min} = c.(b$-$1) + 1$. This comes from the fact that the maximum number of cells affected in any possible combination so that they always lie within a maximum of b-adjacent symbols is given by $(c.(b$-$1) + 1)$ and thus is the worst possible scenario. Similarly, the maximum number of adjacent memory cells being protected is $C_{max} = c.b$. This is the best possible scenario wherein all the cells affected in a b-symbol burst lie within b-adjacent symbols.

IV. EVALUATION

The proposed codes were synthesized on Synopsys Design Compiler using NCSU FreePDK45 45nm library for DAsEC and TAsEC for different information symbol lengths. Both these codes were implemented using Dataflow model in Verilog and errors were injected to ensure that all double adjacent errors and triple adjacent errors were corrected. Exhaustive testing was done for different magnitudes and locations of the symbols. For triple adjacent errors, the codes were tested to ensure that all

errors within any 3 adjacent columns were also corrected.

[Revirieo 15] proposed a method to correct double adjacent errors using OLS codes for binary bits by augmenting the parity check matrix and the decoding procedure of general OLS codes. This method was extended to non-binary symbols using the method proposed in [Namba 15a], wherein each bit in the symbol had its own independent parity check matrix as well as its own independent decoder, to correct double adjacent symbol errors. This double adjacent symbol error correcting OLS (DAsEC OLS) code was implemented using Dataflow model in Verilog and synthesized on Synopsys Design Compiler using NCSU FreePDK45 45nm library. A comparison was made to the proposed DAsEC codes.

[Namba 15b] proposed an adjacent symbol error correcting code based on Reed-Solomon Codes. For an adjacent error in the i-th and $(i+1)$-th bits, the syndrome appears as:

$$S_0 = e_i + e_{i+1} \qquad (10)$$

$$S_j = e_i\alpha^{ij} + e_{i+1}\alpha^{(i+1)j} \ \forall \ (1 \leq j \leq 2t - 1) \quad (11)$$

The adjacent error pattern is calculated from equation (10) and equation (11) for $i = 1,2,3$ and the error pattern is specified using these four equations. This double adjacent symbol error correcting Reed-Solomon (DAsEC RS) code was implemented using Behavioral model in Verilog and synthesized on Synopsys Design Compiler using NCSU FreePDK45 45nm library in order to compare it to the proposed DAsEC codes.

[Reviriego 13] proposed a method for correcting triple adjacent errors for binary bits using OLS codes with the same number of check bits as a double error correcting OLS code but with an augmented decoding procedure. This was also extended using the procedure in [Namba 15a] to correct triple adjacent symbol errors. This triple adjacent symbol error correcting OLS (TAsEC OLS) code was implemented using Dataflow model in Verilog and also synthesized on Synopsys Design Compiler using NCSU FreePDK45 45nm library to compare it to the proposed TAsEC codes. The evaluation of the implemented codes and the proposed codes was done on the basis of area of the decoder, decoder latency and redundancy (i.e. number of check symbols required for a given information symbol length per codeword).

A. Redundancy

Table I shows the comparison of redundancy between, the DAsEC OLS codes [Reviriego 15] [Namba 15a], the DAsEC RS codes in [Namba 15b] and the proposed DAsEC codes for $k = 8$, 16, 32 and 64. k refers to the number of information symbols in the codeword, r refers to the number of check symbols required and k_{max} refers to the maximum number of information symbols the codeword can have for the given r number of check symbols. It is seen that the redundancy of the proposed codes and DAsEC RS codes are the same. It is also seen that the proposed codes have much better redundancy compared to the DAsEC OLS codes. Table II shows the comparison between the TAsEC OLS codes [Reviriego 13] [Namba 15a] and the proposed TAsEC codes for $k = 8$, 16, 32 and 64. As seen from the table, the proposed codes have a much better redundancy compared to the TAsEC OLS codes.

B. Hardware Complexity

Table I also shows the comparisons of area of the decoder and the decoder latency between the DAsEC OLS codes [Reviriego 15] [Namba 15a], the DAsEC RS codes in [Namba 15b] and the proposed DAsEC codes for $k = 8$, 16, 32 and 64. The operations involved in the DAsEC RS codes were of $GF(2^m)$, where m = number of bits per symbol, which increases the complexity of the decoder. The complex operations result in a much greater circuit depth compared to the proposed codes. Due to this higher circuit depth, these codes have high decoder latency as well. The use of a companion matrix makes all operations binary for the proposed DAsEC codes. As a result, there are no complex operations involved for the proposed codes. The decoder circuit is comprised mainly of XOR gates, with additional AND and OR gates. As a result, the decoder latency and the decoder area of the proposed codes are much less compared to the DAsEC RS codes as seen in Table I. The area and decoder latency of the DAsEC OLS codes are better than the proposed codes, because the circuit depth is higher for the proposed codes due to the low redundancy. The DAsEC OLS codes have lesser complexity and parallel decoding logic. This reduces the decoder latency of the DAsEC OLS codes compared to the proposed codes. But this reduced decoder latency comes at an expense of very high redundancy.

Table II also shows the comparison of area of the decoder and the decoder latency between the TAsEC OLS codes [Reviriego 13] [Namba 15a] and the proposed TAsEC codes. Similar to the DAsEC code comparison, the proposed codes have higher complexity due to the higher circuit depth arising from the low redundancy. The low decoding latency of the TAsEC OLS codes is due to a low complexity parallel decoding logic, but comes at an expense of very high redundancy. The proposed codes can provide much better redundancy for a slightly higher cost of decoder latency and decoder area.

V. CONCLUSION

In this paper, b-adjacent symbol error correction for Reed-Solomon codes are proposed which are systematic by design and have low complexity parallel one step decoding. The proposed codes have better decoding latency and decoder area compared to existing adjacent error correction for Reed-Solomon codes. The proposed codes are also compared to symbol error correcting OLS codes and it is shown that the proposed codes achieve much better redundancy, but at a cost of slightly higher decoder latency and decoder area. The proposed codes thus provide a balanced tradeoff between the amount of redundancy required and the decoder complexity and latency. As a result, the proposed codes can be used to increase the reliability of latest memory technologies like MLC PCM and STT-MRAM in lower technology nodes.

TABLE I. REDUNDANCY, DECODER AREA AND DECODER LATENCY COMPARISON FOR DAsEC CODES

Bits/ Symbol	k	DAsEC OLS Codes [Namba 15a] [Reviriego 15]				DAsEC RS Codes [Namba 15b]				Proposed Codes			
		k_{max} (symbols)	r (symbols)	Area (μm^2)	Latency (ns)	k_{max} (symbols)	r (symbols)	Area (μm^2)	Latency (ns)	k_{max} (symbols)	r (symbols)	Area (μm^2)	Latency (ns)
4	8	9	9	659	0.27	11	4	42711	3.09	15	4	2328	0.96
5	16	16	12	1634	0.31	27	4	165222	4.76	31	4	8420	1.47
6	32	49	21	3847	0.42	59	4	1314387	6.61	63	4	18379	1.64
7	64	64	24	9225	0.42	123	4	16563101	9.89	127	4	43541	2.38

TABLE II. REDUNDANCY, DECODER AREA AND DECODER LATENCY COMPARISON FOR TAsEC CODES

Bits/Symbol	k	TAsEC OLS Codes [Namba 15a] [Reviriego 13]				Proposed Codes			
		k_{max} (symbols)	r (symbols)	Area (μm^2)	Latency (ns)	k_{max} (symbols)	r (symbols)	Area (μm^2)	Latency (ns)
4	8	-	-	-	-	15	6	2943	1.09
5	16	16	16	6174	0.77	31	6	12572	1.74
6	32	49	26	15456	1.22	63	6	41045	2.25
7	64	64	32	37679	1.64	127	6	128011	3.21

978-1-5386-3775-3/18 $31.00 © 2018 IEEE

ACKNOWLEDGMENTS

This research was supported in part by the National Science Foundation under Grant No. CCF-1617665.

REFERENCES

[Bossen 70] D. C. Bossen, "b-Adjacent Error Correction," *IBM Journal of Research and Development*, vol. 14, no. 4, pp. 402-408, 1970.

[Carrasco 08] R. A. Carrasco and M. Johnston, *Non-binary Error Control Coding for Wireless Communication and Data Storage.* Chichester, West Sussex, UK: Wiley, 2008.

[Dutta 11] R. Datta and N.A. Touba, "Generating Burst-Error Correcting Codes from Orthogonal Latin Square Codes -- A Graph Theoretic Approach," in *Proc. of IEEE Symposium on Defect and Fault Tolerance*, pp. 367-373, 2011.

[Fujiwara 06] E. Fujiwara, *Code Design for Dependable Systems: Theory and Practical Applications.* Hoboken, NJ, USA:Wiley-Interscience, 2006.

[Jiang 14] L. Jiang, Y. Zhang and J. Yang, "Mitigating Write Disturbance in Super-Dense Phase Change Memories," in *Proc. of Annual IEEE/IFIP International Conference on Dependable Systems and Networks*, pp. 216-227, 2014.

[Klockmann 17] A. Klockmann, G. Georgakos and M. Goessel, "A new 3-bit burst-error correcting code," in *Proc. of IEEE International Symposium on On-Line Testing and Robust System Design (IOLTS)*, pp. 3-5, 2017.

[Namba 15a] K. Namba and F. Lombardi, "Non-Binary Orthogonal Latin Square Codes for a Multilevel Phase Charge Memory (PCM)," in *IEEE Transactions on Computers*, vol. 64, no. 7, pp. 2092-2097, Jul. 2015.

[Namba 15b] K. Namba and F. Lombardi, "A Single and Adjacent Symbol Error-Correcting Parallel Decoder for Reed-Solomon Codes," in *IEEE Transactions on Device and Materials Reliability*, vol. 15, no. 1, pp. 75-81, Mar. 2015.

[Papandreou 10] N. Papandreou, A. Pantazi, A. Sebastian, M. Breitwisch, C. Lam, H. Pozidis and E. Eleftheriou, "Multilevel Phase-Change Memory," in *Proc. of IEEE International Conference on Electronics, Circuits and Systems*, pp. 1017-1020, 2010.

[Reviriego 13] P. Reviriego, S. Liu, J.A. Maestro, S. Lee, N.A. Touba and R. Datta, "Implementing Triple Adjacent Error Correction in Double Error Correction Orthogonal Latin Square Codes," in *Proc. of IEEE Symposium on Defect and Fault Tolerance*, pp. 167-171, 2013.

[Reviriego 15] P. Reviriego, S. Pontarelli, A. Evans and J.A. Maestro, "A Class of SEC-DED-DAEC Codes Derived from Orthogonal Latin Square Codes," in *IEEE Transactions on very Large Scale Integration (VLSI) Systems*, vol. 23, no. 5, pp. 968-972, May 2015.

[Yoon 17] I. Yoon and A. Raychowdhury, "Modeling and Analysis of Magnetic Field Induced Coupling on Embedded STT-MRAM Arrays," in *IEEE Transactions on Computer-Aided Design of Integrated Circuits and Systems*, vol. 37, no. 2, pp. 337-349, Feb. 2018.

978-1-5386-3775-3/18 $31.00 © 2018 IEEE

2018 IEEE 36th VLSI Test Symposium (VTS)

Circuit-Level Reliability Simulator for Front-End-of-Line and Middle-of-Line Time-Dependent Dielectric Breakdown in FinFET Technology

Kexin Yang, Taizhi Liu, Rui Zhang, Linda Milor

School of Electrical and Computer Engineering, Georgia Institute of Technology, Atlanta, GA USA

kyang70@gatech.edu

Abstract— **This paper presents a lifetime simulator for both Front-End-of-Line (FEOL) time dependent dielectric breakdown (TDDB) and the newly emerging Middle-of-Line (MOL) time dependent dielectric breakdown for FinFET technology. A lifetime assessment flow for digital circuits and microprocessors is proposed for the target wearout mechanisms, and its associated vulnerable feature extraction algorithms are discussed in detail. Our simulator incorporates the detailed electrical stress, temperature, linewidth of each standard cell within the digital circuit and microprocessor. Also, FEOL TDDB and MOL TDDB lifetimes are combined in the calculation of TDDB lifetime. Circuit designers can use the resulting lifetime information to guide and improve their circuits to make them more robust and reliable.**

Keywords—time-dependent dielectric breakdown; lifetime simulator; wearout; frontend-of-line dielectric breakdown; middle-of-line breakdown; digital circuit; microprocessor; reliability

I. INTRODUCTION

Traditional FEOL time-dependent dielectric breakdown (FEOL TDDB) is one of the main concerns for advanced CMOS technology. Accurate circuit lifetime assessment due to TDDB has become an significant part of the circuit design process. A new source of breakdown is Middle-of-Line (MOL) dielectric breakdown, which is breakdown between the polysilicon/high-k control gate (PC) and diffusion contacts (CA) [1]. MOL TDDB is a growing concern for semiconductor device reliability. It is necessary to investigate and perform detailed lifetime analysis of MOL TDDB in state-of-art FinFET technology.

This paper presents a simulator that can be used to assess logic circuit lifetime due to not only the traditional FEOL TDDB, but also the above mentioned MOL TDDB in FinFET technology. There are studies of MOL TDDB on dielectric materials [2], [3]. A budget-based MOL reliability management in FinFET technology is proposed in [4]; however, the authors declare that voltage does not have a strong impact on MOL TDDB, and assume a fixed voltage for the lifetime calculation. The assumption may be broken by a compact standard cell layout and a high switching activity, where the voltage difference between two segments plays an important role. In addition, no study has been conducted to investigate the vulnerable feature extraction algorithms in MOL TDDB in both digital circuits and microprocessors.

In previous FEOL system-level reliability studies, researchers have studied bias temperature instability (BTI) [5], [6], hot carrier injection (HCI) [7], [8], FEOL TDDB [9], [10]. None have considered MOL TDDB at the system-level which

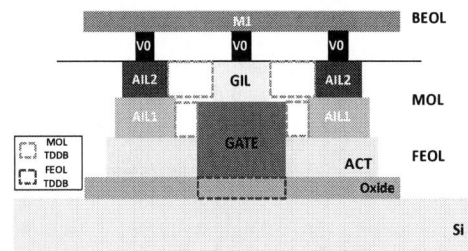

Fig. 1. FinFET cross-section.

involves the extraction of vulnerable features of MOL TDDB in a circuit layout. In addition, most of the studies are conducted with traditional CMOS technology and fail to consider the state-of-art FinFET technology.

A methodology to link device level wearout models of MOL TDDB and FEOL TDDB to circuit lifetime is introduced in this work. The MOL TDDB vulnerable features in a FinFET are presented. The corresponding algorithms to extract such vulnerable features in a standard cell are discussed in detail. Our simulator runs in three steps. First, it characterizes the standard cell library corresponding to a given FinFET technology library (generating vulnerable features for each standard cell). After that, the simulator combines the vulnerable features with cell activity and the temperature profile to calculate TDDB lifetime of standard cells. The last step combines the lifetime of vulnerable features caused by both FEOL and MOL TDDB.

To demonstrate our simulator's functionality, the lifetime simulator is used to study the lifetime distribution for an 8-bit FFT circuit and a Leon3 processor implemented with FinFET technology. For the Leon3, we also consider the impact of use scenarios on the lifetime of the microprocessor.

The rest of this paper is organized as follows. Section 2 presents the wearout models used for TDDB. Section 3 describes the extraction algorithm in detail. Section 4 presents the lifetime simulator using an FFT circuit and the Leon3 processor as examples. The paper is concluded in Section 5.

II. DEVICE-LEVEL WEAROUT MODELS

Fig. 1 shows the breakdown paths of FEOL TDDB and MOL TDDB. FEOL TDDB is the breakdown between the gate and source or drain; whereas MOL TDDB is the breakdown between the gate and its adjacent contact or active interconnect layer.

A. FEOL TDDB Model

FEOL TDDB is described as the build-up of traps in the gate oxide as a function of time under voltage and thermal stress. We

978-1-5386-3775-3/18 $31.00 © 2018 IEEE   127

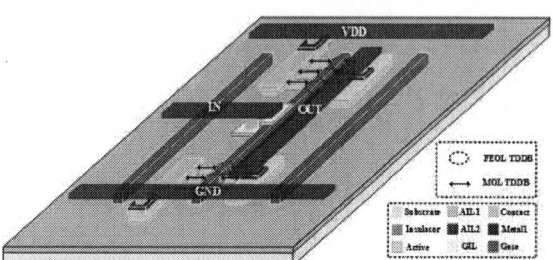

Fig. 2. 3D inverter view of MOL TDDB and FEOL TDDB.

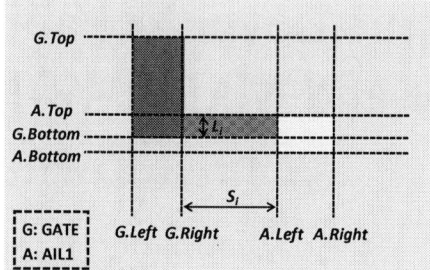

Fig. 3. Illustration of MOL TDDB vulnerable feature.

use the hard breakdown (HBD) model to characterize the transistor lifetime distribution. For ultra-thin (<5nm) gate dielectrics, the time-to-failure due to gate-oxide degradation can be derived by connecting the oxide degradation model to the Weibull failure distribution function [11] which is described by a shape parameter, β and a characteristic lifetime η, which is the time-to-failure at the 63% probability point, i.e.,

$$\eta = A_{ox}\left(\frac{1}{WL}\right)^{\frac{1}{\beta}} e^{-\frac{1}{\beta}} V^{a+bT} \exp\left(\frac{c}{T} + \frac{d}{T^2}\right) s^{-1} \quad (1)$$

where W and L are the device width and length, respectively, s is the probability of stress, T is temperature, V is gate voltage, and a, b, c, d, and A_{ox} are fitting parameters, which include the activation energy between 0.6 and 0.9 eV. The constants in (1) are determined using test structure data at high temperatures and voltages [9].

B. MOL TDDB Model

Although Back-End-Line TDDB (BTDDB) is not discussed in this paper, the device-level lifetime model for MOL TDDB is similar to that of BTDDB [12] as follows:

$$\eta = A_{MTDDB} L_i^{-1/\beta_i} \exp(-\gamma E^m + E_a/kT) \quad (2)$$

where $A_{MOL\ TDDB}$ is a constant that depends on the material properties of the dielectric, γ is the field acceleration factor; electric filed is a function of voltage, V and the linespace, S_i, i.e., $E=V/S_i$, and m is 1 for the E model [13]. L_i is the vulnerable length and E_a is the activation energy (~0.5eV). The temperature dependence is modelled with the Arrhenius relationship [14], where k is the Boltzmann constant. The parameters are obtained from experimental data from [1], [14].

III. VULNERABLE FEATURE EXTRACTION

In this section, the algorithms that are used to extract vulnerable features in each standard cell for TDDB are introduced. As shown in Section 2, the device-level models for FEOL TDDB and MOL TDDB are different; thus, we need to develop a unique algorithm to extract vulnerable features due to each type of TDDB. In this study, FreePDK15 [15] was implemented and used as a case study for FinFET technology.

In Fig. 1, a cross-section view of a FinFET transistor is presented. The blue dashed squares stand for locations for MOL TDDB, while FEOL TDDB is represented by the purple dashed squares. A detailed 3D illustration of TDDB in a layout in FinFET technology can be found in Fig. 2, which uses an inverter's layout as an example.

A. FEOL TDDB Vulnerable Feature Extraction

To characterize device's FEOL TDDB lifetime, we only need to obtain the transistor's width (W) and channel length (L). We can get the transistor's size information for each standard cell from the spice netlist. Notice instead of using the width (W)

directly from the netlist, which represents the drawn width of the source and drain, we should calculate the effective width [16] as follows,

$$W_{eff} = T_{fin} + 2H_{fin} \quad (3)$$

where T_{fin} is the fin thickness and H_{fin} is the fin height.

To take the number of fins into account, we obtain the total effective width,

$$W_{eff,total} = n_{fin} \cdot W_{eff} \quad (4)$$

where n_{fin} is number of fins in the transistor.

B. MOL TDDB Vulnerable Feature Extraction

We need to analyze and extract MOL TDDB vulnerable features from each standard cell layout. From Fig 1, there are two types of MOL TDDB features; one is the GATE-AIL1 pair and the other is the GIL-AIL2 pair. Our simulator needs to find all the existing vulnerable features.

As shown in Fig 3, L_i and S_i are indicated. The yellow square represents the AIL1 layer; the red square stands for the gate. The blue dashed square is the vulnerable feature and our goal to extract these vulnerable features. In our simulator, only the nearest vulnerable features are extracted, since the further ones are separated by poly segments in the middle and the electric field between them will be shielded. For layout generation, we have used the NanGate 15nm Open Cell Library [17]. To identify the vulnerable features in a standard cell layout, we should find the pin to which the corresponding segment (GATE or AIL1 layer) is connected. This is because the layout needs to be linked to the netlist and activity information from running benchmarks.

In the layout file, the pin connection is stored as a single point coordinate (x, y), while a polygon is stored as its vertices' coordinates: bottom left corner (*Left*, *Bottom*), and the upper right corner (*Right*, *Top*). Therefore, we need to start by finding the top layer to which the pin is directly connected and continue the process downward in the stack. In a standard cell, in most cases, the top layer will be a metal layer (M1 or M2) depending on the type of cell being analyzed.

In our simulator, the ray-casting algorithm [18] has been implemented to find the connected layer. If the number of crossings is odd, the point is inside a polygon. Fig 4 gives a set of example points which we need to test. The Python implementation of the ray-casting algorithm to determine whether a point is inside a polygon is presented in Fig. 5.

After finding the pin's directly connected layer's segment, we start to process downward to find all the layers to which the pin is connected. There are two types of overlap we could find in a standard cell layout, which are illustrated in Fig 6. Fig 6(a) shows the overlap situation that only happens between GIL and

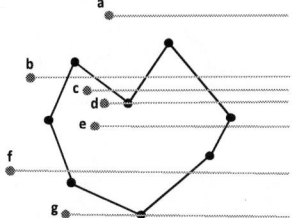

Fig. 4. Point inclusion problem.

Algorithm 1: point inclusion algorithm

Input: *polygon (poly), point (p)*
Output: whether the test point is inside the polygon
def PinPoly(*poly, p*):
    nvert = number of vertex in the polygon
    testx = p.x, testy = p.y, result = false, i = 0, j = nvert - 1
    while (i < nvert):
        if (((poly.Vertex[i].y > testy) != (poly.Vertex[j].y > testy))
        and *(testx < (poly.Vertex[j].x - poly.Vertex[i].x)*
        * (testy-poly.Vertex[i].y) / (poly.Vertex[j].y-poly.Vertex[i].y)
        + self.Vertex[i].x)):
            result = !result
        j = i, i = i + 1
    return *result*

Fig. 5. Point inclusion algorithm.

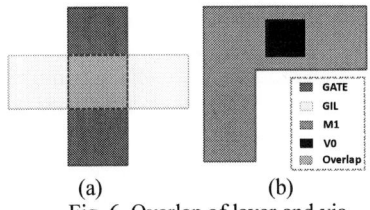

(a)   (b)
Fig. 6. Overlap of layer and via.

Algorithm 2: rectangle overlap determination

Input: rectangle_A (*RectA*), rectangle_B (*RectB*)
Output: whether the two rectangles overlap
if (*RectA.Left < RectB.Right* **and** *RectA.Right > RectB.Left* **and**
    RectA.Top > RectB.Bottom **and** *RectA.Bottom < RectB.Top*):
    return *true*
else:
    return *false*

Fig. 7. Overlap determination algorithm.

GATE, while most of the overlap situations are presented in Fig 6(b).

The overlap between two rectangles is easy to implement and its corresponding algorithm is shown in Fig. 7. The proof of the "if statement" is by contradiction. Any one of the following four cases guarantee that no overlap exists between rectangles A and B:

Case #1: If A's left edge is to the right of B's right edge.
Case #2: If A's right edge is to the left of B's left.
Case #3: If A's top edge is below B's bottom.
Case #4: If A's bottom edge is above B's top.

As for the second situation in Fig. 6(b), we can utilize the poly inclusion algorithm to do the job. That is, if one of the via's vertices is inside the polygon, then the two overlap. Once the layer connection determination is finished, we store the pin's connected layers for each standard cell in a Python dictionary named "*std_cell_info*".

Algorithm 3: vulnerable feature extraction

Input: standard cell layout information (.txt file)
Output: vulnerable feature for each standard cell
for each *pin* in standard cell:
    for each *pin* in std_cell_info[*pin'*]   (*pin != pin'*):
        if max(G.Bottom, A.Bottom) < min(G.Top, A.Top):
            L_i = min(G.Top, A.Top) - max(G.Bottom, A.Bottom)
            X_coord = sort([G.Left, G.Right, A.Left, A.Right])
            S_i = X_coord[2] - X_coord[1]
#*std_cell_info* is a dictionrary which stores each pin connected layers

---

Fig. 8. Vulnerable feature extraction algorithm.

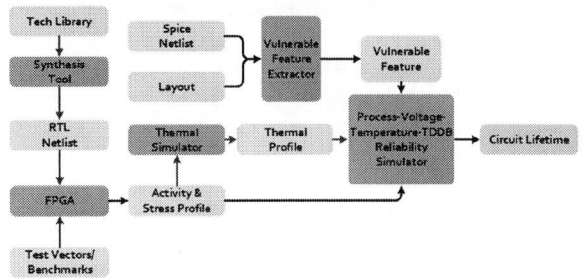

Fig. 9. Framework for the reliability simulator. Yellow boxes are data and blue boxes are tools.

As mentioned before, the cell layout is composed of multiple polygons with their corresponding vertices. To extract the vulnerable features in Fig. 3, we use the vulnerable feature extraction algorithm which is presented in Fig. 8.

We perform vulnerable feature extraction only if the "overlap" configuration of a GATE-AIL1 pair exists. If an GATE-AIL1 pair overlaps, the maximum y value of the bottom left corner of the GATE and AIL1, max(*G.Bottom, A.Bottom*) must be less than the minimum y value of the upper right corner of GATE and AIL1, min(*G.Top, A.Top*).

The vulnerable length L_i is computed as follows,
$$L_i = \min(G.Top, A.Top) - \max(G.Bottom, A.Bottom) \quad (5)$$

To extract the linespace S_i between the GATE-AIL1 pair, we sort the horizontal coordinates and put them into an array X_coord[] first. After sorting, the linespace can be easily computed by the subtraction of the middle two elements,
$$S_i = X\_coord[2] - X\_coord[1] \quad (6)$$

IV. LIFETIME SIMULATOR

The framework of our reliability simulator is presented in Fig. 9. This figure describes the tool flow needed to compute lifetime. The left most part of the figure includes the tools needed to determine operating profiles, such as activity, duty cycle, and temperature for each net, while the circuit is supplied with a set of random input vectors. The blocks on the right combine the operating profiles together to determine the lifetime. The lifetime is first computed for individual standard cells, and then these lifetimes are combined to find the lifetime of the whole circuit.

For FEOL TDDB and MOL TDDB, the significant factors are activity, voltage (VDD), temperature, and the vulnerable features. For activity tracking, the circuit netlist is loaded onto an FPGA for emulation [19]. The resulting state probabilities and toggle rates of the I/O ports are recorded. By using

978-1-5386-3775-3/18 $31.00 © 2018 IEEE   129

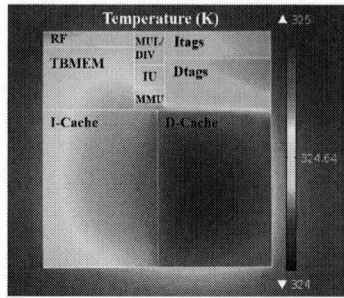

Fig. 10. The average temperature distribution of Leon3 while running a standard benchmark.

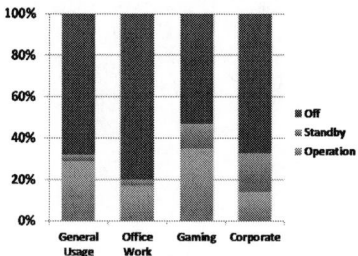

Fig. 11. The use scenarios provided by Intel [23].

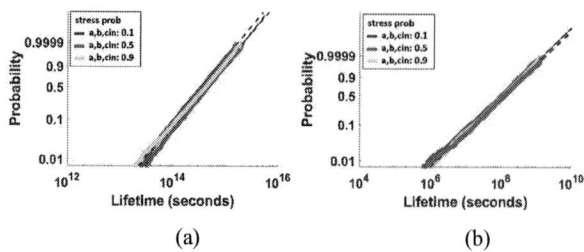

Fig. 12. Standard cell lifetime distribution for the FA (full adder) in the FFT circuit: (a) FEOL TDDB and (b) MOL TDDB.

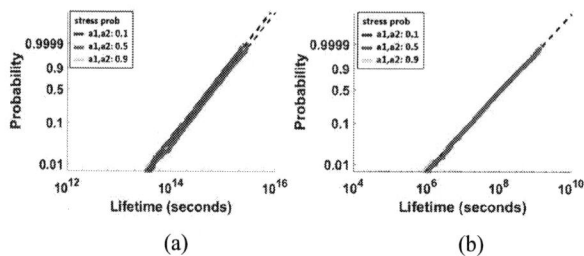

Fig. 13. Standard cell lifetime distribution for the OR2_X1 cell in Leon3: (a) FEOL TDDB and (b) MOL TDDB.

PrimeTime [20] activity propagation, we obtain the state probabilities and toggle rates for all the internal nets. The state probabilities are the key parameters to determine the lifetime of each layout feature, since signal states determine the time that each layout feature is under stress.

An 8-bit FFT circuit and a Leon3 microprocessor are implemented in FreePDK15 in this paper to demonstrate the functionality of our simulator. Synthesis is done with Synopsys Design Compiler [21], with the standard library with 69 different standard cells. After synthesis, the FFT circuit is composed of 38 types of standard cells with a total cell count of 112K. The Leon3 has 18 types of standard cells, with a total cell count of 312K.

Using the net activity and RC information from the layout, we can find the power consumed by each component of the Leon3 microprocessor. To determine the thermal distribution, we consider the self-heating effects of FinFETs [22] and supply the power consumption data to COMSOL. The temperature distribution when the Leon3 is running a standard benchmark is shown in Fig. 10. We associate this temperature profile with every standard cell in the microprocessor to calculate lifetime.

The FFT circuit is supplied with randomly generated inputs and the circuit continuously performs the Fast Fourier Transformation on the data. For the Leon3, we consider the degradation under different use scenarios, as shown in Fig. 11. The use scenarios have different fractions of time when the system is in three modes: operation, standby, and off [23]. The activity profiles during the operation mode are determined by running benchmarks [24]. Our experimental results use a combination of standard benchmarks.

First, we calculate the FEOL TDDB lifetime of a single device using (1) and the MOL TDDB lifetime of a single feature using (2). To combine different device lifetimes in a standard cell, we assume a standard cell is composed of n devices (n features for MOL TDDB), each modelled with a Weibull distribution, for each wearout mechanism. The characteristic

lifetime of the cell, η_{cell}, is a combination of Weibull distributions and is the solution of [25], [26]:

$$1 = \sum_{i=1}^{n}(\eta_{cell}/\eta_i)^{\beta_i} \qquad (7)$$

where $\eta_i, i = 1, \dots, n$ are the characteristic lifetimes of all of the devices; and $\beta_i, i = 1, \dots, n$ are the corresponding shape parameters. Similarly [26]:

$$\beta_{cell} = \sum_{i=1}^{n}\beta_i(\eta_{cell}/\eta_i)^{\beta_i} \qquad (8)$$

To calculate the FEOL TDDB lifetime of a standard cell, we need to obtain the gate-source voltage, V_{gs}, for each transistor and analyze each transistor's gate stress probability p. The lifetime of a transistor is a function of its stress probability, which in turn depends on the input pattern probabilities.

If the shape parameter is the same for each device (feature), which is typically assumed,

$$\eta_{cell} = \left(\sum_{i=1}^{n}\eta_i^{-\beta}\right)^{-1/\beta} \qquad (9)$$

As for the MOL TDDB lifetime calculation, we should analyze the circuit's layout. For each adjacent GATE-AIL1 pair, we extract the linespace S_i and vulnerable length L_i. After that, for each pair, the vulnerable feature pair (S_i, L_i) is associated with the poly-contact voltage difference V. If the two pins' stress probability are independent, the stress probability of a single dielectric segment feature is calculated as follows (if not, we evaluate its correlation and get the corresponding stress probability):

$$p_{total} = p_1 \cdot (1 - p_2) + p_2 \cdot (1 - p_1) \qquad (10)$$

where p_1 and p_2 are the probabilities of the poly and contact being at logic "1", respectively.

By using (7) – (10), we get the characteristic lifetime of FEOL TDDB and MOL TDDB for every standard cell in the circuits, which is shown in Figs. 12 and 13. The standard cell lifetime distributions are simply combinations of the transistor/layout feature lifetimes of all of the transistors and layout features in the standard cell. The input probabilities for each logic state for the standard cells propagate to internal nodes within the cell and determine the stress of transistors and layout features. As we can see in the figures, the lifetimes of the full

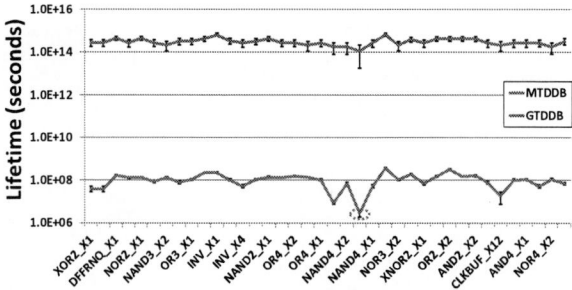

Fig. 14. FFT circuit FEOL TDDB and MOL TDDB characteristic lifetimes for each type of standard cell and its lifetime limiting cell (shown in the red dashed circle). The confidence bound indicate variation in characteristic lifetime due to activity.

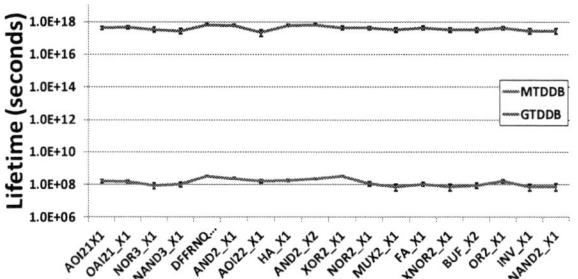

Fig. 15. Leon3 FEOL TDDB and MOL TDDB characteristic lifetimes for each type of standard cell. The confidence bounds indicate variation in characteristic lifetime due to activity.

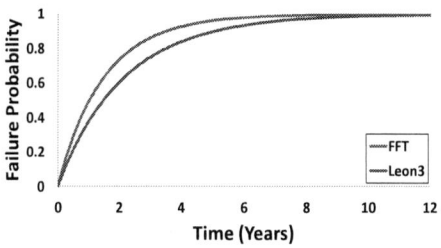

Fig. 16. FFT and Leon3 failure probability.

adder cell and the OR gate are partitioned by probability ranges for logic "1" at the inputs and plotted with a lognormal plot. We can partition the other standard cells in the same way and fit their corresponding lifetimes with the lognormal distribution.

Regression is used to determine the parameters of the lognormal distribution. After regression, each standard cell is shown as a mean and standard deviation of the characteristic lifetime in Figs. 14 and 15.

Since the lifetime of each type of standard cell is calculated, we can proceed to calculate the circuit's failure probability at time t. We model each of the standard cell lifetimes with a lognormal distribution, for each wearout mechanism and its corresponding activity range. For n standard cells, there are n failure rates, F_i, $i = 1, \ldots, n$, where F_i is the cumulative probability of the lognormal distribution. These n failure rates contribute a reliability defect density d_i, which must be added together to find the failure probability of the whole circuit. The failure probability of the whole circuit F_{total} is a combination of lognormal distributions and is calculated as follows [27]:

$$d_i = -\ln(1 - F_i(t)) \quad (11)$$
$$F_{total} = 1 - exp^{-\Sigma d_i} \quad (12)$$

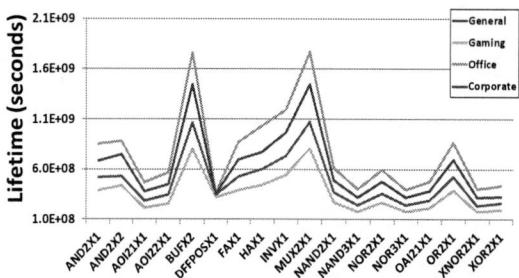

Fig. 17. Leon3 MOL TDDB lifetime for different use scenarios.

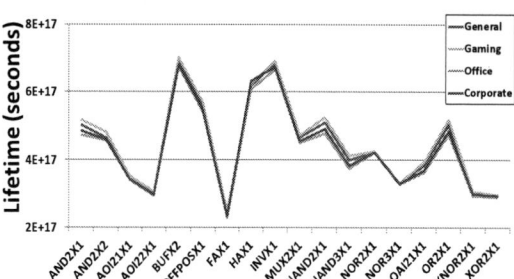

Fig. 18. Leon3 FEOL TDDB lifetime for different use scenarios.

For each (μ_i, σ_i) pair, we obtain F_i by looking up the cumulative probability for a normal distribution. Using (11) and (12), we combine standard cell lifetimes for FEOL TDDB and for MOL TDDB together, as they are independent mechanisms. The calculated failure probabilities are shown in Fig. 16.

From Figs. 14 and 15, we can see in the result that MOL TDDB is dominant. This result is technology dependent, since in FinFET technology, the layout is more compact, and thus the linespace and vulnerable length become smaller, which translates into severe degradation due to MOL TDDB. On the other hand, FEOL TDDB is more sensitive to voltage, and voltage scaling helps to alleviate the impact of FEOL TDDB.

To consider the use scenario impact on the Leon3, the stress during operation is computed based on activities when the Leon3 runs a standard benchmark. It idles with a random state in standby mode and powers down for the off mode. Shown in Fig. 17, we can conclude that MOL TDDB is more sensitive to use scenarios while, as observed in Fig. 18, FEOL TDDB is not sensitive. The vulnerable features in MOL TDDB are associated with two pins in a standard cell, while for FEOL TDDB, each device is only associated with its gate voltage; and thus, the disturbance of two pins will have a larger impact than just one.

In addition, one can find from Fig. 17 that not all standard cells have the same lifetime under different use scenarios; and thus, it is possible for a circuit designer to choose specific cells over others to ensure a longer lifetime in certain applications. One can replace the lifetime limiting standard cells with ones with a longer lifetime. As can be seen in Fig. 14, for the FFT circuit, the "INV_X16" cell is the lifetime limiting cell, and thus, we could use 16 "INV_X1" cells to replace the "INV_X16" cell to achieve the same functionality with a higher lifetime. The layout of "INV_X16" is more compact and has more vulnerable features in the layout, which causes a lower lifetime. An illustration is shown in Fig. 19; if the inverter is operated at 50% duty cycle, we can increase the lifetime by 6.75X, while

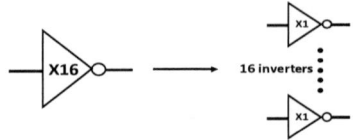

Fig. 19. Using cells with longer lifetime to replace a lifetime limiting cell to improve the lifetime of the circuit.

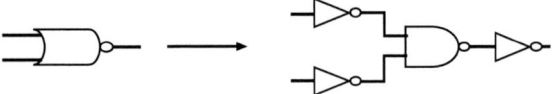

Fig. 20. Using an inverter and a NAND gate to replace a NOR. gate.

increasing the area by 2.67X, with similar power consumption. In addition, further optimization is possible if we characterize each standard cell in the technology library and find combinations of cells that have the same functionality with a longer lifetime, at the expense of possibly more area and power. Fig. 20 gives another possible example which uses a combination of cells to replace a NOR gate in the circuit; the area increases by 3.25X and power increases by 3.87X.

V. CONCLUSION

This paper investigates not only the traditional reliability concern, FEOL TDDB, but also the newly emerging wearout mechanism, MOL TDDB. A novel lifetime simulator for FinFET technology is proposed for the target wearout mechanisms. The shrinking feature size leads to severe degradation caused by MOL TDDB because of its sensitivity to alignment errors. On the other hand, voltage scaling alleviates the impact of FEOL TDDB and MOL TDDB.

With reliability simulation, a circuit designer can use the information to redesign a circuit or redraw the layout in a more robust and reliable way; also, a circuit designer can use application specific information to choose certain cells that have longer lifetimes than others. It is also possible to use the lifetime information to add some constraints on circuits to ensure the circuit's performance over the product lifetime.

This work provides a framework to identify the lifetime limiting cells in a circuit; further optimization on the trade-off between power, area and lifetime of a circuit needs further investigation.

ACKNOWLEDGEMENT

The authors would like to thank the NSF for support under Award Number 1700914.

REFERENCES

[1] F. Chen, Carole Graas, Michael Shinosky, Kai Zhao, Shreesh Narasimha, Xiao Hu Liu, Chunyan Tian, "Breakdown data generation and in-die deconvolution methodology to address BEOL and MOL dielectric breakdown challenges," Microelectronics Reliability, vol. 55, no. 12, pp. 2727–2747, 2015.

[2] F. Chen, et al."New breakdown data generation and analytics methodology to address BEOL and MOL dielectric TDDB process development and technology qualification challenges," IEEE Int. Reliability Physics Symp., 2014, pp. 3A.1.1-3A.1.11.

[3] E. Wu, J. Stathis, B. Li, B. Linder, K. Zhao, and G. Bonilla, "A critical analysis of sampling-based reconstruction methodology for dielectric breakdown systems (BEOL/MOL/FEOL)," IEEE Int. Reliability Physics Symp., 2015, pp. 2A.2.1-2A.2.11, 2015.

[4] J. C. Ahn, Jae-Gyung, Ming Feng Lu, Nitin Navale, Dawn Graves, Gamal Refai-Ahmed, Ping-Chin Yeh, "Product-level reliability estimator with budget-based reliability management in 16nm technology," IEEE Int. Reliability Physics Symp., 2017, pp. 3A–3.1–3A–3.6.

[5] W. Wang, S. Yang, S. Bhardwaj, S. Vrudhula, F. Liu, and Y. Cao, "The impact of NBTI effect on combinational circuit: modeling, simulation, and analysis," IEEE Trans. on Very Large Scale Integration Systems, vol. 18, no. 2, pp. 173–183, 2010.

[6] T. Liu, C.-C. Chen, and L. Milor, "Comprehensive Reliability-Aware Statistical Timing Analysis Using a Unified Gate-Delay Model for Microprocessors," IEEE Trans. on Emerging Topics in Computing, 2016.

[7] W. Wang, V. Reddy, A. T. Krishnan, R. Vattikonda, S. Krishnan, and Y. Cao, "Compact modeling and simulation of circuit reliability for 65-nm CMOS technology," IEEE Trans. on Device and Materials Reliability, vol. 7, no. 4, pp. 509–517, 2007.

[8] Y. Wang, S. Cotofana, and L. Fang, "A unified aging model of NBTI and HCI degradation towards lifetime reliability management for nanoscale MOSFET circuits," IEEE/ACM Int. Symp. on Nanoscale Architectures, 2011, pp. 175–180.

[9] X. Li, J. Qin, and J. B. Bernstein, "Compact modeling of MOSFET wearout mechanisms for circuit-reliability simulation," IEEE Trans. on Device and Materials Reliability, vol. 8, no. 1, pp. 98–121, 2008.

[10] K. Yang and L. Milor, "Impact of stress acceleration on mixed-signal gate oxide lifetime," IEEE Int. Mixed-Signals Testing Workshop, 2015.

[11] C.-C. Chen and L. Milor, "System-level modeling and microprocessor reliability analysis for backend wearout mechanisms," Design, Automation & Test in Europe Conf. & Exhibition, 2013, pp. 1615–1620.

[12] G. S. Haase and J. W. McPherson, "Modeling of interconnect dielectric lifetime under stress conditions and new extrapolation methodologies for time-dependent dielectric breakdown," IEEE Int. Reliability Physics Symp., 2007, pp. 390–398.

[13] K.-Y. Yiang, H. W. Yao, and A. Marathe, "TDDB Kinetics and their Relationship with the E-and√ E-models," Int. Interconnect Technology Conf., 2008, pp. 168–170.

[14] T. Kauerauf, A. Branka, G. Sorrentino, P. Roussel, S. Demuynck, K. Croes, K. Mercha, J. Bommels, Z. Tokei, and G. Groeseneken, "Reliability of MOL local interconnects," IEEE Int. Reliability Physics Symp., 2013, pp. 2F.5.1-2F.5.5.

[15] NCSU, "FreePDK15." [Online]. Available: https://www.eda.ncsu.edu/wiki/FreePDK15:Contents.

[16] J.-W. Yang and J. G. Fossum, "On the feasibility of nanoscale triple-gate CMOS transistors," IEEE Trans. on Electron Devices, vol. 52, no. 6, pp. 1159–1164, 2005.

[17] NanGate, "NanGate FreePDK15 Open Cell Library." [Online]. Available: http://www.nangate.com/?page_id=2328.

[18] R. J. Segura and F. R. Feito, "An algorithm for determining intersection segment-polygon in 3D," Computers & Graphics, vol. 22, no. 5, pp. 587–592, 1998.

[19] C.-C. Chen, S. Cha, T. Liu, and L. Milor, "System-level modeling of microprocessor reliability degradation due to BTI and HCI," IEEE Int. Reliability Physics Symp., 2014, pp. CA.8.1-CA.8.9.

[20] "PrimeTime." [Online]. Available: https://www.synopsys.com/

[21] "Design Compiler." [Online]. Available: https://www.synopsys.com/

[22] "C.O.M.S.O.L. Multiphysics, Heat Transfer Module User's Guide Version 5.2, COMSOL," 2015.

[23] R. Kwasnick, A. E. Papathanasiou, M. Reilly, A. Rashid, B. Zaknoon, and J. Falk, "Determination of CPU use conditions," Int. Reliability Physics Symp., 2011, pp. 2C.3.1-2C.3.6.

[24] "Mibench benchmark." [Online]. Available: http://vhosts.eecs.umich.edu/mibench//.

[25] M. Bashir and L. Milor, "Towards a chip level reliability simulator for copper/low-k backend processes," Design, Automation and Test in Europe, 2010, pp. 279–282.

[26] M. Bashir, L. Milor, D. H. Kim, and S. K. Lim, "Methodology to determine the impact of linewidth variation on chip scale copper/low-k backend dielectric breakdown," Microelectronics Reliability, vol. 50, no. 9, pp. 1341–1346, 2010.

[27] L. Milor and C. Hong, "Area scaling for backend dielectric breakdown," IEEE Trans. on Semiconductor Manufacturing, vol. 23, no. 3, pp. 429-441, 2010.

978-1-5386-3775-3/18 $31.00 © 2018 IEEE

2018 IEEE 36th VLSI Test Symposium (VTS)

On-Line Monitoring and Error Correction in Sensor Interface Circuits Using Digital Calibration Techniques

Sascha Heinssen, Theodor Hillebrand, Maike Taddiken, Steffen Paul and Dagmar Peters-Drolshagen

Institute of Electrodynamics and Microelectronics (ITEM.me)

University of Bremen, Bremen, Germany, +49 421/218-62535

Email: {heinssen, hillebrand, taddiken, steffen.paul, peters}@me.uni-bremen.de

Abstract—**The functionality of modern integrated circuits is affected by numerous non-ideal effects. Especially during the design of analog circuit components, much effort is invested in compensating these effects by various methods. In this work, an alternative approach is presented, which is based on digital calibration techniques and used in a sensor interface circuit. A suitable test signal is interleaved with the sensor signal, whereby the interface is permanently monitored. Static and dynamic errors are detected in this way without interrupting normal sensor activities. Afterwards, these errors are determined and corrected in the digital domain by adaptive filters. Hereby, an on-line error correction is realized, which allows the sensor interface to react to changing ambient conditions and makes it more robust. Furthermore, the design requirements of the analog circuit components can be drastically reduced since not all errors need to be fully compensated in the analog domain. The introduced approach is demonstrated by correcting different gain- and offset-errors in a 65 nm CMOS sensor interface.**

Index Terms—**sensor interface, on-line testing, on-line error correction, monitoring, digital calibration**

I. INTRODUCTION

Numerous safety-critical applications (e.g. aerospace, automotive) require the precise measurement of various physical quantities. In order to perform this measurement and the following signal processing, the components of the utilized sensor interface circuit need to be highly accurate. Unfortunately, several non-ideal effects as well as environmental influences affect the functionality of these components. This is especially true for the analog part of the system. Therefore, a lot of work is spent on the design of precise and reliable analog circuit components to conquer typical non-idealities such as device mismatch, charge injection, clock feedthrough, noise as well as process variation, the dependency on supply voltage or temperature and aging (PVTA influences) [1]. Usually, the effects resulting from these non-idealities are counteracted by different design techniques, circuit topologies or additional components. While complex topologies for operational amplifiers or common-mode feedback structures are used to compensate offset-errors and noise [2], [3], capacitor arrays can compensate mismatches between the capacitors of the interface [4]. In switched-capacitor (SC) circuits, techniques like correlated double sampling or chopper stabilization can also be applied to suppress offset-errors and noise [5]–[7]. Furthermore, process variation as well as the dependency on

temperature and supply voltage can be counteracted by special layout techniques or by a complementary circuit design [1], [8], [9]. But even though all of these methods are very effective in compensating different errors, they significantly increase the design time and costs of the whole sensor system and it can not be guaranteed that the interface works properly in changing environmental conditions. For these reasons, a completely different approach is chosen in this work, which is based on digital calibration [10].

With this approach, errors are no longer completely compensated in the analog domain but detected and corrected digitally in a two-step method. Firstly, a suitable test signal is fed into the analog part of the interface circuit to make different errors visible. Subsequently, these errors are determined and corrected to a certain extent by digital signal processing algorithms and adaptive filters. As a consequence, not all errors need to be compensated directly in the analog domain and the design requirements of the analog circuit components can be relaxed. Since the error determination and correction is operating continuously, it is not only possible to correct static non-idealities but also time-variant errors resulting from temperature dependency or aging with only a low overhead on digital circuitry. Hereby, the whole sensor system gains the ability to monitor itself and becomes more robust.

The proposed concept of on-line error detection and correction is demonstrated in this work by considering a capacitive sensor interface circuit. This interface suffers from typical non-ideal effects, which lead to different linear gain- and offset-errors. In order to detect these errors, a sinusoidal test signal is transmitted through the interface alternating with the actual sensor signal by using time-division multiplexing. In the digital domain, variations of the test signal are evaluated and the required correction parameters are determined. Adaptive filters are used afterwards to eliminate present gain- and offset-errors from the sensor signal.

This work is structured as follows. Section II describes the sensor interface circuit and the non-ideal effects, which can occur inside the circuit components. In Section III the procedure of error detection and correction is presented as well as the applied digital algorithms. Section IV provides different simulation results to illustrate the performance of the proposed approach. The final Section V concludes this paper.

978-1-5386-3775-3/18 $31.00 © 2018 IEEE

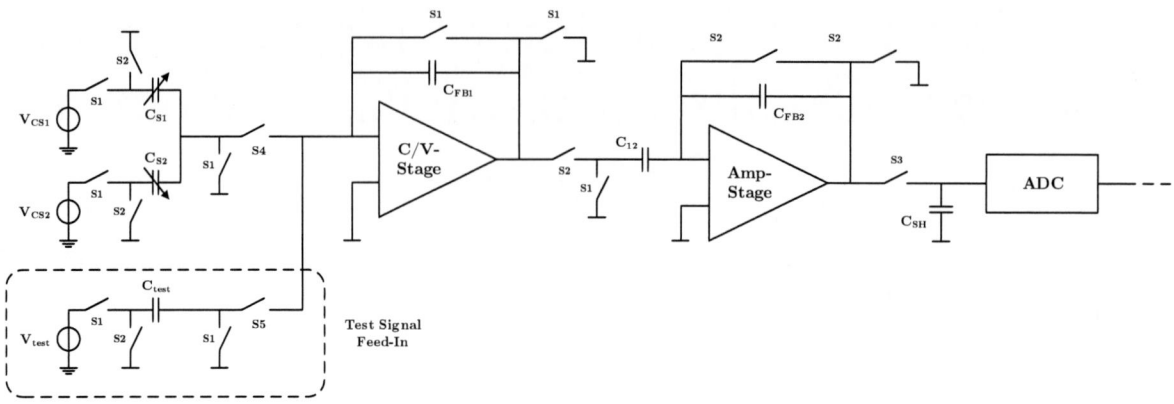

Fig. 1. Analog part of the sensor interface with electrical model of a capacitive sensor, C/V-stage, amplifier stage, ADC and the test signal feed-in, which is used for error detection.

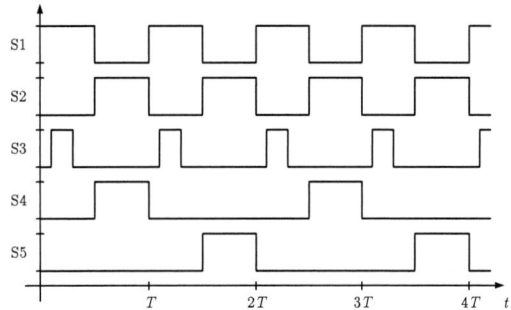

Fig. 2. Clock-phases of the different SC-switches inside the sensor interface.

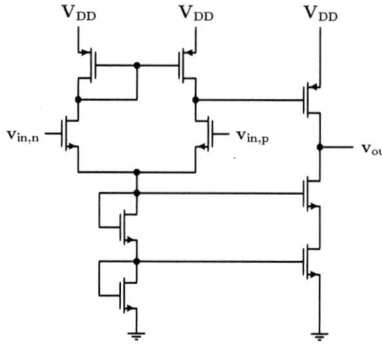

Fig. 3. Schematic of the operational transconductance amplifier, which is used to implement the C/V-stage and the amplifier stage.

II. SENSOR INTERFACE CIRCUIT

A. Topology and Function

A capacitive sensor interface circuit is analyzed in this work and used to illustrate the approach of on-line error detection and correction. The topology of its analog part is shown in Fig. 1 together with the electrical model of a differential capacitive sensor. Moreover, the test signal feed-in is shown, which is needed for error detection and will be described in Section III. The interface is implemented in SC-technology and based on comparable interface circuits, which are typical for evaluating the output signal of a capacitive sensor [4], [6]. Its switches are controlled by the different clock phases presented in Fig. 2. While the clock phases $S1$, $S2$ and $S3$ are required to operate the sensor interface, the clock phases $S4$ and $S5$ are only needed to feed in the test signal. If the test signal feed-in is non-existent, these two clock phases are no longer necessary and the clock phase $S4$ is replaced by $S2$.

Initially, the sensor interface is considered without the test signal feed-in. It consists of a Capacitance-to-Voltage converter (C/V-stage), a second amplifier stage and an Analog-to-Digital Converter (ADC). The input signal of the interface is generated by the electrical model of the capacitive sensor [11]. In this model, the two capacitances C_{S1} and C_{S2} represent the variable capacitances of the sensor element. They are charged and discharged periodically by two voltage sources, which

apply opposite square-wave signals to their outer electrodes in the clock phase $S1$. If there is a difference between the two sensor capacitances, this difference leads to a charge imbalance at the middle electrode of C_{S1} and C_{S2}. The resulting charge is then transferred to the C/V-stage of the interface in clock phase $S2$ and converted into an equivalent voltage by the operational amplifier (Op-Amp). In the following clock phase $S1$, this voltage is further processed by the second amplifier stage while the capacitances around the C/V-stage are resetted. The output voltage of the second stage is finally stored on the sample-and-hold capacitance C_{SH} during clock phase $S3$ before it is digitized by the ADC.

The analog part of the interface circuit is implemented in a 65 nm low power CMOS process with a supply voltage of 1 V. A two-stage operational transconductance amplifier (OTA) is used to implement the Op-Amps of the C/V- and amplifier stage. The schematic of this OTA is shown in Fig. 3. Due to its simple topology, it suffers from many non-idealities and is therefore well-suited to show the performance of the on-line error correction. The OTA also offers the advantage of consuming only 185 nW of power. Furthermore, the switches of the interface are realized by transmission gates and all voltage sources as well as the ADC are assumed to be ideal since these components are not analyzed in this paper.

978-1-5386-3775-3/18 $31.00 © 2018 IEEE   134

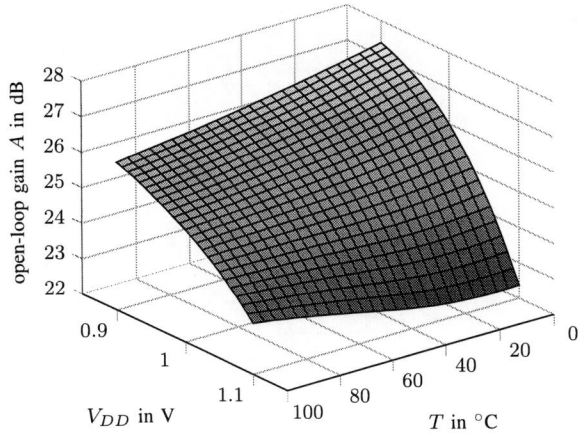

Fig. 4. Open loop gain A of the utilized operational amplifiers depending on the temperature T and the supply voltage V_{DD}.

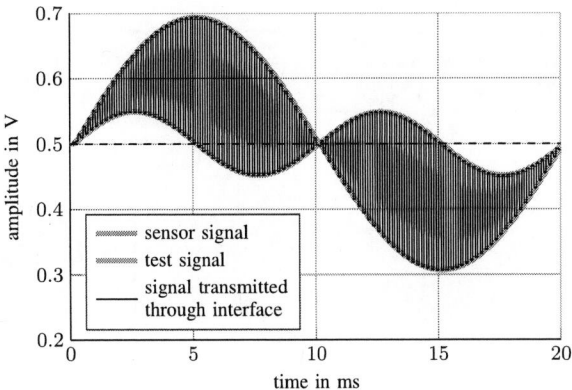

Fig. 5. Sensor signal, test signal and signal that is transmitted through the sensor interface. For the sake of clarity only the clock phase $S2$ is shown.

B. Non-Idealities

As already mentioned, numerous non-ideal effects can affect the behaviour of the interface components and need to be considered during circuit design. Important non-idealities that mainly influence the Op-Amps are VTA influences, which can lead to gain- and offset-errors or changes in the transfer characteristics of the Op-Amps. The reason for this is that numerous transistor parameters (e.g. threshold voltage, charge carrier mobility) are dependent on the temperature or the supply voltage [12]. Aging effects like Negative Bias Temperature Instability (NBTI) or Hot Carrier Injection (HCI) can also alter the transfer function of the Op-Amps and need to be considered as well [13]. The open-loop gain of the two Op-Amps used in this work is displayed in Fig. 4. It is shown depending on the temperature and the supply voltage for a temperature range from $0°\,C$ to $100°\,C$ and a voltage range from $0.9\,V$ to $1.1\,V$.

Additional non-idealities concerning the switches of a SC-circuit are charge injection and clock feedthrough. Both effects can occur when a switch is opening and the transistors inside the switch are turning off. At this time, the charge that is stored in the conductive channel of the transistor is transferred into the signal path. If a capacitor is connected to the output of a switch, a portion of this charge is transferred onto this capacitor and added to the stored signal value. Afterwards, this offset remains in the signal path and distorts the transmitted sensor signal.

Further non-ideal influences are charge leakage and device mismatch. Both influences can appear at all transistors and capacitors or between them. While charge leakage usually results in negative offset-voltages, device mismatch can lead to offset-voltages as well as gain-errors.

III. ON-LINE MONITORING AND ERROR CORRECTION

For the on-line detection and correction of the aforementioned non-ideal effects, a novel concept based on digital calibration is developed. In this approach, a well-known test signal is used to detect different linear gain- and offset-errors

before these errors are corrected digitally by adaptive filters. A similar approach using a correlation-based error correction was already presented in [14] but had several drawbacks since the test signal was directly added to the sensor signal. Thus, the sensor signal was distorted by the test signal and present errors were even increased if the correction parameters were not chosen correctly. These drawbacks are eliminated in this work by permanently separating the test signal from the sensor signal and evaluating only the well-known test signal to determine linear gain- and offset-errors.

The complete procedure of error detection and correction can be separated into three parts: 1) the error detection in the analog domain, 2) the determination of present errors and correction of the test signal in the digital domain and 3) the final digital correction of the sensor signal.

A. Error Detection in the Analog Domain

The error detection in the analog domain has to fulfill several requirements. First of all, a suitable and well-known test signal is needed to make present errors visible and to identify them correctly. Furthermore, the test signal must neither distort the sensor signal nor interrupt normal sensor activities. The first requirement is fulfilled by using a sinusoidal test signal in this work, which is generated by an additional and configurable voltage source. However, other well-known signals can be used as well, e.g. a pseudo random noise sequence. As it can be seen from Fig. 1, the test signal is fed into the sensor interface by a capacitance C_{test}, whose value equals the nominal value of the sensor capacitances C_{S1} and C_{S2}.

In order to fulfill the other requirements on error detection, the switching scheme of the SC-circuit is modified and the two clock phases $S4$ and $S5$ are added to realize time-division multiplexing. From Fig. 2, it can be seen that the clock phases $S4$ and $S5$ are both synchronized with $S2$ but only have half of its frequency. In addition, the clock phase $S5$ has an offset of half a period compared to $S4$. Relating to the interface circuit in Fig. 1, this means that the sensor signal is fed into the interface during the clock phase $S4$ while the test signal is fed into the interface during $S5$. Since a clock phase $S1$ follows both phases $S4$ and $S5$, the capacitances

978-1-5386-3775-3/18 $31.00 © 2018 IEEE   135

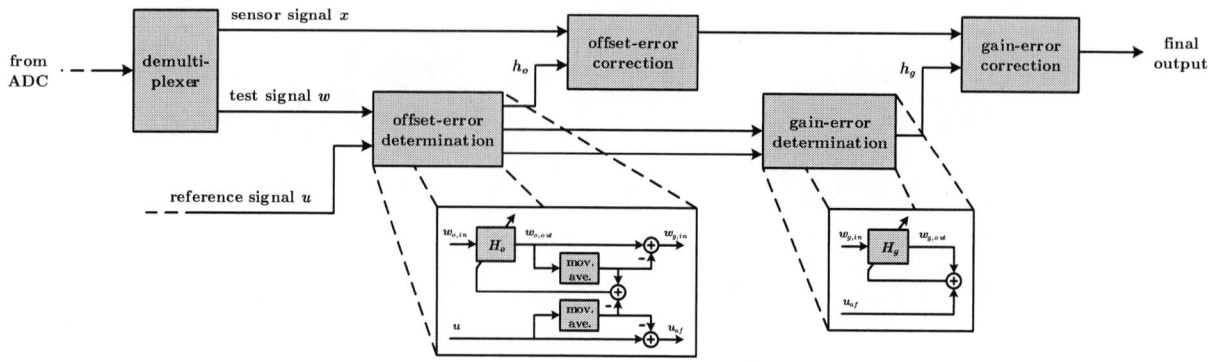

Fig. 6. Block diagram of the digital on-line error determination and correction.

around the C/V-stage are always resetted after a signal value was fed into the interface. Thus, the sensor signal and the test signal can be transmitted through the interface separately and without interfering each other. The signal that is created in this way is shown in Fig. 5 for the case of a sinusoidal sensor signal and a sinusoidal test signal with one forth of the amplitude and twice the frequency of the sensor signal. With this method, a continuous error detection is realized fulfilling the requirements named above. However, the drawback of this approach is that the effective sampling rate of the interface is reduced by one half. But this problem can be solved by increasing the overall sampling frequency of the interface, which is possible in many applications due to the relatively low sampling frequencies in most capacitive interfaces [4], [6].

B. Error Determination and Correction of the Test Signal in the Digital Domain

After the sensor signal x and the test signal w have passed the sensor interface, the present gain- and offset-errors can be determined in the digital part of the circuit. In order to perform this task, the signals x and w are separated again by a demultiplexer and are further transmitted on different signal paths. Afterwards, the test signal w is compared to a reference signal u, which corresponds to the output signal of the error-free interface circuit with the test signal as an input. Thus, differences between the actual and the nominal behavior of the interface can be detected and errors can be determined. Fig. 6 illustrates the complete process of on-line error determination and correction.

The first step in this process is the determination of offset-errors. For this purpose, the moving averages of the test signal and the reference signal are continuously calculated and compared. The difference between these two values is the present offset-error. In order to determine and to correct it, an adaptive filter H_o is inserted into the signal path of the test signal. Due to this and the use of a suitable adaptation algorithm, it becomes possible to permanently adapt the test signal and thereby eliminate the offset-error from the test signal. Since it can be implemented very efficiently, the least-mean-square (LMS) algorithm is utilized in this work [15]. With this algorithm, a filter coefficient h_o is calculated, which

corresponds to the present offset-error. Starting from its initial value of zero, h_o is updated with every test signal value by the update equation:

$$h_o(n+1) = h_o(n) + \mu_o \cdot (\overline{w_{o,out}}(n) - \overline{u}(n)) \cdot w_{o,in}(n), \quad (1)$$

where μ_o is the step-size of the algorithm, $\overline{w_{o,out}}$ the moving average of the test signal after the adaptive filter H_o, $\overline{u}$ the moving average of the reference signal and $w_{o,in}$ the current test signal value before it is filtered by H_o. Inside the adaptive filter H_o, the filter coefficient h_o is used to correct the test signal by the formula:

$$w_{o,out}(n) = w_{o,in}(n) - h_o(n). \quad (2)$$

With this approach, the offset-error in the test signal is continuously decreasing after the error determination process has started. After a certain amount of time, the offset-error is completely eliminated from the test signal. At this point, the present offset-error is fully determined and the correct filter coefficient is found.

Following the detection of offset-errors, linear gain-errors are determined. Since the algorithm used for this task only works correctly with average-free signals, it must be ensured that the test signal is zero-mean after the offset-error determination. Therefore, the moving average of the test signal, which is calculated to determine the offset-error, is subtracted from the test signal. The moving average of the resulting signal is calculated again afterwards and used to set an offset-flag if the mean value of the test signal exceeds a pre-defined threshold. Only after this flag is resetted, it can be guaranteed that the test signal is nearly zero-mean and the determination of linear gain-errors starts.

Basically, the gain-error determination follows the same procedure as the offset-error determination. An adaptive filter H_g is also inserted into the test signal path and used to correct gain-errors by a filter coefficient h_g. But in contrast to the previous case, the present gain-error is calculated by comparing the current value of the average-free test signal with the current value of the average-free reference signal. In order to correct this error, the LMS-algorithm is utilized again. Its update equation is now given by:

$$h_g(n+1) = h_g(n) + \mu_g \cdot (w_{g,out}(n) - u_{af}(n)) \cdot w_{g,in}(n), \quad (3)$$

978-1-5386-3775-3/18 $31.00 © 2018 IEEE   136

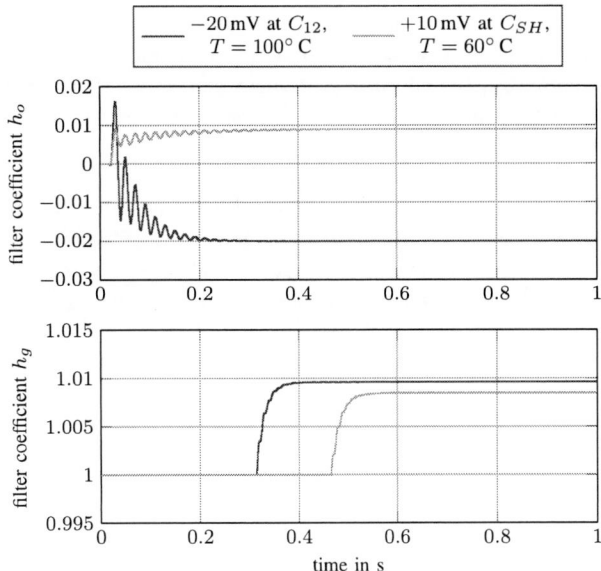

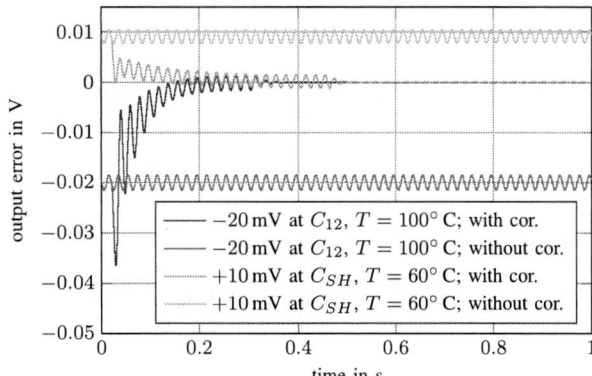

Fig. 8. Error in the digital output signal of the sensor interface for two different static gain- and offset-errors with and without error correction.

Fig. 7. Filter coefficients h_o and h_g for two different static gain- and offset-errors, which result from charge leakage, charge injection and temperature dependency.

where μ_g is the step-size of the algorithm, $w_{g,in}$ the current value of the average-free test signal before, $w_{g,out}$ the current value of the average-free test signal after the filter H_g and u_{af} the current value of the average-free reference signal. The correction of the test signal inside H_g is performed by the equation:

$$w_{g,out}(n) = h_g(n) \cdot w_{g,in}(n). \qquad (4)$$

At the beginning of the gain-error determination, the initial value of h_g is one. It is then continuously adapted as in the case of the offset-error detection until the correct filter coefficient h_g is found. Afterwards, the process of error determination is continuing and any further variations of the gain- or offset-errors are monitored. Consequently, this approach can be applied to determine and correct static gain- and offset-errors as well as time-varying ones.

C. Digital Error Correction of the Sensor Signal

The two filter coefficients h_o and h_g, which are determined by evaluating the test signal w, can not only be used to correct the test signal but also to correct the sensor signal x. For this reason, two more adaptive filters are added to the digital part of the sensor interface and inserted into the signal path of the sensor signal. These two filters use the coefficients h_o and h_g again to continuously adapt the sensor signal and thus eliminate any gain- and offset-errors from it. The signal correction inside these filters is performed by the equations (2) and (4), whereby the test signal w is replaced by the sensor signal x.

IV. SIMULATION RESULTS

The performance of the introduced concept of on-line error determination and correction is demonstrated by two different simulation scenarios. In both scenarios, the capacitive sensor

is excited by an external sinusoidal acceleration, which leads to an equivalent sinusoidal input voltage with an amplitude of 200 mV and a frequency of 50 Hz. In order to detect any errors, a sinusoidal test signal with an amplitude of 50 mV and a frequency of 100 Hz is used. Both signals are shown in Fig. 5. Furthermore, the utilized ADC has a resolution of 14 Bit, which corresponds to a voltage resolution of 61.035 μV in the digital part of the interface.

In the first scenario, the sensor interface is assumed to suffer from two different static gain- and offset-errors. In the first case, charge leakage leads to a constant negative offset-error of 20 mV at the capacitance C_{12} and the ambient temperature of the interface is set to 100° C. The result is an additional gain-error since the open-loop gain of the Op-Amps is altered according to Fig. 4. In the second case, charge injection causes a positive offset-error of 10 mV at the capacitance C_{SH} and the ambient temperature of the interface is set to 60° C.

After an initialization time, which is needed to calculate the moving average of the test signal, the LMS algorithm starts to adapt the filter coefficients h_o and h_g. The resulting convergence process is shown in Fig. 7 for the two static gain- and offset-errors. It can be seen that in both cases the errors are determined correctly within a simulation time of less than 1 second. It can also be recognized that initially only the offset-error is corrected and no gain-error correction is performed until this error is fully determined. Afterwards, the gain-error correction starts and the value of h_g tends towards the existing gain-error. Although the detected errors are slightly different from the assumed error values, they are determined correctly since differences between them result from the nominal transfer characteristics of the interface circuit.

The performance of the on-line error correction is also shown by considering the error in the digital output signal of the sensor interface. This error is calculated by comparing the output signal of the interface with an error-free reference signal. It is illustrated in Fig. 8 for the two static gain- and offset-errors with and without error correction. It becomes clear that the present gain- and offset-errors are determined correctly and that they can be drastically reduced by the on-line error correction. With this approach, the error in the output

978-1-5386-3775-3/18 $31.00 © 2018 IEEE

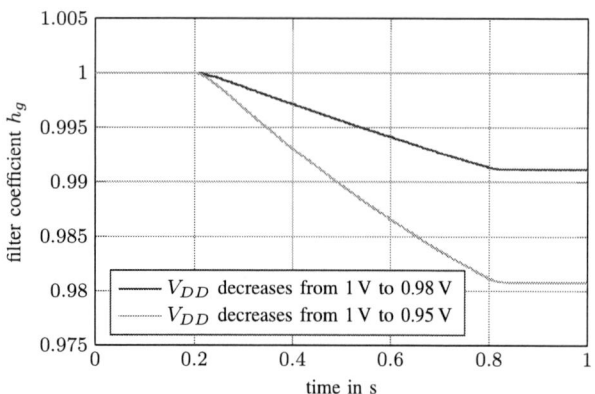

Fig. 9. Filter coefficient h_g for two different time-variant gain-errors, which result from a linearly decreasing supply voltage.

signal of the interface is decreased to a value of ± 2 least significant bits (LSB) in both cases.

In the second simulation scenario, the on-line error determination and correction is used to eliminate time-variant gain-errors. For this reason, the supply voltage of the sensor interface is assumed to decrease linearly over time in two cases, which leads to variations of the Op-Amps open-loop gain. A voltage drop of $20\,\text{mV}$ is assumed in the first case and a voltage drop of $50\,\text{mV}$ in the second case. In both cases, the voltage starts to decrease after a simulation time of 0.2 seconds and ends to decrease at a simulation time of 0.8 seconds.

Since no offset-errors are assumed in this scenario, only the filter coefficient h_g is considered. It is depicted in Fig. 9 over the simulation time of 1 second. It can be seen that in both cases the filter coefficient h_g remains at its initial value until the supply voltage starts to decrease. Subsequently, h_g is continuously re-adapted to track the present gain-error until the supply voltage reaches a stationary value again. The gain-error is therefore determined and eliminated correctly at all times. Moreover, the error in the output signal of the interface do not exceed a value of ± 3 LSB during the complete simulation.

V. CONCLUSION

A concept of on-line error determination and correction is presented in this work, which continuously monitors a sensor interface circuit and eliminates existing and arising linear gain- and offset-errors. Time-division multiplexing is used to insert a test signal into the analog part of the sensor interface. In the digital domain, distortions of the test signal are evaluated and two error correction parameters are determined. These parameters are used afterwards to eliminate different errors, which are typical for their appearance in sensor interface circuits and are usually conquered by analog compensations methods. Since the correction parameters are updated continuously, static as well as time-variant errors can be corrected. Furthermore, the gain- and offset-errors are corrected independently of their origin. Thus, the proposed approach is not limited to sensor interface circuits, but can also be applied to other mixed-signal systems.

The successful application of the developed on-line error correction is shown in two different simulation scenarios. It is demonstrated that combined static gain- and offset-errors can be fully determined and corrected in less than 1 second and that time-variant errors can be tracked and eliminated correctly. Moreover, the error in the digital output signal of the sensor interface is not significantly increased and kept in a range of ± 2 LSB for static errors and ± 3 LSB for dynamic errors. It can therefore be concluded that the used on-line error correction is well-suited to monitor a sensor interface and to correct various gain- and offset-errors.

ACKNOWLEDGMENT

This research project (RoMulus) is supported by the Federal Ministry of Education and Research, Germany, under reference number 16ES0362-72K.

REFERENCES

[1] R. J. Baker, *CMOS Circuit Design, Layout, and Simulation*, 3rd ed. Wiley-IEEE Press, Sep 2010.

[2] M. Lemkin and B. E. Boser, "A Three-Axis Micromachined Accelerometer with a CMOS Position-Sense Interface and Digital Offset-trim Electronics," *IEEE Journal of Solid-State Circuits*, vol. 34, no. 4, pp. 456–468, Apr 1999.

[3] V. Petkov, G. Balachandran, and J. Beintner, "A Fully Differential Charge-Balanced Accelerometer for Electronic Stability Control," *Solid-State Circuits, IEEE Journal of*, vol. 49, no. 1, pp. 262–270, Jan 2014.

[4] M. Kamarainen, M. Saukoski, M. Paavola, J. A. M. Jarvinen, M. Laiho, and K. A. I. Halonen, "A Micropower Front End for Three-Axis Capacitive Microaccelerometers," *IEEE Transactions on Instrumentation and Measurement*, vol. 58, no. 10, pp. 3642–3652, Oct 2009.

[5] H. Kulah, N. Yazdi, and K. Najafi, "A CMOS Switched-Capacitor Interface Circuit for an Integrated Accelerometer," in *Proceedings of the 43rd IEEE Midwest Symposium on Circuits and Systems*, vol. 1, 2000, pp. 244–247.

[6] W. Bracke, P. Merken, R. Puers, and C. V. Hoof, "Ultra-Low-Power Interface Chip for Autonomous Capacitive Sensor Systems," *IEEE Transactions on Circuits and Systems I: Regular Papers*, vol. 54, no. 1, pp. 130–140, Jan 2007.

[7] H. L. Chen, P. S. Chen, and J. S. Chiang, "A Low-Offset Low-Noise Sigma-Delta Modulator with Pseudorandom Chopper-Stabilization Technique," *IEEE Transactions on Circuits and Systems I: Regular Papers*, vol. 56, no. 12, pp. 2533–2543, Dec 2009.

[8] Y. Zhang and J. S. Yuan, "CMOS Transistor Amplifier Temperature Compensation: Modeling and Analysis," *IEEE Transactions on Device and Materials Reliability*, vol. 12, no. 2, pp. 376–381, June 2012.

[9] Y. Wang, P. K. Chan, and K. H. Li, "A Compact CMOS Ring Oscillator with Temperature and Supply Compensation for Sensor Applications," in *2014 IEEE Computer Society Annual Symposium on VLSI*, July 2014, pp. 267–272.

[10] B. Murmann, "Digitally Assisted Data Converter Design," in *ESSCIRC (ESSCIRC), 2013 Proceedings of the*, Sept 2013, pp. 24–31.

[11] X. Li and G. C. M. Meijer, "An Accurate Interface for Capacitive Sensors," *IEEE Transactions on Instrumentation and Measurement*, vol. 51, no. 5, pp. 935–939, Oct 2002.

[12] K. Tscherkaschin, T. Hillebrand, M. Taddiken, S. Paul, and D. Peters-Drolshagen, "Degradation and Temperature Analysis of Voltage-Controlled Ring Oscillators for Robust and Reliable Oscillator Designs in a 65nm Bulk CMOS Process," in *2016 MIXDES*, June 2016, pp. 353–358.

[13] T. Hillebrand, M. Taddiken, K. Tscherkaschin, S. Paul, and D. Peters-Drolshagen, "Online Monitoring of NBTI and HCD in Beta-Multiplier Circuits," in *2016 IEEE 22nd International Symposium on On-Line Testing and Robust System Design (IOLTS)*, July 2016, pp. 209–210.

[14] S. Heinssen, N. Hellwege, N. Heidmann, S. Paul, and D. Peters-Drolshagen, "Robust Digital Calibration Engine for MEMS Inertial Sensor Systems," in *2015 IEEE SENSORS*, Nov 2015, pp. 1–4.

[15] S. Haykin, *Adaptive Filter Theory*, 5th ed. Addison Wesley Pub Co Inc, 2013.

978-1-5386-3775-3/18 $31.00 © 2018 IEEE

Special Session on BIST/Calibration of A/MS devices

Hans-Mart von Staudt, Dialog Semiconductor, Germany
James Izon, Texas Instruments, High accuracy trim, USA
Sule Ozev, Arizona State University, USA
Peter Sarson, Dialog Semiconductor, UK (organizer)

I. Introduction

This special session will focus on new ways of performing calibration and test of on chip circuits that historically were performed on ATE. The presentation in this special session will demonstrate how these new test and calibration techniques help to reduce cost and increase the quality of semiconductor shipped into the field.

II. Trimming: The Challenge for Yield and Test Cost (Hans-Martin von Staudt)

Ever increasing accuracy and precision requirements push designers often beyond the capability of their circuits because of manufacturing induced variability. Digital trimming, where a trim DAC provides a compensation value to adjust the circuit under trim, looks to be a simple and attractive solution to the problem. It defers the production of accuracy and precision from the fab to the test floor. In theory, wide distributions can be trimmed to become narrow. However, various error sources prevent the expected narrow rectangular distribution. Noise and trim DAC non-linearity can cause distribution tails that in turn may cause yield loss, as the desired trim result cannot be achieved. The presentation shows the error sources and mitigation techniques. Besides the technical challenge trimming comes organizationally at a cost. Not only does it require test time. From an engineering management perspective, trimming only defers the engineering efforts from Design to Test Engineering, where usually less resource is available. Self-trimming is a technique to address both issues. In addition, it can be turned into built-in self-test.

III. High Accuracy Trim, Calibration and Testing of Integrated On-Chip Op-Amps (James Izon)

Achieving very high performance in motor control applications has traditionally required expensive on-board op-amps to meet low offset, high gain accuracy system requirements. To achieve performance near this standard for an integrated programmable gain amplifier (PGA) solution requires utilizing software correction which in turn adds noticeable latency creating substantial stability issues for real-time control. A new design, using analog trim of the PGA offset and gain along with a new interpolation based gain trim architecture can achieve an embedded amplifier design requiring no software overhead while enabling low offset and high gain accuracy across PVT. The introduction of such high performance PGAs into a system-on-chip (SoC) necessitates stringent testing to guarantee design spec, but must also maintain low impact to test cost, reflecting in the end cost per PGA. This presentation will cover the trim and test infrastructure developed to achieve these high performance metrics with low cost manufacturing requirements.

IV. Dynamic Testing and Trimming for Embedded DC-DC Converters (Sule Ozev)

Embedded power converters are generally not tested for their dynamic response. This process may miss some defects that affect the overall performance of the converters. Built-in self-test is a viable option to ensure high quality products. BIST also helps with in-field calibration as the devices ages to extend the lifetime. For dynamic testing, DC-DC converters need to be placed in their normal operation mode in closed loop. Small perturbations can be used to approximate the linearized transfer function, or large perturbations can be used to approximate the step response. Linear transfer function provides a way to approximate the dynamic characteristics as well as the available margin and therefore is preferred. This can be done with a sine-wave excitation and sweeping the frequency or with the excitation of a noise-based stimulus. The talk will be focused on the implementation, at the board level of using noise-based stimulus and experimental data obtained from commercially available parts

Innovative Practices on Machine Learning for Emerging Applications

Kareem Madkour, Mentor, A Siemens Business, Egypt
Zhaobo Zhang, Huawei Technologies Co. Ltd, USA
Alfred L. Crouch, Peter L. Levin, and Eve Hunter, Amida Technology Solutions, Inc., USA
Yu Huang, Mentor, A Siemens Business, USA (Organizer)

I. INTRODUCTION

The IP session focuses on using Machine Learning (ML) techniques on several emerging applications. The first contribution discusses hotspot detection by using ML. The second presentation then talks a data-driven health monitoring solution. The last contribution discusses using ML to emulate hardware Trojans.

II. OVERCOMING THE CHALLENGES OF HOTSPOT DETECTION USING DEEP LEARNING (KAREEM MADKOUR)

Identifying layout hotspots during the design stage is an important yet a challenging verification step in the state-of-art technologies. Because of its success in the image based classification applications, Machine Learning (ML) promises an alternative that is faster than rigorous model-based simulation and can extend the prediction capability beyond known hotspots in the training set. The problem of hotspots detection is different from standard classification problems because the hotspot patterns are very few relative to the good patterns. And because of this highly unbalanced "needle-in-the-hay" problem, it is challenging to maintain high accuracy and low false-alarm.

Another challenge specific to the problem of hotspots detection is the very small variations between hotspot and non-hotspot patterns. This challenge requires a high fidelity feature representation so that small differences are detectable.

In this presentation we propose a framework that trains a Deep Neural Network ML-system for hotspot detection based on a training set of known hotspots and good patterns that accommodates for the high imbalance between hotspots and good patterns. We will use this system to predict new hotspots that were not part of the training set. We also propose a method to enrich the training set with synthetically generated patterns to improve the accuracy. Our results outperform previous reported results for accuracy and false-alarm rate.

III. DATA-DRIVEN HEALTH MONITORING SOLUTION FOR NETWORK DEVICES (ZHAOBO ZHANG)

Network outage has significant impact in modern internet-connected world. The demand of high reliability and availability on network is ever-increasing from carrier and enterprise. This work introduces the health monitoring research and practice we have done on network devices, e.g. core routers. The goal is to detect health issues by monitoring and analyzing system runtime data. Therefore, reliability improvement strategies can be applied in time, in order to prevent failures from occurring or minimize the root cause identification time. Our data-driven health monitoring solution includes data collection, data analysis, and feedback learning. Data under analysis are mainly time series. We describe common techniques in time series analysis like smoothing, auto regression, change point detection etc., and how those techniques are used for anomaly detection in the network domain. As for the self-learning capability, we describe feature extraction, training set preparation and the usage of supervised learning algorithms like random forest and support vector machines. Therefore previous knowledge can be learnt and integrated into the solution automatically. At the end, some real cases are demonstrated.

IV. DATA COLLECTION OF A TROJAN INSERTION HARDWARE EMULATOR FOR MACHINE LEARNING (ALFRED L. CROUCH, PETER L. LEVIN, AND EVE HUNTER)

Trojan insertion in a virtualized emulation environment against a golden model can be used to explore the behaviors and impacts of Trojans and their associated triggers in a similar way that defect/fault insertion into hardware can be used to generate a "failure modes and effects analysis" (FMEA). The Trojan behavior and effects analysis (TBEA) can be used to help harden a design under development (Design-for-Security); can be used to explore detection strategies and embedded detection instruments; and can be used to explore potential countermeasures. If the TBEA is accomplished in an automated fashion in an emulation environment with Trojan payloads treated as library elements to be inserted in key locations, then Trojan signatures can be learned and used to assess anomalous behavior on existing hardware. Symptom matching allows observations and data collection from real hardware to be input and compared to learned behaviors in the data analysis engine if the minimal goals of data collection and analysis are: detection, identification, location, threat impact, signature development, and detection/countermeasure recommendation. Amida has developed a proof-of-concept hardware-software appliance, has inserted Trojans, triggers and detection instruments, has collected data and is building the analysis engine.

IC Layout Weak Point Effectiveness Evaluation based on Statistical Methods

Fang Lin*, Ali Ahmadi[†], Kannan Sekar[†], Yan Pan[†] and Ke Huang*

*Department of Electrical and Computer Engineering, San Diego State University, San Diego, CA 92182
[†]GLOBALFOUNDRIES Inc., Malta, NY 12020

Abstract—Design hotspots, a.k.a. layout weak points are layout patterns that are susceptible to systematic failure and yield loss in high-volume manufacturing (HVM). Therefore, understanding the yield impact of layout weak point is a crucial step for yield learning. Layout weak points can be identified by layout analysis based on simulation or silicon learning from past failures. In advanced technology nodes, the interaction between design and manufacturing process is increasingly significant, making it paramount to quantify which layout weak points are causing yield loss for a specific manufacturing process. This can be achieved by collecting volume scan diagnosis results and then map the callouts to suspected weak points for defect root-cause analysis. One major challenge in this process is the often-false assumptions that correlation is causation, especially as scan diagnosis resolution is typically not ideal and random correlation hits can be common as the weak point count grows. In this work, we propose a novel approach for quantitatively assessing the impact of layout weak points on IC failure using statistical methods. We develop a new weak point effectiveness metric to help guide the decision on whether a specific weak point is a root cause in a population of volume scan diagnosis results. Experimental results show that our approach is able to interpret high count weak points for "meaningful" root-cause and rank the relative contribution of different weak point sets.

I. Introduction

The increasing complexity of modern semiconductor technology with aggressive scaling leads to various design robustness concerns. The high performance expectations lead to aggressive design techniques that are continuously pushing the limit of manufacturing process. To overcome the challenges faced in the modern integrated circuit (IC) design and manufacturing process, various design for manufacturability (DFM) techniques were proposed to improve the process at the design stage [1]–[3]. Typical DFM techniques include design rule enforcement, identification of minimum area of physical layers and single contact/via, lithography and chemical & mechanical polishing (CMP) hotspot detection and repair [4], [5], etc. The DFM-based process improvement typically passes through many silicon cycles before a final yield enhancement can be achieved.

During HVM, some layout patterns are susceptible to causing systematic failures and result to yield loss. These patterns are often called hotspots [6], a.k.a. weak points. Therefore, a commonly-used DFM approach is to continuously identify and evaluate potential high-risk weak points and try to mitigate them through both design changes and manufacturing process improvement. Defect-prone layout patterns such as

single contact/via and lithography hotspot can be identified by electronic design automation (EDA) layout analysis tools at the design stage, or by silicon learning from HVM historical failure data. Then, this information can be used in several ways: designers utilize it to improve the DFM features during design/manufacturing iterations by focusing improvement efforts only on the high-risk patterns; Process engineers can identify the systematic root-cause of these failure mechanisms.

Various techniques exist today to identify and simulate weak points at the design stage such as true physical modeling, fitted behavioral modeling, or rule-based techniques [3], etc. For a specific design, the yield impact of various weak points is usually assessed by correlating them to scan diagnosis call-out data collected from wafer or package level test. This correlation analysis allows us to establish a relationship between weakpoints and yield loss. When evaluating weak points for a specific design, the fundamental question on the "effectiveness" of these weak points was often raised. Indeed, not all weak points will result in an actual failure in manufacturing. Historical PFA and logic diagnosis data show that only a subset of weak points are correlated with diagnosis call-out data. Then the next question is: are the overlap between weak points and diagnosis call-out locations a result of random hits, or effective causation?

In this work, we propose a methodology to statistically assess the "effectiveness" of design weak points using silicon diagnosis data. We define the effectiveness of a weakpoint as the likelihood that the weakpoint is representing underlying causation for silicon failure. To do so, first we compute the hit rate between diagnosis call-outs and weak points. However, a simple hit count may be misleading as it heavily depends on the number of weak points. To address this, we introduce a statistical metric to take into consideration the possibility of random weakpoint hits. Our conjecture is that if the hits occur as a result of random coincidence, then shifting the weak points would not necessarily result in a decrease of hit number. On the other hand, "effective" weak points are expected to have a significant decrease in hit number when they are shifted. Moreover, we use random and bootstrapping resampling methods to assess the confidence level of our metric. Using a set of generated weak points and diagnosis callout data from HVM, we show that i) we can demonstrate that effective weak points behave differently from a set of randomly generated weak points, and ii) our metric can be

978-1-5386-3775-3/18 $31.00 © 2018 IEEE

used to rank the relative contribution of different weak point sets.

The remainder of this paper is organized as follows. In Section II, we discuss the background and rational of IC layout weak point identification. In Section III, we describe the proposed framework for weak point quality evaluation in detail. Experimental results from HVM data are shown in Section IV and the conclusion is drawn in Section V.

II. IC LAYOUT WEAK POINT IDENTIFICATION AND EFFECTIVENESS EVALUATION

A. Weak point identification methods

In an effort to ramp IC manufacturing yield, the weak points are often identified by EDA tools based on process simulation models or past silicon learning. These weak points are specific structures on the layout which are susceptible to causing manufacturing defects such as short or open circuits [7]. Several simulation methods such as true physical modeling, fitted behavioral modeling [3], or rule-based techniques [8] can be used to simulate these weak points. However, accurate modeling of yield killing weak points still remains a formidable challenge. It is not uncommon to find that a true defect is not covered by any modeled weak points when logic diagnosis/PFA is performed on a failed chip. In some cases, design rules are modified as new design marginalities are uncovered during yield ramp. One can perform diagnosis by locating logic failures using layout-aware diagnosis tools [9], and correlate these diagnosis call-out data to weak points to identify yield limiting factors.

B. Evaluation of weak point effectiveness

Assessing the likelihood that weak points can cause actual failure is crucial in the yield enhancement process. Layout weakpoints serve as hypothesis for potential systematic yield loss mechanisms. The sooner a definitive relationship can be established between layout weakpoints and a population of scan diagnosis results, the faster corrective actions can be taken, either to improve relative manufacturing process or to carry out targeted design modifications to avoid the occurence of the weakpoints in a design. In order to establish such relationship, a straightforward approach would be to compute the percentage of "hit" numbers among all diagnosis call-out locations, which is known as hit rate [5]. As an example, given a set S of M generated weak points for a specific design, and let N denote the total number of diagnosis defects from failed devices of a certain wafer/lot, we can define a hit rate value $hit_r = P/N$, where P is the number of diagnosis defects with defect locations overlaid with at least one weak point from the set S. For a set of failed devices under investigation, larger hit_r value signifies higher likelihood that the weak point set S is causing actual silicon failure.

While hit rate computation is a fast way to assess the causality of weak point set S for silicon failure, it may not be a reliable indicator and its absolute value cannot be properly interpreted by itself. Fig. 1 shows a scenario of two sets of weak points, namely S_1 and S_2, shown by the blue squares,

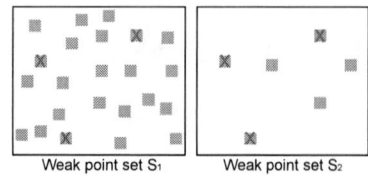

Fig. 1: Example of two weak point sets S_1 and S_2 with weak points shown by blue squares and diagnosis callout locations shown by red X signs.

and a set of diagnosis call-out locations superimposed on one device layout shown by the red X signs. It can be observed that the hit rate values hit_r for both sets S_1 and S_2 are 100%. However, these 2 sets differs significantly as the set S_1 contain higher number of potential locations. Moreover, it is difficult to assess if the hits from the set S_1 are a result of meaningful systematic defect caused by layout feature expressed by the weakpoint, or solely due to random hits due to the larger size of the weak point set.

To analytically compute the random hit rate of weak point set, we can assume a box of size $d_1 \times d_2$ as the bounding box of the entire device layout, a set S of M random weak points with the size $a_1 \times a_2$, and a set Q of N diagnosis callout locations with the size of the i-th location being $b_{i1} \times b_{i2}$. We can then estimate the probability of having exactly k hits between S and Q using Bernoulli trials:

$$P(k) = \binom{M}{k} p^k q^{M-k} \tag{1}$$

where p denotes the probability of success at each trial, i.e., the probability that the j-th sample in S hits at least one sample in Q: $p = (A_1 \cap \cdots \cap A_N)/(d_1 \times d_2)$ with A_i being the total possible hit area of the i-th sample in Q: $A_i = (b_{i1} + a_1) \times (b_{i2} + a_2)$, and q denotes the probability of failure at each trial: $q = 1 - p$. In the same way, we can compute the probability of having at least k hits $P_a(k)$ by adding up $P(l)$ for all $l \geqslant k$: $P_a(k) = \sum_{l=k}^{M} P(l)$. However, this straightforward approach has several limitations. For example, it assumes random weak point generation process has a uniform distribution across the entire chip layout, while there are always empty areas on the layout where no weak point would be generated or the scan test is not covering. In order to assess the effectiveness of weak points by taking into account all the practical aspects, we propose a shifting approach and we evaluate the effectiveness of the weak point based on statistical methods, which will be shown in the next section.

III. PROPOSED WEAK POINT EFFECTIVENESS EVALUATION SCHEME

The goal of the proposed scheme is to answer two fundamental questions in weak point quality evaluation: i) given a set S of M weak points for a particular design, how to interpret their effectiveness, i.e., "meaningful" root-causes for diagnosis call-out? ii) For R different weak point sets, $S_1, \ldots, S_R$, how do we quantitatively compare the effectiveness of these sets in

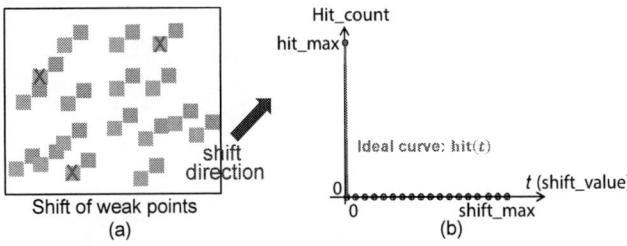

Fig. 2: Weak point shift method with (a) an example of the shift process and (b) ideal weak point efficiency curve.

terms of root-cause identification quality? In this section, we show the proposed scheme using a weak point shift approach and statistical methods to evaluate/compare weak point sets.

A. Weak point shifting method

Fig. 2(a) illustrates the weak point shifting method. We reuse the set S to denote the M original weak points, as shown by the blue squares in Fig. 2(a): $S = [(x_1, y_1), \ldots, (x_M, y_M)]$, where (x_j, y_j) denotes the Cartesian coordinates of the j-th sample in S. Let P denote the number of diagnosis call-out locations collected from scan diagnosis that are overlaid with at least one sample from S, as shown by the red X signs in Fig. 2(a). In order to assess the trustiness of these P hits, we propose to shift all samples in S by (x_{sft}, y_{sft}). Note that the range of (x_{sft}, y_{sft}) can be determined based on the metal pitch value (i.e., the center to center distance between the metals having minimum width and spacing) of a particular layer. The orange squares in Fig. 2(a) show the shifted weak points with shift direction shown by the blue arrow. We use S_{sft} to denote the new set of shifted weak points: $S_{sft} = [(x_1 + x_{sft}, y_1 + y_{sft}), \ldots, (x_M + x_{sft}, y_M + y_{sft})]$.

The rationale behind this shifting approach is: if the original set S is indeed representing a root-cause for the scan diagnosis call-out set, the P hits with the diagnosis call-out locations should be significantly higher than the hit count that can be obtained by overlaying the same diagnosis call-out set with a nonsensical set of locations of the same size M. By shifting the original set S by a small amount, we essentially create such a nonsensical set of locations. As the shift is relatively small (e.g., in um range), the shifted set should still land in a region of similar design style or test coverage as the original set S As we can see from Fig. 2(a), no red X sign is overlaid with the orange squares, which signifies a 0 hit number for the shifted set S_{sft}.

B. Weak point effectiveness metric

Based on the previous discussion on the impact of weak point shifting on the hit count, we can quantitatively define a new metric for evaluating the effectiveness of weak points. Fig. 2(b) illustrates the weak point effectiveness (WPE) metric by showing the ideal hit count curve expressed as a function of shift value t by assuming equal horizontal and vertical absolute shift values $t = |x_{sft}| = |y_{sft}|$. The x-axis in Fig. 2(b) denotes the shift value t, and the y-axis denotes the total number of hits between the diagnosis call-out set Q and the shifted weak point set S_{sft} with the shift amount t. As can be observed by the red curve in Fig. 2(b), the best scenario of this shifting process is to have the maximum number of hits hit_{max} when the shift value is 0. When the shift value t is larger than a user-defined tolerance value, e.g. the pitch value, then the ideal hit count should drop to 0. The ideal hit curve can be expressed as:

$$hit(t) = \begin{cases} hit_{max}, & t \leqslant Tol \\ 0, & t > Tol \end{cases} \tag{2}$$

where hit_{max} denotes the maximum achievable hit number for a given diagnosis call-out data set, and Tol denotes the tolerance considered when computing the hits. Based on the definition in (2), we can then evaluate the effectiveness of any weak point sets that produce the hit curve $hitn(t)$ by the following equation:

$$wpe = \int_0^{shift_{max}} |hit(t) - hitn(t)| dt \tag{3}$$

where wpe denotes the weak point effectiveness evaluation metric and $shift_{max}$ denotes the maximum shift value. Equation (3) allows us to quantitatively evaluate the efficiency of any weak point set by comparing its hit curve to the ideal one. Smaller wpe values signify higher effectiveness of weak point sets in revealing failure root causes. Obviously, any randomly generated weak point sets would result in considerably higher hit curve values even when t is much larger than the tolerance Tol.

C. Effectiveness evaluation for a given weak point set S

Using the weak point effectiveness metric defined in Equation (3), we can evaluate the effectiveness of a given weak point set S by comparing its wpe value to a random, nonsensical weak point set S_r. The efficiency of the set S can be inferred as how better it can perform in terms of wpe than a randomly generated set S_r. In other words, if the wpe value obtained using S is significantly lower than that obtained from S_r, then we can conclude that S is generated with meaningful root cause information as compared to a randomly generated set.

In order to generate a random set S_r for fair comparison, as mentioned before, we need to consider the practical constraints of the IC layout patterns. A uniform random weak point generation across the entire chip may not accurately represent the layout hotspot patterns as there are always empty areas on the layout where no weak point would be generated. Thus, we need to generate samples from locations susceptible to causing physical defects. For this purpose, we propose to generate new random samples in the proximity of original samples from S. The goal is to generate new samples with a distance from original points large enough to avoid duplicating points, while still in the proximity of original points. To achieve this, we use an exclusive/inclusive generation box as illustrated in Fig.

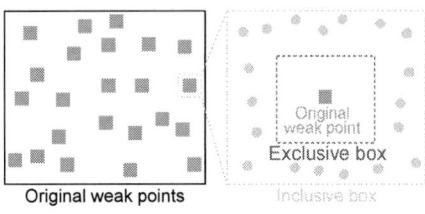

Fig. 3: Exclusive/inclusive box for random weak point generation.

3. As can be shown by the red and green dotted boxes on the right side of Fig. 3, for each original weak point, we generate new samples uniformly at random inside the inclusive box (green dotted box) by excluding all samples from the exclusive box (red dotted box). The new generated samples around one original point are shown by the green dots in Fig. 3. The sizes of exclusive/inclusive boxes can be set such that new points are not too close to the original ones (e.g. 2 pitches for exclusive box size), while they are still in the proximity of the original ones (e.g. 4 pitches for inclusive box size).

Algorithm 1 Generation of a new weak point set S_r of M_r samples from an original set S of M samples.

1: **procedure** GENERATION
2: /*Original weak point set S contains M samples*/
3: /*M_r is the desired sample number in the new set S_r*/
4: /*The widths of exclusive/inclusive boxes: d_{ex}/d_{in}*/
5:     $c \leftarrow 1$
6:     $i \leftarrow 1$
7:     $S_r \leftarrow \emptyset$
8:     **while** $c \leqslant M_r$ **do**
9:         **if** $i == M$ **then**
10:            $i \leftarrow 1$
11:         Generate a random point (x_c, y_c) around the i-th
12:         sample in S using $d_{ex} \times d_{ex}$ and $d_{in} \times d_{in}$
13:         $S_r \leftarrow \{S_r, (x_c, y_c)\}$
14:         $i = i + 1$
15:         $c = c + 1$
16:     **return** S_r
    end procedure

Algorithm 1 shows the pseudocode for generating random weak points. Specifically, it generates a new set S_r of M_r samples from an original set S of M samples. Once S_r is generated, the effectiveness of weak point set S can be evaluated by comparing its wpe value defined in Equation (3) to that obtained from S_r. If the wpe value of S is significantly lower than that obtained from S_r, then we can conclude that S has a higher effectiveness than a randomly generated set with its weak points more likely to reveal root cause failure mechanisms of a given design. Note that we can repeat the procedure outlined in Algorithm 1 N times to assess the robustness of the results. We can also apply statistical resampling approaches such as bootstrapping to resample the data set S_r K times with replacement and compute the corresponding mean and standard deviation of wpe values to

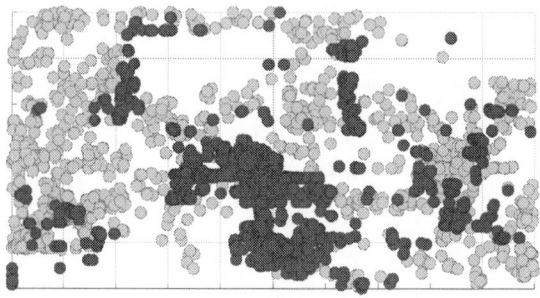

Fig. 4: Original weak points (green dots) and diagnosis callout locations (red dots) in one block of the design.

estimate the robustness of the results.

The proposed approach can be also used to compare effectiveness of different weak point sets by directly comparing their respective wpe values. The set that has the lowest wpe value is then considered as the most efficient weak point set. One thing to note is that the highest hit count among all weakpoints should be used as hit_{max} in wpe calculation for all weakpoint sets.

IV. EXPERIMENTAL RESULTS

In this section, we show the effectiveness of the proposed approach using weak point data generated from EDA analysis at the design stage and failed location data obtained through logic diagnosis from a set of defective chips.

Our first case study is a large-scale device. The EDA weak point analysis resulted in a set S_1 of $M = 158,171$ weak points spread out of the device layout. The green dots in Fig. 4 show these weak points in one block of the layout. We also collected a set Q_1 of $N = 431,522$ diagnosis callout locations from logic diagnosis results of defective devices from various wafers/lots, which are shown by the red dots in Fig. 4. It should be noted that the weak points and diagnosis call-out locations shown in Fig. 4 represent the center of their corresponding bounding boxes, since failure locations are usually represented by a rectangular area such as an entire area of a contact/via. A diagnosis call-out location is considered as being "hit" by a weak point when there exists an overlap area of the two bounding boxes. The initial hit number between S_1 and Q_1 is $1,140$, i.e., $1,140$ diagnosis locations from Q_1 were overlaid with at least 1 weak point in S_1, and we use this number as the maximum achievable hit number hit_{max} for this case study as defined in (2).

To assess the effectiveness of our weak point set S_1, we applied the shifting method as describe in Section III-A. In this work, we consider one shift direction, i.e., the values of x_{sft} and y_{sft} in (x_{sft}, y_{sft}) are always positive and set to be equal: $t = x_{sft} = y_{sft}$. Our experiments with other shift values/directions show very similar results and we only demonstrate the results of one shift direction for the purpose of presentation. Table I summarizes the hit count for different shift values. The ideal hit count numbers for various shift values for this case study are shown in the 2nd row of Table

---

978-1-5386-3775-3/18 $31.00 © 2018 IEEE

TABLE I: Hit count for sets S_1, S_r, and S_2

Shift value	$(0,0)$	(t_1,t_1)	(t_2,t_2)	(t_3,t_3)
Ideal	$1,140$	0	0	0
Set S_1	$1,140$	17	0	0
Set S_r	5	7	117	16
Set S_2	107	204	4	104

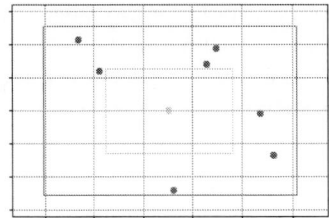

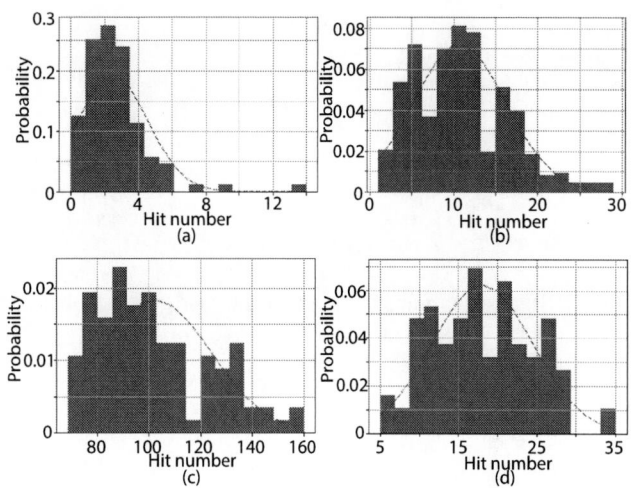

Fig. 5: Example of newly generated weak points (red dots) around an original weak point (green dot).

Fig. 6: Histogram and PDF plots of hit numbers from the $1,000$ new random sets for various shift values: (a) $(0,0)$, (b) (t_1,t_1), (c) (t_2,t_2), and (d) (t_3,t_3).

I as computed from (2). The 3rd row of Table I shows the actual hit count numbers obtained from S_1. The minimum shift value (t_1,t_1) is determined based on the pitch value of the considered layer. It can be observed that when the original weak points are shifted, the hit number decreased significantly, which corroborates our conjecture in Section III-B regarding the weak point efficiency.

As discussed in Section III-C, a random set of weak points S_r can be generated to evaluate the effectiveness of the original set S_1. For this purpose, we randomly generated a new set of 10^6 weak points by setting the exclusive/inclusive bounding box sizes as $d_{ex} \times d_{ex}/d_{in} \times d_{in}$ as described in Algorithm 1. The values of exclusive/inclusive bounding box sizes are chosen according to the particular pitch value of the considered layer such that we avoid duplicating original weak points while new points are still in the proximity of original ones. As the original set S_1 has $M = 158,171$ samples, according to Algorithm 1, each original point is associated with 6-7 new random points. Fig. 5 shows an example of newly generated weak points shown by the red dots around an original point shown by the green dot. The exclusive/inclusive boxes are shown by the green/red boxes. The same shift values were then applied to the set S_r. The hit numbers for various shift amounts between S_r and Q_1 are shown in the 4th row of Table I. It can be observed that the hit count drops significantly for S_r when no shift was applied, while this number exhibits random trend when we increase the shift amount. The maximum number of hits occurred with the shift amount of (t_2,t_2), which is anther indicator of random behavior of the set S_r.

To quantitatively assess the quality of S_1 as compared to S_r, we applied the wpe metric defined in (3). Based on the limited number of shifts we performed, we used the trapezoidal rule to compute the integral approximation of wpe value in (3):

$$wpe_a = \sum_{i=1}^{n-1} \left[\left(wpe(t_i) + wpe(t_{i+1}) \right)/2 \right] \Delta t_i \qquad (4)$$

where n is the total number of shift values, $wpe(t_i)$ is the wpe

value for the shift value t_i: $wpe(t_i) = |hit(t_i) - hitn(t_i)|$ with $hit(t_i)/hitn(t_i)$ representing the ideal/actual hit count number of the shift value t_i, and Δt_i is the interval between t_i and t_{i+1}: $\Delta t_i = t_{i+1} - t_i$. According to the experimental results from Table I, it can be calculated that the wpe_a value for S_1/S_r is 0.85/96.6. We can then readily conclude that the set S_1 performs much better than S_r and it contains meaningful failure root-cause information worth further investigation.

In order to assess the robustness of the results, we have repeated the analysis by re-sampling the random weak point set S_r $1,000$ times with Algorithm 1 re-performed and wpe_a recalculated each time. Fig. 6 shows the histogram plots of the hit count numbers of all the $1,000$ new sample sets for various shift values as shown in Table I. We also plotted an estimated probability density function (PDF) of hit numbers for each shift value by assuming a normal distribution. It can be observed that the mean of hit numbers for each shift value is very close to the one shown in the 4th row of Table I. To further evaluate the robustness of the results, we computed the wpe_a value as defined in (4) for each of the $1,000$ new random sample sets. Fig. 7(a) shows the histogram and PDF plots of the wpe_a values obtained from the new $1,000$ samples of S_r. It can be observed that the wpe_a values range from 60 to 120, which are always significantly larger than the wpe_a value 0.85 obtained from S_1. To further confirm this observation, we have performed a one-sample one-tailed t-test by considering the null hypothesis that the average wpe_a value from the new $1,000$ samples of S_r is equal to that obtained from S_1. The t-test result suggested that this null hypothesis is rejected even at 1% significance level, which confirms that the set S_1 is more efficient than the random set S_r with statistical significance.

As mentioned in Section III-C, we can also use bootstrapping resampling approach to evaluate the robustness of the results. For this purpose, we resampled the 10^6 samples

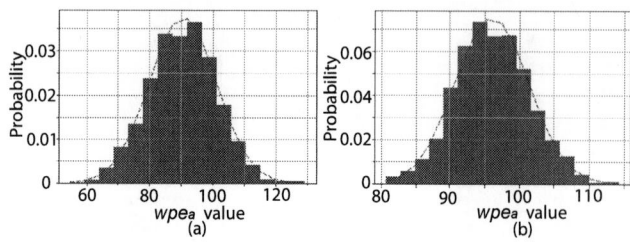

Fig. 7: Histogram and PDF plots of the wpe_a values obtained from (a) the new $1,000$ random samples of S_r, and (b) the $1,000$ boostrapping samples of S_r.

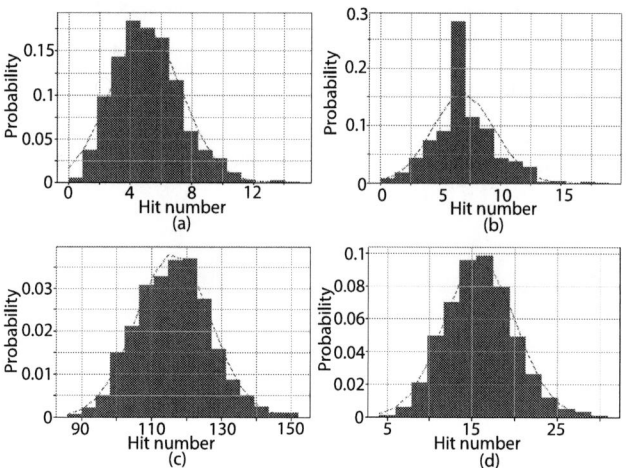

Fig. 8: Histogram and PDF plots of hit numbers from the $1,000$ bootstrapping samples of S_r for various shift values: (a) $(0,0)$, (b) (t_1, t_1), (c) (t_2, t_2), and (d) (t_3, t_3).

contained in S_r with replacement to form a bootstrapping sample set S'_r of the same size. The new wpe_a value was then computed for S'_r as defined in (4). We repeated the process $1,000$ times by computing the hit numbers for various shift values and the value of wpe_a each time. Fig. 8 shows the histogram and PDF plots of hit numbers from the $1,000$ bootstrapping samples for various shift values, and Fig. 7(b) shows the histogram and PDF plots of wpe_a values from the $1,000$ bootstrapping samples. It can be observed that the original set S_1 still outperforms the $1,000$ bootstrapping samples as well. We then performed the same t-test as before to confirm that S_1 is more efficient than the bootstrapping samples of S_r with statistical significance.

As we discussed in Section III, the second goal of the proposed approach is to quantitatively compare different weak point sets to determine which one is more efficient in revealing real failure root causes. For this purpose, we used another weak point set S_2 of $M_2 = 1,908,806$ samples generated from EDA analysis for another layout layer, and we also collected a set Q_2 of $N_2 = 171,225$ diagnosis callout locations from logic diagnosis results of defective devices from

various wafers/lots. We performed the same shifting analysis for this second case study and we obtained the hit count numbers for various shift values, which are shown in the 5th row of Table I. It can be observed that although the size of S_2 is an order of magnitude larger than that of S_1 (1,908,806 vs 158,171), its efficiency is worse as the hit count numbers are still significant even with large shift values. Indeed, a wpe_a calculation can easily confirm this observation: the wpe_a value for S_2 is 89.87, which is significantly larger than that of S_1 0.85. This analysis confirms the ability of our approach to compare the efficiency of different weak point sets.

V. CONCLUSION

In this work, we presented a new metric to quantitatively evaluate the effectiveness of layout weak point sets generated as part of the DFM process. Using diagnosis call-out data collected from testing, we assessed the effectiveness of weak points by shifting them and observing the resulting hit numbers between weak points and diagnosis call-out locations. A new weak point effectiveness (WPE) metric was then used to measure and compare the efficiency of different weak point sets. Experimental results showed that our approach is able to interpret high count weak points for "meaningful" root-cause analysis and compare the effectiveness of different weak point sets.

REFERENCES

[1] W. C. Tam and S. Blanton, "To DFM or not to DFM?," in *ACM/IEEE DAC*, 2011, pp. 65–70.

[2] W. C. Tam and S. Blanton, "Design-for-manufacturability assessment for integrated circuits using radar," *IEEE Transactions on Computer-Aided Design of Integrated Circuits and Systems*, vol. 33, no. 10, pp. 1559–1572, 2014.

[3] R. Radojcic, D. Perry, and M. Nakamoto, "Design for manufacturability for fabless manufactuers," *IEEE Solid-State Circuits Magazine*, vol. 1, no. 3, pp. 24–33, 2009.

[4] W. Wen, J. Li, S. Lin, J. Chen, and S. Chang, "A fuzzy-matching model with grid reduction for lithography hotspot detection," *IEEE Transactions on Computer-Aided Design of Integrated Circuits and Systems*, vol. 33, no. 11, pp. 1671–1680, 2014.

[5] D. Ding, B. Yu, J. Ghosh, and D. Pan, "EPIC: Efficient prediction of IC manufacturing hotspots with a unified meta-classification formulation," in *ASPDAC*, 2012, pp. 263–270.

[6] M. Redford, J. Sawicki, P. Subramaniam, C. Hou, Y. Zorian, and K. Michaels, "DFM - don't care or competitive weapon?," in *ACM/IEEE DAC*, 2009, pp. 296–297.

[7] Y. Pan, R. Desineni, J. Lambert, E. Teoh, T. Berndt, V. Lim, G. Huat, J. Kim, and S. Kekare, "Spotme effective co-optimization of design and defect inspection for fast yield ramp," in *SEMI Advanced Semiconductor Manufacturing Conference*, 2013, pp. 200–205.

[8] A. Vikram, V. Agarwal, and A. Agarwal, "Lithography technology for advanced devices and introduction to integrated cad analysis for hotspot detection," *IET Circuits, Devices & Systems*, vol. 11, no. 1, pp. 1–9, 2017.

[9] R. Desineni, O. Poku, and S. Blanton, "A logic diagnosis methodology for improved localization and extraction of accurate defect behavior," in *IEEE ITC*, 2006, pp. 1–10.

2018 IEEE 36th VLSI Test Symposium (VTS)

Analyzing and Mitigating the Impact of Permanent Faults on a Systolic Array Based Neural Network Accelerator

Jeff (Jun) Zhang   Tianyu Gu   Kanad Basu   Siddharth Garg

Department of Electrical and Computer Engineering, New York University, Brooklyn, New York, 11201
Email: {jeffjunzhang,tg1553,kb150, sg175}@nyu.edu

Abstract—Due to their growing popularity and computational cost, deep neural networks (DNNs) are being targeted for hardware acceleration. A popular architecture for DNN acceleration, adopted by the Google Tensor Processing Unit (TPU), utilizes a systolic array based matrix multiplication unit at its core. This paper deals with the design of fault-tolerant, systolic array based DNN accelerators for high defect rate technologies. To this end, we empirically show that the classification accuracy of a baseline TPU drops significantly even at extremely low fault rates (as low as 0.006%). We then propose two novel strategies, fault-aware pruning (FAP) and fault-aware pruning+retraining (FAP+T), that enable the TPU to operate at fault rates of up to 50%, with negligible drop in classification accuracy (as low as 0.1%) and no run-time performance overhead. The FAP+T does introduce a one-time retraining penalty per TPU chip before it is deployed, but we propose optimizations that reduce this one-time penalty to under 12 minutes. The penalty is then amortized over the entire lifetime of the TPU's operation.

1. Introduction

Deep neural networks (DNN) have, in the past few years, surpassed the performance of traditional machine learning algorithms and are now considered as state-of-the-art for a range of applications, including image and video recognition, text classification [1], [2] and language translation [3]. One drawback of DNNs, however, is their complexity. State-of-the-art DNNs have millions of parameters and are computationally expensive, both to train and execute. GPU acceleration is one way to mitigate this computational burden. Nonetheless, there is a growing interest in *special purpose hardware accelerators* for DNN execution to achieve even greater performance and energy efficiency. Examples of recent special-purpose DNN accelerators include [4]–[7].

DNNs contain multiple layers of computation. Each layer comprises of a computationally expensive matrix multiplication or convolution operation, the latter for convolutional neural networks (CNN), followed by an inexpensive non-linear "activation." Work dating back to the 1980s has shown that matrix multiplications and convolutions can be efficiently implemented using *systolic arrays* [8]. A systolic array is a grid of connected processing elements (PE) that only communicate with their neighbors. For matrix multiplications and convolutions, each PE simply performs a

multiply-and-accumulate (MAC) operation and passes the accumulated value downstream. Data is streamed through the array in a synchronized manner such that each PE received its inputs at just the right time, thus obviating the need for input buffering or complex routing. Further, the cost of reading inputs from memory is amortized over several compute cycles, providing high energy efficiency.

An example of a systolic array based DNN accelerator is the Google Tensor Processing Unit (TPU), that uses 256×256 grid of MAC units at its core, and provides between $30\times$ to $80\times$ times greater performance than CPU or GPU based servers [6]. The TPU is widely deployed in Google datacenters to accelerate DNN inference while reducing the total datacenter energy consumption. In this paper, we use the TPU architecture from Google as the baseline design, but our proposed techniques can apply to any systolic array based DNN accelerators. In the rest of the paper, we will use TPU to refer to the general class of systolic array based DNN accelerators.

An important challenge with future technology scaling is the increase in fault rates, including both permanent (hard errors) and temporary faults (soft errors) [9]–[12]. Temporary faults might occasionally impact the DNN's classification results, but their overall impact on classification accuracy is small, even at high soft error rates. On the other hand, permanent faults, as we show in this paper, can affect the result of *every* DNN execution and significantly reduce the classification accuracy. While permanent faults, at least those that are related to manufacturing defects, can be identified during post-fabrication testing, discarding every chip with a permanent fault reduces yield. This paper addresses the design of *fault-tolerant TPU*, with a focus on manufacturing defects or process variation induced permanent faults.

Prior work has suggested that several machine learning (ML) applications, including DNNs, are inherently resilient to errors; One potential solution for addressing permanent faults in a TPU, therefore, is to simply ignore the permanent faults in the hope that the application itself is inherently resilient to errors. Indeed, prior work has suggested that this is the case for several machine learning (ML) applications, including DNNs [13]. However, using detailed gate-level simulations of stuck-at permanent faults, we show that the accuracy of the TPU drops sharply even if as few as four out of the approximately 64K MAC units are faulty.

Next, we propose two new solutions: *fault-aware prun-*

978-1-5386-3775-3/18 $31.00 © 2018 IEEE   147

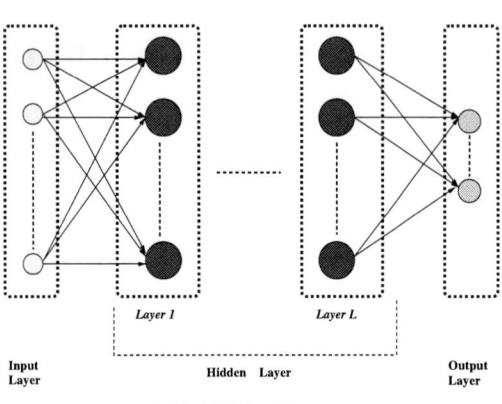

(a) DNN Architecture

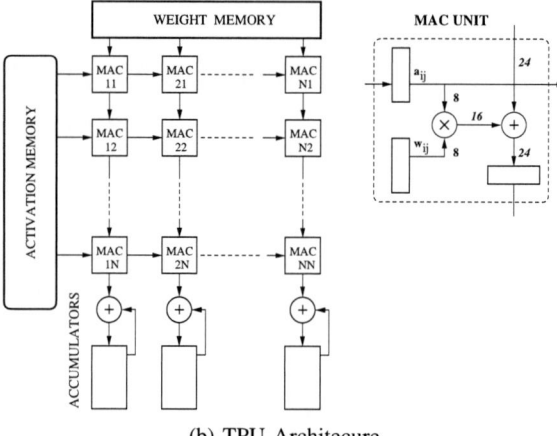

(b) TPU Architecure

Figure 1: Illustration of DNN Architecture and TPU.

ing (FAP) and *fault-aware pruning plus retraining* (FAP+T) that enable the TPU to operate at high fault rates of up to 50% with a negligible impact on classification accuracy and no run-time performance overhead compared to the defect-free baseline. Our proposed solutions build on the recent work that shows that a significant fraction of a DNN's connections can be *pruned* with no (or limited) impact on accuracy. However, while the prior work using pruning to reduce DNN execution time and memory usage [14]–[18], we do so to enable *fault tolerance*. FAP prunes all connections in a DNN that map to faulty MACs using simple bypass circuitry that requires only minor modifications to the baseline TPU. FAP+T additionally *retrains* the DNN after pruning to restore classification accuracy back or close to its baseline, but comes at the expense of extra "test time" per TPU chip. To the best of our knowledge, ours is the *first* work that analyzes and proposes solutions to mitigate the impact of permanent faults on TPU (or TPU-like) DNN accelerators.

2. Related Work

Our work falls in the area of fault-tolerant digital system design. The prior work in this area has a rich history, starting with early work on addressing permanent faults in memory systems using error correcting codes (ECC) [19], and redundancy based approaches for addressing faults in logic [20]–[23] and in the routing fabric [24], [25]. However, these techniques do not directly apply to systolic arrays.

Kung et al. [26] were the first to describe fault-tolerant systolic array designs, which were later improved upon by [27]. The basic idea is to view a systolic array with faults as a smaller systolic array with *fewer* rows and columns. These solutions, in addition to requiring complex bypassing and additional registers inside PEs, can have a significant performance penalty at high-defect rates since an entire column/row is eliminated for each faulty PE. More sophisticated solutions can reduce the performance penalty, but at even higher design complexity [28], [29]. It is important to note that, as opposed to FAP and FAP+T (our proposed solutions), all of these techniques are *application agnostic*, that

is, they seek to correctly execute the original task/application in the presence of faults. FAP and FAP+T, on the other hand, are application-aware, that is, we modify the underlying DNN architecture using pruning and re-training to adapt to faults in the TPU. Consequently, FAP and FAP+T have no performance penalty and negligible area overhead.

Another line of related work is on the design of crossbar-based DNN accelerators leveraging emerging technologies like phase change memory (PCM), resistive RAM (RRAM) and spin-based devices [30], [31]. Specifically, prior work has proposed techniques to cope with non-ideal device characteristics including device non-linearities, limited dynamic range and process variations. However, these techniques are specific to the devices used and to crossbar architectures, and do not apply to TPU designs [32], [33].

3. Background and Preliminaries

In this section, we breifly describe the relvant background on DNNs and the design of our baseline hardware accelerator for DNN execution.

3.1. Deep Neural Networks

A DNN consists of L stacked layers of computation, as shown in Figure 1a. Layer l has N^l *neurons* whose outputs are referred to as *activations*, represented by an N^l dimensional vector a^l. Each layer multiplies the vector of activations from the previous layer with a weight matrix w^l of dimensions $N^{l-1} \times N^l$ and adds constant biases represented by an N^l dimensional vector b^l, followed by an element-wise activation function ϕ. Mathematically, each layer of a DNN performs the following operation:

$$ y_i^l = \sum_j w_{i,j}^l a_j^{l-1} + b_i^l, \quad a_i^l = \phi(y_i^l) \quad \forall l \in [1, L], \quad (1) $$

In our description so far, we have assumed that each neuron's activation depends on the activations of all neurons in the previous layers. Such layers are referred to as *fully connected* (FC) layers. A commonly used special case of an FC layer is a convolutional (conv) layer. The activations of a

978-1-5386-3775-3/18 $31.00 © 2018 IEEE   148

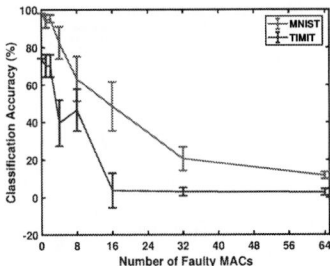
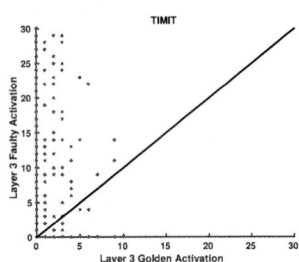

(a) Classification Accuracy Drop Due to Stuck-at-Fault MACs.

(b) TIMIT Output Regression for Layer 3 Activation with 8 Faulty MACs.

Figure 2: Impact of TPU Stuck-at-Faults on DNN Applications

convolutional neural network are represented as 3-D *tensors* of dimensions $W^l \times H^l \times D^l$ which refer to the width, height and number of channels in layer l of the network. Also, the weights are represented as 4-D tensors of dimensions $F^l \times F^l \times D^{l-1} \times D^l$, where F^l is the filter size in layer l. While the mathematical expression of the convolutional layer is omitted for brevity; for the purposes of this paper it suffices to know that the outputs of channel $d \in D^l$ is determined by convolving the 3-D input tensor with a 3-D weight tensor obtained by setting the last dimension of the 4-D weight tensor to d.

3.2. DNN Acceleration on TPU

Figure 1b shows a block-diagram of the TPU architecture, at the heart of which is a systolic array containing $N \times N$ MAC units that is used to perform the computationally expensive matrix multiplication and convolution operations. To understand how, consider a fully-connected layer with N input neurons and N output neurons, and consequently an $N \times N$ weight matrix. The weight matrix is first loaded into the systolic array, with weight $w_{i,j}$ being loaded into the MAC in row j and column i, which we refer to as $\text{MAC}_{i,j}$.

Now consider the first column of the TPU. $\text{MAC}_{1,1}$ computes $w_{1,1}a_1$ in the first clock cycle and sends the result to $\text{MAC}_{1,2}$ which adds the product $w_{1,2}a_2$ to its input and forwards the result further downwards. In clock cycle N, the $\text{MAC}_{1,N}$ unit outputs $y_1 = \sum_{i=1}^{N} w_{1i}a_i$, which is the the first element of the output matrix.

The second column receives the same stream of inputs as the first column, but delayed by one clock cycle. This column is loaded with weights from the second row of the weight matrix, and outputs y_2, and so on. In practice, the TPU operates in *batches*, where each batch has B inputs. A batch of B inputs is multiplied by an $N \times N$ weight matrix in $2N + B$ clock cycles.

4. Motivational Analysis: Impact of Permanent Faults on TPU

To motivate the proposed fault-tolerant TPU design, we begin by empirically analyzing the impact of permanent faults

on the classification accuracy of the TPU. To do so, we first synthesized an RTL description of the TPU into the gate-level netlist using the 45 nm OSU PDK, and then inserted stuck-at faults at internal nodes in the gate-level netlist. For this analysis, we focused only on faults in the data-path and ignored faults in the memory components (since they can be addressed using ECC) and the control logic since it consumes an insignificant fraction of the design.

We then mapped DNNs for two classification tasks, MNIST digit classification and TIMIT speech recognition, on the faulty TPU and plot the classification accuracy versus number of faulty MAC units in Figure 2a. For TIMIT, we observe that even with only four faulty MACs (i.e., with only $\sim 0.005\%$ MACs faulty), the classification accuracy drops from the 74.13% to 39.69%.

The reason for the drop in accuracy is that stuck-at faults frequently affect the higher order bits of the MAC output, resulting in large absolute errors in the matrix-vector product. Figures 2b scatters the golden (fault-free) activations of the final layer of the TIMIT DNNs with the corresponding faulty outputs. Observe that for TIMIT, the faulty outputs have much higher magnitudes than the golden outputs.

To mitigate the permanent faults on the TPU, one simple idea is to bypass the entire columns where the faulty MACs reside. However, with the number of permanent faults increasing, the performance penalty would be unacceptable.

These observations together motivates the fault aware design of this paper.

5. Proposed Fault-Tolerant TPU Design

From the motivational example, it is clear that TPUs with even a relatively small number of faulty MAC units cannot be used unless the fault impact is mitigated. Our proposed fault-tolerant TPU design starts with an important observation: note from the description of the TPU operation in Section 3.2 that *each weight in the DNN maps to exactly one MAC unit*. In other words, there is a static mapping between DNN weights and MAC units. We then exploit the static mapping to determine which weights to prune.

This observation can be formalized using mapping functions $r()$ and $c()$ that take as input the indices of a DNN weight and output the row and column, respectively, of the MAC unit on which the weight is mapped. Specifically, based on the discussion in Section 3.2, the mapping functions for weight $w_{i,j}$ in a fully-connected layer are $r(i,j) = j\%N$ and $c(i,j) = i\%N$, where $\%$ is the modulo operator. Note that implicit in this mapping function is the fact that weight matrices that do not fit fully in the systolic array are first *blocked* into smaller $N \times N$ sub-matrices.

Our mapping strategy for conv layers sums along input channels along rows of the TPU, and each column computes outputs for a different output channel. Consequently, the mapping functions for weight $w_{i,j,k,l}$ in a conv layer are $r(i,j) = k\%N$ and $c(i,j) = l\%N$. We now describe our two proposed fault-tolerance methodologies: FAP and FAP+T.

978-1-5386-3775-3/18 $31.00 © 2018 IEEE

5.1. Fault-Aware Pruning (FAP)

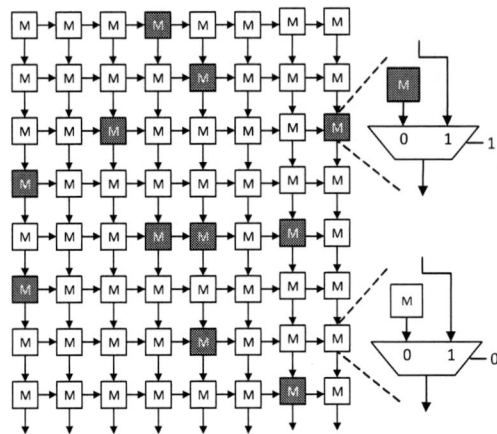

Figure 3: Diagram illustrating FAP.

The proposed FAP and FAP+T techniques both assume that standard post-fabrication tests are used on each TPU chip to determine the location of faulty MACs. Given this information, the idea behind FAP is to set (or prune) any weight that maps to a faulty MAC to zero. That is, for all pairs of (i, j) values such that $\text{MAC}_{c(i,j),r(i,j)}$ is faulty, we set the corresponding $w_{i,j} = 0$ (this is for fully connected layers, a similar strategy is used for conv layers). Note that multiple weights can map to one MAC unit; correspondingly, even a single faulty MAC can result in multiple weights being pruned.

In hardware, pruning is achieved by introducing a separate bypass path for faulty MAC units, as shown in Figure 3. With the bypass path being enabled, the faulty MAC unit's contribution to the column sum is skipped, which is equivalent to setting the faulty MAC's weight to zero. Note that actually loading a zero weight into a faulty MAC is *not* equivalent to setting it's weight to zero. The area overhead due to the new bypass path is only about 9%.

5.2. FAP+T Solution

The FAP+T approach starts with FAP based on each TPU's fault map, but additionally *retrains* the unpruned weights in the DNN while forcing all pruned weights to zero during the re-training process. The FAP+T algorithm (see Algorithm 1) returns new, optimized values for the unpruned weights that improve the classification accuracy compared to the FAP solution. One drawback, however, is that re-training needs to be performed for each TPU chip based on its unique fault map; however, this needs to be done *only once* per TPU chip and the cost of doing so is amortized over the entire lifetime of the TPU chip.

Note that the FAP+T algorithm has a parameter MAX_EPOCHS that determines the number of iterations of the re-training algorithm. Setting this parameter to zero is equivalent to FAP. As MAX_EPOCHS is increased, the re-training time increases in return for increased classification accuracy.

Algorithm 1: FAP+T Training Algorithm

1 **Algorithm** FAP+T()
2 Load the pre-trained DNN weights and TPU fault map;
3 Determine indices of pruned weights from TPU fault map;
4 Set all pruned weights to zero;
5 **for** *Training epochs* $\leq MAX\_EPOCHS$ **do**
6   *Update weights using back-prop.;*
7   *Set all pruned weights to zero;*
8 **end**
9 **return** *Retrained model;*

Multi-Layer Perceptrons (MLPs)		
Name	MNIST	TIMIT
# Neurons	784-256-256-256-10	1845-2000-2000-2000-183

AlexNet Convolutional Feature Extraction				
layer	filter	stride	padding	activation
conv1	96x3x11x11	4	0	ReLU+LRN
pool1	max, 3x3	2	0	/
conv2	256x96x5x5	1	2	ReLU+LRN
pool2	max, 3x3	2	0	/
conv3	384x256x3x3	1	1	ReLU
conv4	384x384x3x3	1	1	ReLU
conv5	256x384x3x3	1	1	ReLU
pool5	max, 3x3	2	0	/

AlexNet Fully-connected Layers		
layer	#neurons	activation
fc6	4096	ReLU
fc7	4096	ReLU
fc8	4096	/

TABLE 1: Benchmark DNNs architecture

6. Empirical Evaluation

In this section, we introduce our evaluation framework and discuss the results on three popular DNN applications.

6.1. Setup

Benchmarks The three benchmarks we used are: MNIST digit classification, TIMIT speech recognition [34] and AlexNet for image recognition using the PASCAL VOC2007 dataset [35]. We trained all DNNs from scratch and achieve the-state-of-art classification accuracy. Table 1 shows the detail parameters of the three DNN architectures. The MNIST and TIMIT DNNs have only fully connected layers, while AlexNet has five conv layers and three fully connected layers.

Systolic Array Parameters A prototype of the systolic array of a grid of 256×256 MACs was developed in synthesizable Verilog and synthesized with OSU FreePDK 45 nm standard cell library using the Cadence Genus synthesis tool. The systolic array is synthesized to be run at 658 MHz with a nominal supply voltage of $1.1V$ and consumes 19.7W of dynamic power.

Simulation Methodology We scheduled our benchmark DNNs on the systolic array using the mapping techniques described in Section 5. To analyze the impact of permanent

978-1-5386-3775-3/18 $31.00 © 2018 IEEE

faults on the TPU, we use stuck-at fault injection methodology described in Section 4. We then conduct a detailed gate-level simulation in Modelsim to collect the final accuracy for the benchmarks. Each experiment is repeated 10 times with faults injected in different locations, picked uniformly at random.

6.2. Results

Figure 4 shows the classification accuracy versus the number of permanent faults using the FAP and FAP+T techniques for our three benchmark DNNs. Note that we have about 65K MAC units in the TPU, and show results for high defect rates of up to 50%, i.e., with upto half of the MAC units faulty. For MNIST and TIMIT, both FAP and FAP+T are able to go up to fault rates of 25% with negligible classification accuracy reduction. However, with 50% faulty MACs, only FAP+T is able to provide close to the baseline classification accuracy, while that of the FAP technique reduces.

Compared to MNIST/TIMIT, the classification accuracy for AlexNet as seen in Figure 4b, drops markedly with increasing fault rates with FAP. FAP+T, however, performs much better and only incurs an 8% accuracy drop even with a 50% fault rate. One reason for the accuracy drop comes from the scheduling policy for the CNNs. In the current policy, one permanent faulty MAC would lead to a whole channel of the filter to be pruned. A more sophisticated scheduling may show better results but it is out of the scope of this paper.

As mentioned before, FAP+T incurs a one-time re-training overhead per TPU chip. While the retraining overhead (about 1 hour for AlexNet in the worst case, 25 epochs) is negligible compared to the lifetime of the TPU, we show in Figure 5 that it can be reduced even further by setting the value of MAX_EPOCHS appropriately. Specifically, Figure 5 shows the classification accuracy of FAP+T with increasing number of training epochs. Observe, for instance, that for AlexNet with 25% faulty MACs, the classification accuracy after the first 5 epochs only marginally lower than the classification accuracy after 25 epochs. Consequently, the re-training time for this example of AlexNet can be reduced by $5\times$, from 1 hour to only 12 minutes.

7. Conclusion

In this paper, we address the problem of designing fault-tolerant hardware accelerators for DNN execution for high defect rate technologies. Using the systolic array based Google TPU architecture as a baseline, we first showed using detailed gate-level simulations that the TPU's classification accuracy drops significantly even in the presence of very low fault rates. Then, we proposed two novel techniques, FAP and FAP+T, that allow TPUs to operate even with fault rates as high as 50%, with negligible to tolerable drops in classification accuracy (from 0.1% drop for TIMIT to 8% drop for AlexNet). While neither technique has any run-time performance overhead, FAP+T introduces a one-time retraining overhead of about 12 minutes per TPU chip for

AlexNet. However, this one-time cost is amortized over the entire TPU's lifetime. As future work, we plan to address the impact of aging-related faults on DNN accelerators. Also, we are planning to apply our proposed methodology on larger neural-networks.

References

[1] A. Krizhevsky et al., "Imagenet classification with deep convolutional neural networks," in *Advances in Neural Information Processing Systems (NIPS)*, pp. 1097–1105, 2012.

[2] A. Karpathy et al., "Large-scale video classification with convolutional neural networks," in *Proceedings of the IEEE conference on Computer Vision and Pattern Recognition*, pp. 1725–1732, 2014.

[3] I. Sutskever et al., "Sequence to sequence learning with neural networks," in *Advances in Neural Information Processing Systems (NIPS)*, pp. 3104–3112, 2014.

[4] Y.-H. Chen et al., "Eyeriss: An energy-efficient reconfigurable accelerator for deep convolutional neural networks," *IEEE Journal of Solid-State Circuits*, vol. 52, no. 1, pp. 127–138, 2017.

[5] Z. Du et al., "Shidiannao: Shifting vision processing closer to the sensor," in *ACM SIGARCH Computer Architecture News*, vol. 43, pp. 92–104, 2015.

[6] N. P. Jouppi et al., "In-datacenter performance analysis of a tensor processing unit," in *Proceedings of the 44th Annual International Symposium on Computer Architecture*, pp. 1–12, ACM, 2017.

[7] S. Park et al., "4.6 a1. 93tops/w scalable deep learning/inference processor with tetra-parallel mimd architecture for big-data applications," in *IEEE International Solid-State Circuits Conference (ISSCC)*, pp. 1–3, 2015.

[8] H.-T. Kung, "Why systolic architectures?," *IEEE computer*, vol. 15, no. 1, pp. 37–46, 1982.

[9] S. S. Sahoo et al., "Design and evaluation of reliability-oriented task re-mapping in mpsocs using time-series analysis of intermittent faults," in *Design, Automation & Test in Europe Conference & Exhibition (DATE)*, pp. 798–803, IEEE, 2016.

[10] J. Zhang et al., "Enabling extreme energy efficiency via timing speculation for deep neural network accelerators," 2017.

[11] S. Borkar, "Design perspectives on 22nm cmos and beyond," in *Proceedings of the 46th Annual Design Automation Conference*, pp. 93–94, ACM, 2009.

[12] C. Constantinescu, "Trends and challenges in vlsi circuit reliability," *IEEE Micro*, vol. 23, no. 4, pp. 14–19, 2003.

[13] A. Gebregiorgis et al., "Error propagation aware timing relaxation for approximate near threshold computing," in *Proceedings of the 54th Annual Design Automation Conference (DAC)*, pp. 1–6, IEEE, 2017.

[14] S. Han et al., "Deep compression: Compressing deep neural networks with pruning, trained quantization and huffman coding," *arXiv preprint arXiv:1510.00149*, 2015.

[15] J. Yu et al., "Scalpel: Customizing dnn pruning to the underlying hardware parallelism," in *Proceedings of the 44th Annual International Symposium on Computer Architecture*, pp. 548–560, ACM, 2017.

[16] H. Li et al., "Pruning filters for efficient convnets," *arXiv preprint arXiv:1608.08710*, 2016.

[17] S. Anwar et al., "Structured pruning of deep convolutional neural networks," *ACM Journal on Emerging Technologies in Computing Systems (JETC)*, vol. 13, no. 3, p. 32, 2017.

[18] P. Molchanov et al., "Pruning convolutional neural networks for resource efficient inference," 2016.

[19] L. Levine et al., "Special feature: Semiconductor memory reliability with error detecting and correcting codes," *Computer*, vol. 9, no. 10, pp. 43–50, 1976.

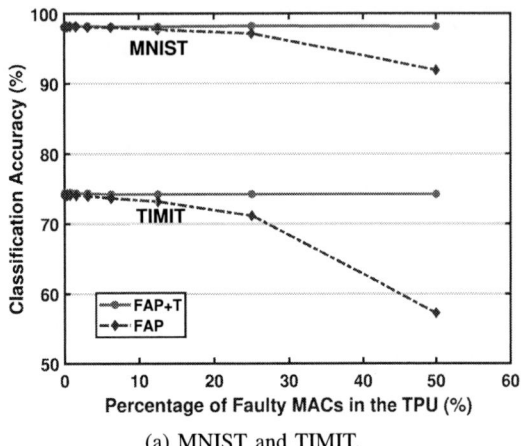

(a) MNIST and TIMIT.

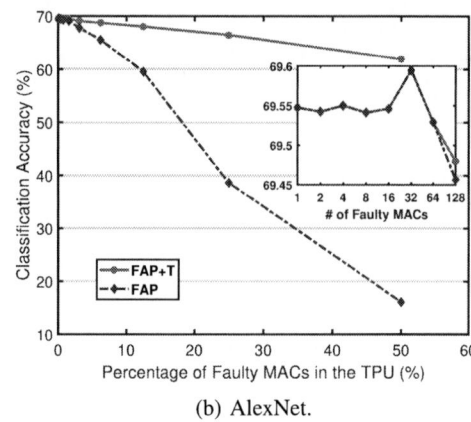

(b) AlexNet.

Figure 4: Classification accuracy vs. Percentage of Faulty MACs using FAP and FAP+T for (a) MNIST and TIMIT and (b) AlexNet.

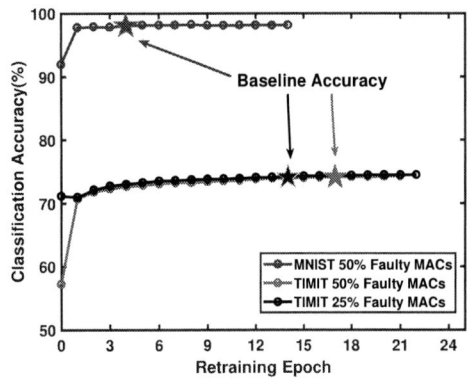

(a) MNIST and TIMIT.

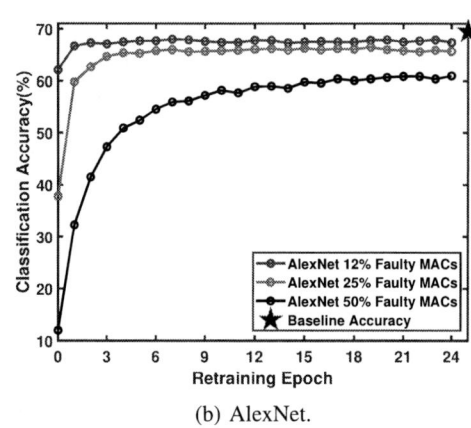

(b) AlexNet.

Figure 5: Classification accuracy vs. MAX_EPOCHS for (a) MNIST and TIMIT, and (b) AlexNet.

[20] J. H. Patel *et al.*, "Concurrent error detection in alu's by recomputing with shifted operands," *IEEE Transactions on Computers*, vol. 31, no. 7, pp. 589–595, 1982.

[21] N. Oh *et al.*, "Error detection by duplicated instructions in super-scalar processors," *IEEE Transactions on Reliability*, vol. 51, no. 1, pp. 63–75, 2002.

[22] A. Meixner *et al.*, "Argus: Low-cost, comprehensive error detection in simple cores," in *40th Annual IEEE/ACM International Symposium on Microarchitecture (MICRO)*, pp. 210–222, 2007.

[23] M. Zhang *et al.*, "Sequential element design with built-in soft error resilience," *IEEE Transactions on Very Large Scale Integration (VLSI) Systems*, vol. 14, no. 12, pp. 1368–1378, 2006.

[24] Y.-C. Chang *et al.*, "On the design and analysis of fault tolerant noc architecture using spare routers," in *Proceedings of the 16th Asia and South Pacific Design Automation Conference (ASPDAC)*, pp. 431–436, 2011.

[25] W.-C. Tsai *et al.*, "A fault-tolerant noc scheme using bidirectional channel," in *Proceedings of the 48th Design Automation Conference (DAC)*, pp. 918–923, ACM, 2011.

[26] H. Kung *et al.*, "Fault-tolerance and two-level pipelining in vlsi systolic arrays," tech. rep., Carnegie-Mellon UNIV, 1983.

[27] J. H. Kim *et al.*, "On the design of fault-tolerant two-dimensional systolic arrays for yield enhancement," *IEEE Transactions on Computers*, vol. 38, no. 4, pp. 515–525, 1989.

[28] M. Esonu *et al.*, "Fault-tolerant design methodology for systolic array architectures," *IEE Proceedings-Computers and Digital Techniques*, vol. 141, no. 1, pp. 17–28, 1994.

[29] H. F. Li *et al.*, "Restructuring for fault-tolerant systolic arrays," *IEEE Transactions on Computers*, vol. 38, no. 2, pp. 307–311, 1989.

[30] G. W. Burr *et al.*, "Neuromorphic computing using non-volatile memory," *Advances in Physics: X*, vol. 2, no. 1, pp. 89–124, 2017.

[31] C. Liu *et al.*, "Rescuing memristor-based neuromorphic design with high defects," in *Proceedings of the 54th Annual Design Automation Conference (DAC)*, pp. 87:1–87:6, ACM, 2017.

[32] G. W. Burr *et al.*, "Experimental demonstration and tolerancing of a large-scale neural network (165 000 synapses) using phase-change memory as the synaptic weight element," *IEEE Transactions on Electron Devices*, vol. 62, no. 11, pp. 3498–3507, 2015.

[33] D. Chabi *et al.*, "Robust neural logic block (nlb) based on memristor crossbar array," in *Proceedings of the IEEE/ACM International Symposium on Nanoscale Architectures*, pp. 137–143, 2011.

[34] J. Ba *et al.*, "Do deep nets really need to be deep?," in *Advances in Neural Information Processing Systems (NIPS)*, pp. 2654–2662, 2014.

[35] M. Everingham *et al.*, "The PASCAL Visual Object Classes Challenge 2007 (VOC2007) Results." http://www.pascal-network.org/challenges/VOC/voc2007/workshop/index.html.

978-1-5386-3775-3/18 $31.00 © 2018 IEEE   152

2018 IEEE 36th VLSI Test Symposium (VTS)

IR Drop Prediction of ECO-Revised Circuits Using Machine Learning

Shih-Yao Lin[1], Yen-Chun Fang[1], Yu-Ching Li[1], Yu-Cheng Liu[1], Tsung-Shan Yang[1], Shang-Chien Lin[1], Chien-Mo Li[1]
Eric Jia-Wei Fang[2]
[1]Graduate Institute of Electronics Engineering
National Taiwan University, Taipei 106, Taiwan
[2]MediaTek Inc., Hsinchu 300, Taiwan

Abstract — Excessive *power supply noise* (PSN), such as IR drop, can cause timing violation in VLSI chips. However, simulation PSN takes a very long time, especially when multiple iterations are needed in IR drop signoff. In this work, we propose a machine learning technique to build an IR drop prediction model based on circuits before ECO (engineer change order) revision. After revision, we can re-use this model to predict the IR drop of the revised circuit. Because the previous circuit(s) and the revised circuit are very similar, the model can be applied with small error. We proposed seven feature extractions, which are simple and scalable for large designs. Our experiment results show that prediction accuracy (average error 3.7mV) and correlation (0.55) are very high for a three million-gate real design. The run time speedup is up to 30X. The proposed method is very useful for designers to save the simulation time when fixing the IR drop problem.

Keywords — *power supply noise, IR drop analyzer, machine learning*

I. INTRODUCTION

Power supply noise (PSN) has become an important concern for VLSI system design and test [1, 2]. Excessive PSN degrades circuit performance, which even leads to timing failure [3, 4]. It is a well-known problem that excessive PSN can induce significant yield loss (overkill) [5, 6, 7]. PSN include IR drop and *Ldi/dt* noise. Since IR drop is more significant than the *Ldi/dt* noise for on-chip power integrity analysis, this paper will focus on the IR drop effect only.

Traditional dynamic IR drop analyzer solves large linear equation systems to obtain the IR drop of every node in the circuit, and then simulate critical paths to verify if there is any IR drop violation [8]. However, this process is very slow, especially when multiple iterations are needed in IR drop signoff. For an industry scale design (~3M gate count), IR drop analysis can take up to one day. Every time a minor revision is made, the whole process has to be repeated, even if the revised circuit just changed a small number of cells.

It has been shown that machine learning prediction of circuit speedpath [9] and timing signoff [10] is feasible. Recently, Ye et al.[11] developed an SVM-based regression method to predict circuit delay at runtime

without PSN consideration. However, it has been shown that IR drop analysis is inaccurate if PSN is ignored [12]. Unfortunately, realistic large circuits are difficult for machine learning since the dimension is very large. Power-aware dynamic IR drop prediction of cells can be found in [13]. They used linear model to predict the IR drop of cells. However, the prediction rule is based on designer's experience, which cannot be generalized and automated. So far, there is still no good machine learning technique available to predict PSN for large circuits.

Fig. 1 shows the traditional flow of IR drop analysis. After each circuit revision, we need to rerun the IR drop analyzer to make sure there is no violation. The source of patterns can be either functional patterns or test patterns. Because real design process needs many revisions, repeated IR-drop analysis during each iteration can be very time consuming.

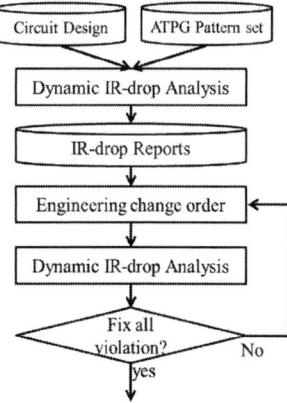

Fig. 1. Traditional IR drop analysis flow

In this work, we propose to use machine learning to build an IR drop prediction model for the circuit(s) before revision. After a circuit revision, we can re-use this model to predict the IR drop of the revised circuit. After the predicted IR drop meets our specification, we need to rerun the dynamic IR drop analyzer again to make sure there is indeed no violation before the final signoff. This work has three major contributions. We take advantage of the similarity between the original circuit and the revised circuit to learn a model to speed up the signoff process so very few dynamic IR drop analyses are needed. This new

978-1-5386-3775-3/18 $31.00 © 2018 IEEE   153

flow saved a lot of iterative simulation time during revision. Second, we propose to sample a small portion of cells to predict IR drop of all cells. This greatly reduces the size of input data so that machine learning of realistic industrial design is feasible. Third, we propose seven simple but important feature extraction methods to greatly reduce the dimension, so the proposal is scalable for large designs. Our experiment results on a three million-gate GPU show that average error of prediction IR drop is 3.7mV and correlation is 0.55. The run time speedup is up to 30X compared to a commercial tool *Ansys RedHawk*. The proposed method is very useful for designers to save the simulation time during ECO to fix IR drop problems.

The rest of this paper is organized as follows. Section II provides previous research papers in PSN-aware IR analysis. Section III presents the proposed machine learning technique. Section IV shows experimental results on benchmark circuits. Finally, Section V concludes this paper.

II. PAST RESEARCH

A. Statistical IR drop Prediction

Many different metrics have been proposed as alternatives to IR drop, such as *weighted switching activity* (WSA) [14, 15], *switching cycle average power* (SCAP) [16], *flip-flop toggle count* (FFTC) [17], and *etc*. Although some metrics show good correlations with actual IR drop values, there is no known model to translate the proposed metrics to the actual IR drop values. It is not clear what is the pass/fail threshold for these metrics. Therefore, it is impossible to use these alternative metrics to sign off a design. A recent paper used a linear model to predict the IR drop values [13]. For each cell, they calculated a linear model to predict the IR drop based on the power consumption. The problem of the linear model is that it may not be good enough for complex designs. In addition, it is computationally expensive to calibrate a linear model for each cell in large designs. A paper tried to identify high power area (*hot-spot*) using switching probability and logic level [18]. Although we see a correlation between real hot-spot and the predicted area, it is still not clear what is the pass/fail threshold for design sign off.

B. Machine Learning IR drop Prediction

Machine learning has been applied to identify speedpath outliers [9]. Various feature extractions have been performed based on topology, dynamic effects, static effects, statistical effects, and random effects. Nevertheless, it did not consider IR drop effects. Support vector machine has been applied to predict IR drop [11]. This technique was implemented on FPGA to dynamically adjust the CPU operation frequency. Their technique used only input patterns, no feature extraction, to predict IR drop. The number of dimensions is very large and therefore it is not scalable for large designs. Another previous work is IR-drop-aware timing prediction using machine learning

[19]. This work proposed feature extraction so it is scalable for large designs. However, it did not consider the ECO revision issue. Every time a new revision is made, a new model is needed.

C. Dynamic IR drop Analyzer

Our proposed machine learning technique can be applied to speed up any circuit IR drop analyzer. In this paper, we use a PSN-aware dynamic IR drop analyzer, *IDEA* (*IR drop-aware Efficient timing Analyzer*) as our benchmark simulator [12]. This technique is very scalable because they model the voltage-delay characteristic function in a simple analytical function, which just require limited simulation of library cells. Experimental results showed that, for small circuits, the error is less than 5% compared with HSPICE. Although IDEA is up to 272 times faster than a commercial tool, NANOSIM, it still takes days to simulate million-gate designs.

III. PROPOSED TECHNIQUE

A. Proposed Flow

Fig. 2 shows the proposed flow of our work. During the design phase, we have several ECO-revised circuits, including *previous versions* (..., E_{n-2}, E_{n-1}) and *the current version* (E_n). Suppose that we have performed dynamic IR drop analysis on previous versions, using dynamic IR drop analyzer, such as *Ansys Redhawk*[8]. We can then extract important features from a small number of *sampled cells*. After that, we run a machine learning to build a *model* for this circuit so we can re-use this model to predict IR drop of the current version (E_n). Designers can use our prediction results to quickly evaluate whether IR drop of the current version meets the specification or not. Our machine learning prediction can save a lot of IR drop analysis runtime during iterations. Finally, when the predicted IR drop all meet our specification, we need to run the dynamic IR drop analyzer again to make sure there is indeed no violation before the final signoff. Compare Fig. 2 with Fig. 1, we can save simulation time during the prediction phase.

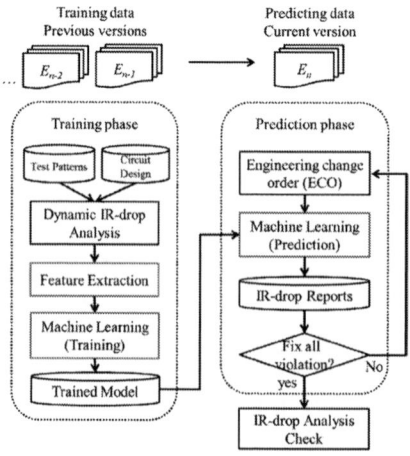

Fig. 2. Proposed IR Drop Prediction flow

B. Cell Sampling and Feature Extraction

Because there are many cells in a real design, it is impractical to use all cells to build a machine learning model. In this research, we propose to sample a portion of cells to build a model. Two factors should be considered when we take samples: (1) physical location and (2) IR drop values. For (1), we divide the chip layout into $M \times N$ windows. Based on our experience, we take 5% ~ 10% sampling cells randomly from each window. For (2), we sort all cells by their IR drops. We take sampled from three categories of cells: serious IR drop, medium IR drop cells, and low IR drop.

Table I shows the features we consider in this work. Given a sampled cell, there are three categories of features. *Power features* of a sampled cell include the power of this cell, toggle rate of this cell, and type of this cell. *Physical features* include this cell location (*i.e. X, Y* coordination), toggle rate of neighbor cells, and neighbor count (number of cells in the neighborhood). Finally, the *via feature* is the distance to via. Each feature is explained as follows.

TABLE I.   SEVEN FEATURES OF A SAMPLED CELL

categories	1	2	3
Power features	Cell power	Cell toggle rate	Cell type
Physical features	Cell location	Neighbor toggle rate	Neighbor count
Via feature	Distance to via		

Cell power is the power consumption of the sampled cell given a set of input patterns. *Cell toggle rate* measures the switching activity of the sampled cell. *Toggle rate* is defined as the number of toggles over the number of clock cycles and it is a number between 0% and 200%. The reason for 200% toggle rate is because clock buffers toggle twice in each cycle. Both cell power and cell toggle rate are scalar variables that can be obtained by a dynamic IR drop analyzer, such as *Redhawk*. *Cell type* is the logic gate type of the sampled cell, such as NAND, NOR, and *etc*. This is a categorical scalar variable, which can be obtained from the netlist or the IR drop analysis report.

Fig. 3 shows how to define neighbors for a given sampled cell. We draw a rectangle window, centered at the given sampled cell. The window height and width can be adjusted by the user. Different technology may have different setting. In this work, we increasingly enlarge the window size and observed the prediction accuracy under different window width and height. After several experiments, our ANN model reached the highest prediction accuracy when window height is set to three row heights and window width is 50.

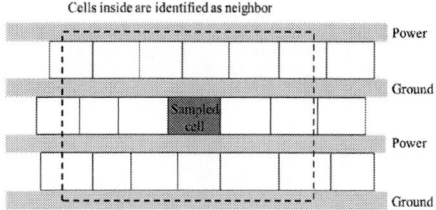

Fig. 3.   Neighbors of a sampled cell

Neighbor toggle rate (NTR) is the toggle rate among all neighbor cells. Because different cell types have different impact on IR drop, so we need to count *NTR* according to cell types. For each cell type, toggle rates of the same cell type are summed up. This feature is a vector, whose dimension equals to the number of cell types. *NTR* of a sampled cell s is shown as the following equation (1).

$$NTR(s) = [\sum_{k \in W} TR_k^{type=t_1}, \dots, \sum_{k \in W} TR_k^{type=t_N}] \quad (1)$$

, where TR_k^t indicates the toggle rate of the k_{th} cell of type t, and W is the neighbor window of the sample cell s.

Neighbor count (NC) means the total count of neighbor cells. This feature is a scalar, defined in the following equation (2), where $Cell_k$ is an indicator variable (1 means the presence of the k_{th} cell, and 0 otherwise).

$$NC(s) = \sum_{k \in W} Cell_k \quad (2)$$

Distance to via (D) is the distance to the closest power via, and this via must be in the same row as the sampled cell. Fig. 4 shows the definition of D. The sampled cell is in the middle of the window and the red rectangles represent power vias. Number D represents the resistance value from the sampled cell to the power network.

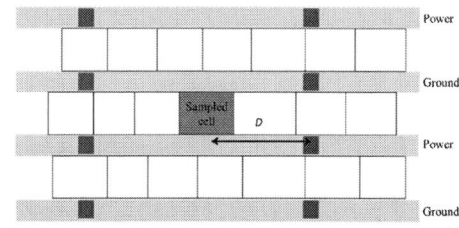

Fig. 4   Distance to via

Totally, we propose seven features of dimension $(T+7)$, where T is the total number of cell types used in the design. This is a very small dimension and scalable for large designs.

C. Machine Learning Prediction Model

Artificial neural network (ANN) [20] imitated the neural structure of human's brain. Figure 5 shows an example ANN model with one hidden layer, where x_n is the input features of n_{th} sampled cell, and t_n is the target IR drop value of the n_{th} sampled cell. w is the weight of neurons in ANN and N is the number of training data.

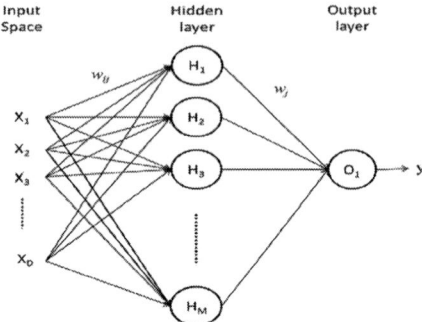

Fig. 5   ANN Model (with one hidden layer)

Our goal is to find a function $y(x_n, w)$ to minimize the following error function,

$$E(w) = \frac{1}{2}\sum_{i=1}^{N}\|y(x_n, w) - t_n\|^2 \qquad (3)$$

IV. EXPERIMENTAL RESULTS

Three ITC'99/IWLS'05 benchmark circuits (*b18, b19, leon3mp*) in 45nm and one real GPU (*Graphic Processor Unit*) in 16nm technology from the industry have been evaluated by our proposed method. Profiles of these four circuits are shown in Table II. The first column shows the number of cells. Given a commercial ATPG test pattern set, three ITC/IWLS benchmark circuits have been simulated by our own dynamic IR drop simulator, IDEA [12]. The second column shows the number of patterns simulated. The max IR drop and the average IR drop of the circuit are shown in the third and fourth columns. Dynamic IR drop analysis of the GPU was performed by a commercial tool, *Ansys RedHawk*. All the machine learning experiments use the artificial neural network open source FANN [21]. The experiments were run on Intel Xeon CPU E5520 @ 2.27GH with 32GB RAM.

TABLE II.   PROFILE OF BENCHMARK CIRCUITS (E_1, BEFORE ECO)

Circuit	Cells	Patterns	V_{DD} (V)	Avg. IR drop(mV)	Max IR drop(mV)
b18	64K	50	1.1	29	39
b19	128K	50	1.1	59	83
leon3mp	638K	50	1.1	92	241
GPU	3,006K	240	0.9	25	190

A. IR Drop Prediction before ECO

We first evaluate the effectiveness of the IR drop prediction for the circuit before engineering change order (ECO). Prediction accuracy is measured by *Normalized root mean square error (NRMSE)*, which is defined in equation (4) and (5). In these equations, $\hat{y}_i$ is the simulated IR drop of the i_{th} sample cell, and y_i is the predicted IR drop of the i_{th} sample cell. N is the number of data.

$$RMSE = \sqrt{\frac{\sum_{i=1}^{N}(\hat{y}_i - y_i)^2}{N}} \qquad (4)$$

$$NRMSE = \frac{RMSE}{mean(y)} * 100\% \qquad (5)$$

First, we want to know how many samples we need to build a model with high prediction accuracy. In this experiment, we sampled a small portion of cells and predict IR drop of all cells. The training data and predicting data are from the same design (E_1). There is no ECO-revision in this experiment. The prediction accuracy for three benchmark circuits is plotted in Fig. 6. We observe that *NRMSE* drops quickly when sampling cells increase. We can see that *NRMSE* remain constant when the percentage of sampled cells is more than 10%. These experiments show that 10% sampling is enough for our designs. Please note that IR drop is highly design-dependent. Each design has a unique model, even if they are the same technology.

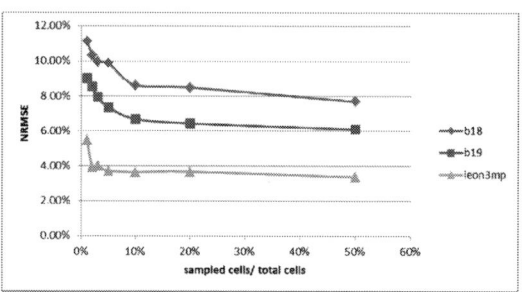

Fig. 6.   Prediction accuracy vs. number of samples (before ECO)

Table III displays prediction results of four benchmark circuits. The machine learning model is trained by data from 10% sampled cells of the first edition E_1. Both *NRMSE* and *CC* are very good. As shown in the table, machine learning can predict IR drop accurately, without any ECO-revision, compared to simulation results. To evaluate the ANN technique, we also tried the *extra tree* technique [22]. Results of three benchmark circuits are very similar to those of ANN.

TABLE III.   EXPERIMENT RESULTS (TRAINING=PREDICTION=E_1)

Circuit	Feature Dimension	NRMSE ANN, Tree	CC ANN, Tree
b18	42	8.7%, 6.8%	0.94, 0.95
b19	44	6.6%, 6.1%	0.94, 0.94
leon3mp	55	3.3%, 4.4%	0.98, 0.98
GPU	1,201	6.7%, NA	0.78, NA

Correlation coefficient (CC) is defined in equation (6). Smaller *NRMSE* and bigger *CC* indicates better results.

$$CC = \frac{\sum_{i=1}^{N}[y_i - mean(y)][\hat{y}_i - mean(\hat{y})]}{\sqrt{\sum_{i=1}^{N}[y_i - mean(y)]^2 \sum_{i=1}^{N}[\hat{y}_i - mean(\hat{y})]^2}} \qquad (6)$$

B. IR Drop Prediction after ECO

We evaluate the effectiveness of the IR drop prediction for circuits after ECO. First, we use the original circuit as edition E_1. Then, we use *Cadence SOC Encounter* to move 13 and 7 serious IR drop cells in benchmark circuits b18 and b19, respectively, to produce a new edition E_2. For benchmark circuit leon3mp, we add one power stripe to produce edition E_2. Then we move 32 cells to produce edition E_3. Three editions of GPU are real data from MediaTek. Table IV shows the prediction accuracy of four circuits after ECO. Machine learning model is trained by data from 10% sampled cells of the first edition (E_1). And then we use the model to predict the second (E_2) and the third edition circuits (E_3).

TABLE IV.   PREDICTION RESULTS OF FOUR CIRCUITS AFTER ECO

	E_1		E_2		E_3	
Circuit	NRMSE	CC	NRMSE	CC	NRMSE	CC
b18	8.7%	0.94	11.2%	0.88	-	-
b19	6.6%	0.94	9.7%	0.93	-	-
leon3mp	3.3%	0.98	6.1%	0.98	7.7%	0.98
GPU	6.7%	0.78	9.0%	0.59	11.2%	0.61

We can see from Table IV that our machine learning model has the best prediction accuracy when predicting the first edition circuit. E_1. As the number of revision increases, prediction accuracy becomes worse. Therefore, it is important to train the model using the most recent revision. Table V and Table VI use both previous editions (E_1 and E_2) data to improve the prediction accuracy of the third edition (E_3). Table V shows the prediction results of randomly sampled 10% cells in E_3. Table VI shows the prediction results of top 10% serious IR drop cells in E_3. With both E_1 and E_2 data in the training, the prediction accuracy is much better than that of using E_1 data only (Table IV). Average error is defined in equation (7). Max Error is defined in equation (8).

$$\text{Average Error} = \frac{\sum_{i=1}^{N} \|\hat{y}_i - y_i\|}{N} \quad (7)$$

$$\text{Max Error} = \max(\hat{y}_i - y_i), \ i = 1 \text{ to } N \quad (8)$$

where $\hat{y}_i$ and y_i are simulated IR drop and predicted IR drop of the i_{th} sample cell, respectively. A positive error means under-prediction but a negative error means over-prediction. The average error of leon3mp is 5mV, 5% of the average IR drop values. The average error of GPU is 3.7mV, which is 15% of the average IR drop values. The max error is 40mv, which is about 20% of the worst case IR.

TABLE V.   PREDICTION RESULTS OF E_3 CIRCUIT (TRAINED BY E_1+E_2)

	E_3		
Circuit	NRMSE	CC	Avg. Error
leon3mp	3.4%	0.98	3.8mV
GPU	6.8%	0.81	3.3mV

TABLE VI.   PREDICTION OF TOP 10% SERIOUS IR DROP CELLS OF E_3

	E_3					
Circuit	NR MSE	CC	Avg. IR drop (mV)	Avg. Error (mV)	Max IR drop (mV)	Max Error (mV)
leon3mp	3.7%	0.54	92	5.0 (5%)	241	49.0 (20%)
GPU	7.4%	0.55	25	3.7 (15%)	190	39.3 (21%)

Fig. 7 is error distribution of leon3mp and GPU (top 10% worst cells in Table VI). Totally 60K and 300K cells for leon3mp and GPU, respectively. 99.9% of errors are smaller than 15% of max IR drop (36mV to leon3mp and 28.5mV to GPU). Red lines mean the 15% boundary. Only 10 cells (out of 60K) in leon3mp and 22 cells (out of 300K) in GPU are under-predicted.

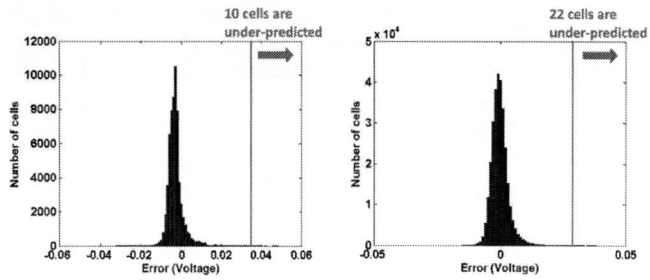

Fig. 7a. Leon3mp error distribution (60K cells)

Fig. 7b. GPU error distribution (300K cells)

Fig. 8 shows the plot of simulation IR drop results versus predicted IR drop for leon3mp and b18. Training data are E_1 plus E_2 and prediction data is E_3. Y axis represents simulated IR drop. X axis represents predicted IR drop. Correlation of simulated IR drop and predicted IR drop is 0.98 for leo3map and 0.88 for b18.

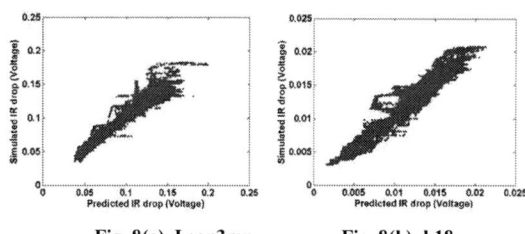

Fig. 8(a). Leon3mp   Fig. 8(b). b18

Fig. 9 shows the IR drop map of leon3mp E_3 circuit. Fig. 9a is simulated IR drop map and Fig 9b is predicted IR drop map. Green area is low IR drop area, yellow area is medium IR drop area, and orange area is serious IR drop area. Red dots are high IR drop cells. Correlation between simulated IR drop map and predicted IR drop map is high.

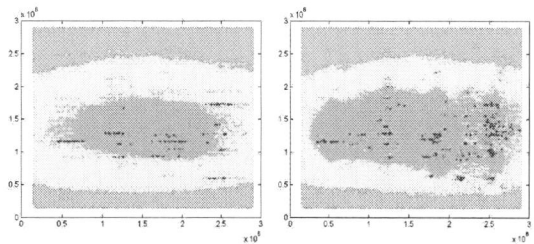

Fig. 9a. Leon3mp simulated IR drop map

Fig. 9b. Leon3mp predicted IR drop map

C. Runtime

Table VII shows the runtime comparison between proposed technique and commercial tools. In the proposed flow, we only need one feature extraction plus one training in the training phase. In the prediction phase, we need one feature extraction plus one (or more) prediction. Total time of proposed technique is two feature extraction time plus one training time plus one prediction time. Although we cannot save time for small circuits, we can save a significant amount of simulation time for large circuits. We need only 13 minutes to predict IR drop for GPU whereas *RedHawk* needs almost one day to simulate the circuit. The run time speedup is shown in the parenthesis (including feature extraction and training). Our technique significantly reduces the IR drop simulation time.

TABLE VII.   RUNTIME COMPARISON

Circuit	b18	b19	leon3mp	GPU
Feature Extraction	12s	32s	147s	11m27s
Training	51s	106s	204s	24m57s
Prediction	1s	2s	19s	1m29s
Total time	76s	172s	517s	49m20s
NANOSIM	46s (0.6X)	95s (0.55X)	734s (1.4X)	-
RedHawk	-	-	-	1 day (30X)

V. DISCUSSION

For ANN to work well, both the number of hidden layers and neurons should be carefully tuned. Using too few neurons will result in *underfitting*. It occurs when there are too few neurons to detect important information in a large data set. Too many neurons may lead to *overfitting*. In this work, we tried two, three up to four hidden layers. We found that two hidden layers with twenty neurons for each hidden layer are enough for our data set. Too many layers would not improve the accuracy, and too many neurons would lead to overfitting.

Our proposal is good for the design sign-off stage, when the revised circuit is very similar to its previous version. Every time we add a new version, we would need to add this new version to our training so that this assumption can be valid.

VI. CONCLUSIONS

In this work, we have proposed an IR drop prediction for ECO-revised circuits using artificial neural network. We sampled a small portion of cells on a die to train the neural network. We proposed seven feature extractions, which are simple and scalable for large designs. Our experiment results show that prediction accuracy (average error 3.7mV) and correlation (0.55) are very high for a 3 million-gate real design. The run time speedup is up to

30X. The proposed method is very useful for designers to save the simulation time to fix the IR drop problem.

REFERENCES

[1] Kenneth L. Shepard, and Vinod Narayanan. "Noise in deep submicron digital design." *Proc. international conference on Computer-aided design.* IEEE Computer Society, 1997, pp. 524-531.

[2] Mohammad Tehranipoor, and Kenneth M. Butler. "Power supply noise: A survey on effects and research." *IEEE Design & Test of Computers* 2.27 (2010): 51-67.

[3] Howard H. Chen, and David D. Ling. "Power supply noise analysis methodology for deep-submicron VLSI chip design." *Proc. Design Automation Conference.* ACM, 1997, pp. 638-648.

[4] Yi-Min Jiang, and Kwang-Ting Cheng. "Analysis of performance impact caused by power supply noise in deep submicron devices." *Proc. Design Automation Conference.* ACM, 1999, pp. 760-765.

[5] L.-C. Wang, D.M.H. Walker, A. Majhi, B. Kruseman, G. Gronthoud, L.E. Villagra, P. van de Wiel,S. Eichenberger, "Power supply noise in delay testing," *Proc. International Test Conference*, pp. 1-10, 2006.

[6] Yi-Hua Li, et al. "Capture-power-safe test pattern determination for at-speed scan-based testing." *IEEE Transactions on computer-aided design of integrated circuits and systems* 33.1 (2014): 127-138.

[7] P. Girard, CW Wu, X Wen, *Power-aware testing and test strategies for low power devices*, 2010.

[8] *Apache RedHawk User Manual*, 2015.

[9] P. Bastani, K. Killpack, Li.-C. Wang, E. Chiprout, "Speedpath prediction based on learning from a small set of examples," *Proc. Design Automation Conference*, 2008, pp. 217 - 222.

[10] Andrew B. Kahng, Mulong Luo, and Siddhartha Nath. "SI for free: machine learning of interconnect coupling delay and transition effects." *System Level Interconnect Prediction (SLIP), 2015 ACM/IEEE International Workshop on.* IEEE, 2015, pp. 1-8.

[11] Fangming Ye, Firouzi, F., Yang Yang, K. Chakrabarty, M.B. Tahoori, "On-chip voltage-droop prediction using support-vector machines," *Proc. VLSI Test Symposium*, 2014.

[12] C.-Y. Han, Y.-C. Li, H.-T. Kan, and James C.-M. Li, "Power-Supply-Noise-Aware Test Pattern Analysis and Regeneration for Yield Improvement," IEICE, Vol.E99-A,No.12,pp.-,Dec. 2016.

[13] Yuta Yamato, "A Fast and Accurate Per-Cell Dynamic IR drop Estimation Method for At-Speed Scan Test Pattern Validation," *Proc. International Test Conference* 2012.

[14] Jeremy Lee, et al. "Layout-aware, IR-drop tolerant transition fault pattern generation." *Proc. Design, automation and test in Europe.* ACM, 2008, pp. 1172-1177.

[15] Junxia Ma, Jeremy Lee, and Mohammad Tehranipoor. "Layout-aware pattern generation for maximizing supply noise effects on critical paths." *2009 27th IEEE VLSI Test Symposium.* IEEE, 2009, pp. 221-226.

[16] N. Ahmed and M. Tehranipoor, "Transition Delay Fault Test Pattern Generation Considering Supply Voltage Noise in a SOC Design," *Proc. Design Automation Conference*, 2007, pp. 533-538.

[17] Xiaoqing Wen, et al. "Low-capture-power test generation for scan-based at-speed testing." *IEEE International Conference on Test*, IEEE, 2005, pp. -1028.

[18] Kohei Miyase, et al. "Identification of high power consuming areas with gate type and logic level information." *2015 20th IEEE European Test Symposium (ETS).* IEEE, 2015, pp. 1-6.

[19] Yu-Cheng Liu, Cheng-Yu Han, Shih-Yao Lin, and James Chien-Mo Li, "PSN-aware Circuit Test Timing Prediction using Machine Learning," *Proc. IET Computers & Digital Techniques*, 2016.

[20] Warren S. McCulloch, and Walter Pitts. "A logical calculus of the ideas immanent in nervous activity." *The bulletin of mathematical biophysics* 5.4 (1943): 115-133.

[21] Fast Artificial Neural Network. Available: http://libfann.github.io/fann/docs/files/fann-h.html

[22] P. Geurts, and *et. al.* "Extremely Randomized Trees," Machine Learning, Springer, 2006.

Special Session on Machine Learning for Test and Diagnosis

Krishnendu Chakrabarty, Duke University, USA
Li-C. Wang, University of California, Santa Barbara, USA
Gaurav Veda, Mentor, A Siemens Business, USA
Yu Huang, Mentor, A Siemens Business, USA (Organizer)

I. Introduction

The special session focuses on using Machine Learning (ML) techniques on different applications in test and diagnosis. The first contribution discusses how to close the gap between working silicon and a working system by using ML. The second presentation then talks an alternative ML view and its various applications such as functional verification, Fmax prediction, and production yield optimization. The last presentation discusses using supervised ML on volume diagnosis to further improve the accuracy of identifying root causes.

II. Data-Driven Resiliency Solutions for Boards and Systems (Krishnendu Chakrabarty)

This presentation will describe how data analytics and real-time monitoring can be used to ensure that boards and systems operate as intended. In the first part of the talk, the speaker will focus on the resilience problem for complex boards; we are seeing a significant gap today between working silicon and a working board, which is reflected in failures at the board level that cannot be duplicated at the component level. The speaker will describe how machine learning, statistical techniques, and information-theoretic analysis can be used to close the gap between working silicon and a working system. Next, the presenter will describe how time-series analysis can be used to detect anomalies in complex core router systems. The effectiveness of proactive fault tolerance depends on whether anomalies can be accurately detected before a failure occurs. However, traditional anomaly detection techniques fail to detect "outliers" when the monitored data involves temporal measurements and exhibits significantly different statistical characteristics for its constituent features. The speaker will describe a feature-categorization-based hybrid method to overcome the difficulty of detecting anomalies in features with different statistical characteristics. A correlation analyzer will be described to remove irrelevant and redundant features. A comprehensive set of experimental results will be presented for data collected during 30 days of field operation from over 20 core routers deployed by customers of a major telecom company.

III. Machine Learning For Feature-Based Analytics in Some Test Applications (Li-C. Wang)

Applying machine learning in Electronic Design Automation (EDA) and Test has received growing interests in recent years. One approach to analyze data from design and test processes can be called feature-based analytics. In this context, the talk explains the inadequacy of adopting a traditional machine learning problem formulation view. Then, an alternative machine learning view is presented where learning from data is treated as an iterative feature search process. The concept of learnable is explained between the traditional big data learning and the small/limited dataset learning commonly encountered in EDA and Test applications. The theoretical and practical considerations for implementing such an iterative feature search process are discussed in the context of various applications such as functional verification, Fmax prediction, and production yield optimization.

IV. Supervised Techniques for Volume Diagnosis (Gaurav Veda)

Traditionally, determining defect distributions in digital logic has relied heavily on diagnosis of scan based test failures followed by failure analysis (FA) techniques. More recently, to address the challenge of increasing fabrication complexity, and at the same time, market pressure to reduce cost/cycle times, unsupervised machine learning approaches to identify root causes from volume diagnosis results, have emerged as a powerful methodology. Supervised machine learning techniques offer an exciting new way of further improving the accuracy of identifying root causes from volume diagnosis results. In this talk, we will present some supervised learning techniques that have shown a lot of promise in our internal research. These techniques provide accurate results, as well as a reliable path to accommodate new defects in new technologies such as cell internal root causes. We've only begun to scratch the surface and this is an area of active research. We will also mention promising future research directions.

Innovative Practices on Challenges, Opportunities, and Solutions to Hardware Security

Sohrab Aftabjahani, Intel, USA
Jason Oberg, Tortuga Logic, USA
Michael Chen, Mentor, A Siemens Business, USA
Huawei Li, SKLCA, Institute of Computing Technology, CAS, China (Organizer)

I. INTRODUCTION

The IP session focuses on challenges, opportunities, and solutions to hardware security. The first contribution discusses design for security and security validation challenges needed to be considered by EDA industry. The second presentation then talks about security verification technologies throughout the design lifecycle. The last contribution discusses a solution that can improve the security of electronic hardware manufacturing.

II. IS EDA INDUSTRY READY FOR DESIGN FOR SECURITY AND SECURITY VALIDATION CHALLENGES? (SOHRAB AFTABJAHANI)

Digital systems speed up the evolution of many aspects of our lives including health, medicine, finance, transportation, defense, education, and even recreation. As our modern life increasingly rely on them, their security becomes of utmost importance.

While it is now well-known the necessity of the security built into digital systems and its verification, security validation is still a big challenge due to their exponential complexity growth and time-to-market pressures. Because EDA tools are not mature enough to provide fast, accurate, and fully-automated analysis to determine the vulnerability of systems to various types of security attacks while attackers demonstrate the feasibility of more advanced attacks such as side-channel and physical attacks. For these attacks, even the basic capability to analyze the systems (semi) automatically to determine if they are vulnerable to such attacks to mitigate the vulnerabilities does not exist in commercial EDA tools.

We conclude that current EDA still need to go a long way with respect to Design for Security and Security Analysis (DFSASV) to keep up with ever increasing demand of industry. Finally, we describe our effort to accelerate realizing an echo-system of security-aware EDA tools for DFSASV and call academia, industry, and government for participation and support.

III. ENABLING FULL SOC HARDWARE/SOFTWARE SECURITY VERIFICATION (JASON OBERG)

Today's SoCs are produced as massive compositions of constituent parts that include new hardware designs, legacy cores from previous designs, and third-party IP cores. Stakeholders responsible for the development of each of these parts may either inadvertently design their hardware in an insecure way, or be malicious - both which lead to hidden design flaws that can be exploited once the SoC is deployed in the field. Because of this disparate design strategy and questionable stakeholder intentions, successfully validating system-wide security properties becomes an intractable and daunting task that results in security vulnerabilities such as Meltdown and Spectre going undetected. To properly identify these types of security vulnerabilities before tapeout, the appropriate security verification technologies need to be in place throughout the design lifecycle. In this talk, we discuss how each stage of the design process (including formal, simulation, and emulation) can be augmented to perform security review without interruption to current design flow standards. We will also present real world security review case studies ranging from small IP blocks to larger SoC subsystems.

IV. FINDING OPPORTUNITIES TO APPLY HARDWARE SECURITY (MICHAEL CHEN)

What needs to be done when a probe touches a wafer-die for the first time? When should the hardware configuration be set? Just how many device variances should there be? How will trust impact global electronic manufacturing and its supply-chain? Are there security vulnerabilities in Industry 4.0 (I-IoT)? Is AWS IoT or 5G a game changer for edge-node protection?

This presentation presents a solution that can improve the security of electronic hardware manufacturing. End-to-end solutions based on a reliable pre-boot Root-of-Trust (RoT) are needed to minimize vulnerability and limit attacks in today's complex supply chain. The solution enables the connection of SoCs to a secure server, tracks the chip at each step in the manufacturing process, and securely configures and updates them in the field. As a result, IC suppliers will have better visibility of the SoC lifecycle from design to chip birth to proliferation in the supply chain to decommissioning.

Fast Fault Coverage Estimation of Sequential Tests Using Entropy Measurements

Sarmad Tanwir
Virginia Tech
Blacksburg, VA 24060, USA
Email: sarmadt@vt.edu

Michael S. Hsiao
Virginia Tech
Blacksburg, VA 24060, USA
Email: hsiao@vt.edu

Abstract—**In this paper, we explore the relationship between fault coverage for sequential circuits and the observed entropy during test application. We find that the entropy measured at a subset of the primary and pseudo-primary outputs is highly and significantly correlated to the regions of the design that are exercised by the test vectors. Consequently, they can be used for estimating the coverages for the newly generated test sets. As computing the entropy requires only logic simulation, it can greatly speed up the fault grading process. In our experiments on the ISCAS'89 benchmarks, we are able to achieve an average root mean squared error of 2.1% in fault coverage estimation for test sets of varying lengths, together with an order of magnitude speedup over fault simulation.**

I. INTRODUCTION

Functional test suites consist of sequences of test vectors. A set of effective sequential tests can help not only for functional testing, but also in Logic Built-in Self Test (LBIST), system level testing, post silicon validation, pre-silicon validation, etc. Determining whether a test suite is effective requires some form of simulation. One popular method is fault simulation. However, fault simulation of functional vectors can be computationally intensive, as it involves not only logic simulation, but also injection and simulation of the faulty events induced by modeled faults. The cost of non-scan (or sequential) fault simulation is exacerbated over the scan (or combinational) counterpart because the simulator also needs to store the faulty values for every fault for multiple time frames as long as the fault has not been detected. With the increasing gate count in industrial circuits, this cost is becoming prohibitive. Therefore, there has been a considerable interest in fast fault coverage estimation without fault simulation.

Initial work in this area such as [1]–[6] was mostly done for the fault coverage estimation of combinational and full scan circuits. In [1], Jain et al. introduce STAFAN, which uses the signal probabilities obtained from logic simulation to estimate the fault coverage. Nakazawa et al. [2] replace the empirical sensitization probabilities in STAFAN with computable propagation probabilities to achieve similar results and Cui et al. further improve the method in [5]. The authors of [3], [4] present a fault sampling paradigm to statistically estimate the coverages for the entire population of faults. Whereas, in [6], the authors demonstrate a correlation between an RTL level coverage metric and the gate level fault coverage and then use it to estimate the fault coverage for the test sets.

supported in part by NSF grant 1422054 and an Intel grant.

In [7], Pomeranz et al. propose a method for ranking the functional test vectors by computing a metric that represents the achieved state traversal during the test application. Instead of estimating the actual fault coverage, it is aimed for ranking the test sets. Mirkhani et al. in [8] use stand-alone fault simulation of modules and some probabilistic analysis to estimate the fault coverage of the entire circuit. In [9] and [10] Mirkhani et al. propose another method that uses Gate Input Combinations (GIC) measured during logic simulation and partial fault simulation to provide better estimates of the fault coverage.

In this paper, we present a different method that uses the entropy measurements at a subset of the primary outputs (POs) and pseudo primary outputs (PPOs) of a sequential circuit during test application to estimate the fault coverage of the applied test set. Another key difference with [10] is that we build a single regression model that captures the circuit specific characteristics by training on many different random test sets. It, therefore, has a better generalizability across other totally different test sets and is also independent of the test vector ordering within the test set. While [10] requires partial fault simulation for each new test set, our model, once trained, requires only logic simulation to estimate the fault coverage of the unseen test sets of varying lengths. The time complexity of our method during prediction is $O(G)$ (G is the number of gates) as compared to the $O(G^2)$ complexity of the full blown fault simulation.

The rest of the paper is organized as follows. In Section II, we define the preliminaries for subsequent discussion. In Section III and IV, we explain the dataset creation and evaluation methodology for our method respectively. In Sections V and VI, we respectively present the experimental results and include a note on a promising application. Finally, we conclude the paper in Section VII.

II. PRELIMINARIES

A. Output Partitions

For the purpose of generating data for this paper, we divide the primary and pseudo primary outputs of a sequential circuit into multiple *partitions* of a pre-defined size. For example, if a benchmark has 6 primary outputs and 12 state elements (resulting in a total count of 18), a partition size of 4 would mean four partitions of size 4 and one partition of size 2. We create these partitions by serially following the order in which the outputs are numbered in the circuit netlist. Later in

our analysis, we compute the correlation of metrics computed over these partitions with the fault coverage of the test set.

B. Observed Count

The *observed count* is the number of unique binary numbers observed at a particular partition during the application of the test set. For example, if three vectors were applied at the circuit inputs and we saw 0000, 0010, and 0000 at a 4-bit output partition, we would say that the observed count at the partition is 2 as only two unique patterns (0000, 0010) are observed.

C. Observed Entropy

Let us say that the proportion in which a binary number i is seen at a partition is p_i, then we define the observed entropy H at that partition for the applied test set using the following equation.

$$H = -\sum_{\forall i} p_i \log p_i$$

For the above example, we have

- $i : 0000:\ p_i = 2/3$

- $i : 0010:\ p_i = 1/3$

Here, the numerator is the number of times the number is observed whereas the denominator is the total number of applied vectors. With the abovementioned observations, the entropy comes out to be:

$$H = -(0.67 \times \log(0.67) + 0.33 \times \log(0.33)) = 0.915.$$

D. Linear Regression

Linear regression is a statistical method that is used to model the relationship between a dependent variable and one or more independent variables. It is called single or multiple linear regression depending on whether there are single or multiple independent variables. It is customary to represent the dependent variable as y and the collection of independent variables by a vector x. Linear regression assumes a linear relationship between y and x such that: -

$$y_i = x_i^T \beta + \epsilon_i$$

Here T represents the transpose of the vector and ϵ represents the noise in the assumed linear relationship. We are interested in computing the values of β. The above equation can be written in vector form to cater for all values of x and y in a dataset as follows.

$$y = X\beta + \epsilon$$

where, $X = [x_1, x_2, ..., x_n]^T$, $y = [y_1, y_2, ..., y_n]^T$ and $\beta = [\beta_1, \beta_2, ..., \beta_n]^T$ respectively.

Given a dataset, the values of X and y are known, whereas the values of β need to be determined, which constitutes the training of the linear regressor. The simplest method to estimate the values of β is known as Ordinary Least Squares estimation (OLS), which gives the following estimate of β.

$$\hat{\beta} = (X^T X)^{-1} X^T y$$

Once we have $\hat{\beta}$, we can use it to estimate y for any unseen set of independent variables x. In this paper, we use the OLS to train the linear regression model.

E. Residuals

We can gauge the quality of the linear regression by studying the statistical properties of the residuals . The residual is simply the error between the actual and the predicted value. Mathematically,

$$r_i = y_i - \hat{y}_i$$

Here y_i is the actual measured value and $\hat{y}_i$ is the predicted (or estimated) value by the linear regression model. The OLS procedure stated above minimizes the $\sum r_i^2$ for the training dataset. The following conditions are closely met when there exists a high linear correlation between the dependent variable and the set of independent variables.

1) The residuals are normally distributed with mean of zero.

2) The residuals are uncorrelated with each other.

In our experiments, we'll test these conditions to validate our assumed linear relationships.

F. Explained Variance

We also look at the explained variance score for our trained regression models. The explained variance is defined as:-

$$explained\_variance(y, \hat{y}) = 1 - \frac{\sigma_{y-\hat{y}}^2}{\sigma_y^2}$$

where, $\sigma_{y-\hat{y}}^2$ is the variance of the residuals and σ_y^2 is the underlying variance of the variable of interest (fault coverage in our case) in the population. Ideally, we'd like the variance of the residuals to be significantly smaller as compared to the variance of the variable of interest. That will lead to a near perfect score of 1 for the explained variance. In practice, the closer the explained variance is to 1, the higher is the prediction quality of the linear regression model.

G. Multicollinearity

Multicollinearity happens if two or more independent predictor variables are highly correlated with each other. This is not a good situation when training a linear regression model. The worst case of multicollinearity happens when two variables are perfectly correlated. In that case, the matrix X has less than full rank and therefore, $X^T X$ cannot be inverted. Resultantly, the OLS estimator $\hat{\beta}$ cannot be computed.

In less than perfect cases of multicollinearity, we see other adverse effects such as:

a) The regression coefficient for a particular feature is highly dependent on whether other features are included in the model.

b) The precision of the regression coefficients decreases as more features are added to the model.

This has a direct impact on model interpretability and generalizability. The partial derivatives of the multicollinear independent variables computed from the model will not correctly capture the relationship between those and the predicted dependent variable. Also, spurious relationships in the model (such as two variables that are negatively correlated with each other and have a negligible combined effect on the predicted variable) may increase the prediction error over the unseen (or test) data.

III. DATASET CREATION

A. Test Generation

For each of the considered benchmark circuit, we generate a population of at least 1000 randomly generated test sets as well as at least one ATPG-generated test set. To achieve a good variation in the fault coverage of these test sets, we compute a good seed by shuffling together the contents of the process id (pid), time of day and the system clock ticks value. We use a different seed for each of the randomly generated test set. Besides a different seed, we also define a different probability of generating a 0 for each test set. This means that the proportion of zeros and ones is randomly varied in the generated tests. This increases the variation in the test sets and their fault coverages. Besides these characteristics, we also randomly vary the size of the generated test sets.

For the ATPG-generated test sets, we use the test sets generated by STRATEGATE [11] in our population. Both the original and compacted test sets from STRATEGATE are used.

B. Feature Computation

We logic simulate the benchmark circuit using each of the generated test sets and measure the observed count and entropy values for each partition. We use a constant partition size of 2 in all of our experiments. We save this data in a text file for analysis.

C. Fault Simulation

We fault simulate each of the test sets using a parallel-fault fault simulator and record the actual fault coverage achieved in the above-mentioned text file. These recorded values help us train the regression model during the training time and help evaluate the performance of the model during the test time.

D. Final Dataset

The final dataset for each benchmark circuit contains 1000+ rows, one row for a separate test set under consideration. Each column in the dataset corresponds to a measured feature for a particular partition such as entropy or the observed count.

TABLE I. BENCHMARK CHARACTERISTICS

Benchmark	Gates	Inputs	States	Outputs
s298	125	6	14	7
s344	210	12	15	12
s382	209	6	21	7
s400	215	6	21	7
s444	232	6	21	7
s526	244	6	21	7
s641	476	38	19	25
s820	336	21	5	20
s832	334	21	5	20
s1196	593	17	18	15
s1423	827	20	74	6
s1488	692	11	6	20
s1494	686	11	6	20
s5378	2779	35	49	179
s35932	19876	38	1728	320

IV. INVESTIGATION METHODOLOGY

Once we have the dataset for each circuit, we first analyze each dataset to find the features (i.e., the metrics computed at different partitions) that are most correlated to the fault coverage of a particular circuit. Then, we build a linear regression model over the selected features to predict the fault coverage for some randomly selected test sets. Finally, we evaluate the predictive power of our developed model by making a comparison with the fault coverage measured by fault simulation. We also repeat the experiment for different number of selected features and study the variation in prediction accuracy. Table I provides the total number of gates, inputs, states and outputs of the considered sequential benchmarks.

A. Feature Selection

We predict the fault coverage, which is a quantitative value using quantitative metrics computed over a small number of partitions in the design. The total number of partitions can be quite large depending on the number of POs and PPOs of the circuit. We, however, select only a few to perform our calculations. We perform this selection by computing the Pearson's correlation a feature f_w computed at partition w and the fault coverage fc, which is defined as follows.

$$\rho_{f_w, fc} = \frac{cov(f_w, fc)}{\sigma_{f_w}\sigma_{fc}} = \frac{\sum_{i=1}^{n}(f_{w,i} - \mu_{f_w})(fc_i - \mu_{fc})}{\sigma_{f_w}\sigma_{fc}}$$

Here, μ represents the mean and σ the standard deviation of the subscripted quantities and n is the total number of records in the dataset. The correlation coefficient can have a value between -1 and 1. A value of 0 implies no correlation, whereas a value of 1 (-1) implies strong positive (negative) correlation. We use the *selectKBest()* function from the Sci-kit Learn's feature selection library to carry out this step. This selects the features sequentially to come up with a set of variables that together are best able to predict the fault coverage.

An alternate approach to feature selection could have been performing a Principal Component Analysis (PCA) over the entire set of features. We, however, want to predict the fault coverage using as little of the circuit outputs (or partitions) as possible. The PCA features, on the other hand, are a linear combination of many circuit outputs, which might improve our fault coverage prediction accuracy, but is contrary to the objective of our investigation.

B. Avoiding Multicollinearity

The library function *selectKBest()* selects the features sequentially. This inherently avoids the occurrence of multicollinearity. For example, let's assume that we have three features x_1, x_2, x_3 where both x_1 and x_2 have a high correlation with the output y, while also being highly correlated to each other. The third variable x_3, on the other hand, is only mildly correlated with y. The sequential feature selector will first pick the best feature say x_2 and then will look for other features. When selecting the second feature, it will find that x_1 is highly correlated with x_2, which means that its contribution is already incorporated in the model. At this point, it may select x_3 because even though it is mildly correlated to y, it can reduce the residuals by explaining the variance of y that is unexplained by the existing variable x_2.

Here, we must also discuss the multicollinearity that might occur *within* the predefined partitions, which are nothing but sets of output bits. This generally will not cause a multicollinearity issue because the correlated circuit outputs are tied together in a single feature and share a single coefficient in the linear regression model.

There does exist a possibility that two inversely correlated outputs end up in the same partition. In that case, they might render the metric computed on that partition useless for predictive purposes. This situation is compensated in practice by the existence of many other partitions that can alternatively explain the variance in the fault coverage.

C. Test and Train Split

To train and test our prediction model, we split the dataset of each benchmark circuit into training and test portions. To perform this split, we randomly select 30% of the rows and put them in the test portion. The ATPG-generated test sets are also used in the test portion. The remaining dataset becomes the train portion. The train portion is then used to train the prediction model, whereas the test portion is used to evaluate the predictive power of the trained model.

We use the *train_test_split()* function from Sci-kit Learn's model selection library to perform this split over the features of the randomly generated test sets. At this point, we ensure that the high coverage deterministic test sets for the ISCAS'89 circuits are only included in the test portion so that the model is trained only on the random test tests.

D. Training the Model

After we have split the dataset into train and test portions, we build a linear regression model over the selected features from the train portion.

TABLE II.   BENCHMARK-WISE RESULTS FOR FAULT COVERAGE PREDICTION

Ckt.	# Features	RMS Error	Expl. Var	FS Time	Est. Time
s298	5(23%)	1.91%	99%	0.0422	0.0074
s344	6(21%)	2.90%	95%	0.0421	0.0094
s382	5(18%)	1.12%	98%	0.1138	0.0077
s400	5(18%)	1.22%	98%	0.0145	0.0009
s444	5(18%)	1.04%	98%	0.0184	0.0010
s526	6(21%)	2.77%	98%	0.0197	0.0011
s641	4(9%)	2.90%	88%	0.0536	0.0202
s820	8(31%)	1.30%	97%	0.1215	0.0183
s832	15(58%)	0.99%	99%	0.1330	0.0180
s1196	6(18%)	2.77%	98%	0.2791	0.0630
s1423	39(49%)	2.36%	96%	1.3004	0.0717
s1488	9(35%)	2.67%	97%	0.4048	0.0606
s1494	9(35%)	2.64%	97%	0.3899	0.0622
s5378	16(7%)	1.82%	97%	2.8469	0.1985
s35932	10(0.5%)	3.50%	95%	28.2520	2.9238
Avg		**2.13%**	**97%**		

E. Evaluation

For evaluation purposes, we use the trained model to predict the fault coverage of the test sets whose features comprise the test portion of the dataset. We measure the root mean squared error and the explained variance for test sets of each benchmark circuit. We also draw scatter plots for the predicted and measured fault coverages and residuals to visualize the quality of our predictions.

V. RESULTS

Table II presents the fault coverage prediction results for the considered benchmark circuits. Column 1 states the benchmark circuit name, column 2 shows the minimum number of features selected for a good fault coverage prediction and within the parentheses, the number of features as a percentage of the total number of features in the dataset. Column 3 gives the Root Mean Squared (RMS) error between the predicted fault coverage and the actual measured fault coverage. Column 4 lists the explained variance for each benchmark. Column 5 shows the runtime of fault simulation, whereas column 6 includes the cumulative runtime for logic simulation, feature computation and execution of the trained prediction model. As expected, the estimation takes an order of magnitude lesser time than the actual measurement by fault simulation. The difference clearly grows for the larger circuits. It is interesting to note that for the largest circuits, the percentage of features used to make the prediction is also quite smaller.

We observe that we are able to achieve an average prediction error of 2.13% as compared to 2.53% achieved by [10] at the saturation level of 0.95, for the common sequential benchmarks, even though we do not use any part of the evaluation test sets to learn the correlation.

978-1-5386-3775-3/18 $31.00 © 2018 IEEE

For a better appreciation by the reader, we present the scatter plots between the predicted and measured values, residual plots and the distribution of the residuals for select IS-CAS'89 circuits of different size. The first column of Figure 1 includes the scatter plots between the predicted fault coverage and the measured fault coverage for the test set populations included in the test portion of the dataset for each considered benchmark. The test portion of the dataset not only contains a representative collection of the random test sets but also the ATPG-generated test sets from [11]. The ATPG-generated tests are the highest fault coverage test sets and can be seen as the rightmost dots in the scatter plots. As may be seen, there are two dots with the highest fault coverage in each scatter plot. These are the compacted and non-compacted test sets from [11].

At this point, we'd like to comment that such accurate prediction of fault coverage of ATPG-generated tests is rather unexpected. What we see here is that the detection of random-resistant faults in the circuit have effected the entropy at the partitions that are not mutually exclusive from the partitions effected by the non random-resistant faults. This may be a consequence of the design of the considered benchmarks and/or the assumptions in naming or numbering the circuit outputs and states. In our experiments, we have simply partitioned the outputs in the defined numeric order and that has worked well for the sake of these experiments. We believe that the explanation of these results requires further research on different circuit designs and numbering conventions so that we are better able to formulate a theory on output groupings. However, the findings are interesting from an empirical standpoint.

The second column Figures 1 includes the corresponding residual plots for the fault coverage estimations. A dot in these graphs represents the residual against the measured true value of the fault coverage.

To visualize how the residuals are distributed, we plot the distributions of the residuals in the third column of Figures 1. We observe that all the distributions are centered at zero and appear similar to the normal distributions with the same mean and variance. These analytical normal distributions are plotted as dotted red curves.

VI. APPLICATION TO TEST GENERATION

We believe that the entropy fault coverage relation is a useful finding that can also be utilized for sequential test generation using advanced numerical optimization techniques. For example, we employed reinforcement learning (the policy gradients method) to generate sequential tests for the circuit s820. Reinforcement learning is an area of machine learning, where the agents that interact with an environment try to learn an action policy that maximizes some notion of cumulative reward. For acting optimally, the agents must 'reason' about the long term consequences of their actions (i.e., maximize total future reward), although the immediate reward of their actions may be suboptimal. It is particularly well-suited to 'sequential' problems that involve a long-term versus short-term reward trade-off. In the case of sequential test generation, the agents are the inputs to the circuit and the reward is the estimated fault coverage computed as a function of the observed entropy during logic simulation. In our initial proof of concept implementation, we were able to consistently improve the fault coverage for circuit s820 using reinforcement learning over multiple iterations. We acknowledge that there does exist a significant room for improvement in the test generation problem, which may be achieved for example by parallel rollouts in a cloud environment. Such enhancements are currently the subject of another study. We, however, do share our initial implementation at the url https://github.com/sarmadt/TestGenRL, as a proof of concept demonstration for the readers.

VII. CONCLUSION

In this paper, we have used entropy measurements at selected primary outputs and state elements to estimate the fault coverages of the sequential test sets for ISCAS'89 circuits. We observe that the linear regression model trained over the random test sets is able to accurately predict the fault coverage for the unseen random as well as ATPG generated test sets. As the method relies only on logic simulation, it can greatly speed up the fault grading process as compared to the expensive sequential fault simulation. We believe that this is a useful finding that can not only be used for fast fault grading of randomly generated test sets, but also, in the future, for the generation of high quality sequential tests through numerical optimization of the extracted features using advanced machine learning techniques.

REFERENCES

[1] S. K. Jain and V. D. Agrawal, "Stafan: An alternative to fault simulation," in *Papers on Twenty-five years of electronic design automation*. ACM, 1988, pp. 475–480.

[2] M. Nakazawa, S. Nitta, and K. Hirabayashi, "Probabilistic fault grading based on activation checking and observability analysis," *Journal of Electronic Testing*, vol. 1, no. 3, pp. 235–238, 1990.

[3] V. D. Agrawal, "Fault sampling revisited," *IEEE Design & Test of Computers*, vol. 7, no. 4, pp. 32–35, 1990.

[4] H. A. Farhat and S. G. From, "A beta model for estimating the testability and coverage distributions of a vlsi circuit," *IEEE transactions on computer-aided design of integrated circuits and systems*, vol. 12, no. 4, pp. 550–554, 1993.

[5] H. Cui, S. C. Seth, and S. K. Mehta, "Modeling fault coverage of random test patterns," *Journal of Electronic Testing*, vol. 19, no. 3, pp. 271–284, 2003.

[6] S. Park, L. Chen, P. K. Parvathala, S. Patil, and I. Pomeranz, "A functional coverage metric for estimating the gate-level fault coverage of functional tests," in *Test Conference, 2006. ITC'06. IEEE International*. IEEE, 2006, pp. 1–10.

[7] I. Pomeranz, P. K. Parvathala, and S. Patil, "Estimating the fault coverage of functional test sequences without fault simulation," in *Asian Test Symposium, 2007. ATS'07. 16th*. IEEE, 2007, pp. 25–32.

[8] S. Mirkhani, J. A. Abraham, T. Vo, H. Jun, and B. Eklow, "Falcon: Rapid statistical fault coverage estimation for complex designs," in *Test Conference (ITC), 2012 IEEE International*. IEEE, 2012, pp. 1–10.

[9] S. Mirkhani and J. A. Abraham, "Eagle: A regression model for fault coverage estimation using a simulation based metric," in *Test Conference (ITC), 2014 IEEE International*. IEEE, 2014, pp. 1–10.

[10] ——, "Fast evaluation of test vector sets using a simulation-based statistical metric," in *VLSI Test Symposium (VTS), 2014 IEEE 32nd*. IEEE, 2014, pp. 1–6.

[11] M. S. Hsiao, E. M. Rudnick, and J. H. Patel, "Sequential circuit test generation using dynamic state traversal," in *Proceedings of the 1997 European conference on Design and Test*. IEEE Computer Society, 1997, p. 22.

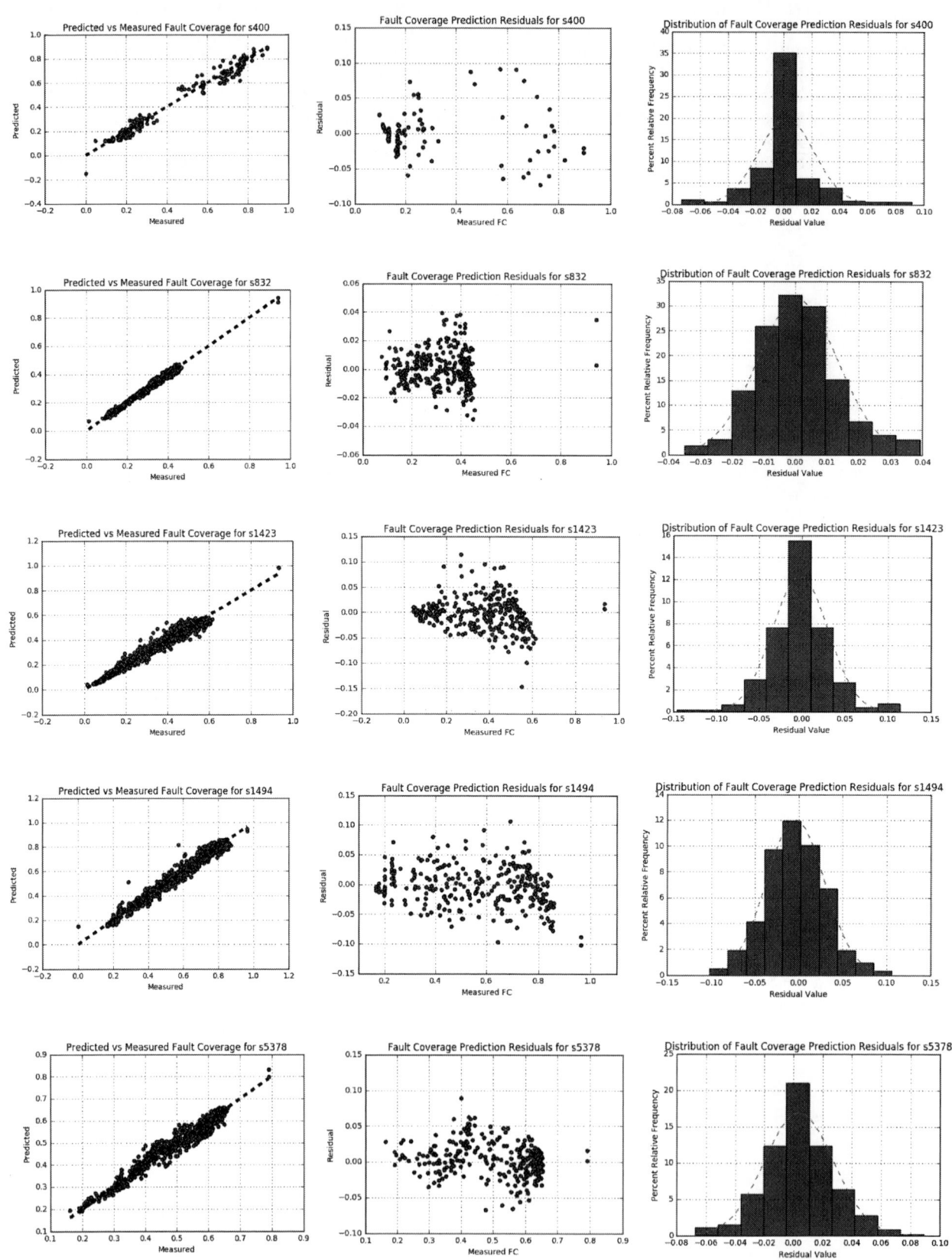

Fig. 1. Scatter plot, residual plots and distribution of residuals for select ISCAS'89 circuits

978-1-5386-3775-3/18 $31.00 © 2018 IEEE

Real-Time Monitoring of Test Fallout Data to Quickly Identify Tester and Yield Issues in a Multi-Site Environment

Qutaiba Khasawneh
Electrical Engineering Department
Southern Methodist University
Dallas, TX USA
qkhasawneh@smu.edu

Jennifer Dworak
Computer Science & Engineering Department
Southern Methodist University
Dallas, TX USA
jdworak@smu.edu

Ping Gui
Electrical Engineering Department
Southern Methodist University
Dallas, TX USA
pgui@smu.edu

Benjamin Williams
Statistical Science Department
Southern Methodist University
Dallas, TX USA
benjamin@smu.edu

Alan C. Elliott
Statistical Science Department
Southern Methodist University
Dallas, TX USA
aelliott@smu.edu

Anand Muthaiah
VP Test Engineering
Tessolve Semiconductor Pvt. Ltd
Bangalore, India
anand.muthaiah@tessolve.com

Abstract— Variations in test fallout during the testing of high-volume devices may arise from a variety of sources. Some of these, such as process variations and design marginalities, correspond to actual problems with the devices being tested and should lead to those devices being scrapped and/or changes being made to the fabrication process. However, in other cases, the test equipment itself could be unreliable or require an alteration in the test procedure, leading to either bad chips tested as good or good chips failing the test. Thus, a real-time monitor of the yield data and the quality of the test result data is essential for assuring high test quality and for ruling out possible issues arising from the test hardware. The real-time monitoring of the yield can save the chip manufacturer time and cost by finding issues with the testing system in the early stages of production (or as soon as those issues arise), avoiding the scrapping of good devices, and preventing the penalty of sending bad units to the customer. In this paper, a low-cost algorithm to monitor the yield of multiple test sites is presented. The method is capable of being implemented in the test program of a standard tester, and the ability of the approach to provide early warning of test site problems during wafer test is demonstrated through an industrial case study.

Keywords—Multi-site, ATE, Yield Loss, Test Quality, Test Variation, Chi-Square, Null Hypothesis, Site Issue, Production Issues.

I. INTRODUCTION

It is important to guarantee that only working chips are sent to customers. This requires testing every chip before shipping it. As a result, it is important to confirm that the system used to test the chips is in good condition to guarantee the quality of the test results. Test systems generally consist of some or all of the following: testers, probe cards, final test boards, probers, handlers, and burn-in boards. Variations in test results may occur due to variations or faults in any of these components in addition to variations or defects in the devices under test. For example, testers usually contain many internal boards. Different testers on the same production line may contain different board revisions and/or operate with different software drivers. All of these factors could cause variations in the test results, especially when many testers are used to accommodate the testing of high-volume devices. The final test and probe boards could also introduce variations due to the design of the boards, the number and type of components used in the boards, the number of sites, the use of sockets, and how long the boards have been in use, which may lead to mechanical and electrical stress. The mechanical damage in the probe needles and

the wafers, its effect on the yield, and proposed solutions are all discussed in [1], [2], and [3].

Site-to-site variations may also exist in multi-site testing due to variations arising from differences in the version of the board layout, tester resource arrangements, and handler/prober alignment. For example, in the case of analog test, the analog waveform generator, voltmeter, DC sources, and other sources/meters come to the test head in a certain arrangement. With a high number of sites, it is impossible to have the same resource arrangement for all sites, and some sites may have longer traces than others due to their arrangement. In addition, the handler/prober alignment affects the Z-direction, where the dies/devices are pushed to the prober/socket. In multi-site testing, the pressure on all sites will not be equal, and the resulting measurements may be affected.

The number of external components on the test board also increases as the number of sites increases. Therefore, the possibility of component degradation and faulty components increases with multi-site testing as well. A faulty component may affect the apparent yield and test fallout from a specific site on that board. This means that the variation of measured data from this site would be different from that obtained from the other sites and the other boards. This would affect the quality of the devices tested in that site and would cause the results to be questionable.

In spite of the possible increase in the test measurement variation as the number of sites increases, the demand is to have more sites in production due to the pressure to fulfill the high-volume demand for tested parts at low cost. In fact, it was shown in [4] that the multi-site solution is one of the most effective approaches for reducing the cost of testing. Hence, it is very important to find any multi-site tester-based hardware issues early on to avoid any quality issues arising during production.

To some extent, this need is addressed by the fact that each production line has a hardware diagnostic flow where the hardware is tested to make sure it is in good condition. The diagnostic flow usually operates when the production line has a problem and when the test program gets loaded to test production material. However, when testing complex devices that need a high number of external components such as relays, resistors, capacitors, inductors and sometimes other chips in the load board, it is very difficult to achieve one hundred-percent full coverage of all those components because it would complicate the load board

978-1-5386-3775-3/18 $31.00 © 2018 IEEE

design and the diagnostic program. This may cause the diagnostic program to miss some faulty marginal components in addition to any faulty untested components. As a result, this paper explores an approach that uses real-time monitors to flag possible yield issues and changes in the distribution of the data. Such monitoring is needed to avoid delays and device losses in production and to meet customers' demands.

II. BACKGROUND OF USING DATA ANALYSIS IN THE SEMICONDUCTOR INDUSTRY

Statistical process control goes back to Western Electric's 1956 rules in [24], which flag a process as being out of control based on trends of measured values (e.g. the number of data points falling outside a given number of standard deviations of the mean).

At the same time, statistical data analysis has been used in the semiconductor testing industry for many years. For example, reference [5] discussed how the design modeling, verification, and testing of analog ICs improved when their variations were understood. It summarized the use of statistical tests to replace the specification-based tests by alternative fast and cheap measurements in the test program. Finally, it discussed the use of adaptive tests to reduce the test time by dynamically changing the limits, test flow, or test content based on historical and real-time data. However, the paper did not explore the effect of any possible issues in the tester hardware or software on the yield and quality of the test results.

Statistical analysis has also been used extensively for outlier identification and to prevent field returns. For example, such analysis was used to screen the outlier units from analog products during manufacturing test in [6]. In [7], statistical analyses of the parametric measurements in the corners of the test were used to identify latent defects. In addition, moving limits and multiple parameter correlation techniques were used to screen the outliers in [8]. In [9] and [10], different adaptive testing approaches and the establishment and application of dynamic test limits based on the measurements obtained from the neighboring dies were explained. References [11] and [12] discussed the virtual probe technique, where the test results of a small number of dies in the wafer can predict the test results of the other dies. This can be done by learning the spatial correlation of the test data across the wafer. The virtual probe in [12] was developed based on compressed sensing [13], [14], and [15], and the algorithm was adopted from maximum a posterior (MAP) estimation. In [15], the part average analysis (PAA) tool was discussed and compared to the part average test (PAT) discussed in [16]. PAA is a tool used by automotive suppliers for early detection of latent defects in the components, and it is based on the PAT. The PAT is used in the semiconductor industry to mark conspicuous dies in the wafer as possible outliers.

In this paper, we further consider statistical methods in test and will use a case study to explore the effectiveness of an "easy-to-calculate" low-cost method that uses a small window of recent test data to identify problems with any site due to the test equipment, load board, or tester hardware. The method is designed to be easy to implement on a tester to target problem detection in real time (i.e. as opposed to post-processing) in a modern multi-site electronic test environment. The goal is to provide opportunities to find and debug problems quickly before a significant amount of the product has been tested on an unreliable site.

III. GOODNESS OF FIT TEST AND HYPOTHESIS TEST

The goodness of fit test is a statistical test used to measure how well the observed data fits the given statistical model as discussed in [20], [21] and [22]. This will be used to determine if the distribution at different sites fits the distribution of the expected yield. Those measurements are used in statistical hypothesis testing to find whether the samples are drawn from identical distributions, such as with the Kolmogorov-Smirnov test or whether the outcome frequencies follow a specific distribution, as with Pearson's Chi-squared test. Therefore, a hypothesis of a statistical relationship existing between the two data sets is tested against the null hypothesis that proposes no statistical relationship between the two sets of data. The test of the hypothesis determines if the null hypothesis will be rejected for a pre-specified threshold probability. This threshold is defined as the level of significance.

Two kinds of errors happen in hypothesis testing. A Type-I error is a false positive error where the true null hypothesis is rejected incorrectly. In contrast, a Type-II error is a false negative where a false null hypothesis is retained incorrectly. The probability of a Type-I error occurring is denoted by α and is the probability of rejecting the null hypothesis when it is true. The probability of a Type-II error occurring is denoted by β and is the probability of failing to reject the null hypothesis when the alternative hypothesis is true.

A contingency table or association table is built by tabulating the actual frequency distribution of the variables in the observed samples. The degrees of freedom (d.f.) is calculated using Equation (1):

$$\text{d.f.} = (r-1)(c-1) \qquad (1)$$

where r is the number of rows in the contingency table and c is the number of columns in the contingency table. For example, a data set of eight categories, where each category has two levels, can be represented in a contingency table of eight columns by two rows. So, the degree of freedom is seven.

Pearson's Chi-squared test (χ^2 test), called simply Chi-squared test for short, is a statistical hypothesis test where the null hypothesis is that the distribution of the sample data follows some hypothesized distribution. The Chi-squared test calculates the deviation of the data from the expected value as χ^2 using (2):

$$\chi^2 = \sum_{i=1}^{i=N} \frac{(O_i - E_i)^2}{E_i} \qquad (2)$$

where O_i is the frequency of the observed value in the contingency table for cell i, E_i is the expected frequency of cell i assuming that the null hypothesis is true, and N is the number of cells in the contingency table.

After calculating the Chi-squared (χ^2) test statistic using (2), the probability of obtaining this specific χ^2 value is found using the published Chi-squared distribution table using the degrees of freedom of the data. This probability is compared with the specified significance level. In the case of multi-site testing, the null hypothesis for a Chi-squared test is: the test result is independent of the site. Thus, rejecting the null hypothesis implies there is significant evidence that the test result depends on the site. The null hypothesis is rejected when the given probability is less than the significance level. Equivalently, the null hypothesis is rejected when the observed χ^2 value is greater than the critical value. The critical value is a χ^2 value at which the probability of obtaining the specific χ^2 statistic is equal to the stated significance level. The observed χ^2 value is compared to the critical value to determine if the result is statistically significant.

978-1-5386-3775-3/18 $31.00 © 2018 IEEE

IV. TEST VARIATION

Test measurement variations are a well-known issue in the semiconductor testing industry. In general, different aspects of the design itself and the process used for manufacturing are the main sources of test result variations. This is true for both digital and analog devices. However, there are many additional potential sources of variations in test measurements, especially for analog tests. For example, testing sensitivity due to program instability could cause variations in the test results, where testing one unit many times using the same test program gives different results. Usually, this instability is due to a short wait time in the test program before making the analog measurement, an insufficient number of samples in the measurement, or a glitch due to a programming mistake that could cause the device to be in the wrong mode or that resets the device. Site-to-site variations and within-site variations that arise due to the wide variety of load board layouts, alignment, and component issues listed earlier are common. Tester-to-tester and load-board-to-load-board variations add to the overall test variations.

The process capability is the variation of the process used to manufacture a product and is defined as the ratio of the spread between the process specifications over the spread of the process values as measured by six process standard deviation units. The process potential index (C_p) is defined in (3). The process capability index C_{pk} is defined in (4), and it is a statistical measurement of the capability of the process to produce output within the specification limits.

$$C_p = (USL—LSL)/(6\sigma) \quad (3)$$

$$C_{pk} = \min [(\mu—LSL)/3\sigma, (USL-\mu)/3\sigma] \quad (4)$$

where: μ is the mean of the data, USL and LSL are the upper specification limit and the lower specification limit, and σ is the overall standard deviation. The standard deviation is due to reproducibility errors and repeatability errors in the system and is defined in (5):

$$\sigma = \sqrt{(\sigma_R^2 + \sigma_r^2)} \quad (5)$$

where σ_r is the standard deviation of the repeatability data, and σ_R is the standard deviation of the reproducibility data. Note that repeatability refers to testing the same unit on the same setup multiple times and obtaining repeatable data. Reproducibility refers to testing the unit on different setups (sites, boards, or testers) and being able to reproduce the same data.

Gauge repeatability and reproducibility (GRR) is one of the methods used to study the variance of the test systems in the semiconductor industry, and GRR is defined in [19] as

$$\%GRR = 100/C_p \quad (6)$$

References [17] and [18] explain the process of finding GRR and the different equations used. The lower the GRR value is, the more repeatable and reproducible the test is. According to the Automotive Industry Action Group (AIAG) in [18], the GRR should be less than 10% to have a system with acceptable repeatability and reproducibility.

V. THE NEED FOR YIELD MONITORING

The GRR calculation is done before releasing the board, the program, and the device to production and in the early stages of production. However, after some time in production, some of the components on the load boards might become faulty or damaged. This would affect the test results and reduce the quality of the testing, where the data could have outliers, level shifts, and variance changes. Outlier devices or even bad units could be sent to the customer if the changes in the test system were not addressed, and good units could be discarded. Therefore, it is very important to have a real-time monitor to detect these issues as early as possible.

Additionally, if only one site or one load board is affected, the reliability of the measurements taken from the affected site or board would be less than that achieved with the other boards and sites. This presents both an opportunity and a challenge. First, if we can identify the different quality of the responses coming from that particular site or board, we can not only identify the sources of the problem, but also correct them. This is the opportunity. The challenge lies in finding a low-cost, effective way to identify such a problem in real time.

Today, post processing the yield data is the most common way to monitor variations in the production data across sites, wafers, etc. For example, this is usually done for outlier identification in probe testing, where a wafer map of the test results is saved in the database. Thus, it is possible to use data mining to find the suspicious units on the wafer and mark them as outliers, as seen in [7]. However, in the final test, where the packaged devices are tested, it is time consuming to use post processing because each unit has to be retested to find its ChipID, if available. Most of the time, if a problem in the yield or test setup is suspected, all of the suspected units have to be retested.

VI. PROPOSED APPROACH FOR REAL-TIME MONITORING

When a high-volume device is running in production, it is important to deliver the high quality units on time. Therefore, it is crucial to monitor the quality of the units as they are being tested. It is not easy to judge the quality of the tests as soon as the units start to fail. The use of statistics helps in making the decision about the quality of the testing and how many units are needed to make this decision. Statistically speaking, a small sample size may not produce accurate decisions, and the larger the sample size, the more accurate the decisions are. However, sample sizes that are too large consume a lot of time and resources. In [19], it is suggested that thirty is the least number of samples that should be used to get a meaningful decision without special handling. The algorithm proposed here monitors the yield of each site in the board using the Chi-squared test to check if the test results are independent of the site on which the tests were performed. The goal is to quickly find that a site in a multi-site setup has begun to fail.

Fig. 1 shows the proposed algorithm to monitor the yield. The algorithm keeps track of the total number of units tested, the number of passing units in each site, the number of failing units in each site, and how many alarms were generated to report a problem. The total number of tested units is compared with the predefined window size, which represents the number of units tested across all of the sites, whose test results will be analyzed together to identify possible dependency between the sites and the test results. If the test results are dependent on the site used to perform the test, a problem with one or more sites may be present.

If we used the minimum recommended sample size of thirty for an eight-site solution, the window size is 240 units in total. The number of failing and passing units at each site is needed to calculate the yield of each site and compare it to the expected value in the Chi-squared test. If test time is especially critical, the Chi-squared test can be run only when the number of failing units in any site is more than expected (RunDependency in Fig. 1).

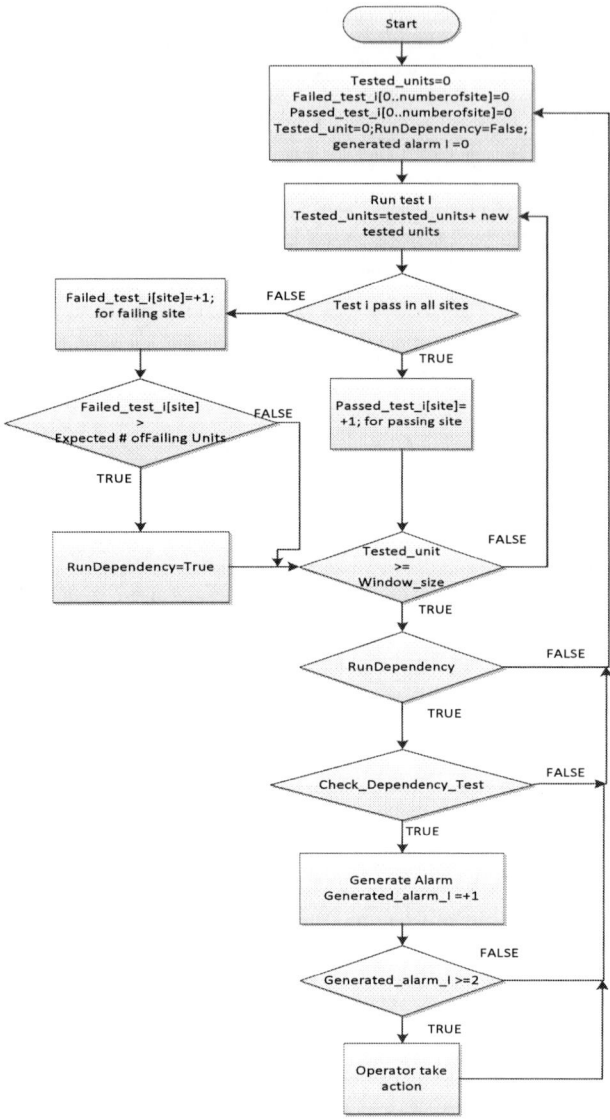

Fig. 1. *Real-time yield monitoring algorithm.*

The degrees of freedom is needed to calculate the Chi-squared test value. In an eight-site solution, there are seven degrees of freedom. This is because the 8 sites constitute eight "categories" (i.e. columns) in the contingency table, and there are two possible test results: pass and fail (corresponding to the rows of the table). According to (1), the corresponding degrees of freedom can thus be calculated as $(8-1)*(2-1) = 7*1=7$. The critical value for d.f. of 7 is 14.07 if the significance level is 0.05 and 18.48 if the significance level is 0.01. The critical value will be used to reject or accept the null hypothesis, where the null hypothesis is that the test results are independent of the site used to test the part. In the algorithm shown in Fig. 1, once an appropriate number of parts have been tested to fill the desired window, a dependency check function may be called. The function is defined in Fig. 2. The code is used if any site has more failing units than the expected yield.

Fig. 2 shows the pseudo-code of the function to check the dependency test using the Chi-squared test. The function inputs are the number of passing and failing units in each site, the expected yield ratio, and the critical Chi-squared value. The

function calculates the Chi-squared value of the results and returns true if it is greater than the critical Chi-squared value provided. Also, the function finds the residual of all the sites and the site with the highest residual. The site with the highest residual is used to identify the site with a possible issue. As shown in Fig. 1, if the result is determined to be dependent, the algorithm generates an informational alarm to be sent to the list of the operators warning them of a possible issue in the production the first time an issue appears. The alarm will include the calculated residual of each site, the site with the maximum residual, and the test name. The residual information is necessary to find the site with the issue. However, if the issue happens again, another alarm is sent to the list with the same information, and an action can be specified depending on the device and the production engineer. The action could be a pause on the production to give the operators a chance to check the status and decide if there is a need to stop the production or a need to stop using that site only. Two alarms identifying the same site will indicate that that site likely has an issue. The dual alarm requirement before action is taken is used to reduce the chance of false alarms stopping production. It also allows more data to be collected before stopping the test to help indicate whether the problem is affecting a single site or multiple sites. However, if different sites are identified with each alarm, this may indicate issues with multiple sites, a board issue, another tester issue, or even a new issue with the underlying processed parts, and more investigation is needed.

```
1   function Check_Dependency_Test(Passedunits,Failedunits,
2                                  CHI_SQUARE_CRITICAL_VALUE,
3                                  ExpectedYield, var max_Res):bool
4   {
5       CHI_SUM = 0
6       for {i = 1 to NmbrSites)
7       {
8           if ((Passedunits[i]+Failedunits[i])>0)
9           {
10              Y[1,i] = Passedunits[i];
11              Y[2,i] = Failedunits[i];
12              E[1,i] = (Y[1,i]+Y[2,i])*ExpectedYield
13              E[2,i] = (Y[1,i]+Y[2,i])*(1-ExpectedYield)
14              RESIDUAL [1,i] = (Y[1,i]-E[1,i])^2/E[1,i]
15              RESIDUAL [2,i] = (Y[2,i]-E[2,i])^2/E[2,i]
16              CHI_SUM = CHI_SUM+RESIDUAL [1,i]+RESIDUAL [2,i]
17          }
18      }
19      max_Res = find_max_location(RESIDUAL)
20      if (CHI_SUM > CHI_SQUARE_CRITICAL_VALUE)
21          return true
22      else
23          return false
24  }
```

Fig. 2. *Dependency Check pseudo-code.*

VII. CASE STUDY

The algorithm was implemented using R, an open source programming language for statistical computing and graphics as defined in [23]. The algorithm was verified using data from four different lots, where the yield of many lots was more than 97%. The goal was to discover whether the algorithm could have identified the presence of a problem more quickly during production, had it been applied at that time.

The production program had 378 tests, and each lot had 5432 units. The four lots investigated included two lots with the expected yield and two lots with lower than expected yield. Specifically, the yield of the third lot was 48.8%, and the biggest failures were in test T1 and test T2. The yield of T1 in lot 3 was 49.7%. Only fifty units from lot 3 failed the T2 test, but all the failing parts were in site 2. The fourth lot's yield was 91.39% and all of the failures were coming from test T3 in site 1. The T3 yield

978-1-5386-3775-3/18 $31.00 © 2018 IEEE

in lot 4 and site 1 was 57.76%. The proposed algorithm was able to quickly find the issues in the 2 lots with low yield without flagging any errors in the other two good lots. Furthermore, it was able to find whether the issue was coming from a specific site.

Fig. 3 has the results for T1 in lot 3 using a significance level of 0.01. The figure contains two plots. The top plot shows the dependency check for each window, and the bottom plot shows the site with the highest residual for the window if the test results are dependent on the site. Clearly, multiple sites have issues.

In Table 1, the residuals of all sites in first window are shown for T1 of lot 3 and T3 of lot 4. The residual is in red if it is larger than the critical Chi-squared level. T1 had an issue on all sites, indicating a possible load board, tester, or lot issue. The debug showed that the problem was a resistor tolerance issue, where the resistor was used to control the analog input to the DUT amplifier. The solution was to use a resistor with lower tolerance.

Although the graph for T3 and lot 4 is not shown due to space limitations, running the proposed algorithm identified a problem in all the windows in the lot. Table 1 clearly identifies Site 1 as the problematic site. In this case, the failing site was out of calibration. The calibration issue, if not fixed, would have caused even more issues if other lots continued to be tested with the same miscalibrated setup.

Fig. 4 has results for test T2 from lot 3 using a significance level of 0.05. Five windows out of the 24 windows showed a site dependence: specifically windows 1, 2, 20, 22 and 24. However, the other windows had at least one failing device in site two even though the other sites had no failures. The source of the failures was a wait time issue, and it was fixed by increasing the wait time.

Using the proposed algorithm allows the problem to be flagged earlier than could occur with a post-processing analysis based on the yield. For example, it would take two windows and a total of 480 tested parts to detect the issue with our real-time monitoring algorithm for T1 and T2 in lot 3 and T3 in lot 4.

Identifying the problem earlier allows debug to start earlier. It also reduces the number of parts that need to be re-tested, as well as the corresponding time needed to perform the re-test. For example, finding the problem after two windows corresponded to 249 failing units in T1. In the case of post processing, the problem would be identified only after the entire lot was tested, and 2699 units had failed the test. In the case of T3 and lot 4, if the issue were found later using post processing, all 256 failing parts in the lot would need to be retested. They would need to be tested one die at a time because they are not physically adjacent on the wafer. The test time for a single unit would be very close to the 25.7 seconds of the octal sites' test time. Also, the setup time for the prober and the tester would add more time to retest those units. In contrast, using the proposed algorithm would flag the site issue with only 27 failing units in the first 2 windows, not 256 failing units in the lot. Thus, only 27 units would need to be retested. Of course, if all parts (not just the failing ones) need to be retested, the number of re-tests we save is even greater.

The proposed approach may also allow problems to be found that would otherwise remain hidden in relatively high yields. For example, in the case of T2, with only fifty total failures in the lot, the failures might have been ignored because the loss is less than 1%. However, with the proposed windowing algorithm the problem would be flagged with the sent alarms, and the failing site would have been automatically identified. This is important because failing to identify a test issue may lead to poor test quality. This could lead not only to yield loss, but also to less

trustworthy test results overall. Note that when good parts fail and are scrapped, the customer delivery commitments may be affected, and the cost may increase due to requirements for testing more lots with more resources (e.g. testers, load boards, and operators).

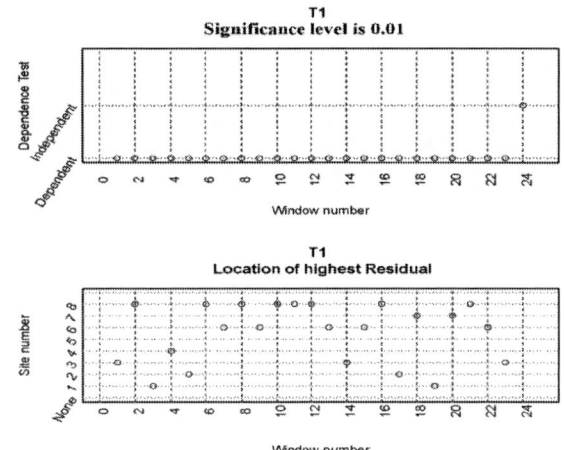

Fig. 3. Dependency Check Function Results for T1.

TABLE I. THE CALCULATED RESIDUAL FOR THE FIRST WINDOW.

	Site	1	2	3	4	5	6	7	8
T1	F	484	285	535	351	271	283	342	206
	P	15	8.8	16.5	10.8	8.38	8.7	10.6	6.37
T3	F	95	0.8	0.8	0.8	1.1	1.0	0.9	0.9
	P	2.9	0.02	0.03	0.02	0.03	0.03	0.03	0.03

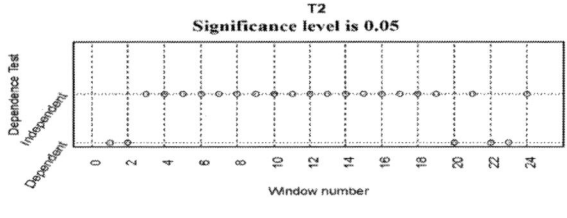

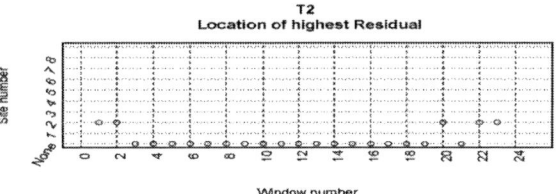

Fig. 4. Dependency Check Function Results for T2.

Once an issue is flagged, several responses are possible. For example, flagging a problem and identifying the site in question will enable the operators to disable that site and run the material without causing more damage to the units that would otherwise be tested at that site. The wafer, the probe card or the load board in the final test could be damaged if testing is not stopped. Alternatively, the operators may be able to switch the load board to a backup board and continue testing to meet the commitment to customers while debugging the board with the issue. Another option is to switch to the diagnostic program for the test board and the tester to better identify the problem. In any case, flagging the issue early is necessary to give a timely warning to the test engineer and the management so that the issue can be resolved and any customer commitment addressed.

VIII. CONCLUSIONS

A real-time monitor of the yield and the quality of the test results in the production test can significantly reduce the time required to find problematic sites in a multi-site environment. The faster an issue is found, the faster it can be fixed, and the less money is lost in retesting, recalling bad units, and scrapping good units. Thus, it is desired to develop a method of quickly identifying a problem at a site using simple and inexpensive modifications to the test procedure. The yield monitoring algorithm presented here checks the yields of each site and compares them to the expected yield using the Chi-squared test, and generates an alarm if the yield is not statistically independent.

The proposed algorithm can catch an issue and indicate whether it affects all the sites equally (e.g. due to aging) or if it affects only one site at a time. In the algorithms described here, an alarm is sent when a possible issue is found, and a threshold of two failing windows is used to spur action, such as pausing the testing to avoid any damaging effect on the hardware or the units under test. The threshold and the action taken could be changed depending on the type of the device, the customer commitment, the production engineer's experience, and the history of the device. For some parts and production environments, the "pause the production" approach is preferable to the common response of immediately scrapping the yield loss. For example, in the case of high current devices, it is a better solution to pause the production instead of possibly damaging the devices, test board, or tester.

The proposed algorithm can be used for both probe testing of the wafer level units and for the final test that tests the packaged units. Thus, it has advantages over approaches that are only used for wafer-level test. Also, the number of failing units in the algorithm is less than the numbers of failing units in the post processing method. This will save time in retesting the questionable units. The proposed algorithm is also easy to implement as a real time monitor in the test program of standard testers used in the semiconductor industry without the need for new infrastructure or significant expenditures. There are companies in the semiconductor testing industry that sell services to monitor the yield and send reports about the yield almost in real time. While such services can provide detailed and valuable information for yield learning and addressing other process and test problems, the proposed approach can provide a simple solution that can be employed to identify problems with the tester at very low cost if implemented early in production program if outside services are not contracted. The time and training required to setup the code is likely to be minimal, and a library could be prepared by the test engineer and reused with every new program. The Chi-squared test could also be added to the operating system that comes with the tester. This would reduce the complexity of the code and will make it faster to implement for new devices.

Future work will explore the ability of the proposed approach to handle environments with low expected yields and more subtle/early failures of the test equipment. The sensitivity of the proposed approach to changes in the expected yield and the significance level will also be explored.

ACKNOWLEDGMENT

Thanks to Tessolve Semiconductor Pvt. Ltd for providing the test data used in this paper.

REFERENCES

[1] G. W. Mann, F. Taber, P. Seitzer, and J. Broz, "The leading edge of production wafer probe test technology," in Proc. Int. Test Conf., 2004, pp. 1168–1195.

[2] G. Hotchkiss, G. Ryan, J. Broz, R.M. Rincon, S. Mitchell, R. Rolda, R. "Effects of Probe Damage on Wire Bond Integrity", IEEE ECTC, 2001.

[3] W. Sauter, T. Aoki, T. Hasida, H. Miyai, K. Petrarca, F. Beaulieu, S. Allard, J. Power, and M. Agbesi, "Problems with Wire Bonding on Probe Marks and Possible Solutions", IEEE ECTC, 2003.

[4] Volkerink, E.H.; Khoche, A.; Rivoir, J.; Hilliges, K.-D., "Test economics for multi-site test with modern cost reduction techniques," VLSI Test Symposium, 2002. (VTS 2002). Proceedings 20th IEEE, vol., no., pp.411, 416, 2002.

[5] De Jonghe, D.; Maricau, E.; Gielen, G.; McConaghy, T.; Tasic′, B.; Stratigopoulos, H., "Advances in variation-aware modeling, verification, and testing of analog ICs," Design, Automation & Test in Europe Conference & Exhibition (DATE), 2012 , vol., no., pp.1615,1620, 12-16 March 2012.

[6] Krishnan, S.; Kerkhoff, H.G., "Exploiting Multiple Mahalanobis Distance Metrics to Screen Outliers From Analog Product Manufacturing Test Responses," Design & Test, IEEE, vol.30, no.3, pp.18, 24, June 2013.

[7] Turakhia, R.P.; Daasch, W.R.; Lurkins, J.; Benware, B., "Changing test and data modeling requirements for screening latent defects as statistical outliers," Design & Test of Computers, IEEE, vol.23, no.2, pp.100,109, March-April 2006.

[8] Liquan Fang; Lemnawar, M.; Yizi Xing, "Cost Effective Outliers Screening with Moving Limits and Correlation Testing for Analogue ICs," Test Conference, 2006. ITC '06. IEEE International, vol., no., pp.1, 10, Oct. 2006.

[9] Mingjing Chen; Orailoglu, A., "Test cost minimization through adaptive test development," Computer Design, 2008. ICCD 2008. IEEE International Conference on, vol., no., pp.234, 239, 12-15 Oct. 2008

[10] Marinissen, E.J.; Singh, A.; Glotter, D.; Esposito, M.; Carulli, J.M.; Nahar, A.; Butler, K.M.; Appello, D.; Portelli, C., "Adapting to adaptive testing," Design, Automation & Test in Europe Conference & Exhibition (DATE), 2010 , vol., no., pp.556,561, 8-12 March 2010.

[11] Hsiu-Ming Chang; Kwang-Ting Cheng; Wangyang Zhang; Xin Li; Butler, K.M., "Test cost reduction through performance prediction using virtual probe," Test Conference (ITC), 2011 IEEE International , vol., no., pp.1,9, 20-22 Sept. 2011.

[12] Hun-Kai Hsu; Fan Lin; Kwang-Ting Cheng; Wangyang Zhang; Xin Li; Carulli, J.M.; Butler, K.M., "Test data analytics — Exploring spatial and test-item correlations in production test data," Test Conference (ITC), 2013 IEEE International , vol., no., pp.1,10, 6-13 Sept. 2013.

[13] C. Bishop, Pattern Recognition and Machine Learning. Upper Saddle River, NJ: Prentice-Hall, 2007.

[14] D. Donoho, "Compressed sensing," IEEE Trans. Inform. Theory, vol.52, no. 4, pp. 1289–1306, Apr. 2006.

[15] Wagner, M., W. Unger, and W. Wondrak. "Part average analysis–A tool for reducing failure rates in automotive electronics." Microelectronics Reliability46.9 (2006): 1433-1438.

[16] AEC-Q001 Rev.D, 12/01/2011, Guidelines for Part Average Testing

[17] Burns, Mark, and Gordon W. Roberts. An introduction to mixed-signal IC test and measurement. Vol. 2001. New York: Oxford University Press, 2001.

[18] Burdick, R. K., Borror, C. M., and Montgomery, D. C. (2003), "A Review of Methods for Measurement Systems Capability Analysis," Journal of Quality Technology, 35, 342–354.

[19] Cohen, Jacob. "Things I have learned (so far)." American psychologist 45.12 (1990): 1304.

[20] Romano, Joseph P., and E. L. Lehmann. "Testing statistical hypotheses." (2005).

[21] Stephens, M. A. "EDF Statistics for Goodness of Fit and Some Comparisons." Journal of the American Statistical Association, vol. 69, no. 347, 1974, pp. 730–737., www.jstor.org/stable/2286009.

[22] Arnold, Taylor B., and John W. Emerson. "Nonparametric goodness-of-fit tests for discrete null distributions." The R Journal 3.2 (2011): 34-39.

[23] https://www.r-project.org/about.html.

[24] Western Electric, Statistical Quality Control Handbook, 1956.

978-1-5386-3775-3/18 $31.00 © 2018 IEEE

Online Information Utility Assessment for Per-Device Adaptive Test Flow[1]

Yanjun Li
Center for Information
Geoscience/School of
Automation Engineering
UESTC
Chengdu, China
yjli@uestc.edu.cn

Ender Yilmaz
NXP Semiconductors
Austin, USA
ender.yilmaz@nxp.com

Peter Sarson
AMS AG
Premstaetten, Austria
peter.sarson@ams.com

Sule Ozev
School of Electrical,
Computer, and Energy
Engineering
Arizona State University
Tempe, USA
Sule.Ozev@asu.edu

Abstract—Per-device adaptive test is a promising direction with the best trade-off between test quality and test time so far. In this work, we propose a method for online assessment of the information content of the next test in the test queue. This assessment can be used to tune the trade-off between test quality and test time of a per-device adaptive test. Since majority of specification parameters are correlated, the overall information content of multiple tests is difficult to extract. We model multivariate correlations among specification parameters and take these correlations into account to estimate the multivariate overall information utility of a given set of tests. The proposed method can be integrated within an existing adaptive test flow (per-device or per-wafer) that runs in the background. Experimental results using 3 distinct industry circuits and sizable data show that the proposed technique can finely tune the trade-off, even achieve zero test escape rates with appreciable test time savings.

I. INTRODUCTION

For mixed-signal/RF circuits, a large number of parameters need to be measured to ensure that the shipped devices perform with respect to their specifications. Due to high levels of integration and the resulting increase in functionality, the number of tests that need to be applied to a mixed-signal device is increasing, which increases the test cost. The prohibitive cost of measuring all specification parameters of a mixed-signal circuit drives researchers to develop new test methods to reduce the test time. Fortunately, most of the specification parameters are mutually correlated, making this full-specification measurement approach unnecessary. It is possible to carefully select a subset of specification parameters to be measured and infer the conformity of the rest of the specifications.

Traditionally, this selection is done based on a small number of initial measurements and statistical tools that evaluate the potential of these measurements to detect failures [1]-[4]. This approach has been very effective as static test compaction with the underlying assumption that manufactured devices behave similarly enough to draw effective conclusions for the entire population based on an initial set of samples. However, there is a fundamental trade-off between test time and test quality. With increasing process variations, the similarity assumption no longer holds. The trade-off between test time and test quality becomes unsustainable where one must increase test time continuously to maintain the same quality level. Thus, the test flow must take this diversity in device behavior into account to provide equally effective solutions.

Adaptive testing is a general term that refers to methods that try to bend the test time-test quality curve by learning from the smaller, localized set of devices, or the device's own behavior. The concept of adaptive test can be applied to learn/optimize within the lot [5], within the wafer [6]-[8], or within the device under test (DUT) itself [9]-[12]. With increasing granularity, there is a better chance to tailor the test, and potentially higher computational effort for learning/optimizing.

A number of static test compaction methods take correlations among specification parameters into account when selecting which tests to apply. A two-class support vector machine is used to model the correlation between pass/fail patterns of devices [13]. SVMs may pass or fail devices based on incomplete information and thus may result in both yield loss and test escapes. In [14], tests are selected based on both their process capability index and correlations among them. In [15], [16], the goal is to select easier tests and infer the pass/fail status of more complex test. Correlation between test suites was analyzed using canonical correlation [17], and a genetic algorithm was applied to optimize for test time reduction.

Since per-device adaptive test provides the best granularity and quality/time trade-off, it is the focus in this work. In [10]-[12], the authors present an adaptive test flow that skips tests based on earlier measurements. The learning process is based on incremental mapping which enables continuous background learning. The prediction of the next test in the queue is based on the joint probability distribution function of the specification parameters, which is modeled using the kernel smoothing method. A fail probability threshold is used to tune the trade-off between the test quality and test time. While adaptive test approaches in general utilize a multi-variate statistical model, the fail probability of

---

[1] Project NO. ZYGX2016J220 Supported by Fundamental Research Funds for the Central Universities, China.

978-1-5386-3775-3/18 $31.00 © 2018 IEEE

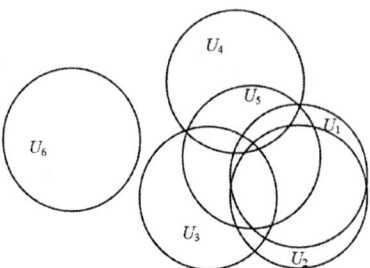

Fig. 1 Demonstration of Information Utility of a 6-parameter case

the next test in the queue is generally computed using pairwise correlations, which skews the information content [10]-[12].

In this paper, we propose a new mathematical model for evaluating the utility of the next test in the queue in terms of detecting defective circuits. The utility function is based on mutual information content between the next test in the queue and the tests that have so far been applied. In order to limit the computation time associated with the utility metric, we propose to use the projection of the test vector to a hyperplane instead of directly calculating the mutual information. The computation of the proposed metric poses a small additional computational burden. However, with the pipelined test/computation flow that can be employed for adaptive test flows [10]-[12], the proposed method poses no additional burden in terms of test time. When the proposed utility assessment method is integrated with an existing adaptive test flow, it provides improved results compared to the state of the art [12].

II. INFORMATION UTILITY

In the domain of information theory [18], the measure of the amount of information one random variable contains about another is defined as mutual information. Let us denote $p(i)$ and $p(j)$ as the probability density functions of test i and test j, and $p(i, j)$ as the joint probability density function of tests i and j. The mutual information I($i;j$) is the relative entropy between the joint distribution and the product of distributions, $p(i)p(j)$, as given in Equation (1).

$$I(i; j) = \sum_{i,j} p(i, j) \log \frac{p(i, j)}{p(i) p(j)} \qquad (1)$$

At one extreme, if tests i and j are independent, the mutual information between them is zero. At the other extreme, if test i is a deterministic function of test j, all information conveyed by test i is shared with test j. In fact, mutual information measures not only the linear dependence as correlation measures, but also nonlinear dependence. Since correlations exist among the circuit parameters, we believe that the applied tests contain information about the not-yet applied ones. We use multi-variate correlations among the specification parameters to estimate the mutual information among them.

In this work, we introduce a metric named as *mutual information utility* to measure the information content of the next test in the queue that has been captured by the tests that have already been applied. If the next test in the queue is a deterministic function of the applied tests, it provides no new

information. In other words, the mutual information utility of the next test based on the already applied tests is one. Conversely, if the next test in the queue is independent of all tests that have already been applied, the mutual information utility equals to zero. When a decision is to be made whether to skip the next test in the queue, the confidence of this decision can be increased with high information utility.

As an example, consider a six-parameter case as in Fig. 1, where each circle represents information each test contains. The overlapped area between two circles represents the mutual information shared by a pair of tests. In this case, after applying the test set {T1, T3, T4}, the information utility of T5 equals one, since the already-applied test set contains the aggregate information of T5. However, after applying test set {T1, T2, T4}, the information utility of T5 is less than 1. Moreover, T6 shares no information with other tests. Thus, after applying any test set excluding T6, the information utility of that set for T6 equals zero. As discussed above, the information utility is related both to the mutual information shared between the concerned test and the applied test set, as well as the mutual information shared inside the applied test set. While this simple example demonstrates the concept of information utility, it is rather involved and tedious to compute the mutual information shared by multiple parameters as the overlapped area in Fig. 1. This computation overhead would raise exponentially with the growing number of dimensions, i.e. the number of specification parameters.

To estimate the mutual information utility, we model tests as vectors in a multi-dimensional space and use multi-variate correlations among the vectors. As an example, consider a 3-parameter case, as illustrated in Fig. 2, where t_k, t_i and t_j represent 3 tests in a 3-dimensional space. Suppose t_i and t_j are applied tests, and t_k is the next test in the queue. If t_k is independent of both t_i and t_j, the vector t_k should be vertical to the space spanned by t_i and t_j, and the test set {t_i, t_j} contains no mutual information with t_k. And if t_k is a deterministic function of t_i and t_j, the vector t_k should be within the space spanned by t_i and t_j, and the test set {t_i, t_j} contains all of the information content of t_k. We use the projection from t_k to the space i-j to estimate the mutual information of the test set {t_i, t_j} with test t_k. The mutual information utility of t_k with the test set {t_i, t_j}, denoted as $U(T_k)$, can be estimated as in Equation (2). We use the square of the vector length for convenience to take both positive and negative correlation into account. Equation (2) can be extended to dimensions more than 3.

$$U(T_k) = \frac{t_{k-proj}^2}{t_k^2} \qquad (2)$$

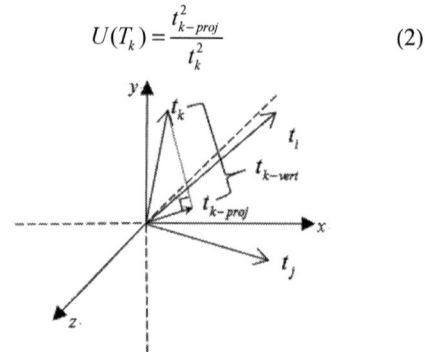

Fig. 2 Information Utility Estimation in a 3-parameter Case

978-1-5386-3775-3/18 $31.00 © 2018 IEEE

III. ONLINE INFORMATION UTILITY ASSESSMENT SCHEME

In general, an adaptive test flow is composed of two stages, the off-line learning stage, and the dynamic learning and prediction stage. In prior work, early characterization data is used to train the parameter model and order the initial test list. In our work, we first utilize the *thin singular value decomposition* [19] to reconstruct the training data as in Equation (3).

$$[\mathbf{x}_1 \quad \cdots \quad \mathbf{x}_p] = \mathbf{U}_p \Sigma_p \mathbf{V}' = [\mathbf{u}_1 \quad \cdots \quad \mathbf{u}_p] \Sigma_p \mathbf{V}' \quad (3)$$

where p is the number of parameters in the test list, n is the number of circuit samples in the training data, the column vector $\mathbf{x}_i = [x_{1i}\, x_{2i} \dots x_{ni}]'$ is the centered measurement results of the ith parameter with n samples, i.e., each $\mathbf{x}_i$ is the result of the measurement minus its empirical mean value. $\mathbf{u}_i = [u_{1i}\, u_{2i} \dots u_{ni}]'\ (i = 1, \dots, p)$ are mutually orthogonal variables with the same variance $\sigma_u = 1/(n\text{-}1)$ [20] and zero mean values, which means $\mathbf{u}_i = (i = 1, \dots, p)$ are mutually uncorrelated. Let us define the test vector matrix $\mathbf{T}$ as Equation (4) and we can rewrite Equation (3) in vector form as Equation (5).

$$\mathbf{T} = \Sigma_p \mathbf{V}' = \begin{bmatrix} t_{11} & \cdots & t_{1p} \\ \vdots & \ddots & \vdots \\ t_{p1} & \cdots & t_{pp} \end{bmatrix} \quad (4)$$

$$\mathbf{x}_i = \sum_{k=1}^{p} t_{ki} \mathbf{u}_k \quad (5)$$

Note that the computation of the test vector matrix is an off-line step, thus adding no test time overhead for adaptive test. Equation (5) implies that the training data can be reconstructed as a linear combination of uncorrelated variables. The correlation coefficient between $\mathbf{x}_i$ and $\mathbf{x}_j$ is given in Equation (5), where, cov($\mathbf{x}_i$, $\mathbf{x}_j$) denotes the covariance of $\mathbf{x}_i$ and $\mathbf{x}_j$, σ_i and σ_j denote the standard variations of $\mathbf{x}_i$ and $\mathbf{x}_j$.

$$\rho_{i,j} = \frac{\text{cov}(\mathbf{x}_i, \mathbf{x}_j)}{\sigma_i \sigma_j} \quad (6)$$

Thus, the correlation between $\mathbf{x}_i$ and $\mathbf{x}_j$ can be calculated as in Equation (6).

$$\rho_{i,j} = \frac{\sigma_{\mathbf{u}}^2 \sum_{k=1}^{p} t_{ki} t_{kj}}{\sqrt{\sigma_{\mathbf{u}}^2 \sum_{k=1}^{p} t_{ki}^2}\sqrt{\sigma_{\mathbf{u}}^2 \sum_{k=1}^{p} t_{kj}^2}} = \frac{\sum_{k=1}^{p} t_{ki} t_{kj}}{\sqrt{\sum_{k=1}^{p} t_{ki}^2}\sqrt{\sum_{k=1}^{p} t_{kj}^2}} \quad (7)$$

From Equation (7), we can see that the correlation between $\mathbf{x}_i$ and $\mathbf{x}_j$ can be represented by the correlation between vectors $\mathbf{t}_i$ and $\mathbf{t}_j$, where $\mathbf{t}_i = [t_{1i}\, t_{2i} \dots t_{pi}]'\ (i = 1, \dots, p)$, is the test vector. By means of this reconstruction, the nxp training data can be represented by a pxp matrix named as test vector matrix. Thus, multi-variate correlations can be analyzed by calculating the projection from the concerned test vector to the hyperplane spanned by the applied ones. This projection is used to estimate the mutual information utility of the given test vector with respect to the tests that have already been applied as in Equation (2).

Estimation of the mutual information utility is an on-line step, which is computed using the projection mentioned above. It is essential that this computation does not exceed the test time of individual tests in the queue to enable background computation. In order to limit the computation time of the projection of the current test vector, we use QR decomposition [19]. Suppose the first i tests are applied, we implement a QR decomposition to the first i vectors in $\mathbf{T}$ as Equation (8) and Equation (9).

$$\mathbf{T}^i = \mathbf{Q}_i \mathbf{T} \quad (8)$$

Where, $\mathbf{Q}_i$ is an orthogonal matrix, which means that any projection from a concerned vector in $\mathbf{T}^i$ to the others is the same as the one in $\mathbf{T}$.

$$\mathbf{T}^i = \begin{bmatrix} t_{1,1}^i & t_{1,2}^i & t_{1,3}^i & \cdots & t_{1,i}^i & t_{1,i+1}^i & \cdots & t_{1,p}^i \\ 0 & t_{2,2}^i & t_{2,3}^i & \cdots & t_{2,i}^i & t_{2,i+1}^i & \cdots & t_{2,p}^i \\ 0 & 0 & t_{3,3}^i & \cdots & t_{3,i}^i & t_{3,i+1}^i & \cdots & t_{3,p}^i \\ \vdots & \vdots & \vdots & \ddots & \vdots & \vdots & \ddots & \vdots \\ 0 & 0 & 0 & \cdots & t_{i,i}^i & t_{i,i+1}^i & \cdots & t_{i,,p}^i \\ 0 & 0 & 0 & \cdots & 0 & t_{i+1,i+1}^i & \cdots & t_{i+1,p}^i \\ \vdots & \vdots & \vdots & \ddots & \vdots & \vdots & \ddots & \vdots \\ 0 & 0 & 0 & \cdots & 0 & t_{p,i+1}^i & \cdots & t_{p,p}^i \end{bmatrix}$$

$$(9)$$

Thus, in $\mathbf{T}^i$, the length of the projection from the $(i+1)$th column vector to the hyperplane spanned by the first i vectors can be calculated as in Equation (10).

$$t_{i+1-proj} = \sqrt{(t_{1,i+1}^i)^2 + (t_{2,i+1}^i)^2 + \cdots + (t_{i,i+1}^i)^2} \quad (10)$$

As a result, the mutual information utility of the $(i+1)$th test with the first (i) tests in the flow can be calculated as in Equation (11).

$$U(T_{i+1}) = U(T_{i+1}^i) = \frac{(t_{1,i+1}^i)^2 + (t_{2,i+1}^i)^2 + \cdots + (t_{i,i+1}^i)^2}{(t_{1,i+1}^i)^2 + (t_{2,i+1}^i)^2 + \cdots + (t_{p,i+1}^i)^2} \quad (11)$$

Fig. 3 outlines the algorithm for the online mutual information utility computation. First, through off-line analysis (Lines 10-30), we construct the test vectors and prepare the mapping model. The mapping model can be based on the kernel based joint probability distribution, as in [12], or any other statistical mapping tool, such as using neural networks.

In the online adaptive test stage, the model is reloaded at the beginning of each device's test flow as in Line 50. For each test in the flow (T_j), the mapping model is used to predict its outcome as in Line 70. If the predicted fail probability of the test T_j is low, it would be skipped. At this point, we also calculate the mutual information utility of the test T_j with the tests that have already been applied, {$T\_applied$}. If this mutual information utility is above the utility threshold (i.e. majority of the information content of T_j is included in {$T\_applied$}), we can skip the test. Otherwise, we will apply this test since it provides us with new information about the device under test (DUT). If test T_j

```
# Static Stage
10      Model Training
20      Initial Test List Ordering
30      Training Data Reconstruction
# Adaptive Test Stage
40      for device i = 1 : N
        {
50      Reload model; T_applied ={}
60      for test Tⱼ, j = 1 : M
            {
70              if (Tⱼ predicted unfaulty & U (j, T_applied) > Uₜₕ)
80                  Skip Tⱼ;
90                  j++;
100         else
110                 Apply Tⱼ;
120                 if (Pass Tⱼ )
130                     j++;
140                     Update model;
150                     T_applied=T_applied∪Tⱼ;
                        Update U(T_applied)
160                 else
170                     device i fail;
180                     Break to next device;
190                 end
200         end
            }
210     if (j=M & device i pass the outlier filer)
220         device i pass
230     else
240         Apply all the unapplied tests
250         if (Pass all the unapplied test)
260             device i pass
270         else
280             device i fail
290         end
300     end
        }
```

Fig. 3 Pseudo-Code of Online Information Utility Assessment Scheme

is applied, its result is used to update the model for the current device as in Line 140. Finally, we update the applied test set, {T_applied} to include the information content of the newly applied test using Equations (8) and (11), as in Line 150. An outlier filter may be added at the end of the test flow to prevent defective devices that do not conform to the learned statistics from escaping the skipped tests, as in Line 210.

At the earlier stages of the test flow for each device, the information utility of unapplied tests is always lower than the threshold, since the small amount of applied tests provide limited information about the unapplied ones. The underlined part in Line 70 guarantees that no test that provides additional information about the device is skipped. This avoids test escapes that are caused by a prediction based on insufficient information. Later in the test flow, the mutual information utility of the unapplied tests with the already applied ones increases, making it easier to skip tests that are predicted to pass. The information utility of the tests highly correlated to the applied ones may exceed the threshold, and the prediction of these tests become reliable. At the later stages, this information utility judgement mechanism has less impact on the test flow as the information utilities of most unapplied tests are higher than the threshold. Moreover, for an extreme case that a test is independent of all the others, the information utility remains zero before it is applied. Of course, this kind of test should never be skipped in an adaptive test flow, even if it is predicted to pass. In such cases, the pass prediction may simply be due to noise caused by a large number of random variables.

The test quality/time trade-off can be tuned by adjusting the mutual information utility threshold. Intuitively, a threshold near zero does little impact on the test flow, and a threshold near one could lead lower test escape rates.

IV. EXPERIMENTAL RESULTS

While the proposed information utility assessment mechanism can be used with any adaptive test flow, we have integrated it into the per-device adaptive test flow of [12]. Since per-device adaptive test provides the most fine-grained tuning of the test list, we evaluate the benefit of the online information utility assessment by comparing the results with those of [12], which is the lowest reported **DPPM** with highest test time reduction in the literature.

We compare the results with and without information utility assessment using the production data of three industry circuits. The first circuit is a large scale circuit with 264 parameters, the second circuit is wireless circuit with 70 parameters, and the third circuit is a diverse mixed-signal circuit with 42 parameters. For all the 3 circuits, the first 2K data are used as the sample set, and the following 48K data are used as the DUT to verify the method.

Test escape and average test time are two important test metrics that determine the test quality and test cost. We use **DPPM** as test escape per million samples, and **Time** as the average number of tests applied. For convenience of comparison, we also use **Test Skipped** as percentage of the tests skipped by the adaptive flow, and **Fault Coverage** as percentage of the faulty devices detected. Because partial initial test list in [12] is randomly generated, all the results are the average simulation results for 10 runs for each circuit.

The test flow with online information utility assessment is coded in MATLAB 2017a, and simulated on a 2.4GHz, 128G RAM server, on which parallel computation is not implemented. Table 1 shows the comparison of the computation time per test of the adaptive test flow with and without online information utility assessment for the 3 circuits. The computation time reported here is the worst case, where **DPPM** target is zero or near-zero, hence the information utility assessment step is used for almost all tests in the flow. We can clearly see that the increased number of specification parameters results in higher computation time for the information utility assessment. However, even with the large SOC circuit with 264 specification parameters (Circuit 1), the computation time is still within a few milliseconds, which is lower than the measurement time for all specification parameters, at DC, mid-band, or RF. Hence, the proposed information utility assessment approach does not pose a burden in terms of test time.

Table 1: Comparison of computation time with respect to [12] which does not take multi-variate correlations into account.

Circuit	# Specification Parameters	Computation Time per Test (ms)	
		[12]	Proposed
1	264	1.40	6.27
2	70	0.75	1.11
3	42	0.68	0.78

978-1-5386-3775-3/18 $31.00 © 2018 IEEE

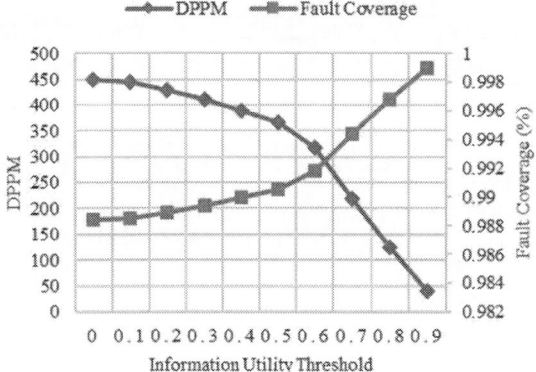

Fig. 4 DPPM and Fault Coverage vs. U_{TH} of Circuit 1

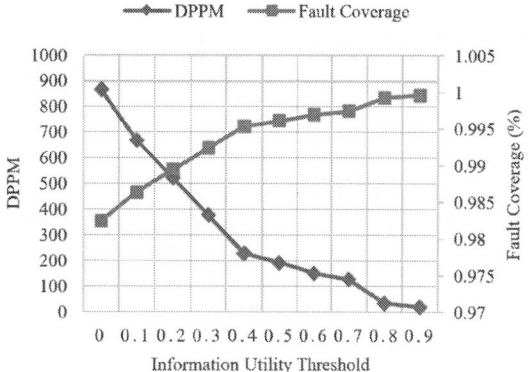

Fig. 6 DPPM and Fault Coverage vs. U_{TH} of Circuit 2

A. Circuit 1

The first circuit is a large-scale SOC with 264 parameters. Fig. 4 shows the **DPPM** and **Fault Coverage** (%) metrics based on the information utility threshold. The **DPPM** decreases from 448 to 40 while **Fault Coverage** increases from 98.84% to 99.90% as U_{TH} increases from 0 to 0.9. Fig. 5 shows the **Time** and **Test Skipped** (%) based on the information utility threshold. The number of tests applied increases from 47 to 114 while **Test Skipped** decreases from 82% to 57% as U_{TH} increase from 0 to 0.9. Moreover, the average **DPPM** reduces to 8 and **Fault Coverage** increases to 99.98% as U_{TH} equals to 0.99, while the **Time** increases up to 184 and **Test Skipped** reduces down to 31%. In fact, among the 10 runs, 6 result in zero **DPPM**. Note that the test flow is equivalent to the previous work [12] when U_{TH} equals to 0. The trade-off between **DPPM** and **Time** can be studied from the results, and an appropriate U_{TH} can be selected for the purpose of tuning the test quality and test time.

B. Circuit 2

The second circuit is a wireless circuit with 70 parameters. Fig. 6 shows the **DPPM** and **Fault Coverage** (%) based on information utility threshold. The **DPPM** decreases from 867 to 21 while **Fault Coverage** increases from 98.24% to 99.96% as U_{TH} increases from 0 to 0.9. Fig. 7 shows the **Time** and **Test Skipped** (%) based on information utility threshold. The number of tests applied increases from 20.3

to 55.1 while **Test Skipped** decreases from 71% to 21% as U_{TH} increase from 0 to 0.9. Moreover all the **DPPM**s of the 10 runs reach 0 with **Fault Coverage** to 100% as U_{TH} equals to 0.91, while the **Time** is 57 tests and **Test Skipped** is 19%.

C. Circuit 3

The third circuit is a diverse mixed-signal circuit with 42 parameters. Fig. 8 shows the **DPPM** and **Fault Coverage** (%) based on information utility threshold. The **DPPM** decreases from 60 to 0 while **Fault Coverage** increases from 98.63% to 100% as U_{TH} increases from 0 to 0.9. Fig. 9 shows the **Time** and **Test Skipped** (%) based on information utility threshold. The number of tests applied increases from 16 to 35 while **Test Skipped** decreases from 63% to 16% as U_{TH} increase from 0 to 0.9. In fact, when we set U_{TH} to 0.88, all the **DPPM**s of the 10 runs reach 0 while the **Time** is to 35 and **Test Skipped** is to 17%.

Table 2 presents a summary of comparison with results of [12], which is state-of-the art for per-device adaptive test of analog/RF circuits. The proposed information utility assessment method enables further reduction in DPPM for every circuit, which was not possible before. In fact, for two of the three circuit, zero DPPM can be achieved with about 20% test time savings. It should be noted that Circuit 2 and Circuit 3 have smaller number of parameters compared to Circuit 1. Since the information utility assessment step forces the application of more tests early on the process, the test time savings for smaller circuits is less. However, achieving zero **DPPM** is required for some applications. Hence, test time savings of about 20% are still significant. For a fair

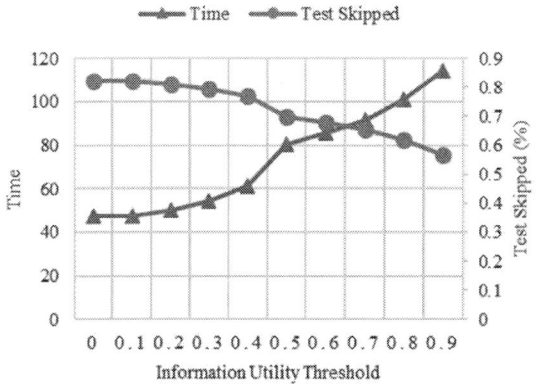

Fig. 5 Time and Test Skipped vs. U_{TH} of Circuit 1

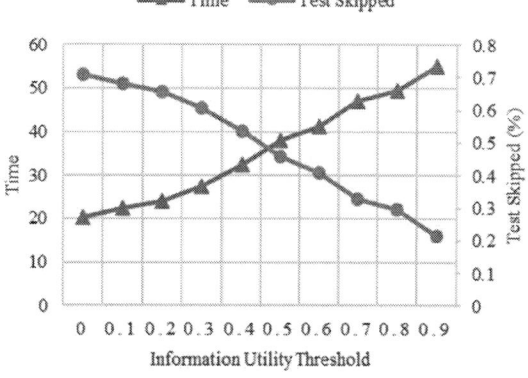

Fig. 7 Time and Test Skipped vs. U_{TH} of Circuit 2

978-1-5386-3775-3/18 $31.00 © 2018 IEEE

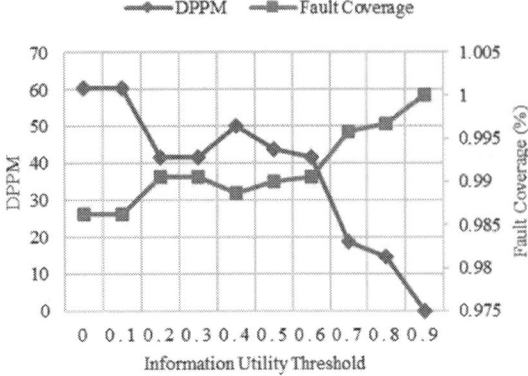

Fig. 8 DPPM and Fault Coverage vs. U_{TH} of Circuit 3

comparison, we did not compare the results of this work with more recent related work, as in [8], which uses adaptive test for a collection of devices on the wafer and reports lower test time savings.

V. CONCLUSIOIN

In this work, we present an online information utility assessment method, which is integrated into a per-device adaptive test flow. The method is demonstrated on three different types and scales of circuits with considerable sizes of production data sets. The trade-off between test escape, and test time can be tuned by controlling the information utility threshold. Compared to the best adaptive test results for analog circuits reported in the literature, the proposed method achieves better **DPPM** and test time trade-off. More importantly, the proposed method enables near- zero or zero **DPPM**. Moreover, the method can be easily integrated into any adaptive test flow.

REFERENCES

[1] P. Drineas and Y. Makris, "Independent test sequence compaction through integer programming," in *IEEE International Conference on Computer-Aided Design*, 2003, pp. 380-386.

[2] H.-G. D. Stratigopoulos, P. Drineas, M. Slamani, and Y. Makris, "Non-RF to RF test correlation using learning machines: A case study," in *IEEE VLSI Test Symposium*, 2007, pp. 9-14.

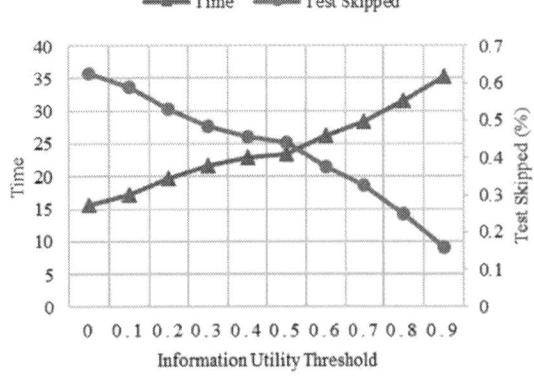

Fig. 9 Time and Test Skipped vs. U_{TH} of Circuit 3

Table 2: Comparison of Test Time Savings and DPPM with [12]

	Proposed Work		[12]	
Circuit	Test Time Savings	DPPM	Test Time Savings	DPPM
1	31%	8	60%	105
2	20%	0	71%	867
3	17%	0	74%	36

[3] S. Biswas and R. D. Blanton, "Test compaction for mixed-signal circuits using pass-fail test data," in *IEEE VLSI Test Symposium*, 2008, pp. 299–308.

[4] S. Biswas and R. D. S. Blanton, "Statistical test compaction using binary decision trees," *IEEE Design and Test of Computers*, vol. 23, no. 6, pp. 452–462, 2006.

[5] S. Benner and O. Boroffice, "Optimal production test times through adaptive test programming," in *IEEE International Test Conference*, 2001, pp. 908–915.

[6] C. Streitwieser, "Real-time adaptive test algorithm including test escape estimation method," in *Proc. International Mixed-Signal Testing Workshop*, 2015

[7] A. Ahmadi, A. Nahar, B. Orr, M. Pas, and Y. Makris, "Wafer-level process variation-driven probe-test flow selection for test cost reduction in analog/RF ICs," in *Proc. IEEE VLSI Test Symposium*, 2016.

[8] H.-G. D. Stratigopoulos and Christian Streiwieser, "Adaptive test flow for mixed-signal ICs", in *IEEE VLSI Test Symposium*, 2017

[9] E. Yilmaz and S. Ozev, "Adaptive test elimination for analog/rf circuits," in *IEEE Design Automation Conference*, 2009, pp. 720–725.

[10] E. Yilmaz, S. Ozev and K. M. Butler, "Adaptive test flow for mixed-signal/rf circuits using learned information from device under test," *IEEE Internatianl Test Conference*, 2010, pp. 1-10.

[11] E. Yilmaz, S. Ozev, and K. M. Butler, "Per-device adaptive test for analog/rf circuits using entropy-based process monitoring," *IEEE Transaction on VLSI Systems*, Vol. 21, No. 6, pp. 1116-1128, June 2013.

[12] E. Yilmaz, S. Ozev, and K. M, Butler, "Adaptive multidimensitional outlier analysis for analog and mixed signal circuits," in *IEEE International Test Conference*, 2011, pp. 1–8.

[13] S. Biswas, P. Li, R. Blanton, and L. Pileggi, "Specification test compaction for analog circuits and mems [accelerometer and opamp examples]," *IEEE Design, Automation and Test in Europe*, vol. 1, pp. 164–169, March 2005.

[14] M. Chen and A. Orailoglu, "Test cost minimization through adaptive test development," in *IEEE International Conference on Circuit Design*, 2008, pp. 234–239.

[15] S. S. Akbay, J. L. Torres, J. M. Rumer, A. Chatterjee, and J. Amtsfield, "Alternate test of RF front ends with IP constraints: Frequency domain test generation and validation," *IEEE International Test Conference*, pp. 1–10, 2006.

[16] N. Kupp, P. Drineas, M. Slamani, and Y. Makris, "On boosting the accuracy of non-RF to RF correlation-based specification test compaction," *Springer Journal of Electronic Testing*, vol. 25, pp. 309–321, 2009.

[17] Ke Huang, Jian Wen and Jim Willmore, "Test-Suite based Analog/RF Test Time Reduction using Canonical Correlation", *IEEE Transactions on Computer-Aided Design of Integrated Circuits and Systems*, Vol. 35, Issue: 12, pp. 2143-2147, 2016

[18] Thomas M. Cover and Joy A. Thomas, "Relative Entropy and Mutual Information", in *Element of Informaiton Theory, 2nd ed*, Wiley-Interscience, New York, 2006, ch. 2, sec. 3, pp. 19-20

[19] Gene H. Golub and Charles F. Van Loan, "The Singular Value Decomposition", in *Matrix Computations, 4th ed*, Baltimore, JHU Press, 2013, ch. 2, sec. 4, pp. 76-80

[20] I. T. Jolliffe, "The Singular Value Decomposition", in *Principal Component Analysis, 2nd ed*, New York, Springer, 2002, ch. 3, sec. 5, pp. 44-46

Special Session on Quantum Systems: Next Challenges in Design, Test, Integration

Carlo Reita, CEA-LETI, France
Jonathan Baugh, University of Waterloo, Canada
Gabriel Poulin-Lamarre, D-Wave Systems, Canada
Bozena Kaminska, Simon Frazer University, Canada (Organizer)
Bernard Courtois, BC Consulting, France (Organizer)

I. INTRODUCTION

In recent years, a clear path towards larger-scale quantum processors has emerged to study and eventually exploit the power of quantum computers at scale. The rapid progress has been observed in quantum computers, semiconductor implementations, and new generation of application development. The speakers will provide a perspective on the numerous challenges and their perspective on the future.

II. CHALLENGES AND OPPORTUNITIES OF SI-BASED QUANTUM BITS INTEGRATION (CARLO REITA)

Since the first demonstration of Si-based QBits by the UNSW group in 2012, the field of Si-based quantum electronics has progressed very quickly. In 2016 CEA-LETI in collaboration with CEA-INAC demonstrated the first quantum bit on 300mm Si wafers and fabricated using a slightly modified CMOS industrial process. More recently, announcements have been made by Intel and others on multi-Qbit systems being developed. However a great amount of work still has to be done before the promise of a quantum computer is materialized. The fabrication of the devices, the handling of decorrelations and errors in the system, the issues with handling the I/O with the chip at cryogenic temperatures, the choice between general computing or dedicated systems are still requiring full answers. The challenge today is both in the physics and in the engineering domains as well as in the computing architecture. Large collaborations are required to bring together all the required competences to arrive at a working system. In the talk will be presented the approach taken by CEA in Grenoble in collaboration with CNRS and other academic partners in order to address all these challenges. A special focus will be done on the specific requirement for characterization and test both in development phase as well as operation phase.

III. CHALLENGES IN SCALING UP SILICON-BASED QUANTUM PROCESSORS (JONATHAN BAUGH)

Several paths to a large scale, universal quantum computer have been proposed, though realization beyond the small scale (~50 qubits) remains a significant challenge. Superconducting qubits, semiconductor quantum dot and donor spin qubits, and topological qubits offer the exciting prospect for quantum analogues of the monolithic computer chip. In particular quantum dots and donors in silicon have the advantage of a CMOS-compatible platform, which can exploit conventional foundry processes to fabricate the quantum device layer and integrate classical control circuitry via cryo-CMOS. In silicon, electron spin coherence times benefit from isotopic purification and a weak intrinsic spin-orbit coupling. Combining these CMOS-compatible qubits with a surface code architecture has been proposed for quantum dots and donor arrays. I will discuss some of the major challenges of such approaches, including density of wiring, integration of control electronics and multiplexing, noise and cross-talk, device yield and variability, and the need for advanced simulation tools to design and characterize real devices. I will describe how a network architecture can realize a surface code quantum computer while reducing wiring density and isolating the qubits most critical to storing quantum information.

IV. BUILDING A SUPERCOMPUTING QUANTUM PROCESSOR AT SCALE (GABRIEL POULIN-LAMARRE)

Quantum mechanics allows us to use completely new ways to process information. There are number of promising architectures exploiting these effects. The company D-Wave Systems stands out in its use of an algorithm called quantum annealing. The company released its 4th generation system in January 2017. The processor successfully integrates more than 2000 superconducting flux qubits and more than 128 000 Josephson junctions on a single chip operated at 12 mK. In this presentation, I will talk about the current processor architecture, putting emphasis on the technical difficulties arising from scaling up the number of qubits.

NOIDA: Noise-resistant Intra-cell Diagnosis

Soumya Mittal and R. D. (Shawn) Blanton
Department of Electrical and Computer Engineering
Carnegie Mellon University, Pittsburgh, PA 15213
https://www.ece.cmu.edu/~actl/

Abstract—The goal of diagnosis is to identify defect locations and subsequently, identify the root cause so as to minimize (and ideally eliminate) the need for physical failure analysis. With advanced technology nodes, there has been an increasing number of front-end (i.e., within a standard cell) defects. Conventional diagnosis approaches typically fail to localize such defects. In addition, circuit-level noise can change the tester response in an unexpected way, and can decrease the quality of diagnosis. This work describes a noise-resistant approach called NOIDA (NOise-resistant Intra-cell DiAgnosis) for effectively diagnosing cell-level defects based on the analysis of the intra-cell physical neighborhoods surrounding likely defect locations. Defect behavior is derived based on the neighborhood, instead of relying on a specific fault model. Experiments demonstrate the effectiveness of NOIDA using a library of standard cells. The results show that for over 16,000 static and sequence-dependent defects, the method achieves an average resolution improvement of 12.1% over prior work with a small accuracy loss (specifically, 1.6%). Additionally, NOIDA is found to be more robust to noise in the tester response. Specifically, in the presence of noisy tester response, NOIDA attains an accuracy of 97.6% with an average resolution improvement of 48.6% over prior work.

I. INTRODUCTION

Diagnosis is a software-based process for determining defect locations within a failing circuit and sometimes, in addition, can characterize the nature of the defects residing in those locations. Suspect defect locations, often called candidates[1], are determined by comparing the observed circuit response with the expected, defect-free response. The outcome of diagnosis is sometimes used to aid physical failure analysis (PFA) in order to identify the root cause of the failure. The quality of diagnosis, which is typically defined by *resolution* (the number of locations reported) and *accuracy* (i.e., are the reported locations correlated to actual defect locations), is a major factor in determining the success rate of PFA.

Besides guiding PFA, the information obtained from volume diagnosis aids the understanding of failure mechanisms, which consequently facilitates yield learning. Moreover, a variety of volume diagnosis approaches aim at using the diagnoses of a statistically-significant number of failing chips to determine if there are any systematic defects [1], [2]. Diagnosis results can also be used to estimate defect distribution for a population of failing chips [3], and grade the effectiveness of fault models [4].

---

[1]A candidate can be (a) an interconnect or a cell in the logic-level representation of the circuit, or (b) an intra-cell net in the transistor-level representation of the circuit. Additional information such as its likely behavior and physical location is also sometimes reported depending on the particular diagnosis approach employed.

Numerous techniques have been proposed over the years to improve the diagnosis of a failed circuit. Approaches such as [5]–[11] use a logic-level description of the circuit to find a possible cause and location of failure. To improve defect localization, techniques have been developed that incorporate design layout information [12]–[18]. All of these techniques focus on diagnosing defects outside the cell and report either an interconnect, a standard-cell pin or the cell itself as a possible candidate. These techniques therefore collectively perform back-end defect diagnosis, which means these approaches cannot locate defects inside a cell. It has been shown in [19] however that the root cause of a significant number of manufacturing defects lies within a cell for advanced nanometer technologies (90nm and below). Because one goal of diagnosis is to eliminate the need for PFA, knowledge of the exact defect location is of utmost importance. Complex standard cells such as full adders, multipliers and scan flip flops contain a large number of transistors that makes PFA challenging [20]. Pinpointing each defect location inside a cell will decrease the physical area that must be examined for failure analysis, resulting in more efficient and cost-effective PFA [19]–[22]. For example, if all the candidates are found to be located within the polysilicon layer, the failing die can be de-layered up to the polysilicon layer and all the higher layers can be ignored during PFA. Moreover, some failure mechanisms including poly-contact shorts, poly-contact opens, and poly-active shorts can only exist within a cell and would not be found by a back-end diagnosis technique. Additionally, diagnoses of intra-cell defects can be correlated to identify yield-limiting layout features within a cell.

A diagnosis approach typically correlates observed defect behavior with one or more fault models to find candidates. Many fault models have been created over the years to capture the variety of observed defect behaviors. While fault models such as single stuck-line, wired logic bridge and transition have been commonly used for diagnosis, their ability to precisely capture the defect behavior is decreasing [23]. Fault models that consider layout information have been shown to be more effective in modeling defects [19], [24]. However, as technology advances, new materials and fabrication steps are employed, which results in new defect types and failure mechanisms. New failure mechanisms (e.g., fin-related defects in FinFET-based circuits [25]) can create misbehaviors that are not sufficiently captured by existing fault models [26]. Thus, the unpredictability of defects necessitates the need for a more generalized approach to defect diagnosis.

Even if fault models accurately captured the typical behaviors of the targeted defects, circuit-level noise can cause deviations between the predicted behavior and the behavior observed on the tester. Various sources of noise exist in digital circuits that include process variation, cross-talk signal noise, power supply noise, and substrate-coupling noise [27]. Deviations between the observed and the predicted behavior can also result from inaccurate SPICE models used for defect extraction, and the transistor-level simulation employed for formulating fault models. Depending on the amount of deviation, a weak logic value at a cell output, due to an intra-cell defect, can be interpreted as a logic-1 or a logic-0 by a receiver cell (depending on its switching threshold, which too is affected by noise), and can result in a tester response that is not predicted by the corresponding fault. Thus, circuit-level noise and the resulting ambiguity in signal logic values warrant the need for a noise-resistant diagnosis approach.

This work describes a novel methodology to diagnose front-end defects that we term NOIDA (NOise-resistant Intra-cell DiAgnosis). Here, the defects are assumed to be localized, i.e., the behavior of a defect is influenced by the circuitry within some radius r surrounding it. Hence, NOIDA derives the defect behavior by analyzing the nets surrounding its location, instead of using a particular fault model. The output of NOIDA is a set of candidates, where each candidate is a tuple consisting of its physical location (x-y coordinate and the physical layer) and its likely behavior (and consequently its defect type). Several defect injection experiments are performed using a 45nm standard-cell library [28] to evaluate NOIDA when noise is introduced to the tester response. Results demonstrate better accuracy and significant resolution improvement for intra-cell defects when compared to [19].

The rest of the paper is organized as follows. Section II provides a brief background on front-end diagnosis and motivates NOIDA. This section also describes NOIDA in detail. Experiment results are presented in Section III. The final section concludes the paper.

II. DIAGNOSIS METHODOLOGY

Many approaches have been put forward over the years to diagnose intra-cell defects. Work in [29] assumes a defect model for defects within the transistor-level description of a cell and maps these defects to logic-level defects using complex transformation rules. Transformation is applied to cells that adhere to the following criteria: the stuck-at fault simulation response at the output of a cell should match the observed circuit response. Logic-level diagnosis tools can then be used on the modified netlist to find the defective cells, and in turn the intra-cell defects. One main drawback of using this approach is that diagnostic accuracy largely depends on the defect models and the transformation rules. Another disadvantage is that a different model is required for each defect type, which means unknown defect types may go undiagnosed.

In [22], possible defective cell locations (i.e., cell candidates) are identified based on the realistic assumption that

the excitation of a cell-internal defect is highly correlated to the input logic values applied to a cell. Logic values at the inputs of each cell candidate are collected for Tester-Fail-Simulation-Fail (TFSF) patterns, that is, patterns that fail on the tester and propagate the error effects from the defect location to the circuit outputs; and Tester-Pass-Simulation-Fail (TPSF) patterns, that is, patterns that pass on the tester but propagate the error effects from the defect location to the circuit outputs. Such input-value combinations will henceforth be referred to as cell-level failing and passing patterns. Cells with inconsistent input conditions, i.e., input conditions that appear in both cell-level passing and failing patterns, are discarded from further analysis. This form of consistency checking is a special case of the approach described in [18]. For the remaining consistent cells, the input conditions are matched with a fault dictionary. The fault dictionary is created by performing a switch-level simulation of various intra-cell defects.

In [19]–[21], [30], [31], a fault dictionary is constructed from extracting realistic defects from physical layout and transistor-level simulation. This has the advantage of generating more accurate responses, though the simulation time may be a limiting factor. Moreover, if the defect extraction or simulation steps are not accurate, then diagnosis will produce inaccurate results.

Methods described in [19]–[22], [30], [31] generate a fault dictionary by extracting intra-cell defects using various techniques. On the other hand, an effect-cause approach is described in [32], where critical path tracing [33] is utilized to trace back from the output of a failing cell to cell inputs. The main drawback of this work is that physical layout information is not considered for diagnosis. For example, bridge defects implicated by this methodology may not be very likely due to lack of proximity.

Prior work on front-end diagnosis discussed up to this point suffers from one major disadvantage – potentially inaccurate defect modeling. NOIDA circumvents this problem by avoiding the use of a specific fault model. Instead, it derives the defect behavior by analyzing the logic activity of the nets surrounding the likely defect location. Here, a defect is assumed to be localized and controlled by the circuitry in close proximity. This assumption holds true for a variety of defects such as bridge, open and transistor defects. The nets near the candidate are collectively referred to as its neighborhood; the logic values applied to the nets in the neighborhood form a neighborhood state. Changes in neighborhood state over time can also be important for sequence- and timing-dependent defects.

Fig. 1 illustrates a typical software diagnosis framework. Given a test response, the first step is to find candidates at the logic level. The result of this logic-level analysis is an initial set of interconnect and cell candidates. Next, the interconnect candidates are examined using back-end diagnosis techniques like [18], [34]. In parallel, each cell candidate identified is further investigated via front-end diagnosis approaches [19]–[22], [29]–[32] to pinpoint defect locations inside a cell.

978-1-5386-3775-3/18 $31.00 © 2018 IEEE

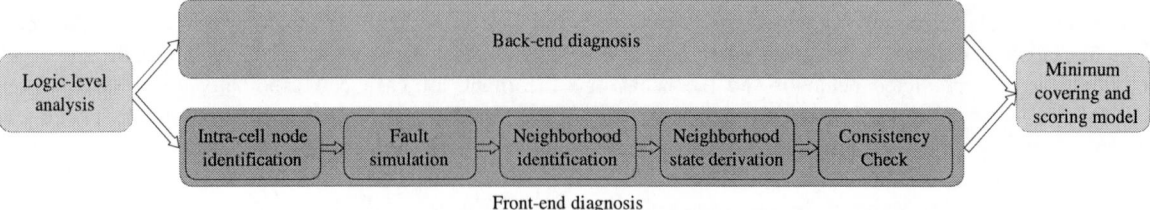

Fig. 1. Overview of a generic diagnosis framework.

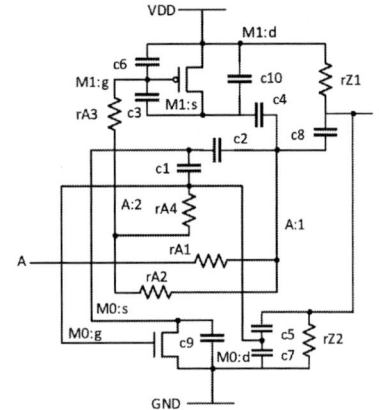

Fig. 2. A schematic view of an inverter cell with parasitics extracted. Parasitics affecting power rails are not shown for clarity.

The resulting set of interconnect and intra-cell candidates are then merged together to constitute a final set of candidates. Minimum set covers are selected from this set and ranked using a scoring model. This paper, in particular, is focused on describing front-end diagnosis. Each step involved in the flow is described next:

1) *Intra-cell node identification*: A transistor-level description of a cell (such as its physical layout, SPICE netlist, etc.) is used to identify intra-cell nodes. A SPICE netlist with parasitics extracted is used here. The inverter schematic is used to illustrate node identification in Fig. 2. The parasitic resistances are denoted using 'r' and the coupling capacitances using 'c'. There are ten internal nodes for this cell, namely, $\{A, A{:}1, A{:}2, Z, M0{:}g, M0{:}d, M0{:}s, M1{:}g, M1{:}d, M1{:}s\}$.

2) *Fault simulation*: Each intra-cell node is faulted at the opposite value of the expected value for the cell-level failing and passing patterns[2]. This is achieved by adding a near-zero resistor between each node and *VDD* (or *GND*, depending on the fault value) in the SPICE netlist. The altered SPICE netlist is then simulated with an analog simulator for the cell-level failing and passing patterns. Each simulation response is digitized using the following criteria: the logic value at the cell output is deemed a logic-1 (logic-0) if the output voltage is more (less) than half of the supply voltage. This logic threshold value depends on the process technology and can be specified by the user. The cell-level passing (failing)

[2]The cell-level failing and passing patterns can be obtained via any logic-level diagnosis technique. However, an exhaustive two-pattern test set is used in Section III because the experiments performed in this work involve diagnosis of individual cells and not a circuit of interconnected cells.

patterns for which a faulted node produces and propagates an error to the cell output are the cell-level TPSF (TFSF) patterns for the faulted node. Each faulted node with at least one cell-level TFSF pattern is deemed an initial diagnosis candidate.

3) *Neighborhood identification*: Neighborhood of each node is found from the extracted SPICE netlist of the cell. Neighbors of a node constitute all nodes that are coupled to it by capacitors. This is a reasonable way to identify neighbors and adheres to the localization assumption. For instance, in Fig. 2, three coupling capacitors, $c2$, $c4$ and $c8$, are associated with node $A{:}1$ and hence the neighbors of $A{:}1$ include $\{Z, M1{:}s, M0{:}s\}$. If parasitic extraction is not possible, then the neighbors can simply be found by identifying the internal nodes that are in close physical proximity to the candidate.

4) *Neighborhood state derivation*: Analog voltage values at each of the internal nodes are stored during fault simulation in step 2. Each value is then converted into its logic equivalent by using the same logic threshold value mentioned in step 2. For static defects, the neighborhood state of a candidate is the set of logical values established in its neighborhood for the pattern applied. For sequence-dependent defects, the neighborhood state tracks the logical values for two or more patterns.

5) *Consistency check*: The neighborhood state of each intra-cell candidate is analyzed for cell-level TPSF and TFSF patterns. If the neighborhood state of a candidate is the same for any pair of cell-level TPSF and TFSF patterns, then the candidate is inconsistent and is removed from the candidate set. It should be noted that the amount of inconsistency can be modulated for candidate elimination. Finally, minimum set covers of the consistent intra-cell candidates are selected to jointly explain the failures observed at the cell output.

III. EXPERIMENTS

This section presents the experiment details of evaluating NOIDA on 79 standard cells within the 45nm standard-cell library of [28].

A population of defective cells is created by injecting cell defects (one at a time) into the layout of each cell. The defects injected include opens, bridges (feedback and non-feedback), stuck-open and stuck-closed transistors with resistance values that range from 1Ω to $20k\Omega$ for bridges, and from $1G\Omega$ to $1k\Omega$ for opens. For each defective cell layout, a corresponding transistor-level netlist is extracted. Because the behavior of the defect is unknown, i.e., whether it is static or sequence-dependent, analog simulation is performed on each altered netlist using an exhaustive two-pattern test set. A defect is

978-1-5386-3775-3/18 $31.00 © 2018 IEEE

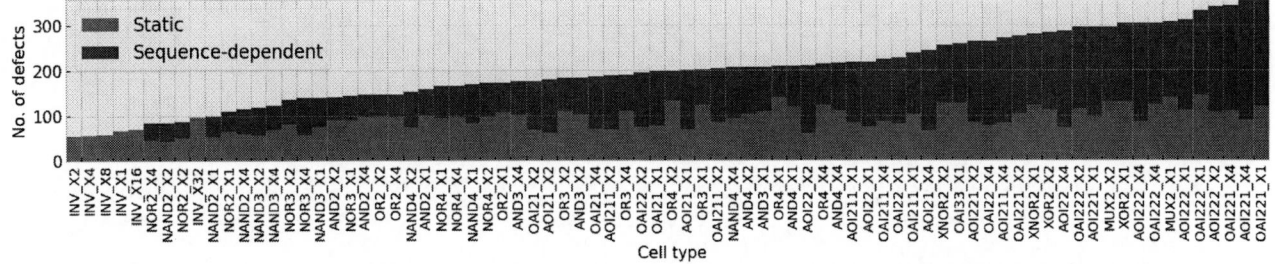

Fig. 3. Distribution of static and sequence-dependent defects for a 45nm standard-cell library [28].

considered detected if the voltage at the cell output deviates from its expected, defect-free value of *VDD* or *GND* by more than 50%. Also, and importantly, patterns that detect the defect are deemed cell-level failing patterns, while the remaining patterns are treated as cell-level passing patterns.

Each defect simulation response is analyzed to determine if it exhibits static or sequence-dependent behavior. A sequence-dependent defect requires a sequence of patterns (two patterns in this case) for detection. Detection of a static defect is independent of the first pattern applied for all two-pattern combinations. In other words, a static defect is detected by a single pattern.

Fig. 3 shows the number of static and sequence-dependent defects for each standard cell in the library. A total of 16,037 defects are injected and simulated, out of which 7,110 (44.3%) are static while 8,927 (55.7%) are sequence-dependent. Out of 7,110 static defects, 449 (6.3%) are open defects, 3,290 (46.3%) are bridge defects, and 3,371 (47.4%) are transistor defects. Similarly, out of 8,927 sequence-dependent defects, 3,711 (41.6%) are open defects, 1,424 (15.9%) are bridge defects, and 3,792 (42.5%) are transistor defects. It is also observed that most of the open defects (89.2%) require a sequence of patterns for detection, and a majority of bridge defects (69.8%) are static.

Each intra-cell defect is diagnosed using NOIDA and [19]. NOIDA returns a set of minimum set covers (step 5 of NOIDA), where each set cover jointly produces a response that exactly matches the observed response at the cell output. Using [19] for diagnosis means that all modeled faults that have a simulation response that matches the observed response are reported as diagnosis candidates. Both diagnosis methodologies are evaluated on two criteria, namely, resolution and accuracy. For NOIDA, resolution is defined as the number of unique intra-cell nodes present in the minimum set covers. For [19], resolution is calculated by counting the number of unique intra-cell nodes corresponding to the faults returned as candidates. For both methodologies, a defect is considered to be accurately diagnosed if the candidates returned include at least one of the intra-cell nodes used for defect injection.

Table 1 shows the number of static and sequence-dependent

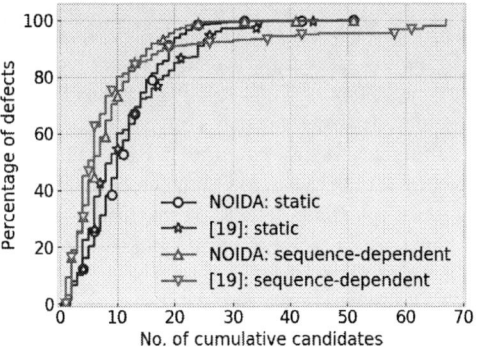

Fig. 4. Cumulative distribution of the number of candidates reported by NOIDA and [19]. Plots marked with "○" and "△" correspond to the resolution obtained by NOIDA for static and sequence-dependent defects, respectively. Plots marked with "*" and "▽" correspond to the resolution obtained by [19] for static and sequence-dependent defects, respectively.

defects accurately diagnosed by NOIDA and [19]. It indicates that while [19] attains perfect accuracy, NOIDA achieves near perfect accuracy for static defects (99.7%) and an accuracy of 97.4% for sequence-dependent defects.

Table 2 categorizes the accuracy attained by NOIDA by defect type for static and sequence-dependent defects. It is observed that 54 transistor defects and 201 open defects are inaccurately diagnosed by NOIDA. These defects are investigated further to determine the reason for inaccuracy. It is discovered that the analog voltage values at the cell output and some of the internal cell nodes lie close to the logic threshold value (i.e., ± 10%) for some patterns. One of the reasons for intermediate voltage at a node is the time at which the voltage of each node is sampled. This observation has two consequences. First, a failing pattern can be interpreted as a passing pattern which can eliminate the correct candidate. Second, voltage (that is close to logic threshold) at an internal node can be inaccurately converted to its logic equivalent, which can change a neighborhood state and in turn, make the correct candidate inconsistent.

Fig. 4 shows the diagnostic resolution achieved by NOIDA

TABLE 1
DIAGNOSTIC ACCURACY ACHIEVED BY NOIDA AND [19].

Diagnosis approach	Static	Sequence-dependent
NOIDA	7090 (99.7%)	8692 (97.4%)
[19]	7110 (100.0%)	8927 (100.0%)

TABLE 2
DIAGNOSTIC ACCURACY ACHIEVED BY NOIDA FOR STATIC AND
SEQUENCE-DEPENDENT DEFECTS.

Defect type	Static		Sequence-dependent	
	Accurate	Inaccurate	Accurate	Inaccurate
Bridge	3290 (100.0%)	0	1424 (100.0%)	0
Open	449 (100.0%)	0	3510 (94.6%)	201
Transistor	3351 (99.4%)	20	3758 (99.1%)	34

978-1-5386-3775-3/18 $31.00 © 2018 IEEE

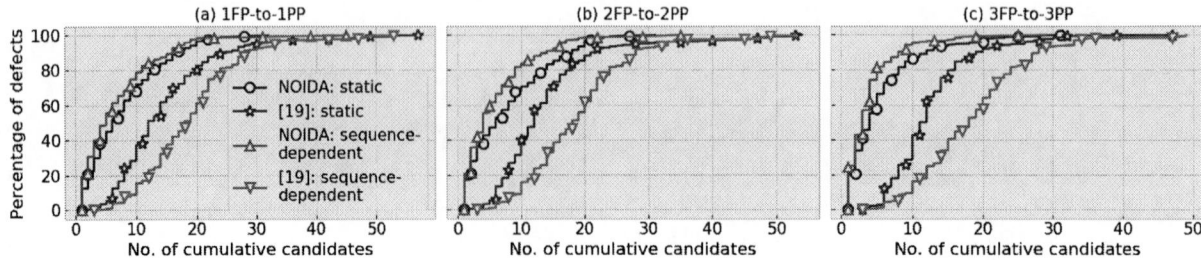

Fig. 5. Cumulative resolution distribution when (a) one (1FP-to-1PP), (b) two (2FP-to-2PP) and (c) three (3FP-to-3PP) cell-level failing patterns are randomly changed to passing patterns. Plot lines marked with "o" and "△" correspond to the resolution obtained by NOIDA for static and sequence-dependent defects, respectively. Plot lines marked with "*" and "▽" correspond to the resolution obtained by [19] for static and sequence-dependent defects, respectively.

and compares it with the resolution achieved by [19]. Fig. 4 contains four plots. Each plot is sorted by the number of candidates. For a plot point x-y, the x-value denotes the number of candidates, and the y-value denotes the percentage of defects y with resolution less than or equal to x. Plots marked with "o" and "△" represent the resolution obtained by NOIDA for static and sequence-dependent defects, respectively. Plots marked with "*" and "▽" is the resolution obtained from the faults derived in [19], for static and sequence-dependent defects, respectively. It is observed that NOIDA returns 12.1% fewer candidates compared to [19]. Moreover, the average resolution obtained from NOIDA is 9.2 candidates per defect, showing an improvement of 1.3 candidates per defect over [19].

However, the resolution improvement achieved by NOIDA is associated with some accuracy loss. When [19] is used for diagnosis, it achieves perfect accuracy. But its important to note here that each defect injected is identical to an instance of the faults extracted in [19] and therefore, it is not at all surprising that the accuracy of [19] is 100.0%.

Another point is that diagnosis here is performed on individual cells, that is, not on a circuit that contains interconnected cells. For circuit-level diagnosis, it is possible that a cell-level failing (passing) pattern can become a circuit-level passing (failing) pattern due to noise. This is because voltage deviation due to noise can change how a driven cell interprets the value. In NOIDA (and other front-end diagnosis approaches [19]–[21], [30], [31]), voltage deviation can also result from inaccurate SPICE models used during defect extraction and analog fault simulation. Thus, a weak logic-1 at a cell output can actually be or interpreted as a weak logic-0 and vice versa, which in turn can transform a cell-level failing pattern to a circuit-level passing pattern. However, a large deviation is required for a strong logic-0 (logic-1) to become a weak logic-1 (logic-0). Thus, for fault models that equate passing with strong logic values, it is less likely for a cell-level passing pattern to become a circuit-level failing pattern. Therefore, to evaluate both NOIDA and [19], "realistic" defect responses are created by randomly changing cell-level failing patterns to passing patterns. Specifically, three new responses are created for each original defect response by changing one, two and three cell-level failing patterns to passing patterns. (Corresponding defect injection experiments will henceforth be referred to as 1FP-to-1PP, 2FP-to-2PP and 3FP-to-3PP.) So as not to modify the behavior of a defect entirely, a defect

response is altered only when the decrease in the number of failing patterns is less than or equal to 50%. For instance, defects with less than six failing patterns are not considered for 3FP-to-3PP. Compared to 16,037 defects considered in the first set of experiments (i.e., when original defect responses are used), the number of defects reduces to 10,891 for 1FP-to-1PP, 8,712 for 2FP-to-2PP, and 5,229 for 3FP-to-3PP.

Table 3 shows the percentage of defects accurately diagnosed by NOIDA and [19] when one, two and three failing patterns are changed to passing patterns. The results indicate that the accuracy of [19] remains at 100.0% while the accuracy of NOIDA slightly increases for 1FP-to-1PP. This is because 102 out of 255 defects inaccurately diagnosed earlier had only one failing pattern and are thus not considered for 1FP-to-1PP. For 2FP-to-2PP and 3FP-to-3PP, it is noticed that the accuracy of [19] drops to 96.5% and 94.2%, respectively while NOIDA performs comparatively better, attaining an accuracy of 97.9% and 96.3%.

The improvement in resolution however is more significant. Specifically, Fig. 5 highlights the improvement in diagnostic resolution from NOIDA. Fig. 5 shows three parts, one each corresponding to 1FP-to-1PP, 2FP-to-2PP and 3FP-to-3PP. Each part contains four plots. Plot lines marked with "o" and "△" show the cumulative diagnostic resolution distribution of static and sequence-dependent defects, respectively, reported by NOIDA. Plot lines marked with "*" and "▽" represent the cumulative resolution distribution of static and sequence-dependent defects, respectively, achieved by [19]. It is observed that NOIDA returns 46.3%, 43.6% and 55.8% fewer candidates compared to [19] for 1FP-to-1PP, 2FP-to-2PP and 3FP-to-3PP, respectively. Moreover, NOIDA shows an improvement of 7.5, 7.3 and 9.5 candidates per defect over [19] for 1FP-to-1PP, 2FP-to-2PP and 3FP-to-3PP, respectively. In addition, the plots reveal that NOIDA achieves a perfect resolution for 13.7%, 17.9% and 20.1% of defects for 1FP-to-1PP, 2FP-to-2PP and 3FP-to-3PP, respectively. On the other

TABLE 3
DIAGNOSTIC ACCURACY FOR NOIDA AND [19] WHEN ONE (1FP-TO-1PP), TWO (2FP-TO-2PP) AND THREE (3FP-TO-3PP) CELL-LEVEL FAILING PATTERNS ARE CHANGED TO PASSING PATTERNS.

Diagnosis approach	1FP-to-1PP	2FP-to-2PP	3FP-to-3PP
NOIDA	98.6%	97.9%	96.3%
[19]	100.0%	96.5%	94.2%

hand, less than 1.0% of defects have a resolution of one when [19] is used. Thus, NOIDA performs significantly better than [19] in terms of accuracy and resolution in the presence of noise.

IV. CONCLUSIONS

This work presents a novel generalized methodology for front-end defect diagnosis we call NOIDA (NOise-resistant Intra-cell DiAgnosis). NOIDA consists of finding defect locations within a cell and deriving defect behavior based on the nets that surround the suspected defect location, instead of correlating the observed defect response with a particular fault model.

Simulation experiments for over 16,000 intra-cell defects are used to evaluate NOIDA on various standard cells. Results indicate that the approach achieves a resolution improvement of 12.1% when compared to [19] with a slight loss in accuracy. Furthermore, when additional defect injection experiments are performed by adding noise to the tester response, it is seen that NOIDA is able to diagnose 97.6% defects accurately compared to 96.9% by [19]. More importantly, NOIDA performs significantly better than [19] in terms of resolution. Specifically, NOIDA returns 48.6% fewer candidates (8.1 fewer candidates per defect, on average) and achieves a perfect resolution for 17.2% of defects. This reduction in the number of potential defective locations within a cell while achieving near perfect accuracy makes PFA efficient and cost-effective, and likely enhances yield learning significantly.

The experiments presented here include running NOIDA on combinational standard cells with a single defect injected at a time. Current work includes running NOIDA on sequential cells. Future work will be focused on extending this approach to include diagnosis of multiple defects, and deriving a scoring model to rank the candidates for further resolution improvement.

REFERENCES

[1] H. Tang et al., "Analyzing Volume Diagnosis Results with Statistical Learning for Yield Improvement," in *IEEE European Test Symposium*, May 2007, pp. 145–150.

[2] R. D. Blanton et al., "Yield Learning Through Physically Aware Diagnosis of IC-failure Populations," *IEEE Design Test of Computers*, vol. 29, no. 1, pp. 36–47, Feb 2012.

[3] X. Yu and R. D. Blanton, "Estimating Defect-Type Distributions through Volume Diagnosis and Defect Behavior Attribution," in *IEEE International Test Conference*, Nov 2010, pp. 1–10.

[4] Y. T. Lin and R. D. Blanton, "METER: Measuring Test Effectiveness Regionally," *IEEE Transactions on Computer-Aided Design of Integrated Circuits and Systems*, vol. 30, no. 7, pp. 1058–1071, July 2011.

[5] I. Pomeranz and S. M. Reddy, "On the Generation of Small Dictionaries for Fault Location," in *International Conference on Computer-Aided Design*, Nov 1992, pp. 272–279.

[6] P. G. Ryan, W. K. Fuchs, and I. Pomeranz, "Fault Dictionary Compression and Equivalence Class Computation for Sequential Circuits," in *International Conference on Computer Aided Design*, Nov 1993, pp. 508–511.

[7] V. Boppana and W. K. Fuchs, "Fault Dictionary Compaction by Output Sequence Removal," in *International Conference on Computer-Aided Design*, 1994, pp. 576–579.

[8] S. Holst and H. J. Wunderlich, "Adaptive Debug and Diagnosis without Fault Dictionaries," in *IEEE European Test Symposium*, May 2007, pp. 7–12.

[9] T. Bartenstein et al., "Diagnosing Combinational Logic Designs using the Single Location at-a-time (SLAT) Paradigm," in *IEEE International Test Conference*, 2001, pp. 287–296.

[10] S. Venkataraman and S. B. Drummonds, "POIROT: A Logic Fault Diagnosis Tool and its Applications," in *IEEE International Test Conference*. IEEE, 2000, pp. 253–262.

[11] J. C. M. Li and E. J. McCluskey, "Diagnosis of Sequence-dependent Chips," in *IEEE VLSI Test Symposium*, 2002, pp. 187–192.

[12] M. Keim et al., "A Rapid Yield Learning Flow Based on Production Integrated Layout-Aware Diagnosis," in *IEEE International Test Conference*, Oct 2006, pp. 1–10.

[13] W. Zou, W.-T. Cheng, and S. M. Reddy, "Bridge Defect Diagnosis with Physical Information," in *IEEE Asian Test Symposium*, Dec 2005, pp. 248–253.

[14] S.-Y. Huang, "Diagnosis of Byzantine Open-segment Faults [Scan Testing]," in *IEEE Asian Test Symposium*, Nov 2002, pp. 248–253.

[15] M. Sharma et al., "Efficiently Performing Yield Enhancements by Identifying Dominant Physical Root Cause from Test Fail Data," in *IEEE International Test Conference*, Oct 2008, pp. 1–9.

[16] Y.-J. Chang et al., "Experiences with Layout-aware Diagnosis - A Case Study," in *Electronic Device Failure Analysis*, vol. 12, no. 12, May 2010, pp. 12–18.

[17] J. Mekkoth et al., "Yield Learning with Layout-aware Advanced Scan Diagnosis," in *International Symposium for Testing and Failure Analysis*, vol. 32, 2006, p. 412.

[18] R. Desineni, O. Poku, and R. D. Blanton, "A Logic Diagnosis Methodology for Improved Localization and Extraction of Accurate Defect Behavior," in *IEEE International Test Conference*, Oct 2006, pp. 1–10.

[19] F. Hapke et al., "Cell-Aware Test," *IEEE Transactions on Computer-Aided Design of Integrated Circuits and Systems*, vol. 33, no. 9, pp. 1396–1409, Sept 2014.

[20] P. Maxwell, F. Hapke, and H. Tang, "Cell-aware Diagnosis: Defective Inmates Exposed in their Cells," in *IEEE European Test Symposium*, May 2016, pp. 1–6.

[21] H. Tang et al., "Diagnosing Cell Internal Defects Using Analog Simulation-based Fault Models," in *IEEE Asian Test Symposium*, Nov 2014, pp. 318–323.

[22] M. E. Amyeen, D. Nayak, and S. Venkataraman, "Improving Precision Using Mixed-level Fault Diagnosis," in *IEEE International Test Conference*, Oct 2006, pp. 1–10.

[23] E. J. McCluskey and C.-W. Tseng, "Stuck-fault Tests vs. Actual Defects," in *IEEE International Test Conference*, 2000, pp. 336–342.

[24] F. J. Ferguson and J. P. Shen, "Extraction and Simulation of Realistic CMOS Faults Using Inductive Fault Analysis," in *International Test Conference, New Frontiers in Testing*, Sep 1988, pp. 475–484.

[25] Y. Liu and Q. Xu, "On Modeling Faults in FinFET Logic Circuits," in *IEEE International Test Conference*, Nov 2012, pp. 1–9.

[26] H. Tang et al., "Diagnosing Timing Related Cell Internal Defects for FinFET Technology," in *VLSI Design, Automation and Test*, Apr 2015, pp. 1–4.

[27] S. Zachariah et al., "On Modeling Cross-Talk Faults," in *Design, Automation and Test in Europe Conference and Exhibition*, Mar 2003, pp. 490–495.

[28] Nangate Inc., "Nangate 45nm Open Cell Library," *http://www.nangate.com*, 2008.

[29] X. Fan et al., "Extending Gate-level Diagnosis Tools to CMOS Intra-gate Faults," *IET Computers Digital Techniques*, vol. 1, no. 6, pp. 685–693, Nov 2007.

[30] A. Ladhar, M. Masmoudi, and L. Bouzaida, "Efficient and Accurate Method for Intra-gate Defect Diagnoses in Nanometer Technology and Volume Data," in *Design, Automation Test in Europe Conference and Exhibition*, Apr 2009, pp. 988–993.

[31] C. W. Tzeng, H. C. Cheng, and S. Y. Huang, "Layout-based Defect-driven Diagnosis for Intracell Bridging Defects," *IEEE Transactions on Computer-Aided Design of Integrated Circuits and Systems*, vol. 28, no. 5, pp. 764–769, May 2009.

[32] Z. Sun et al., "Intra-cell Defects Diagnosis," *Journal of Electronic Testing*, vol. 30, no. 5, pp. 541–555, 2014.

[33] M. Abramovici, P. R. Menon, and D. T. Miller, "Critical Path Tracing - an Alternative to Fault Simulation," in *Design Automation Conference*, 1983, pp. 214–220.

[34] S. Mittal and R. D. Blanton, "PADLOC: Physically-Aware Defect Localization and Characterization," in *IEEE Asian Test Symposium*, Nov 2017, pp. 212–218.

978-1-5386-3775-3/18 $31.00 © 2018 IEEE

2018 IEEE 36th VLSI Test Symposium (VTS)

Multi-faceted Microarchitecture Level Reliability Characterization for NVIDIA and AMD GPUs

Alessandro Vallero[§], Sotiris Tselonis, Dimitris Gizopoulos* and Stefano Di Carlo[§]

[§] Politecnico di Torino, {stefano.dicarlo | alessandro.vallero}@polito.it *University of Athens, dgizop@di.uoa.gr

Abstract – **State-of-the-art GPU chips are designed to deliver extreme throughput for graphics as well as for data-parallel general purpose computing workloads (GPGPU computing). Unlike computing for graphics, GPGPU computing requires highly reliable operations. Since provisioning for high reliability may affect performance, the design of GPGPU systems requires the vulnerability of GPU workloads to soft-errors to be jointly evaluated with the performance of GPU chips. We present an extended study based on a consolidated workflow for the evaluation of the reliability in correlation with the performance of four GPU architectures and corresponding chips: AMD Southern Islands and NVIDIA G80/GT200/Fermi. We obtained reliability measurements (AVF and FIT) employing both fault injection and ACE-analysis based on microarchitecture-level simulators. Apart from the reliability-only and performance-only measurements, we propose combined metrics for performance and reliability that assist comparisons for the same application among GPU chips of different ISAs and vendors, as well as among benchmarks on the same GPU chip.**

Keywords – GPGPU; microarchitecture simulator; reliability; performance; fault injection; throughput.

I. INTRODUCTION

Graphics Processing Units (GPUs) are a powerful computing platform for both graphics and general-purpose, data-parallel and computing-intensive applications. GPUs are increasingly used in applications where reliability and performance are a top tier concern [2]. Trading-off reliability and performance is therefore a key aspect of GPU based systems [1]. Reliability analysis for these systems is challenging. It requires accurate and fast techniques able to carefully trade-off between analysis time and accuracy of the reported measurements. Moreover, it must produce results able to guide the system designers in the selection and development of efficient error protection mechanisms. Apart from the use of physical error injections [3][4], simulation-based techniques are the preferred solution to analyze the reliability of GPU systems. Two reliability estimation methodologies for GPUs have been established in this field similarly to the CPU domain: (i) fault injection [5][6][7][15][21] and (ii) Architectural Correct Execution (ACE) analysis [15][22].

This paper presents the results of an extended study aiming at characterizing the main factors (hardware and software) that influence the reliability of GPU chips in the presence of soft-errors. The study focuses on errors in the register file and in the local/shared memory[1]. These arrays are among the biggest arrays of the GPU and are therefore prone to be affected by radiation induced soft-errors. Even if some of these arrays are often delivered with hardware-protection techniques such as Error Correction Codes (ECC), these can incur performance and energy overheads and hence may not be enabled by users in selected applications [6]. Moreover, they may be protected with an ECC

scheme that cannot cover the expected cardinality of faults. Therefore, evaluating the reliability of even protected arrays is an important task. The paper considers several important factors including: correlation between reliability and performance, size of the hardware structures, resource occupancy and execution scheduling. Different reliability assessment methodologies are used to identify trade-offs between analysis time and accuracy of results. GPUs from different vendors, architectures and programming models are compared: AMD Southern Islands and NVIDIA G80/GT200/Fermi. Reliability of all devices is analyzed running the same set of benchmarks written using the typical development framework for each architecture: OpenCL[2] for the AMD GPU and CUDA[3] for the NVIDIA GPUs. Simulations have been performed using a microarchitecture-level simulation. The framework includes tools to perform both soft-error fault injection campaigns and ACE-analysis. Microarchitecture-level simulators provide a good accuracy for the considered hardware structures as demonstrated for CPUs in [10] and, at the same time, they significantly improve the simulation throughput compared to RTL models that are often not publicly available.

To our knowledge, this is the first work that extensively compares the reliability correlated with the performance of some of most important GPU families with different microarchitecture, Instruction Set Architecture (ISA) and computational model using the same set of benchmarks and employing the two most prominent reliability evaluation methodologies. Preliminary results in this direction have been recently discussed in [11]. Such a multidimensional study delivers significant insights on: differences in GPU vulnerability estimations between fault injection experiments and ACE analysis; variations in the vulnerability of specific hardware components and benchmarks among different GPU architectures; joint evaluation of reliability and performance to support designers and programmers when evaluating different GPUs and workloads.

II. RELIABILITY EVALUATION FRAMEWORK

Simulations on the two GPU families have been carried out using two tools named GUFI and SIFI. GUFI, previously presented in [6], is a micro-architectural level fault injector based on GPGPU-Sim (v3.2.2) [19]. It has been developed to perform reliability analysis on NVIDIA GPUs. GUFI has been extended for the purposes of our study to perform ACE-based analysis. SIFI is a fault injection and ACE analysis tool developed to characterize AMD GPUs [15]. SIFI is built on top of Mult2Sim (v4.2) and supports the Southern Island assembly language [13]. For both tools, reliability is analyzed looking at the low-level assembly code running on the real hardware. Therefore, for NVIDIA GPUs, the SASS assembly is used instead of the intermediate and device-independent PTX assembly. We made this choice to study the

---

[1] Local memory is the AMD terminology while shared memory is the NVIDIA terminology for the same memory type.

[2] https://www.khronos.org/opencl/
[3] https://www.geforce.com/hardware/technology/cuda

978-1-5386-3775-3/18 $31.00 © 2018 IEEE

vulnerability of the actual physical registers (instead of the virtual PTX registers), allowing for a fair comparison of NVIDIA and AMD chips. This study focuses on soft-errors in memory elements. Addressing the impact of these faults is extremely relevant since GPUs include a large number of big memory arrays.

Several reliability metrics are considered. The *architectural vulnerability factor* (AVF) of a hardware structure is the main considered reliability metric. The AVF is the probability that a bit-flip (soft-error) at the target structure manifests as a visible error in the computation (output corruption, crash). It jointly considers hardware and software masking effects. To evaluate the masking properties independently from the occupancy, the concept of AVF$_{Util}$ is used [5]. The AVF$_{Util}$ is the probability that a soft-error in hardware structure that is actually used by the program causes an error in the computation. The AVF can be expressed in terms of the AVF$_{Util}$ and the occupancy, that is the percentage of the hardware structures actually used by the running program, as: AVF = AVF$_{Util}$ × Occupancy. By combining the AVF and the raw soft-error rate (λ_i) of every hardware structure, the GPU failures in time rate (FIT$_{GPU}$) can be computed as: FIT$_{GPU}$ = $\sum_i$ AVF$_i$ × λ_i × #bit$_i$, where #bit$_i$ is the number of memory elements of the hardware structure i. Finally, a system designer or a programmer can be provided with a broader idea of the system performance and reliability for any given workload when combined metrics are used. Such a metric can be the *executions per failure* (EPF) defined as the number of executions of a benchmark before a failure manifests: EPF = EIT/FIT$_{GPU}$ [15], [18]. EIT (*executions in time*) is the number of complete correct executions of a benchmark in 10^9 hours of device operation. Another metric that can be defined to jointly evaluate reliability and performance is the *instructions per failure* (IPF). The IPF measures the instruction throughput of a benchmark before a failure manifests instead of the number of complete program executions per failure like in the EPF case. IPF is defined as: IPF = IIT / FIT$_{GPU}$ where IIT (*instructions in time*) is the number of instructions executed in 10^9 hours.

A. Fault injection campaigns

Both GUFI and SIFI, the tools exploited in this paper, perform fault injection experiments following the flow reported in Fig. 1.A, which is composed of three macro phases.

The *fault generation phase* creates the list of faults (fault pool) to inject. The application is profiled to identify the time intervals in which the GPU is active and to gather information about the executed kernels. This is used to speed up the simulation injecting faults only during active intervals. After that, the fault pool is generated randomly selecting for each fault its time (i.e., the clock cycle it manifests) and the specific memory element to corrupt. During the *fault injection phase*, each fault is injected, and the behavior of the system is observed. Fault injections are simulated in parallel to save time, exploiting concurrent executions. As soon as the simulation of a fault completes, the output of the computation is analyzed (*fault classification*). If the computation ends properly and its output is correct, the fault is classified as masked. Otherwise, it is classified as not masked. To improve the injection throughput, faults that target idle resources (i.e., resources not used in the context of the application) are not simulated since they can be directly classified as masked. These faults are named non-util faults since they do not affect the AVF$_{Util}$

computation. When all faults are classified, the AVF and the AVF$_{Util}$ can be computed as reported in Fig. 1.B.

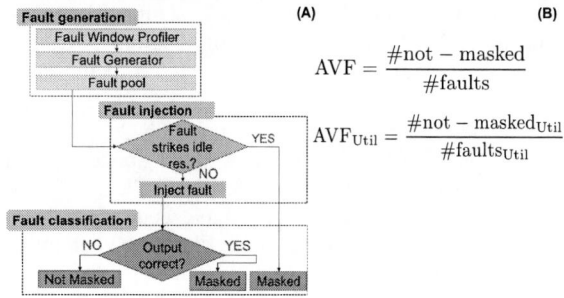

Fig. 1. Fault injection campaign workflow

B. ACE Analysis

ACE (Architecturally Correct Execution) analysis estimates the AVF simulating a single run of the program. It is therefore a very fast analysis that however has reduced accuracy with respect to fault injection. ACE analysis is based on the fact that not all entries of a hardware structure (e.g., registers of the register file) influence the output of the system. The AVF of the structure can therefore be estimated by determining which entry affects the system's output (ACE resources) and which does not (un-ACE resources). Fig. 2.A summarizes the implemented ACE analysis workflow for the register file that follows the approach presented in [8] and [9] for CPU memory arrays. A similar approach is implemented for the other hardware structures.

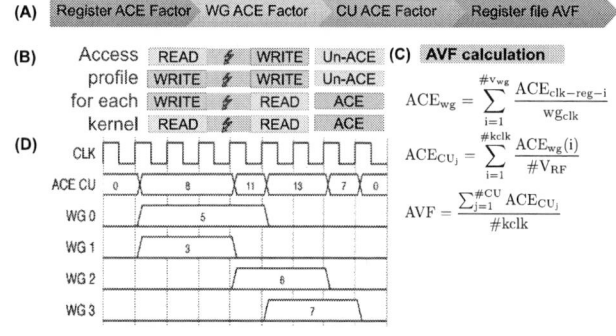

Fig. 2. Register File ACE analysis workflow

Each GPU kernel is analyzed separately and then results are aggregated. For each kernel, the number of registers assigned to each work-group (#v$_{wg}$) is first computed as explained in [20]. All registers not assigned to any work-group are classified as *idle* and marked as un-ACE, while the others are profiled during the execution of the kernel. During the time intervals (i.e., clock cycles) between a read and a write operation (*read-to-write intervals*), and between two consecutive write operations (*write-to-write intervals*) a register can be safely considered un-ACE (Fig. 2.B). In all other cases, it is marked as ACE. The ACE factor of each work-group (ACE$_{wg}$), i.e., the work-group average number of ACE registers per clock cycle, can be computed as reported in Fig. 2.C where wg$_{clk}$ is the number of clock cycles required to execute the work-group and ACE$_{clk-reg-i}$ is the number of clock cycles in which the register i is ACE. At this point, the ACE factors of every work-group of the kernel must be aggregated to obtain the ACE factor of the compute unit (ACE$_{CU}$). To perform this computation, we build a timing diagram representing the time window of every work-group executed by the compute unit (CU).

978-1-5386-3775-3/18 $31.00 © 2018 IEEE

An example is shown in Fig. 2.D, where 4 work-groups (WG1 – WG3) are executed and the number of concurrent work-groups is 2. For each work-group and each clock cycle the diagram reports the related ACE_{wg}. Based on the timing diagram of each work-group, the ACE_{cu} timing diagram is computed by summing up the ACE_{wg} of the scheduled blocks at each clock cycle divided by the number $\#V_{RF}$ of registers per compute unit (Fig. 2.C). The AVF of the entire register file running the selected application can be finally computed as the sum of the AVF of all CUs divided by the number #kclk of clock cycles required to execute the GPU kernel (Fig. 2.C). The AVF_{Util} of the register file can be computed in a similar way, but the ACE_{cu} timing diagram is divided by the number of utilized registers of the running work-groups at each clock cycle (#rf-used), instead of dividing it by the total number of registers of the register file. In the example of Fig. 2.D, considering #rf=32 and #kclk=7, the AVF can be computed as $AVF_{RF} = \frac{\frac{8}{32}+\frac{8}{32}+\frac{8}{32}+\frac{11}{32}+\frac{13}{32}+\frac{13}{32}+\frac{7}{32}}{7} = 0.3$ (30%). Assuming that each work-group consists of 3 work-items and each work-item utilizes 4 registers then a work-group utilizes 3x4=12 registers. Considering the timing of the work-groups the number of utilized registers at each clock cycle is always 24, apart for the last cycle where it is 12. Consequently, the AVF_{Util} can be computed as $AVF_{Util\_RF} = \frac{\frac{8}{24}+\frac{8}{24}+\frac{8}{24}+\frac{11}{24}+\frac{13}{24}+\frac{13}{24}+\frac{7}{12}}{7} = 0.45$ (45%).

The accuracy of the ACE analysis could be increased by taking into account the presence of data processed by dead instructions (i.e., not contributing to the output results) and logic masking (i.e., logical and arithmetic operations resulting in masking of the fault) [8] [9]. From our simulations, we noticed that dead data in the selected benchmarks represent a negligible portion of the application (less than 0.5%). Therefore, neglecting them does not introduce a significant loss of accuracy while reducing the analysis time. Logic masking instead may have an impact on the accuracy. However, its inclusion in the ACE computation significantly increases the complexity of the analysis and therefore has not been implemented in this study. ACE analysis is mainly exploited here for its fast simulation time and we are interested in understanding the difference in accuracy with respect to the fault injection.

III. RELIABILITY EVALUATION

A. Experimental setup

We analyzed 4 GPU chips with different architectures: AMD HD Radeon[TM] 7970 (Southern Islands architecture), NVIDIA Quadro[TM] FX 5600 (G80 architecture), NVIDIA Quadro[TM] FX5800 (GT200 architecture) and NVIDIA Geforce[TM] GTX 480 (Fermi architecture). We used 10 benchmarks: 7 available both in the CUDA SDK[4] and AMD-APP SDK[5] and 3 from the Rodinia benchmarks [16]. All benchmarks have equivalent OpenCL and CUDA implementations. Table 1 summarizes the characteristics of the CUs of the considered GPU chips. Since manufacturing data are not public, the raw bit soft-error rate per cell (λ) has been estimated from literature data [17]. The error rate is normalized to 1E-3 for the 90nm SRAM cell. For all chips, we measured the AVF of the general-purpose register file (RF) and the local memory (LM) using both Fault Injection (FI) and ACE Analysis (ACE). For the FI experiments, we applied statistical fault

sampling [12]. Considering the size of the targeted structures, to reach a 2.88% error margin with 99% confidence level, 2,000 fault injection experiments were conducted for each hardware component of each GPU architecture. Every benchmark was executed with the same input data-set for all considered GPU devices. The data-sets were chosen to maximize and stress the use of the CUs. However, since this significantly increases the simulation time when considering multiple CUs, following the approach used by Farazmand et al. [5], we scaled the analysis considering a single CU.

Table 1: CU details of the target GPU architecture

	Chip name	Quadro[TM] FX 5600	Quadro[TM] FX5800	Geforce[TM] GTX 480	HD Radeon[TM] 7970
	Technology	90 nm	55 nm	40 nm	28 nm
	λ FIT/bit	1E-3	0.72E-3	0.52E-3	0.32E-3
	Vendor	NVIDIA	NVIDIA	NVIDIA	AMD
	Architecture	G80	GT 200	Fermi	S. Islands
	Register file	32KB	64KB	128KB	256KB
	Local memory	16KB	16KB	48KB	64KB
	SIMD Units	1	1	2	4
Max	#wg	8	8	8	40
	#wavefronts	24	32	48	40
	#work-items	768	1024	1536	1840

B. Experimental results

The two charts in Fig. 3 report the AVF for RF and LM computed using both FI and ACE analysis. By analyzing RF (Fig. 3.A), we observe significant differences in the AVF depending both on the hardware platform and on the executed benchmarks. Overall, for all benchmarks the HD Radeon 7970 is the chip with the lowest RF vulnerability while the Quadro FX 5600 is the one with the highest vulnerability. Looking at a single hardware platform, the difference in the AVF for the different benchmarks is the result of the way the software stresses the resource and is able to be resilient to the injected faults. Instead, when looking at the same benchmark executed on the different chips we can note a strong correlation of the AVF with the occupancy (red bullets). Considering RF, based on their average occupancy, the architectures can be ordered as follows: HD Radeon 7970 (12.89%), GTX480 (37%), Quadro FX5800 (50%), Quadro FX 5600 (57%). This trend holds for all benchmarks with the exception of histogram in which the RF occupancy of the GTX480 is 35% while the one of Quadro FX5800 is only 26%. The occupancy of a resource is therefore a strong indicator to predict the AVF trend for a single application executed on different chips. However, it does not provide information on the actual AVF values. In fact, benchmarks with similar occupancy have significantly different AVFs (e.g., backprop and scan in Fig. 3.A). This is an important finding obtained from the analysis of our simulations. It confirms that occupancy is not the only contribution to AVF but the particularities of the access patterns from each benchmark are also important. When considering the AVF of LM (Fig. 3.B), while the correlation between AVF and the occupancy of the resource still holds, a clear AVF variation trend between the four chips valid for all benchmarks cannot be identified. This suggests that the way this resource is used strongly depends on the executed application code. In particular, the AVF of the HD Radeon 7970 is significantly higher than the one of the other architectures for histogram, where occupancy is 100%. In the scan and reduction benchmarks the vulnerability is almost equal for all

---

[4] https://developer.nvidia.com/cuda-toolkit-42-archive

[5] http://developer.amd.com/tools-and-sdks/opencl-zone/amd-accelerated-parallel-processing-app-sdk/

architectures, while Quadro FX 5800 has a slightly higher AVF for backprop and dwt where its occupancy is 22% and 53%, respectively. Not all benchmarks actually use the local memory, this is the case of gaussian, kmeans and vactoadd.

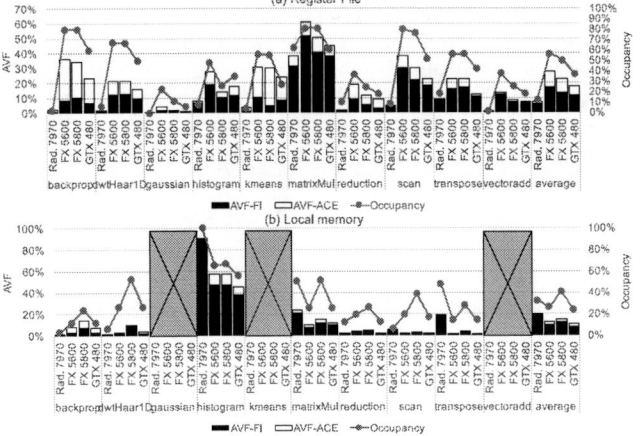

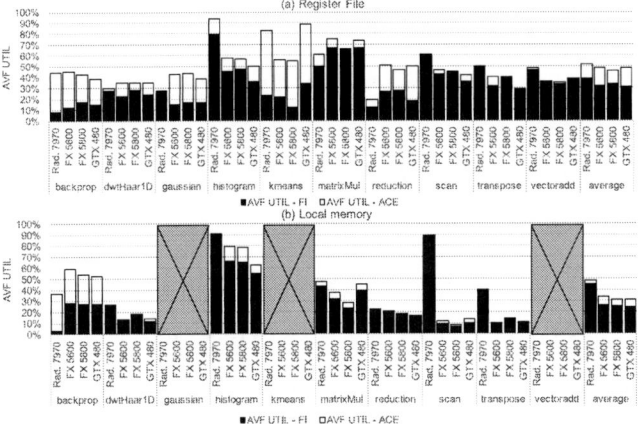

Fig. 3.   AVF measured by fault injection (FI) and ACE.

To discuss vulnerability decoupled from the occupancy of the resources, we can use the AVF$_{Util}$ measurements reported in Fig. 4. Again, both diagrams present AVF$_{Util}$ based on both FI and ACE analysis for all considered GPU models. The AVF$_{Util}$ is mainly influenced by the software logic masking and the different ISAs of the GPU chips. Therefore, it is hard to observe a clear trend or correlation in its value between NVIDIA and AMD architectures. However, it is interesting to note that the three NVIDIA chips, which implement similar native ISAs and use the same programming model, have very similar AVF$_{Util}$ for each benchmark. Apart from the different LM and RF size, which in turn influence the number of vulnerable resources leading to different vulnerabilities (Fig. 6), the NVIDIA GPUs employ different wavefront scheduling mechanisms (warps in NVIDIA/CUDA terminology). Each SIMD unit can accommodate a different number of wavefronts. Changing the number of resident wavefronts changes the scheduling process. This leads to a variation of the vulnerability timing windows for both registers and memory words. An increment of the time a wavefront has to wait before being scheduled leads to a longer exposure of a critical resource to a fault. Moreover, the number of wavefronts that a

single CU can concurrently execute is another factor that influences the wavefront scheduling. The CUs of NVIDIA Quadro FX 5600 and NVIDIA Quadro FX 5800 can schedule a single wavefront at a time while the CUs of NVIDIA GeForce GTX 480 process two wavefronts in parallel. This difference significantly influences the ACE analysis results for the AVF$_{Util}$ of RF (Fig. 4.A), which measures reliability on the basis of the vulnerable timing windows of the used memory elements. This does not apply to LM since it is shared among all wavefronts of the same CU. In particular, for some benchmarks, the two NVIDIA Quadro chips show very close values while the NVIDIA GeForce chip features a different value in benchmarks gaussian, histogram, kmeans and transpose.

Fig. 3 and Fig. 4 also allow us to compare AVF estimations obtained with FI and ACE analysis. Such a comparison must consider two main aspects: the measurement accuracy and the time required to perform the analysis. Regarding the accuracy, it is well-known from the literature that the error margin and the confidence interval of statistical fault injection are determined by the number of injected fault. Differently form FI, the accuracy of the ACE analysis cannot be quantified even if it is well known that it delivers pessimistic evaluations. This trend is confirmed for RF, where FI estimates lower AVF compared to ACE analysis (Fig. 3.A and Fig. 4.A). The error is strongly benchmark dependent. In particular, in our ACE analysis, to be very fast, we do not consider the program logical masking of faults. Fig. 3.B shows that ACE analysis AVF overestimation is lower for LM than for RF. This can be explained since RF and LM are used in a different way by the work-items of an application. LM is significantly slower than RF. For this reason, LM is mainly used to move data from/to RF, where logic and arithmetic operations take place. Consequently, logic masking effects not detected by ACE analysis are more relevant in RF than in LM. The only exceptions to this trend are histogram and backprop. For these two benchmarks, the ACE analysis introduces a higher error. This can be explained looking at the AVF$_{Util}$ (Fig. 4). Their AVF$_{Util}$ significantly changes depending on the evaluation method (FI or ACE analysis). Although the error between AVF$_{Util}$ based on FI and AVF$_{Util}$ based on ACE analysis is higher for backprop than histogram, histogram occupies more intensely the local memory compared to backprop. Thus, even if histogram features the second highest difference in LM AVF$_{Util}$ (for different methods of evaluation), its high local memory occupancy leads to the highest difference in AVF depending on the evaluation method. Among the different benchmarks, backprop, is the one presenting the highest AVF overestimation between FI and ACE analysis for both RF and LM, respectively 3.3 and 1.1 times higher (Fig. 3). On average, the overestimation of the AVF and AVF$_{Util}$ made by ACE analysis with respect to FI is respectively 95% and 80% for RF, while 15.9% and 17.4% for LM. It is also interesting to remark that, for some combinations of benchmarks and architectures, we observe that ACE analysis slightly underestimates AVF. This applies to HD Radeon for dwtHaar1D (0.35 p.u. – percentile units), histogram (0.38 p.u.) and reduction (0.93 p.u.) and to Quadro FX 5800 for reduction (0.31 p.u.). Finally, in case of the scan benchmark we observe a singularity: ACE analysis underestimates the AVF$_{Util}$ (88.9% for FI against 85.2% of ACE analysis). Nevertheless, this difference is very close to the 2.88% error margin of fault injection.

In terms of simulation time, the single-run ACE analysis offers significantly better performance compared to FI. Table 2

978-1-5386-3775-3/18 $31.00 © 2018 IEEE   189

quantifies this benefit comparing the simulation time of ACE analysis with the number of fault Injections Per Hour (IPH) that we were able to simulate for each benchmark employing both GUFI and SIFI. However, it is important to remember that this benefit must be traded-off with the reduced accuracy delivered by ACE analysis and with the capability of FI to precisely quantify the error margin of the computed metrics. Nevertheless, looking at the results provided in Fig. 3 and Fig. 4, it is clear that, despite its lower accuracy, ACE analysis is very efficient in providing a rough idea about the vulnerability of a hardware component or the differences between benchmarks in a very short simulation time. Overall, from our analysis we can conclude that ACE analysis represents a good characterization technique for LM, since it combines good accuracy and low computation time, whereas it is less suitable for RF given the higher inaccuracy. In the remaining of this section discussions will focus on results obtained resorting to fault-injection experiments.

Table 2. Simulation time required to perform the reliability analysis

Benchmark	SIFI		GUFI	
	ACE time (s)	IPH	ACE time (s)	IPH
backprop	3	1200	13	277
dwt	9	400	1	3600
gaussian	29	124	37	97
histogram	173	21	44	82
kmeans	24	150	90	40
matrixMul	21	171	20	180
reduction	4	900	4	900
scan	5	720	2	1800
transpose	2	1800	6	600
vectoradd	39	92	5	720
AVERAGE	30.9	557.8	22.2	829.6

Fig. 5 combines the AVF of the different hardware structures with the raw bit soft-error rate of the technology (λ) used to build the different chips (see Table 1). The result is the global FIT_{GPU}. The figure breaks down the contribution of RF and LM to the global FIT_{GPU}. Interestingly, the contribution of LM is significantly lower than the one of RF for most benchmarks in which this memory array is used. Understanding the motivations for the variation of the FIT rate is not simple since it depends on the fabrication technology, the size of the hardware structures and their vulnerability for each benchmark. In general, the combination of the technology node and the size of the structure seems to be the predominant factors. In our experimental setup, the size of RF is bigger than the one of LM and we observe a higher number of failures caused by faults in RF for all benchmarks and architectures except for histogram. In the histogram benchmark, which has an intense use of LM, the LM AVF is much higher than the one of RF and represents the main contribution to the FIT. In histogram, LM is used as a read-only memory, so all the Util resources are always ACE. Moreover, the GPU architectures whose hardware components are larger have the potential to better exploit their intrinsic parallelism, since the number of parallel work-items is also influenced by the availability of resources for the execute kernels. Executing more work-items concurrently increases the number of potentially vulnerable resources. More specifically, this trend can be evicted from Fig. 6 for NVIDIA architectures. However, Radeon 7970, featuring the largest components, shows an opposite behavior in some cases, highlighting the influence of compilers on vulnerable resources.

The bigger size of a GPU hardware component naturally makes it more vulnerable to soft-errors. However, it increases the execution parallelism and thus improves performance. Therefore, as discussed in Section II, to combine the reliability evaluation

with the performance profile of each benchmark and GPU chip we analyzed the EPF (Fig. 7) and the IPF (Fig. 8) metrics because FIT alone (Fig. 5) does not take into account the amount of work carried out by the GPUs before a failure arises. EPF incorporates the execution time and FIT for a program, while IPF also includes information about the instruction throughput of GPUs when executing an application.

Table 3 shows for each benchmark (rows) and architecture (cols) the execution time (cycles) as well as the number of executed instructions required to compute EPF and IPF.

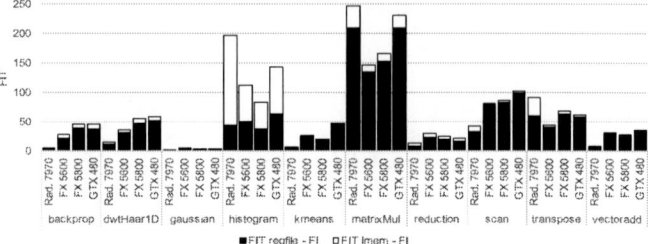

Fig. 5. Breakdown of Failures in Time (FIT) rate using the AVF measurements from Fault Injection.

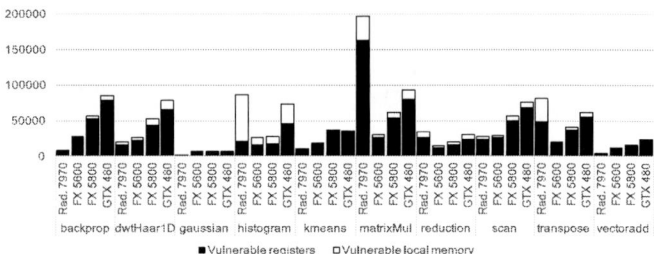

Fig. 6. Vulnerable resources in bits.

The IPF for a particular benchmark is proportional to the EPF and to the instruction throughput. Since this throughput strongly depends on the target execution device, to fairly compare different GPU architectures we must look at the EPF instead of IPF. The IPF is instead useful for evaluating the reliability of different programs on the same GPU chip. On the one hand, the EPF metric is useful to the architects who can quantify the effectiveness of a hardware-based error protection technique which can be applied to their designs (if needed) along with a performance cost at the early design stages. Larger EPF numbers show a larger number of executions before a failure and different protection mechanisms can deliver different improvements in the FIT rates and can also have different impact on performance. Combining performance and reliability measurements in the EPF metric delivers a broader view for decision-making. This could be for instance important when evaluating real-time applications that are not continuously executed, but they are scheduled once every time period. On the other hand, IPF is useful to the programmers who want to quantify the effectiveness of software redundancy-based protection techniques which can be applied to their programs running on the same architecture, thereby correlating the error resilience of their applications at a performance cost. IPF summarizes both the performance cost and the resilience improvement.

In Fig. 7, using the EPF to compare the different architectures, we can identify that the HD Radeon is in general the best choice for the selected benchmarks with the exception of histogram, scan and transpose. Using the IPF to compare different benchmarks on the same architecture (Fig. 8), we can instead notice that for HD Radeon, backprop and gaussian have higher IPF than the other benchmarks while gaussian has the higher IPF when executed on Quadro FX5600, Quadro FX 5800 and GTX 480. Metrics that combine reliability and performance have also the potential to help comparing CPUs and GPUs. In particular, while the IPF is a raw throughput of work (instructions per failure occurrence) the EPF is a complete execution rate per failure occurrence, which is very useful if one wants to compare different processing elements like CPUs and GPUs [18].

Table 3. Execution time and instructions

Benchmark		HD Radeon 7970	Quadro FX 5600	Quadro FX 5800	GeForce GTX 480
	Freq (MHz)	925	337.5	325	700
backprop	cycles/	94376/	423594	369855	206834
	inst.	2108160	10312032		
dwt	cycles	41072	44998	39412	25859
	inst.	1839075	1180042		
gaussian	cycles	5862543	561060	555732	541687
	inst.	7308189	5488224		
histogram	cycles	3198537	1031394	1029746	885491
	inst.	20029440	21784328		
kmeans	cycles	913526	1278216	1267144	1397604
	inst.	31930960	35984844		
matrixMul	cycles	269591	439591	400594	299346
	inst.	10924032	15007744		
reduction	cycles	27836	47377	47086	27231
	inst.	312736	854719		
scan	cycles	123763	18707	16572	19721
	inst.	3025801	468720		
transpose	cycles	50862	98911	82821	49942
	inst.	733184	2818048		
vectoradd	cycles	31687	29219	21603	30225
	inst.	1523712	638976		

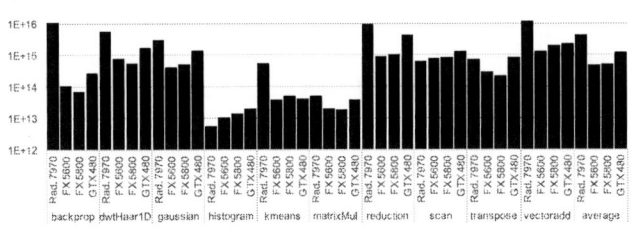

Fig. 7. Executions per Failure (EPF).

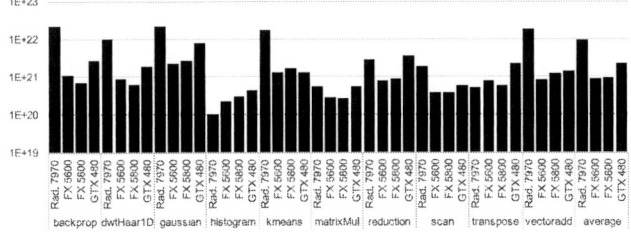

Fig. 8. Instructions per Failure.

IV. CONCLUSIONS

We have presented a multi-faceted comprehensive reliability assessment framework for state-of-the-art AMD and NVIDIA GPUs. Reliability measurements have been performed using both fault injection and ACE analysis to reveal the differences between the two approaches. We used 10 benchmarks to compare the vulnerability of the AMD/OpenCL versions and the NVIDIA/CUDA versions. We also proposed two combined

performance/vulnerability metrics (EPF and IPF) that report the throughput of complete executions per failure or the throughput of individual instructions per failure. These metrics provide a wider picture of the GPU quality of execution and can be employed to compare different GPU chips for the same application or different programs on the same GPU chip. The proposed framework can be flexibly to jointly assess reliability and performance of further GPU configurations, any OpenCL and CUDA workload and, of course, to assist designers in making decisions for hardware- based or software-based error protections techniques in GPUs.

ACKNOWLEDGMENT

This research has been supported by the 7th Framework Program of the European Union through the CLERECO Project, under Grant Agreement 611404.

REFERENCES

[1] G. H. Asadi et al. "Balancing Performance and Reliability in the Memory Hierarchy," *ISPASS 2005*, pp. 269-279

[2] P. Rech et al. "Impact of GPUs Parallelism Management on Safety-Critical and HPC Applications Reliability," DSN 2014, pp. 455-466.

[3] P. Rech *et al.*, "Neutron-Induced Soft Errors in Graphic Processing Units," *REDW 2012*, pp. 1-6.

[4] I. S. Haque et al. "Hard Data on Soft Errors: A Large-Scale Assessment of Real-World Error Rates in GPGPU," *CCGrid 2012*, pp. 691-696.

[5] N. Farazmand et al. "Statistical fault injection-based AVF analysis of a GPU architecture," *SELSE, 2012*.

[6] B. Fang et al. "GPU-Qin: A methodology for evaluating the error resilience of GPGPU applications," *ISPASS 2014*, pp. 221-230.

[7] S. Tselonis et al. "GUFI: A framework for GPUs reliability assessment," *ISPASS 2016*, pp. 90-100.

[8] S. S. Mukherjee et al. "A systematic methodology to compute the architectural vulnerability factors for a high-performance microprocessor," *MICRO-36*, pp. 29–40.

[9] A. Biswas et al. "Computing architectural vulnerability factors for address-based structures," *ISCA '05*, pp. 543.

[10] A. Chatzidimitriou et al. "RT Level vs. Microarchitecture Level Reliability Assessment: Case Study on ARM Cortex-A9 CPU", *DSN 2017*.

[11] A. Vallero et al. "Microarchitecture Level Reliability Comparison of Modern GPU Designs: First Findings", *ISPASS 2017*.

[12] R. Leveugle et al. " *Design, Automation & Test in Europe Conference & Exhibition. DATE 2009*, pp. 502-506.

[13] R. Ubal et al. "Multi2Sim: A simulation framework for CPU-GPU computing," *PACT 2012*, pp. 335-344.

[14] N. J. Wang et al., "Examining ACE analysis reliability estimates using fault-injection," *ISCA 2007*, pp. 460-469.

[15] A. Vallero et al., "SIFI: AMD Southern Island GPU Microarchitectural Level Fault Injector". *IOLTS 2017*.

[16] S. Che *et al.*, "Rodinia: A benchmark suite for heterogeneous computing," *IISWC 2009*, pp. 44-54.

[17] E. Ibe et al. "Impact of Scaling on Neutron-Induced Soft Error in SRAMs From a 250 nm to a 22 nm Design Rule," *IEEE Transactions on Electron Devices*, vol. 57, no. 7, pp. 1527-1538, July 2010.

[18] A. Chatzidimitriou et al. "Performance-aware reliability assessment of heterogeneous chips," *IEEE 35th VLSI Test Symposium 2017. VTS 2017*, Las Vegas, NV, USA, pp. 1-6.

[19] A. Bakhoda et al. "Analyzing CUDA workloads using a detailed GPU simulator," *ISPASS 2009*, pp. 163-174.

[20] AMD accelerated parallel processing opencl optimization guide available. [Online]. Available: http://amd-dev.wpengine.netdna-cdn.com/wordpress/media/ 2013/12/AMD OpenCL Programming Optimization Guide.pdf

[21] S. K. S. Hari et al., "Sassifi: An architecture-level fault injection tool for gpu application resilience evaluation," *ISPASS 2017*, pp. 249–258.

[22] J.Tan et al. "Analyzing soft-error vulnerability on GPGPU microarchitecture," *IISWC 2011*, pp. 226-23.

2018 IEEE 36th VLSI Test Symposium (VTS)

RF Circuit Authentication for Detection of Process Trojans

Fatih Karabacak[1], Richard Welker[2], Matthew J. Casto[3], Jennifer N. Kitchen[1], and Sule Ozev[1]

[1]School of Electrical, Computer, and Energy Engineering, Arizona State University, Arizona, USA, 85281

[2]Alphacore Inc., Arizona, USA, 85281

[3]Air Force Research Lab, Wright-Patterson Air Force Base, Ohio, USA, 45433

Abstract: *Globalized supply chain for electronic circuit manufacturing has reduced the production cost considerably. However, it also presents a challenge since many companies/players contribute to the product and it is not always possible to control or monitor every third-party employee or contractor that takes part in the process. A design house that relies on a foundry for manufacturing needs to ensure that the manufactured devices conform to the agreed-upon process model between the design house and the foundry. Potential deviation from the process model may be due to incidental quality control issues, or due to malicious modifications to the process or circuit layout with the intent of doing harm during in-field operation. In this paper, we present a multivariate methodology to detect even small process and layout level modifications to the circuit by using mission-mode specifications as well as enhanced test modes. We present an algorithm for detecting process/layout modifications and for selection of test inputs to be used in the detection process. Experimental results on an LNA circuit show that the proposed technique can achieve high authentication accuracy even for a single device with a negligible false positive rate.*

1. Introduction

Owning and operating a foundry with advanced capabilities can no longer be afforded by most design companies [1]. The integrated circuit (IC) design industry has adopted a globalized design and manufacturing flow where multiple teams contribute to the design of one IC which gets fabricated in an off-site facility. Globalization of IC design and manufacturing flow has successfully ameliorated the design complexity and fabrication cost challenges, and helped deliver cost-effective products. Splitting design and manufacturing into separate businesses has enabled small and innovative design companies to flourish. While such a distributed flow helps meet the stringent time-to-market deadlines and offer a financially viable model, it is also vulnerable to various forms of security and quality threats in the supply chain that involves designers, foundries, test facilities, and distributors until the end-product reaches the customers. These threats include poor quality control, malicious changes in the process and circuit parameters, intellectual property (IP) piracy, and counterfeit ICs due to overproduction by the foundry or ICs recycled from used devices [2-7].

Emerging hardware security threats have challenged researchers to develop methods for detecting the presence of counterfeit chips [8,9] and unwanted modifications at the process [10-15] and circuit level [16-26].

There has been tremendous research activity in the domain of hardware Trojans, which are defined as modifications to the circuit structures aimed at causing failures or stealing information. Many detection techniques have been proposed for digital circuits. Parametric measurements, such as supply currents [18] and path delays [22,26], can be used to compare the measured values with expected distributions. Repeating test patterns can enable Trojan detection if the Trojan is activated during production testing. Production time or run-time functional testing aims at detecting Trojans by comparing the outcome of the circuit with the expected outcome [20]. Fingerprinting devices through additional parameters, such as heat signatures and electromagnetic emissions, can be used for any generic circuit [17,25] at the expense of increasing test cost. A few examples in the RF domain focus on specific communication standards and run-time detection using functional testing [19,21].

Recently introduced process reliability Trojans are defined as unwanted modifications to process parameters at the foundry aimed at reducing IC lifetime or reducing the fidelity of security mechanisms such as encryption keys [10-15]. These modifications include simple-to-implement, low-level changes, such as changing dopant concentrations or changing annealing temperature during oxidization. At production time, these changes cause minute variations in the parameters of the circuit components (e.g. small shift in the threshold voltage, or increase in generated noise), and are extremely hard to detect. Part of the difficulty lies in the discrete response of the digital circuits; delay and supply current are the only two parametric measurements available. In this work, we focus on process modifications to RF circuits that may result in reliability risks [13,14]. The multitude of performance parameters coupled with the higher sensitivity to process parameters enables better detection of such process modifications. In [15], the authors present compact models that link process modifications to reliability parameters, such as trap generation rate and show that these modifications can indeed reduce the in-field lifetime from years to mere months.

While most research in the arena of authentication and identification focus on digital circuits, analog/RF circuit security is just as important since these devices can result in parametric failures in the system, even if the functionality is met. Techniques for effectively authenticating RF circuits are needed, particularly to detect small modifications to process and circuit that may go unnoticed with current performance testing practices. These small changes may still produce circuits meeting their performance specifications, particularly

978-1-5386-3775-3/18 $31.00 © 2018 IEEE

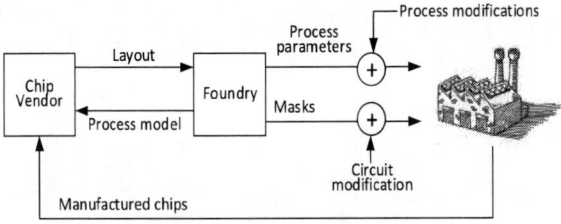

Figure 1: Threat Model

since designers spend extraordinary effort to de-sensitize the circuit performance with respect to process variations. However, the altered circuits may fail and/or exhibit unpredictable behavior shortly after deployment due to accelerated aging or due to environmental triggers, such as temperature and interferers [10-15]. This threat has been recognized for mission critical systems, leading to trusted foundry programs. However, manufacturing at such verifiably trusted foundries is costly and may not be feasible for many commercial applications.

In this work, we present a trust-but-verify framework for RF circuit authentication by combining multi-dimensional statistical analysis with enhanced test modes that sensitize the circuit with respect to process parameters. Essentially, we treat the PDK as a contract between the foundry and the chip designer and use Monte-Carlo simulations to generate an overall statistical model. We present an algorithm to select the inputs and outputs of this model to minimize the rates of false positives. For manufactured devices, which can be sampled randomly, the enhanced test modes (at non-mission-mode supply levels) are used to measure a multitude of circuit parameters. The statistical model is used to determine whether each sample falls within the expected range of distribution for the selected target parameters. If the authentication fails, the circuit may be subjected to invasive diagnostic techniques to confirm the result. We have used an LNA circuit for experimental demonstration. In addition to the baseline LNA circuit layout, two modified circuits were generated, one with line thickness decreased by 8%, and one with an additional wire (capacitive) load of 150fF at the output node. Moreover, additional process deviation for the threshold voltage (40mV shift in the mean) is added in the model file to mimic a process modification. Post layout simulations are used to verify that the modified circuits can be authenticated as such with a very high accuracy while keeping the false positive rate below 2.5%.

2. Threat Model

Figure 1 illustrates our threat model. The IC vendor receives the process model from the foundry and produces the layout of the circuit based on this process model. The threats originate at the foundry due to malicious attackers (third-party consultants, rogue employees) or due to non-malicious quality control issues. Process variables, such as doping concentration and annealing temperature, can be altered, resulting in small global shifts in circuit component parameters. The circuit itself can be altered during mask generation by inserting unwanted circuit

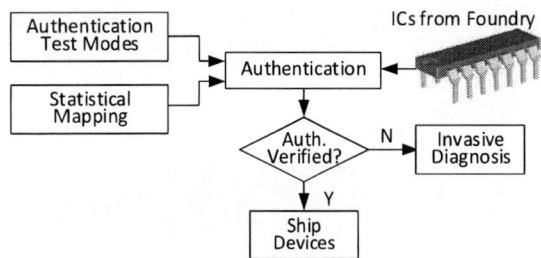

Figure 2: Trust-but-Verify framework

components (e.g. dipole antennas, capacitive loading), or by changing the attributes of connections (e.g. wire thickness). If the modification is nefarious, the goal of the attacker is to weaken the manufactured circuit such that it fails shortly after deployment or becomes easier to tamper with externally. The attacker can make process changes intermittently throughout the production cycle to avoid detection. Hence, all chips manufactured by an untrusted foundry are suspect until verified otherwise.

3. Proposed Authentication Method

Our trust-but-verify vision for authentication is illustrated in Figure 2. During the design process, the IC vendor develops enhanced test modes and statistical authentication models based on the process design kit (PDK) provided by the foundry, which may not be fully trusted. Upon receiving manufactured devices, the IC vendor randomly samples devices from each lot and authenticates each sample with respect to these test modes and authentication models that are not disclosed to the foundry. If a device fails this authentication step, further analysis is conducted to diagnose the potential causes. One cause may be a false alarm due to the statistical nature of the authentication models. However, this failure may be due to an alteration to the circuit or process that is not authorized by the IC vendor. This diagnosis step can involve invasive techniques, such as accelerated aging, over-stress testing, delayering, and X-ray inspection [27-29], and may require more time and resources as it is encountered infrequently.

3.1 Multi-Variate Analysis

Malicious or unintentional changes to process and circuit variables, such as dopant concentration, channel length, metal thickness, annealing time, and temperature can be used to lower resilience to in-field degradation mechanisms, such as hot carrier injection, oxide breakdown, negative bias temperature, and electromigration. These changes in the process result in minor variations in the circuit response that can be hidden within overall process variations, thus making the changes extremely difficult to detect [10-15]. However, in the field, these weaker circuits may fail quickly, due to the exponential relationship between process parameters and degradation. In [15], the authors have developed a model linking modified process parameters to in-field lifetime and have shown that lifetime is reduced drastically. Such

978-1-5386-3775-3/18 $31.00 © 2018 IEEE

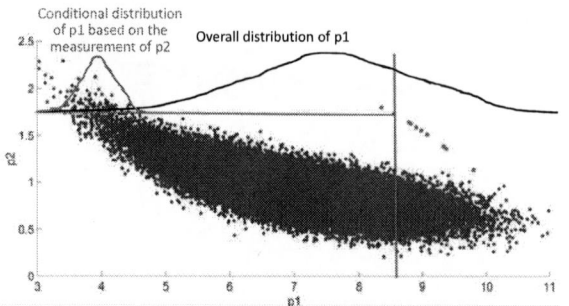

Figure 3: Outlier analysis to detect process and circuit changes that alter relations between performance parameters.

modifications are referred to as Process Reliability Trojans [10-15]. One possible way of addressing the detection of such reliability Trojans is to include additional transistors and extensively test their characteristics for manufactured ICs. However, this would be costly in terms of area/pin overhead.

Fortunately, for RF circuits, changes in almost every process variable will result in some change in performance parameters. If the process shift and/or circuit modification alters the relationship between measurable performances, it can be detected via multivariate analysis, even if the deviation in each performance parameter is within process limits. As an example, Figure 3 illustrates the scatter plot between two performance parameters (p1 and p2) that are correlated through the process. The blue circles are samples that stem from the same functional relation between these two performance parameters, and the red squares show samples that stem from a slightly altered functional relation. In a traditional test flow, p1 and p2 are measured and compared with their limits independently. In such a flow, *none* of the red squares will be flagged as suspicious as both p1 and p2 are well within their 3σ limits. For instance, the overall distribution of p1 is shown as the black probability density function. While checking against limits is ineffective, we can observe, via visual inspection, that these samples are outliers. Of course, for dimensions greater than 3, visual inspection is not possible. We would like to develop an automated method to do the same using more dimensions than 3. If the joint distribution of p1 and p2 are collapsed based on the measurement result of p2 for one of the red samples, the resulting conditional probability is shown as the red distribution curve. The actual location of that sample in the p1 dimension is well outside the expected conditional probability distribution. Hence, we can identify this red sample as not conforming to the expected distribution. Here, the challenge is to select a subset of parameters that will provide good fidelity for detecting process and circuit alterations. For the same example in Figure 3, if p2 is chosen as a target for the conditional probability, the rightmost red square cannot be identified as an outlier.

3.2 Enhanced Test Modes for Increased Process Sensitivity

While designers strive for process robustness at nominal operating conditions, such as supply voltage, noise,

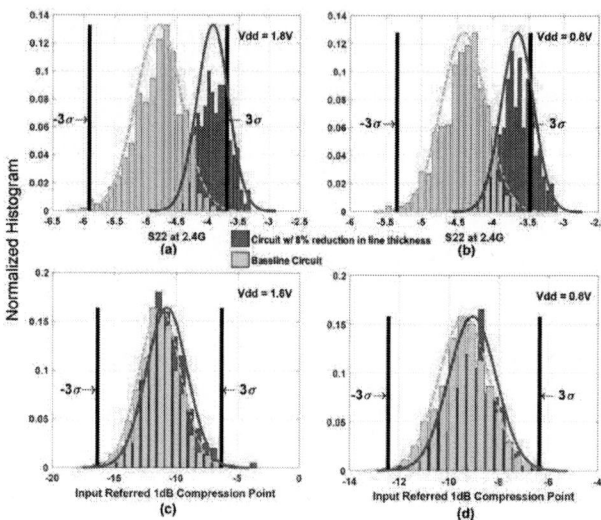

Figure 4: Two LNA performances for nominal process (green histogram) and process modified by decreasing the line thickness by 8% (red histogram) for two supply levels.

temperature, same robustness is generally difficult to maintain over a large variation in operating conditions. By modifying these operating conditions during testing, we can increase sensitivity to process parameters. This sensitivity increase is only temporary and does not result in degradation of circuit performance. Using modified operating conditions for detecting process reliability Trojans has been deemed ineffective for digital circuits [10-15]. However, prior work on this subject has focused only on the analysis of delay based failures, which is a discrete function of path delays and only one-dimensional. For RF circuits with many measurable performance parameters, modified operating conditions enable better detection.

As an example, Figure 4 shows two performance parameters, output reflection coefficient (S22) and 1dB gain compression input point (P1dB), of an LNA circuit under the nominal supply voltage of 1.8V and reduced supply voltage of 0.8V. Green samples are generated via Monte-Carlo sampling using the provided PDK for TowerJazz 180nm technology. Red samples are results obtained when the line thickness of inductors is decreased by 8%, which is a small change. Both performance parameters are affected by the process change, as expected. However, a majority of the samples from the modified process fall squarely within the 3σ limits of the nominal process at both supply voltages. This process modification is not detectable using the two aforementioned measurements at either supply voltage. As a matter of fact, it is not detectable by comparing all performance parameters at any supply voltage. However, by merging the information from all measurements using a multivariate statistical model, such minute changes become detectable.

3.3 Selection of Targets and Inputs

Intuitively, using multiple parameters in the statistical model will lead to better authentication accuracy. However, using all

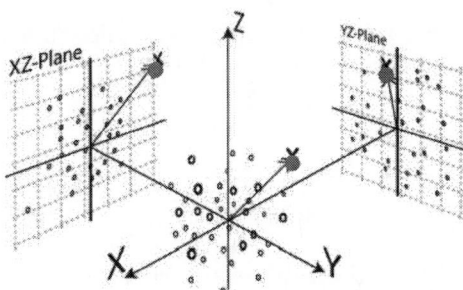

Figure 5: An outlier sample that is not easily detectable in 3 dimensions (XYZ) may become detectable in two selected dimensions (XZ).

performance parameters does not necessarily lead to better detection. Some parametric relations are not altered by modifications to process variables or circuit layout. Including these parameters in the mapping model will only increase uncertainty. The effect of this uncertainty is illustrated in Figure 5. A modified sample with an altered relation between the X and Z axes is not detectable using all 3 dimensions (XYZ) or other 2-dimensional pairs (e.g. YZ). It is only detectable in the XZ-plane. Including the information from the Y-axis only introduces noise and reduces detection capability. Since each process and circuit variable may alter the relations between different sets of performance parameters, we can form multiple statistical models targeting different sets of the process and circuit modifications using the same data.

Potentially, any measured parameter can be a target, and any of the other parameters can be included in the input set. However, this is not the most effective way to detect process/circuit modifications. We need to carefully select the targets and their inputs to minimize false positive rates and maximize the detection capability. In order to select the targets, we conduct Monte-Carlo (MC) simulations under different supply voltage conditions. We then analyze the correlations among the parameters. For each target parameter, we construct a neural network (NN) with inputs comprised of correlated performance parameters. We determine the NN prediction error to set the acceptance threshold. We then evaluate the false positive rate of the prediction using a mutually exclusive set of circuit samples. We remove/add additional inputs to the NN until the false positive rate falls below a pre-determined threshold (we set 1% for each target in this work).

In order to address the flip side of the problem, i.e. maximizing detection capability, we inject a wide range of circuit/layout modifications to the circuit in simulations. The set of modifications is not (and cannot be) complete but they are used as a guide to select the enhanced test modes. Note that for a given set of measurements, there can be more than one target since adding more targets is a post-processing step and does not increase test time or complexity.

3.4 Authentication Algorithm

The proposed authentication framework consists of an off-line stage, where the circuit is analyzed under enhanced test modes and the statistical model is generated, and an on-line stage

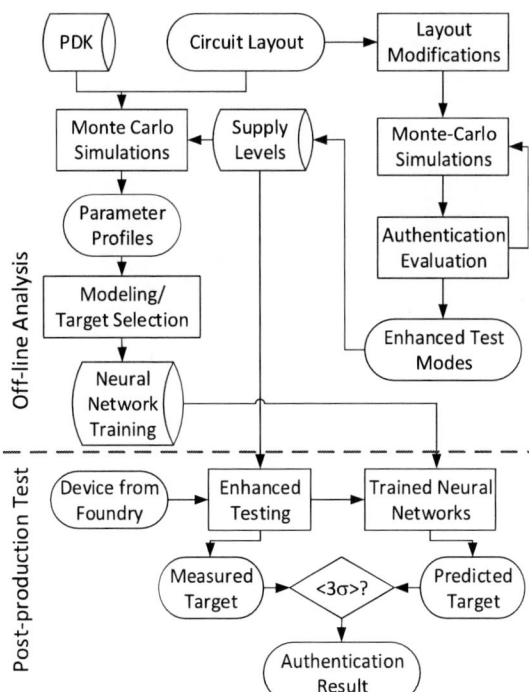

Figure 6: Proposed authentication flow

where the devices that return from the foundry are subjected to enhanced testing. The majority of the computational burden, such as training the NN, takes place during the off-line stage. The online stage post-processing poses negligible computational overhead.

Figure 6 shows the flow of the proposed authentication method. Once the circuit layout is finalized, Monte-Carlo simulations are conducted to form the statistical model and select target/input combinations. Authentication efficacy based on a number of circuit/layout modifications is also evaluated to remove test modes that do not contribute to the coverage. At the end of the off-line stage, a number of NNs are trained for prediction. Once devices are returned from the foundry, samples are taken from each lot for authentication testing (with enhanced test modes). The trained NNs are used to predict the targets and compared with the measurements to determine whether the part is authentic.

In essence, the NN collapses the joint probability distribution around each measurement, except for the target parameter and predicts what the target parameter should be using the learned correlations from simulations. If the target parameter's prediction and actual measurement differ substantially, then the circuit under test is not from the learned distribution. Hence, we can conclude that there has been a modification to either the process or the circuit.

4. Experimental Results

We use a Low Noise Amplifier (Figure 7) circuit to evaluate the proposed authentication methodology for a number of potential process and circuit modifications. As stated earlier,

978-1-5386-3775-3/18 $31.00 © 2018 IEEE

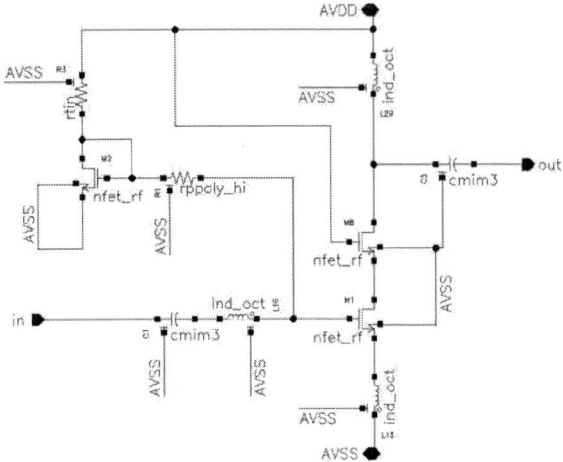

Figure 7: LNA circuit used in the experiments

this set cannot be complete or comprehensive. However, these process modifications have been deemed undetectable by prior work on digital circuits. Moreover, as we have demonstrated in Figure 4, the result of these modifications are minute shifts in the performance parameters under different supply conditions. Hence, by demonstrating the detection capability under these extremely hard to observe

Table 1: Selected Targets with Coverage (Cov) given for line thickness reduction, and False Positive Rates (FP).

Target		# of inputs	Cov. (%)	FP (%)
V_{DD}	Parameter			
0.8V	IDC	28	0	0
	S11 at 2.4G	3	0	0
	S22 at 2.4G	2	46	1
	IIP3	2	0.5	0
	Pin-1dB	1	4	0.5
	S21 at 4.0G	9	2	0
1.0V	S22 at 2.4G	2	46	0
	IIP3	3	0.5	0
	S21 at 4.0G	3	1	0
1.2V	S21 at 2.4G	3	20	0
	S22 at 2.4G	3	18.5	0
	S21 at 4.0G	7	1	0
1.5V	S22 at 2.4G	4	18.5	0
	IIP3	1	2.5	0
	Pin-1dB	10	5	0
	S21 at 2.0G	1	0	0
	S21 at 4.0G	6	5	0
1.8V	S22 at 2.4G	4	42	0
	NF at 2.4G	1	0.5	1
	IIP3	4	1.5	0
	S21 at 4.0G	8	2.5	0

process/circuit modifications, we demonstrate the effectiveness of the proposed technique.

The LNA circuit has been designed at the layout level with TowerJazz 180nm SiGe BiCMOS (SBC18HA) technology. Four versions of this LNA is generated: (a) baseline circuit, as designed, (b) the same LNA circuit where the line thickness is reduced by 8%, which is well within process variations ,(c) the same LNA circuit with a small 150fF capacitive load, and (d) the same LNA circuit where the threshold of the NMOS devices is moved by 40mV. 150fF capacitive load is selected to mimic a dipole antenna placed with the intention of stealing information or jamming the LNA output. The 8% line width reduction and 40mV mean shift in threshold are selected because they are right at the edge of expected process distributions. Hence, these changes would be extremely difficult to detect. The circuits with layout modifications have been taped out for fabrication.

The LNA performances are defined in terms of DC current (IDC), input reflection coefficient (S11) at 2.4GHz and 4GHz, the output reflection coefficient (S22) at 2.4GHz and 4GHz, 1-dB gain compression point at the input (Pin-1dB), third order input intercept (IIP3), noise figure (NF), and output node DC voltage (VoutDC). We use the nominal supply voltage of 1.8V, and reduced supply voltages of 1.5V, 1.2V, 1V, and 0.8V for the measurement of all these parameters.

The baseline circuit is simulated with 1000 MC samples to form the statistical models. After analyzing pairwise correlations, we choose inputs for each target. We used the built-in neural network of MATLAB (feedforwardnet). The network includes 9 hidden layers and employs the Levenberg-Marquardt (LM) algorithm for training. We then analyze false positive rates and add/drop NN inputs for each target to minimize the false positive rate. For targets where the false positive rate cannot be reduced to the pre-determined threshold (1% in this work), the corresponding NN is not used. Note that the NN training process does not involve any information from the modified circuits.

In the evaluation step, we use the modified circuit simulations to obtain the response of the modified circuits. 200 MC samples of each modified circuit are used in this evaluation step. With the selected modifications, a majority of the samples fall within the 3σ limits for all the parameters even when enhanced test modes are used (examples are given in Figure 4). Using only threshold comparison as a criterion, authentication accuracy is less than 30%. This result is in line with the results from prior work on digital circuits, where the researchers concluded that using performance measurements, even at modified supply levels, will not lead to the detection of process reliability Trojans [10-15].

Table 1 shows the selected targets, the number of inputs for the trained NN, the false positive rates that are computed for each of the selected parameters respectively, as well as the coverage of the line thickness modification as an example. The inputs are a subset of the LNA performances. The targets are selected to maximize coverage while keeping the overall false positive rate less than the specified maximum (1%). The false positives

978-1-5386-3775-3/18 $31.00 © 2018 IEEE

Table 2: Authentication results

Supply Level (V)	Authentication Accuracy for Modifications		
	Line width	Load	Vth Shift
1.8	26%	100%	40%
1.8 & 1.5	48%	100%	40%
1.8 & 1.5 & 1.2	71%	100%	50%
1.8 & 1.5 & 1.2 & 1	76%	100%	100%
1.8 & 1.5 & 1.2 & 1 & 0.8	80%	100%	100%

are not fully correlated, and with the selected NNs, the overall false positive rate is less than 2%. Table 2 shows the overall authentication accuracy of the proposed method for the three circuit/process modifications that have been evaluated. The table shows that the majority of the modified circuits can be identified as such even when the process/layout modification is not detectable by comparing individual measurements against their 3σ limits.

5. Conclusion

In this paper, we present a multivariate statistical method to authenticate devices manufactured at a third party foundry with respect to agreed-upon layout and process design kit. We propose to sample a small number of devices from each process lot to subject to authentication testing. The authentication test process will include mission-mode parameters as well as parameters measured under reduced supply voltage. The supply voltage reduction helps in increasing process sensitivity and thus in identifying process/circuit modifications that only cause minute variations in the primary performance of the circuit. We present an algorithm to select the targets and the inputs of the statistical mapping tool to minimize false positive rates. We evaluate our technique on an LNA circuit against three different copies of the same circuit with the process and layout modifications. We find through experiments that even though the evaluated modifications cause only minute deviations in individual performance, using multivariate analysis and enhanced test modes helps achieve high authentication accuracy.

Acknowledgements

This work is supported by US Air Force Research Lab and Alphacore Inc. through a contract with Number FA865016M1788 and the NSF I/UCRC Center for Embedded Systems with grant Number 1361926.

References

1. DIGITIMES Research, "Trends in the global IC design service market." http://www.digitimes.com/Reports/Report.asp?datepublish=2012/3/13n&pages=RSn&seq=400n&read=toc.
2. J. Roy, F. Koushanfar, and I. Markov, "EPIC: Ending Piracy of Integrated Circuits," IEEE/ACM DATE, pp. 1069–1074, 2008.
3. R. Chakraborty and S. Bhunia, "HARPOON: An Obfuscation-Based SoC Design Methodology for Hardware Protection," IEEE TCAD, vol. 28, no. 10, pp. 1493–1502, 2009.
4. SEMI, "Innovation is at risk as semiconductor equipment and materials industry loses up to $4 billion annually due to IP infringement." www.semi.org/en/Press/P043775, 2008.

5. M. Pecht and S. Tiku, "Bogus: electronic manufacturing and consumers confront a rising tide of counterfeit electronics," IEEE Spectrum, vol. 43, no. 5, pp. 37–46, 2006.
6. R. Karri, J. Rajendran, K. Rosenfeld and M. Tehranipoor, "Trustworthy Hardware: Identifying and Classifying Hardware Trojans," IEEE Computer, vol. 43, no. 10, pp. 39-46, Dec. 2010.
7. M. Tehranipoor and F. Koushanfar, "A survey of hardware Trojan taxonomy and detection," IEEE Des. Test Comput., pp. 10–25, 2010.
8. Huang, Ke, John M. Carulli, and Yiorgos Makris. "Counterfeit electronics: A rising threat in the semiconductor manufacturing industry." 2013 IEEE International Test Conference (ITC). IEEE, 2013.
9. Guin, Ujjwal, Daniel DiMase, and Mohammad Tehranipoor. "Counterfeit integrated circuits: detection, avoidance, and the challenges ahead." Journal of Electronic Testing 30.1 (2014): 9-23.
10. Becker, G.T., Regazzoni, F., Paar, C., Burleson, W.P. Burleson, "Stealthy Dopant-Level Hardware Trojans" CHES 2013, vol. 8086, pp. 197–214. Springer, Heidelberg (2013)
11. T. Sugawara, D. Suzuki, R. Fujii, S. Tawa, R. Hori, M. Shiozaki, T. Fujino, "Reversing stealthy dopant-level circuits," Journal of Cryptographic Engineering, vol. 5, no. 2, pp. 85-94, 2015.
12. S. Ghandali, G. T. Becker, D. Holcomb, and C. Paar, "A Design Methodology for Stealthy Parametric Trojans and Its Application to Bug Attacks" CHES 2016, vol. 8086, pp. 197–214. Springer, Heidelberg.
13. R. Kumar, P. Jovanovic, W. P. Burleson, and I. Polian, "Parametric trojans for fault-injection attacks on cryptographic hardware" IACR Cryptology ePrint Archive, p.783. 2014.
14. Shiyanovskii, Y. et.al., W. Process reliability based trojans through NBTI and HCI effects. In IEEE Adaptive Hardware and Systems Conference (pp. 215-222), 2010.
15. Becker, G. T., et.al., Stealthy dopant-level hardware trojans. In International Workshop on Cryptographic Hardware and Embedded Systems (pp. 197-214), 2013.
16. S. Bhasin and F. Regazzoni, "A survey on hardware trojan detection techniques," in IEEE International Symposium on Circuits and Systems pp. 2021–2024, 2015.
17. Balasch, B. Gierlichs, and I. Verbauwhede, "Electromagnetic circuit fingerprints for hardware trojan detection," in IEEE International Symposium on Electromagnetic Compatibility, pp. 246–251, 2015.
18. R. M. Rad, X. Wang, M. Tehranipoor, and J. Plusquellic, "Power supply signal calibration techniques for improving detection resolution to hardware trojans," in IEEE/ACM International Conference on Computer-Aided Design, pp. 632–639, 2008.
19. Y. Jin, D. Maliuk, and Y. Makris, "Post-deployment trust evaluation in wireless cryptographic ics," in IEEE Design, Automation & Test in Europe Conference, pp. 965–970, 2012.
20. K. Xiao, D. Forte, and M. Tehranipoor, "A novel built-in self-authentication technique to prevent inserting hardware Trojans", IEEE Transactions on CAD, vol. 33, no. 12, pp. 1778–1791, 2014.
21. Y. Liu, Y. Jin, and Y. Makris, "Hardware trojans in wireless cryptographic ICs: silicon demonstration & detection method evaluation," in IEEE International Conference on Computer-Aided Design, pp. 399–404, 2014.
22. B. Cha and S. K. Gupta, "Trojan detection via delay measurements: A new approach to select paths and vectors to maximize effectiveness and minimize cost," in DATE, 1265–1270, 2013.
23. R. S. Chakraborty, S. Narasimhan, and S. Bhunia, "Hardware trojan: Threats and emerging solutions," in Proc. of Intl. High Level Design Validation and Test Workshop, 2009, pp. 166–171.
24. S. Jha, "Randomization based probabilistic approach to detect trojan circuits," in High Assurance Systems Engineering Symposium, 2008. HASE 2008. 11th IEEE, 2008, pp. 117–124.
25. D. Agrawal, S. Baktir, D. Karakoyunlu, P. Rohatgi, and B. Sunar, "Trojan detection using IC fingerprinting," in IEEE Symposium on Security and Privacy, pp. 296–310, 2007.
26. Y. Jin and Y. Makris, "Hardware trojan detection using path delay fingerprint," in Intl. Workshop on Hardware-Oriented Security and Trust, 2008, pp. 51–57.
27. Bajura, Michael A., et al. "Verification of integrated circuits against malicious circuit insertions and modifications using non-destructive X-ray microscopy." U.S. Patent No. 8,139,846. 20 Mar. 2012.
28. B. Li, A. Boyer, S. Bendhia, and C. Lemoine, "Ageing effect on electromagnetic susceptibility of a phase locked loop," Microelectron. Rel., vol. 50, pp. 1304–1308, 2010.
29. M. S. Cooper, "Investigation of Arrhenius acceleration factor for integrated circuit early life failure region with several failure mechanisms," IEEE Trans. Compon. Packag. Technol., vol. 28, no. 3, pp. 561–563, Sep. 2005.

Special Session on Machine Learning: How Will Machine Learning Transform Test?

Yiorgos Makris, University of Texas at Dallas, USA
Amit Nahar, Texas Instruments Inc., USA
Haralampos-G. Stratigopoulos, Sorbonne Université, CNRS, LIP6, France
Marc Hutner, Teradyne, Canada (Organizer)

I. INTRODUCTION

This special session will discuss how machine learning can transform test. The first talk will review the key challenges and will argue whether contemporary tools, such as deep learning, could offer any advantages over traditional methods. The second talk will focus on extracting useful information from big test data. The third talk will view adaptive test as running machine learning algorithms in real time on big data.

II. MACHINE LEARNING IN SEMICONDUCTOR TEST: CAN DEEP LEARNING SAVE THE DAY? (YIORGOS MAKRIS)

While many applications of machine learning in various semiconductor manufacturing and testing tasks have been heavily researched over the last two decades, very few have actually seen the light of day in a real production environment. Recently, the popularity of contemporary artificial intelligence methods, such as deep learning, has reignited the enthusiasm and reinvigorated the discussion regarding the potential of statistical and machine learning-based solutions towards reducing test cost, increasing test quality, improving test floor logistics and providing guidance to both designers and process engineers. In this presentation, we will first review the key challenges in transitioning machine learning-based solutions from a research to a production environment; then, we will discuss the main advantages of contemporary methods, such as deep learning, over traditional machine learning solutions. Ultimately, the intention of this presentation is to caution against considering deep learning a panacea to test (and, thereby, prevent disappointment), as well as to steer attention towards the true operational challenges which need to be addressed in order for any machine learning or statistical solution, traditional or contemporary, to be effective in a real production environment.

III. SMALL SIGNALS EXTRACTION IN SEMICONDUCTOR TEST & MANUFACTURING: ROLE OF MACHINE LEARNING (AMIT NAHAR)

Semiconductor manufacturing and testing generates huge amount of data. Identifying small signals from this plethora of data and using it to drive product quality improvements are becoming increasingly important, especially for products going into automotive applications. The big data approaches of data management have increased speed, quality and accessibility of the data. This presentation will discuss challenges and applications of harnessing value from the fab manufacturing and test data. A lot of signals are collected in fab manufacturing from fab tools (e.g. pressures, frequencies, temperature etc). The time interval of collecting these signals varies for different tools (i.e. every second, minute or hour). Analyzing this enormous multidimensional time domain data near real-time and finding signals which deviate from normal is challenging. Machine learning algorithms play a key role in finding these deviations and predicting excursions so that appropriate actions can be taken downstream (most likely at test). Another application that will be discussed is finding small signals from test data. Typically hundreds of tests are performed for every product; hence finding small signals for thousands of such products presents a big challenge. Machine learning algorithms running in a big data environment enables extracting these small signals and using them to correlate with customer returns.

IV. ADAPTIVE TEST: MACHINE LEARNING IN REAL TIME ON BIG DATA (HARALAMPOS-G. STRATIGOPOULOS)

In the standard test approach, the test process is fixed and identical for every circuit and is only occasionally revised, typically every few months, when extensive test data have been collected and confident decisions can be made as to how the test process could be improved to lower the test cost. Adaptive test adheres to the standard test approach, but dynamically adjusts it in real-time to the particularities of each wafer or die instead of having a fixed test process. In this way, significant test time savings and test quality improvements can be achieved. Adjusting the test process may involve adjusting the test limits, the test content, and the test order. The dynamic adjustment may be decided based on available big data: historical (i.e. data from previously tested wafers or lots, customer returns, etc.), near-time (i.e. data from previously tested dies on the same wafer or previously tested wafers in the same lot), and real-time data (i.e. data from already performed tests on the die that is currently under test). The core of an adaptive test flow is the underlying decision-making algorithm. Machine learning algorithms can be very effective in mining available information and identifying meaningful correlations. The challenge is that calculations must be quick such that test application is not delayed, or, if it is delayed, then any such delay is counterbalanced by an average test time reduction per wafer or die or is justified by the improved test quality. This talk will discuss how machine learning can lead to more effective adaptive test flows.

Innovative Practices on Design & Test for Flexible Hybrid Electronics

Tsung-Ching "Jim" Huang, Hewlett Packard Labs, USA
Jason Marsh, NextFlex, USA
Scott H. Goodwin, Micross Components USA, and Dorota S. Temple, RTI International, USA
Tsung-Ching "Jim" Huang, Hewlett Packard Labs, USA (Organizer)

I. INTRODUCTION

The IP session focuses on design and test aspects of emerging flexible hybrid electronics (FHE). The first contribution discusses process design kit for flexible hybrid electronics. The second presentation identifies challenges of FHE design and test as well as application examples and future opportunities. The last contribution discusses design and test considerations for FHE including a proof-of-concept demonstration using proprietary FHE technologies.

II. PROCESS DESIGN KIT (PDK) FOR DESIGN AND TEST CHALLENGES OF FLEXIBLE HYBRID ELECTRONICS (TSUNG-CHING "JIM" HUANG)

Flexible hybrid electronics (FHE) integrating thinned silicon chips and flexile printed components, ranging from sensors to antennas, is emerging for applications such as internet of things (IoT) and wearables. For electronics designers to design FHE systems, however, the entry barrier is high due to lack of trustworthy device models and the design automation infrastructure. In this talk, we present our work in process design kit (PDK) for FHE that provides capabilities of FHE circuit simulations and design verifications with existing electronic design automation (EDA) tools. The key packages of FHE-PDK include technology files for design rule checking (DRC), layout versus schematics (LVS) and layout parasitics extraction (LPE), as well as SPICE models for flexible thin-film transistors (TFT) and passive elements such as resistors and capacitors. With FHE-PDK, the electronics designers can therefore focus on FHE design innovations with guaranteed results with additive manufacturing.

III. CONSIDERATIONS FOR DESIGN AND TEST OF FLEXIBLE HYBRID ELECTRONIC SYSTEMS (JASON MARSH)

The speaker will introduce flexible hybrid electronics (FHE) and relevant application examples. He will then illuminate challenges with design and reliability testing of these devices using current tool sets. He will focus on mechanical failure modes under flexing, twisting, and stretching conditions which are predominantly used in applications for wearables, healthcare, soft robotics and asset monitoring, all of which are areas of development focus at NextFlex. Finally, the speaker will outline critical areas where new development is needed, with emphasis on testing standards, printed active materials, Multiphysics modeling, and high-performance systems.

IV. DESIGN AND TEST CONSIDERATIONS FOR FLEXIBLE HYBRID ELECTRONICS FABRICATED WITH HIGH-PERFORMANCE COTS ICS USING RTI CIRCUITFILM™ TECHNOLOGY (SCOTT H. GOODWIN)

Flexible hybrid electronics is a rapidly growing field which is driven by applications such as wearable sensors and flexible displays. RTI CircuitFilm™ technology incorporates commercial-off-the-shelf (COTS) high performance integrated circuits in flexible electronics systems by leveraging advanced 3D integration processes of wafer thinning, bonding, and direct interconnects. Thinned COTS devices maintain their performance characteristics even when they are 30μm thick and highly flexible. The RTI CircuitFilm™ process incorporates standard semiconductor process technologies to provide good control of the fabrication process. A demonstration system was fabricated and tested at the Micross Advanced Interconnect Technology facility, using the RTI CircuitFilm™ technology. We will discuss design and test considerations of the flexible hybrid electronic system, and present results of the proof-of-concept demonstration. We will conclude with the discussion of potential applications of the RTI CircuitFilm™ technology.

AUTHOR INDEX

Aftabjahani, S. 160
Agrawal, V. 27
Ahmadi, A. 141
Aouini, S. 7
Atwood, E. 120
Awad, T. 73
Barragan, M. 13
Basak, A. 41
Basu, K. 119, 147
Baugh, J. 179
Becker, B. 101
Ben-Hamida, N. 7
Bertozzi, D. 99
Bhattacharya, M. 80
Bhunia, S. 41
Blanton, R. 180
Bosio, A. 39
Bourdel, S. 13
Burchard, J. 101
Casarsa, M. 20
Casto, M. 192
Chakrabarty, K. 61, 81, 93
Chakraborty, K. 159
Chakravarthy, S. 20
Chakravarty, S. 80
Chen, K. 20
Chen, M. 160
Chen, Y. 59
Cho, Y. 7
Commarota, R. 19
Courtois, B. 179
Crepeau, D. 73
Crouch, A. 140
Das, A. 121
Di Carlo, S. 39, 186
Dobbelaere, W. 40
Dulipovici, A. 73
Dworak, J. 87, 167
Elliot, A. 167
Elnaggar, R. 93
Fang, E. 153
Fang, Y.-C. 153
Fernandes, F. 39
Ferrari, P. 13
Garg, S. 19, 147
Ghosh, S. 53
Gizopoulos, D. 186
Goteti, P. 80
Gu, T. 147
Gui, P. 167
Guin, U. 27
Guo, Q. 33
Gupta, S. 87
Harutyunyan, G. 20
Hasib, O. 73

Hatayama, K. 100
Heinssen, S. 133
Hillebrand, T. 133
Hoque, T. 41
Hsiao, M. 161
Hu, Y. 33
Huang, K. 141
Huang, T.-C. 199
Huang, Y. 140, 159
Hunter, E. 140
Hutner, M. 198
Izon, J. 139
Kaminska, B. 179
Karabacak, F. 192
Karam, R. 41
Karimi, N. 19
Karri, R. 93
Keim, M. 20
Khan, M. 53
Khasawneh, Q. 167
Kim, W. 120
Kitchen, J. 67, 192
Koneru, A. 61
Krishnan, A. 80
Leisenberger, F. 47
Levin, P. 140
Li, C.-M. 153
Li, G. 81
Li, H. 33, 99, 160
Li, X. 33
Li, Y. 173
Li, Y.-C. 153
Lin, F. 141
Lin, S.-C. 153
Lin, S.-Y. 153
Liu, C. 59
Liu, T. 127
Liu, Y. 113
Liu, Y.-C. 153
Machida, K. 1
Madkour, K. 140
Maeda, Y. 100
Mak, T. 79
Makris, Y. 107, 198
Margalef-Rovira, M. 13
Marsh, J. 199
Matsushima, J. 100
Matsutani, H. 79
Meade, R. 120
Milor, L. 127
Mittal, S. 180
Muthaiah, A. 167
Nabeel, M. 21
Nahar, A. 198
Neubauer, F. 101

AUTHOR INDEX

Oberg, J. 160
Oliveria, D. 39
Otte, R. 120
Ozev, S. 67, 139, 173, 192
Pan, Y. 141
Pande, P. 79
Parvizi, M. 7
Pasricha, S. 99
Paul, S. 133
Peters-Drolshagen, D. 133
Pistono, E. 13
Podichetty, G. 20
Poulin-Lamarre, G. 179
Qian, J. 81
Raiola, P. 101
Rajendran, J. 19
Rajski, J. 113
Rearick, J. 40
Rech, P. 39
Reddy, G. 107
Reddy, S. 113
Reita, C. 179
Rivoir, J. 101
Roberts, G. 7
Sarson, P. 1, 40, 47, 60, 139, 173
Sauer, M. 101
Savaria, Y. 73
Schaldenbrand, A. 60
Schatzberger, G. 47
Sekanina, L. 39
Sekar, K. 141
Sen, S. 119
Sengupta, A. 21
Shafiee, M. 67
Sharma, E. 13
Sharma, R. 20
Sinanoglu, O. 21
Solecki, J. 113
Stratigopoulos, H. 198
Sunter, S. 60
Taddiken, M. 133
Tanwir, S. 161
Thibeault, C. 73
Touba, N. 121
Traiola, M. 39
Tselonis, S. 186
Tyszer, J. 113
Vallero, A. 186
Van Gelder, J. 80
Vasicek, Z. 39
Veda, G. 159
Violante, M. 40
Von Staudt, H.-M. 139
Wang, L.-C. 159
Wang, X. 41

Welker, R. 192
Wen, W. 59
Wiesner, A. 47
Williams, B. 167
Wu, J. 87
Xanthopoulos, C. 107
Yanagida, T. 1
Yang, K. 127
Yang, Q. 81
Yang, T.-S. 153
Yasin, M. 21
Ye, J. 33
Yilmaz, E. 173
Yu, S. 59
Zhang, J. 147
Zhang, R. 127
Zhang, Z. 140
Zhong, Z. 81
Zhou, Z. 27
Zorian, Y. 60

IEEE
445 Hoes Lane
Piscataway, NJ 08854-4141

ISBN 978-1-5386-3775-3